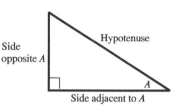
1.4 Definitions of the Trigonometric Functions

$$\sin \theta = \frac{y}{r} \qquad \csc \theta = \frac{r}{y}$$

$$\cos \theta = \frac{x}{r} \qquad \sec \theta = \frac{r}{x}$$

$$\tan \theta = \frac{y}{x} \qquad \cot \theta = \frac{x}{y}$$

$$r = \sqrt{x^2 + y^2}$$

2.1 Trigonometric Functions of Acute Angles in Right Triangles

$$\sin A = \frac{\text{side opposite}}{\text{hypotenuse}}$$

$$\cos A = \frac{\text{side adjacent}}{\text{hypotenuse}}$$

$$\tan A = \frac{\text{side opposite}}{\text{side adjacent}}$$

$$\csc A = \frac{\text{hypotenuse}}{\text{side opposite}}$$

$$\sec A = \frac{\text{hypotenuse}}{\text{side adjacent}}$$

$$\cot A = \frac{\text{side adjacent}}{\text{side opposite}}$$

2.1, 2.2 Values of Trigonometric Functions for Special Angles

Angle θ							
Degrees	**Radians**	$\sin \theta$	$\cos \theta$	$\tan \theta$	$\cot \theta$	$\sec \theta$	$\csc \theta$
0°	0	0	1	0	Undefined	1	Undefined
30°	$\pi/6$	1/2	$\sqrt{3}/2$	$\sqrt{3}/3$	$\sqrt{3}$	$2\sqrt{3}/3$	2
45°	$\pi/4$	$\sqrt{2}/2$	$\sqrt{2}/2$	1	1	$\sqrt{2}$	$\sqrt{2}$
60°	$\pi/3$	$\sqrt{3}/2$	1/2	$\sqrt{3}$	$\sqrt{3}/3$	2	$2\sqrt{3}/3$
90°	$\pi/2$	1	0	Undefined	0	Undefined	1
120°	$2\pi/3$	$\sqrt{3}/2$	$-1/2$	$-\sqrt{3}$	$-\sqrt{3}/3$	-2	$2\sqrt{3}/3$
135°	$3\pi/4$	$\sqrt{2}/2$	$-\sqrt{2}/2$	-1	-1	$-\sqrt{2}$	$\sqrt{2}$
150°	$5\pi/6$	1/2	$-\sqrt{3}/2$	$-\sqrt{3}/3$	$-\sqrt{3}$	$-2\sqrt{3}/3$	2
180°	π	0	-1	0	Undefined	-1	Undefined
210°	$7\pi/6$	$-1/2$	$-\sqrt{3}/2$	$\sqrt{3}/3$	$\sqrt{3}$	$-2\sqrt{3}/3$	-2
225°	$5\pi/4$	$-\sqrt{2}/2$	$-\sqrt{2}/2$	1	1	$-\sqrt{2}$	$-\sqrt{2}$
240°	$4\pi/3$	$-\sqrt{3}/2$	$-1/2$	$\sqrt{3}$	$\sqrt{3}/3$	-2	$-2\sqrt{3}/3$
270°	$3\pi/2$	-1	0	Undefined	0	Undefined	-1
300°	$5\pi/3$	$-\sqrt{3}/2$	1/2	$-\sqrt{3}$	$-\sqrt{3}/3$	2	$-2\sqrt{3}/3$
315°	$7\pi/4$	$-\sqrt{2}/2$	$\sqrt{2}/2$	-1	-1	$\sqrt{2}$	$-\sqrt{2}$
330°	$11\pi/6$	$-1/2$	$\sqrt{3}/2$	$-\sqrt{3}/3$	$-\sqrt{3}$	$2\sqrt{3}/3$	-2
360°	2π	0	1	0	Undefined	1	Undefined

3.1 Conversion of Angular Measure

Degree/Radian Relationship: $180° = \pi$ radians

Conversion Formulas:

From	To	Multiply by
Radians	Degrees	$\dfrac{180°}{\pi}$
Degrees	Radians	$\dfrac{\pi}{180}$

3.2 Applications of Radian Measure

Arc Length: $s = r\theta$, θ in radians

Area of Sector: $A = \dfrac{1}{2} r^2 \theta$, θ in radians

3.4

Angular Velocity	Linear Velocity
$\omega = \dfrac{\theta}{t}$	$v = \dfrac{s}{t}$
(ω in radians per unit time, θ in radians)	$v = \dfrac{r\theta}{t}$
	$v = r\omega$

4.1-4.3 Basic Graphs of Trigonometric Functions

The graph of $y = c + a \sin b(x - d)$ or $y = c + a \cos b(x - d)$, where $b > 0$, has amplitude $|a|$, period $2\pi/b$, a vertical translation c units up if $c > 0$ or $|c|$ units down if $c < 0$, and a phase shift d units to the right if $d > 0$ or $|d|$ units to the left if $d < 0$. The graph of $y = a \tan bx$ or $y = a \cot bx$ has period π/b, where $b > 0$.

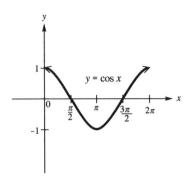

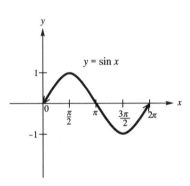

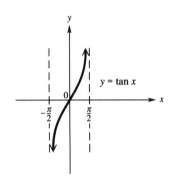

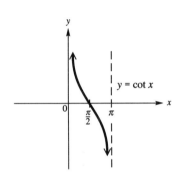

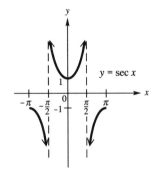

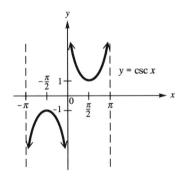

Trigonometry

7th *Edition*

Margaret L. Lial
American River College

John Hornsby
University of New Orleans

David I. Schneider
University of Maryland

Addison
Wesley

Boston San Francisco New York
London Toronto Sydney Tokyo Singapore Madrid
Mexico City Munich Paris Cape Town Hong Kong Montreal

Sponsoring Editor: Bill Poole
Executive Project Manager: Christine O'Brien
Assistant Editor: Jennifer Kerber
Senior Production Supervisor: Rebecca Malone
Project Coordination: Elm Street Publishing Services, Inc.
Executive Marketing Manager: Brenda L. Bravener
Marketing Coordinator: Laura Potter Walton
Senior Prepress Supervisor: Caroline Fell
Manufacturing Buyer: Evelyn Beaton
Text Design: Rebecca Lemna
Cover Design: Barbara T. Atkinson
Cover Photography: © Carr Clifton

Photo Credits:
p. 1, p. 61, p. 103, p. 185, p. 237, p. 281, and p. 387 Images provided by PhotoDisc © 2000.
p. 137 Stone/Hiroyuki Matsumoto. **p. 333** © Digital Art/CORBIS.

Library of Congress Cataloging-in-Publication Data
Lial, Margaret L.
    Trigonometry.–7th ed. / Margaret L. Lial, John Hornsby, David I. Schneider.
      p.   cm.
    ISBN 0-321-05759-7 (alk. paper), ISBN 0-321-06860-2 (annotated instructor's edition), ISBN 0-321-08599-x (NASTA edition)
    1. Trigonometry.   I. Hornsby, E. John.   II. Schneider, David I.   III. Title.
QA531 .L5 2001
516.24–dc21

                                           00-035549
                                           CIP

Reprinted with corrections.

0-321-05759-7   0-321-08599-x
4 5 6 7 8 9 10—CRK—03   8 9 10—CRK— 05

Contents

Preface   vii

1 The Trigonometric Functions

1.1   Basic Concepts   2
1.2   Angles   15
1.3   Angle Relationships and Similar Triangles   24
1.4   Definitions of the Trigonometric Functions   35
1.5   Using the Definitions of the Trigonometric Functions   43

Summary   53
Review Exercises   55
Test   59

2 Acute Angles and Right Triangles

2.1   Trigonometric Functions of Acute Angles   62
2.2   Trigonometric Functions of Non-Acute Angles   71
2.3   Finding Trigonometric Function Values Using a Calculator   77
2.4   Solving Right Triangles   83
2.5   Further Applications of Right Triangles   91

Summary   98
Review Exercises   99
Test   102

3 Radian Measure and the Circular Functions

3.1 Radian Measure   104

3.2 Applications of Radian Measure   109

3.3 Circular Functions of Real Numbers   118

3.4 Linear and Angular Velocity   127

Summary   132

Review Exercises   133

Test   135

4 Graphs of the Circular Functions

4.1 Graphs of the Sine and Cosine Functions   138

4.2 Translations of the Graphs of the Sine and Cosine Functions   154

4.3 Graphs of the Other Circular Functions   166

Summary   179

Review Exercises   180

Test   182

Internet Project   *Modeling Sunset Times*   183

5 Trigonometric Identities

5.1 Fundamental Identities   186

5.2 Verifying Trigonometric Identities   193

5.3 Sum and Difference Identities for Cosine   202

5.4 Sum and Difference Identities for Sine and Tangent   210

5.5 Double-Angle Identities   217

5.6 Half-Angle Identities   226

Summary   233

Review Exercises   234

Test   236

6 Inverse Trigonometric Functions and Trigonometric Equations

6.1 Inverse Trigonometric Functions   238

6.2 Trigonometric Equations I   253

6.3   Trigonometric Equations II   261

6.4   Equations Involving Inverse Trigonometric Functions   268

Summary   275

Review Exercises   276

Test   279

Internet Project   *Modeling a Damped Pendulum*   279

7 Applications of Trigonometry and Vectors

7.1   Oblique Triangles and the Law of Sines   282

7.2   The Ambiguous Case of the Law of Sines   294

7.3   The Law of Cosines   299

7.4   Vectors and the Dot Product   309

7.5   Applications of Vectors   319

Summary   325

Review Exercises   326

Test   331

8 Complex Numbers, Polar Equations, and Parametric Equations

8.1   Complex Numbers   334

8.2   Trigonometric (Polar) Form of Complex Numbers   343

8.3   The Product and Quotient Theorems   350

8.4   Powers and Roots of Complex Numbers   355

8.5   Polar Equations and Graphs   361

8.6   Parametric Equations, Graphs, and Applications   373

Summary   382

Review Exercises   384

Test   386

Internet Project   *The Art of Undersampling*   386

9 Exponential and Logarithmic Functions

9.1   Exponential Functions   388

9.2   Logarithmic Functions   402

9.3   Evaluating Logarithms and the Change-of-Base Theorem   413

9.4   Exponential and Logarithmic Equations   423

Summary   431

Review Exercises   431

Test   434

Internet Project   *Modeling Growth of Internet Hosts*   436

Answers to Selected Exercises   A-1

Index   I-1

Index of Applications   I-7

Preface

The seventh edition of *Trigonometry* reflects our ongoing commitment to providing the best possible text and supplements package to help instructors teach and students succeed. To that end, we have attempted to address the needs of students who will continue their study of mathematics in calculus or other disciplines, as well as those who are taking trigonometry as their final mathematics course. Although we assume that students have had at least one course in algebra, we briefly review algebraic prerequisites as needed. Important concepts from geometry are handled similarly.

A Word About Technology

Technology is part of our lives, and as a result, many students will come to this course with experience using graphing calculators. *While graphing calculators are not required for this text,* we have incorporated a new "windows on graphing" example format that provides a graphing calculator solution alongside the traditional trigonometric solution for selected examples. These graphing calculator solutions are optional and can be easily omitted if desired. On the other hand, we hope this feature will provide an avenue for incorporating graphing technology for those instructors who wish to do so. As in the previous edition, all graphing calculator notes and exercises that require graphing calculators are marked with an icon for easy identification and added course flexibility. Due to rapidly changing technology, we have not in the past nor do we try in this edition to teach students how to use specific models of graphing calculators.

Content Changes

We have worked hard to fine-tune and polish presentations of topics throughout the text based on user and reviewer feedback. Some of the content changes you may notice include the following:

- The presentation of graphing trigonometric functions in Chapter 4 has been expanded. Colored boxes highlighting important characteristics of the graph of each function have been added.

- The algebraic review of inverse functions has been enhanced in Section 6.1. Graphs of the inverse secant, cosecant, and cotangent functions are now presented.

- The presentation of vectors in Chapter 7 now includes operations on vectors and the dot product.

- A helpful summary of polar graphs is included in Section 8.5 for easy student reference.

- Additional examples and exercises on applications of parametric equations are provided in Section 8.6.

New Features

We believe students and instructors will welcome the following new features.

New Real-Life Applications   Wherever possible, we have provided many new or updated applied examples and exercises that focus on real-life applications of mathematics. All applications are now titled, and an index of applications is included at the back of the text.

Increased Emphasis on Modeling and Curve Fitting   Many of the applications feature mathematical models based on real data or real data in table form, thereby providing students with increased opportunities to use, construct, and analyze models. Curve fitting using sine, exponential, and logarithmic models is also covered.

"Windows on Graphing" Example Format   Given the ever-increasing importance of technology in mathematics, selected examples now feature traditional trigonometric as well as optional graphing calculator solutions. We have taken great care to utilize this format only with those examples where the graphing calculator naturally supports and/or enhances the traditional solution.

Looking Ahead to Calculus   These margin notes provide glimpses of how the trigonometric topics currently being studied are used in calculus.

Quantitative Reasoning Problems   Appearing at the end of selected exercise sets, these problems enable students to use trigonometric concepts to explore life issues, such as the relationship between utility costs and average annual temperatures or financial planning for retirement. Others are more whimsical in nature, investigating such topics as the predominant color in the U.S. flag or a method for estimating landmark heights.

Chapter Tests   Each chapter now includes a new end-of-chapter test to help students prepare for examinations.

Internet Projects   Selected chapters conclude with a project for students to complete individually or collaboratively using the material in the chapter and information from the Web site for this text, located at www.awl.com/lhs. Related activities and further readings on the Web are also provided.

Continuing Features

We have retained the popular features of previous editions of the text, some of which follow.

Chapter Themes   These have been enhanced and updated. Each chapter in the text features a particular industry that is presented in the chapter introduction and revisited in examples and exercises throughout the chapter. Identified by special icons (such as), these examples and exercises incorporate real sourced data, often in table or graph form. Featured industries include astronomy, highway engineering, weather, music, and others.

Cautions and Notes   We often give students warnings of common errors and emphasize important ideas in CAUTION and NOTE comments that appear throughout the exposition.

Connections   Retained from the sixth edition, we have included only those Connections boxes that instructors felt were most valuable. They continue to provide connections to the real world and to other mathematical concepts and feature thought-provoking questions for writing, class discussion, or group work. By request, new Connections that highlight historical information have been included.

Ample and Varied Exercise Sets   The text contains a wealth of exercises to provide students with opportunities to practice, apply, and extend the skills they are learning. These include writing exercises and graphing calculator exercises as well as multiple-choice, matching, true/false, and completion problems. Those problems that focus on conceptual understanding, many of which tie together multiple concepts, are now titled *Concept Check*. More illustrations, diagrams, and graphs now accompany exercises.

Relating Concepts   Previously titled Discovering Connections, these exercises appear in selected exercise sets. They tie together topics and highlight the relationships among various concepts and skills. For example, they may show how trigonometry and geometry are related, or how the algebraic concept of symmetry can be applied to trigonometric and polar graphs. Instructors have told us that these sets of exercises make great collaborative activities for small groups of students.

Expanded Review Opportunities   We have retained the popular format of our Chapter Summaries and have added lists of Key Terms and Symbols to the Key Ideas previously included for students. Each Summary is followed by a comprehensive set of Review Exercises and, as mentioned previously, a new Chapter Test.

Supplements

For the Student

Student's Solution Manual ISBN 0-321-05760-0   This helpful supplement contains detailed solutions to odd-numbered Section and Review Exercises and all Chapter Test Exercises, as well as solutions to all Connections, Relating Concepts, and Quantitative Reasoning problems.

Graphing Calculator Manual ISBN 0-321-05762-7   This supplement contains instructions for the TI-82™, TI-83™, TI-83 Plus™, TI-85™, and TI-86™ model graphing calculators. Worked-out examples are taken from the textbook.

Modules for selected Hewlett-Packard and Casio model calculators are also available.

Videotape Series ISBN 0-321-06861-0   Throughout the text, the icon indicates when these would be helpful to students. This videotape series

- Is keyed specifically to the text.
- Features an engaging team of lecturers who provide comprehensive coverage of each section and every topic.
- Uses worked-out examples, visual aids, and manipulatives to reinforce concepts.
- Emphasizes the relevance of material to the real world and relates mathematics to students' everyday lives.
- Can be ordered by mathematics instructors or departments.

Web Site 🌐   The Web site for this text (www.awl.com/lhs) provides additional resources for both students and instructors.

InterAct Math® Tutorial Software Windows/Macintosh (Dual Platform CD-ROM) ISBN 0-321-07703-2   Throughout the text, the icon 🖥 indicates when this software would be helpful to students. InterAct Math Tutorial Software has been developed and designed by professional software engineers working closely with a team of experienced math educators. The software includes exercises that are linked with every objective in the textbook and require the same computational and problem-solving skills as their companion exercises in the text. Each exercise has an example and an interactive guided solution that are designed to involve students in the solution process and to help them identify precisely where they are having trouble. It recognizes common student errors and provides students with appropriate customized feedback. With its sophisticated answer recognition capabilities, InterAct Math Tutorial Software recognizes appropriate forms of the same answer for any kind of input. It also tracks student activity and scores for each section which can then be printed out. The software is free to qualifying adopters or can be bundled with books for sale to students.

InterAct MathXL® Software Text bundled with a 12-month registration coupon ISBN 0-201-71414-0   InterAct MathXL provides diagnostic testing and tutorial help, all on-line using InterAct Math Tutorial Software. Students can take chapter tests correlated to this textbook, receive individualized study plans based on those test results, work practice problems for areas in which they need improvement, receive tutorial instruction in those topics, and take further tests to gauge their progress. With InterAct MathXL, instructors can track students' test results, study plans, and practice work. The software is free when an access code is bundled with a new text.

Digital Video Tutor ISBN 0-201-71459-0   The videotape series for this text is provided on CD-ROM, making it easy and convenient for students to watch video segments from a computer at home or on campus. This complete video set, now affordable and portable for students, is ideal for distance learning or extra instruction.

Addison Wesley Longman Math Tutor Center Text with registration coupon ISBN 0-201-71455-8   Live one-on-one tutoring is available to students who purchase this text. Qualified mathematics tutors are available from 5 P.M. to midnight Eastern Standard Time Sunday through Thursday to answer students' questions on any problem with an answer at the back of the text. Students can contact the Tutor Center via toll free phone, fax, e-mail, or the Internet. This service is provided free when a Tutor Center registration number is bundled with a new Addison Wesley Longman text.

For the Instructor

Annotated Instructor's Edition ISBN 0-321-06860-2   This special edition of the text includes

- All new Teaching Tips appropriately placed in the margins of the text for easy instructor access during lectures and teaching sessions.
- A complete answer section consisting of answers to all text exercises (except writing exercises).

Instructor's Solutions Manual ISBN 0-321-05761-9   Complete solutions for all text exercises, including Section, Connections, Relating Concepts, Quantitative Reasoning, Review, and Chapter Test Exercises (except writing exercises), are given.

Instructor's Testing Manual ISBN 0-321-06859-9   This manual contains two pretests (one open response and one multiple-choice), six tests for each chapter (four open response and two multiple-choice), and additional test items, grouped by section. Answers to all tests and test items are included.

TestGen-EQ with QuizMaster-EQ Windows/Macintosh (Dual-Platform CD-ROM) ISBN 0-321-06863-7   TestGen-EQ's friendly graphical interface enables instructors to easily view, edit, and add questions, transfer questions to tests, and print tests in a variety of fonts and forms. Search and sort features let the instructor quickly locate questions and arrange them in preferred order. Six question formats are available, including short-answer, true/false, multiple-choice, essay, matching, and bimodal formats. A built-in question editor gives the user power to create graphs, import graphics, insert mathematical symbols and templates, and insert variable numbers or text. Computerized testbanks include algorithmically defined problems organized according to each textbook. An "Export to HTML" feature lets instructors create practice tests for the Web.

QuizMaster-EQ enables instructors to create and save tests using TestGen-EQ so students can take them for practice or a grade on a computer network. Instructors can set preferences for how and when tests are administered. QuizMaster-EQ automatically grades the exams, stores results on disk, and allows the instructor to view or print a variety of reports for individual students, classes, or courses. Consult your Addison Wesley Longman sales representative for details.

InterAct Math Plus Software   InterAct Math Plus combines course management and on-line testing with the features of the basic InterAct Math tutorial software to create an invaluable teaching resource. Consult your Addison Wesley Longman sales representative for details.

Acknowledgments

For a textbook to last through seven editions, it is necessary for the authors to rely on comments, criticisms, and suggestions of users, nonusers, instructors, and students. We are grateful for the many responses we have received over the years. We especially wish to thank the following individuals who reviewed this edition of the text:

Carl Anderson	Johnson County Community College	KS
James C. Bishop	Daytona Beach Community College	FL
Robert Brandon	Eastern Oregon University	OR
Hongwei Chen	Christopher Newport University	VA
David Ebert	Peninsula College	WA
Ken Johnston	Hinds Community College	MS
David Longshore	Victor Valley College	CA
Daniel Martinez	California State University, Long Beach	CA
Reynaldo Rivera	Estrella Mountain Community College	AZ
Kathy V. Rodgers	University of Southern Indiana	IN
Al Tinsley	Central Missouri State University	MO
Carol Walker	Hinds Community College	MS

Our sincere thanks go to these dedicated individuals at Addison Wesley Longman who worked long and hard to make this revision a success: Greg Tobin, Bill Poole, Christine O'Brien, Brenda Bravener, Laura Potter Walton, Barbara Atkinson, Becky Malone, and Jennifer Kerber.

While Terry McGinnis has assisted us for many years "behind the scenes" in producing our texts, she has contributed far more to these revisions than ever. There is no question that these books are improved because of her attention to detail and consistency, and we are most grateful for her work above and beyond the call of duty.

Kitty Pellissier did her usual outstanding job checking the answers to the exercises. Steven Pusztai of Elm Street Publishing Services provided excellent production work. We have received positive comments on our indexes thanks to the work of Paul Van Erden. Thanks are also due to Steve Ouellette, who provided the new teaching tips for the *Annotated Instructor's Edition,* and Becky Troutman for preparing the Index of Applications.

To these individuals and all those who have worked in some way on this series over the last 30 years, we are grateful for your contributions. Remember that we could not have done it without you. We hope that you share with us our pride in these books.

Margaret L. Lial

John Hornsby

David I. Schneider

The Trigonometric Functions

1

1.1 Basic Concepts

1.2 Angles

1.3 Angle Relationships and Similar Triangles

1.4 Definitions of the Trigonometric Functions

1.5 Using the Definitions of the Trigonometric Functions

Trigonometry has been used for millennia to solve problems related to astronomy, surveying, and construction. Prior to the fifteenth century, astronomy had the greatest influence on the development of trigonometry. The Greek astronomer Hipparchus is usually given credit for first studying the trigonometric properties of angles.

For centuries, astronomers tried to determine the distances to various stars. It was not until 1838 that the astronomer Friedrich Bessel determined the distance to a star called 61 Cygni. He used a *parallax* method that relied on the measurement of very small angles. This measurement confirmed that the heliocentric model of Copernicus was correct and gave scientists a better understanding of the size and structure of the universe.

You observe parallax when you ride in an automobile and see a nearby object apparently moving backward with respect to more distant objects. The same is true for some stars that are relatively close to Earth. As Earth revolves around the sun, the observed angle θ (the Greek letter *theta**) of some nearby stars changes due to parallax, as shown in the figure on the next page.

*Greek letters are often used to name angles. A list of Greek letters appears inside the back cover of this book.

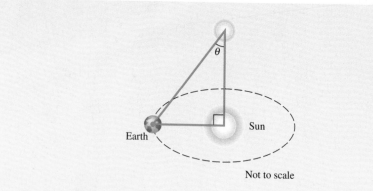

Not to scale

The table lists the size of angle θ in seconds for five stars. (One second is equal to 1/3600th of a degree.)

Star	θ
Alpha Centauri	.763
Barnard's Star	.546
Sirius	.377
61 Cygni	.292
Procyon	.287

Since stars are very distant objects, the parallax of a star is small and always less than one second. How can we use θ to find the distance to these stars? To solve problems like estimating distances to stars, determining the feasibility of a total solar eclipse, or approximating the depth of a crater on the moon, we must understand angles, triangles, and trigonometric functions. These concepts are introduced in this chapter.*

1.1   Basic Concepts

• **The Coordinate Plane**   •   **Interval Notation**   •   **Relations and Functions**

The Coordinate Plane   Many ideas in trigonometry are best explained with a graph in a plane. Recall from algebra that each point in the plane corresponds to an *ordered pair,* two numbers written inside parentheses, such as $(-2, 4)$. Graphs are set up with two axes, one for each number in an ordered pair. The horizontal axis, called the *x-axis,* and the vertical axis, called the *y-axis,* intersect at a point called the *origin.* To locate the point that corresponds to the ordered pair

---

*Sources: Freebury, H. A., *A History of Mathematics,* MacMillan Company, 1968.

Zeilik, M., S. Gregory, and E. Smith, *Introductory Astronomy and Astrophysics,* Second Edition, Saunders College Publishers, 1998.

$(-2, 4)$, start at the origin, and move 2 units left and 4 units up. The point $(-2, 4)$ and other sample points are shown in Figure 1.

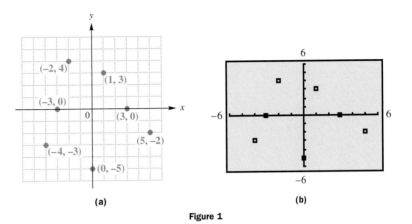

(a)   (b)

Figure 1

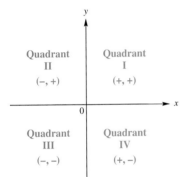

Figure 2

Figure 1(b) shows the points in Figure 1(a) plotted on a typical graphing calculator screen. The minimum and maximum x-values and y-values are -6 and 6. For both axes, the distance between tick marks (the scale) is 1. ∎

The axes divide the plane into four regions called *quadrants*. The quadrants are numbered in a counterclockwise direction, as shown in Figure 2. The points on the axes themselves do not belong to any quadrant. Figure 2 also shows that in quadrant I both the x-coordinate and the y-coordinate are positive; in quadrant II the value of x is negative while y is positive, and so on.

The distance between any two points in a plane can be found by using a formula derived from the **Pythagorean theorem.**

Pythagorean Theorem

The sum of the squares of the legs of a right triangle equals the square of the hypotenuse, so $a^2 + b^2 = c^2$.

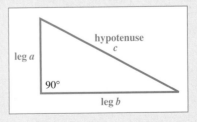

A proof of the Pythagorean theorem is outlined in Exercises 87–91. *The Pythagorean Proposition* by Elisha Scott Loomis (National Council of Teachers of Mathematics, 1968) contains 256 proofs of the Pythagorean Theorem. One of those proofs is attributed to James Garfield, twentieth president of the United States.

● ● ● **Example 1** Finding a Distance from Earth to the Center of the Sun

The maximum distance from the surface of Earth to a point on the sun is 92,955,600 miles. See Figure 3. The diameter of the sun is about 864,930 miles. Find the smallest distance from Earth's surface to the center of the sun.

From the given diameter, we know that the radius of the sun is $864,930/2 = 432,465$ miles. As Figure 3 shows, we are given the lengths of the hypotenuse and one leg of a right triangle, so we use the Pythagorean theorem.

$$c^2 = a^2 + b^2$$
$$92,955,600^2 = 432,465^2 + b^2$$
$$b = 92,954,594$$

The smallest distance is 92,954,594 miles. ● ● ●

Earth

92,955,600 miles

Sun
Not to scale

Figure 3

To find the distance between two points (x_1, y_1) and (x_2, y_2), draw the line segment connecting the points, as shown in Figure 4. Complete a right triangle by drawing a line through (x_1, y_1) parallel to the x-axis and a line through (x_2, y_2) parallel to the y-axis. The ordered pair at the right angle of this triangle is (x_2, y_1).

The horizontal side of the right triangle in Figure 4 has length $x_2 - x_1$, while the vertical side has length $y_2 - y_1$. If d represents the distance between the two original points, then by the Pythagorean theorem,

$$d^2 = (x_2 - x_1)^2 + (y_2 - y_1)^2.$$

Solving for d, we obtain the *distance formula*.

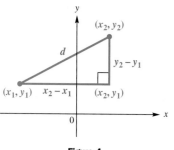

Figure 4

Distance Formula

The distance between the points (x_1, y_1) and (x_2, y_2) is given by the **distance formula,**

$$d = \sqrt{(x_2 - x_1)^2 + (y_2 - y_1)^2}.$$

● ● ● **Example 2** Using the Distance Formula

Find the distance, d, between the points $(-7, 2)$ and $(3, -8)$.

Either point can be used as (x_1, y_1). We choose $(-7, 2)$ as (x_1, y_1) and $(3, -8)$ as (x_2, y_2).

$$d = \sqrt{[3 - (-7)]^2 + (-8 - 2)^2}$$ $x_1 = -7, y_1 = 2, x_2 = 3, y_2 = -8$
$$= \sqrt{10^2 + (-10)^2}$$
$$= \sqrt{100 + 100}$$
$$= \sqrt{200}$$
$$= 10\sqrt{2}$$ $\sqrt{200} = \sqrt{100} \cdot \sqrt{2} = 10\sqrt{2}$ ● ● ●

● ● ● **Example 3** Finding the Distance between a House and a Forest Fire

Firefighters have determined that the coordinates of a house are $(1131.8, 4390.2)$ in feet, and the coordinates of a raging forest fire at the point nearest the house are $(2277.5, -2596.2)$ in feet. Find the distance from the fire to the house.

Use the distance formula.

$$d = \sqrt{(2277.5 - 1131.8)^2 + (-2596.2 - 4390.2)^2}$$
$$= \sqrt{(1145.7)^2 + (-6986.4)^2}$$
$$\approx 7079.7 \quad \approx \text{means "is approximately equal to."}$$

The fire is about 7079.7 feet from the house. ● ● ●

C O N N E C T I O N S

The *midpoint formula* is used to find the coordinates of the midpoint of a line segment, the point equidistant from the endpoints of the segment. Suppose (x_1, y_1) and (x_2, y_2) are any two distinct points in a plane. (Although the figure shows $x_1 < x_2$, no particular order is required.) Assume that the two points are not on a horizontal or vertical line. Let (x, y) be the midpoint of the segment connecting (x_1, y_1) and (x_2, y_2). Draw vertical lines from each of the three points to the x-axis as shown in the figure.

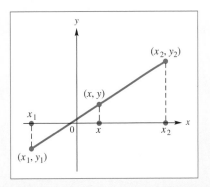

Since (x, y) is the midpoint of the line segment connecting (x_1, y_1) and (x_2, y_2), the distance between x and x_1 equals the distance between x and x_2.

$$x_2 - x = x - x_1$$
$$x_2 + x_1 = 2x$$
$$x = \frac{x_1 + x_2}{2}$$

Thus, the x-coordinate of the midpoint is the average of the x-coordinates of the endpoints of the segment. Similarly, the y-coordinate of the midpoint is the average of the y-coordinates of the endpoints of the segment.

$$y = \frac{y_1 + y_2}{2}$$

For Discussion or Writing

1. Find the midpoint of the segment with endpoints $(8, -4)$ and $(-9, 6)$.
2. A line segment has an endpoint at $(2, -8)$ and midpoint $(-1, -3)$. Find the other endpoint of the segment.
3. Verify the formula given above for the y-coordinate of the midpoint.
4. In the discussion, we used the fact that the x-value of the midpoint is halfway between x_1 and x_2 on the x-axis. Why is this true?

Interval Notation We sometimes specify a set of numbers defined by an inequality using **set-builder notation,** such as $\{x \mid x < 5\}$ (read "the set of all x such that x is less than 5"). We also use another notation, called **interval notation,** for writing intervals. Using interval notation, we write the set $\{x \mid x < 5\}$ as $(-\infty, 5)$. The symbol $-\infty$ does not indicate a number; it is used to show that the interval includes all real numbers less than 5. The right-hand parenthesis indicates that 5 is not included. Since there is no smallest number, and 5 is not included, the interval $(-\infty, 5)$ is an example of an **open interval.** Intervals that include the endpoints, such as $\{x \mid 0 \leq x \leq 5\}$, are **closed intervals.** Closed intervals are indicated with square brackets; we write the interval $\{x \mid 0 \leq x \leq 5\}$ as $[0, 5]$. An interval like $(2, 5]$, that is open on one end and closed on the other, is a **half-open interval.** Examples of other sets written in interval notation are shown in the following chart. In these intervals, assume that $a < b$. Note that a parenthesis is always used with the symbols $-\infty$ and ∞.

Type of Interval	Set	Interval Notation	Graph
Open interval	$\{x \mid x > a\}$	(a, ∞)	
	$\{x \mid a < x < b\}$	(a, b)	
	$\{x \mid x < b\}$	$(-\infty, b)$	
Half-open interval	$\{x \mid x \geq a\}$	$[a, \infty)$	
	$\{x \mid a < x \leq b\}$	$(a, b]$	
	$\{x \mid a \leq x < b\}$	$[a, b)$	
	$\{x \mid x \leq b\}$	$(-\infty, b]$	
Closed interval	$\{x \mid a \leq x \leq b\}$	$[a, b]$	

It is also customary to use $(-\infty, \infty)$ to represent the set of all real numbers.

Relations and Functions A **relation** is defined as a set of ordered pairs. Many relations have a rule or formula showing the connection between the two components of the ordered pairs. For example, the formula

$$y = -5x + 6$$

shows that a value of y can be found from a given value of x by multiplying the value of x by -5 and then adding 6. According to this formula, if $x = 2$, then $y = -5 \cdot 2 + 6 = -4$, so $(2, -4)$ belongs to the relation. In the relation $y = -5x + 6$, the value of y depends on the value of x, so y is the **dependent variable** and x is the **independent variable.**

NOTE A relation is a set of points, often defined by an equation such as $y = -5x + 6$. While precise language would require that we say "the relation defined by the equation $y = -5x + 6$," we will often use the less cumbersome language "the relation $y = -5x + 6$."

Most of the relations we use in trigonometry are also *functions*.

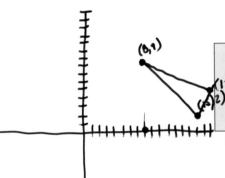

Function

A relation is a **function** if each value of the independent variable leads to exactly one value of the dependent variable. This means that each value of x produces exactly one value of y.

$$d = \sqrt{(17-15)^2 + (4-2)^2}$$

$$d = \sqrt{(15-17)^2 +}$$

It is customary for x to be considered the independent variable and y the dependent variable, and we shall follow that convention.

For example, $y = -5x + 6$ defines a function. For any one value of x that might be chosen, $y = -5x + 6$ gives exactly one value of y. In contrast, $y^2 = x$ defines a relation that is not a function. If we choose the value $x = 16$, then $y^2 = x$ becomes $y^2 = 16$, from which $y = 4$ or $y = -4$. The one x-value, 16, leads to two y-values, 4 and -4, so $y^2 = x$ does not define a function.

Functions are often named with letters such as f, g, or h. For example, the function $y = -5x + 6$ can be written using function notation as

$$f(x) = -5x + 6,$$

where $f(x)$, read "f of x," replaces y. For the function defined by $f(x) = -5x + 6$, if $x = 3$ then

$$f(x) = f(3) = -5 \cdot 3 + 6 = -15 + 6 = -9,$$

or $$f(3) = -9,$$

indicating that the ordered pair $(3, -9)$ belongs to function f. Also,

$$f(-7) = -5(-7) + 6 = 41,$$

so $(-7, 41)$ belongs to f.

Recall that $|a|$ represents the absolute value of a. By definition, $|a| = a$ if $a \geq 0$ and $|a| = -a$ if $a < 0$. Thus $|4| = 4$ and $|-4| = 4$.

● ● ● **Example 4** Using Function Notation

Let $f(x) = -x^2 + |x - 5|$. Find each of the following.

Algebraic Solution

(a) $f(0)$

Use $f(x)$ and replace x with 0.

$$f(0) = -0^2 + |0 - 5|$$
$$= -0 + |-5| = 5$$

(b) $f(-4) = -(-4)^2 + |-4 - 5|$
$$= -16 + |-9|$$
$$= -16 + 9 = -7$$

(c) $f(a) = -a^2 + |a - 5|$ Each x was replaced with a.

(d) Why does f define a function?

For each value of x, there is exactly one value of $f(x)$; therefore f defines a function.

Graphing Calculator Solution

Function notation is available on graphing calculators. If Y_1 is defined as $-x^2 + |x - 5|$, the function values at 0 and -4 are 5 and -7, respectively, as shown in Figure 5. These results agree with the corresponding function values found in parts (a) and (b).

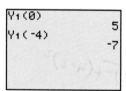

Figure 5

● ● ●

The set of all possible values of the independent variable (x) for which a relation is defined is called the **domain** of the relation. The set of all possible values for the dependent variable (y) is the **range** of the relation. By observing the graph of a relation, we can determine the domain and the range. For example, in Figure 6 the domain is the set of real numbers between -6 and 6 inclusive, and the range is the set of real numbers between -2 and 2 inclusive. Using interval notation, these sets are written $[-6, 6]$ and $[-2, 2]$, respectively.

For a relation to be a function, each value of x in the domain of the function must lead to exactly one value of y. Figure 7 shows the graph of a relation. A point x_1 has been chosen on the x-axis. A vertical line drawn through x_1 intersects the graph in more than one point. Since the x-value x_1 leads to more than one value of y, this graph is not the graph of a function. This example suggests the *vertical line test* for a function.

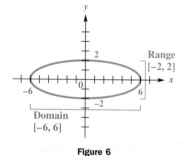

Figure 6

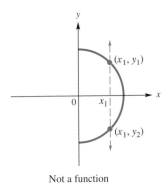

Not a function

Figure 7

Vertical Line Test

If every vertical line intersects the graph of a relation in no more than one point, then the graph is the graph of a function.

● ● ● **Example 5** Identifying Domains, Ranges, and Functions

Find the domain and range of each relation. Identify any functions.

Algebraic Solution

(a) $y = x^2$

Here x, the independent variable, can take on any value, so the domain is the set of all real numbers, $(-\infty, \infty)$. Since the dependent variable y equals the square of x, and since a square is never negative, the range is the set of all nonnegative numbers, $[0, \infty)$.

Each value of x leads to exactly one value of y, so $y = x^2$ defines a function. The graph of $y = x^2$ in Figure 8 shows that it satisfies the conditions of the vertical line test.

Graphing Calculator Solution

(a) To graph a relation with a graphing calculator, the defining equation must first be solved for y. If the equation is in the form $y = f(x)$, where $f(x)$ is some expression in x, then the relation is a function. Therefore, $y = x^2$ defines a function.

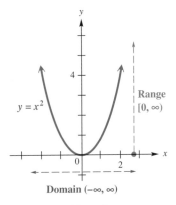

Figure 8

(continued)

(b) $3x + 2y = 6$

In this relation x and y can take on any value, so both the domain and range are $(-\infty, \infty)$. For any value of x that might be chosen, the equation $3x + 2y = 6$ would lead to exactly one value of y. Therefore, $3x + 2y = 6$ defines a function.

(c) $x = y^2 + 2$

For any value of y, the square of y is nonnegative; that is, $y^2 \geq 0$. Since $x = y^2 + 2$, this means that $x \geq 0 + 2 = 2$, making the domain of the relation $[2, \infty)$. Any real number may be squared, so the range is the set of all real numbers, $(-\infty, \infty)$. To decide whether the relation is a function, choose a sample value of x greater than 2 from the domain.

$$6 = y^2 + 2 \quad \text{Let } x = 6.$$
$$4 = y^2$$
$$y = 2 \quad \text{or} \quad y = -2$$

Since one x-value, 6, leads to two y-values, 2 and -2, the relation $x = y^2 + 2$ does not define a function. As can be seen in Figure 9, the graph does not satisfy the vertical line test.

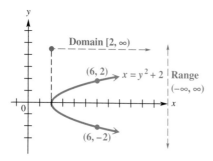

Figure 9

(d) $y = \sqrt{1 - x}$

The quantity under the radical, $1 - x$, must be greater than or equal to 0 for y to be a real number.

$$1 - x \geq 0$$
$$x \leq 1 \quad \text{Add } x \text{ to both sides; } 1 \geq x \text{ means } x \leq 1.$$

Thus, the domain is $(-\infty, 1]$. To determine the range, note that the given radical is nonnegative, so the range is $[0, \infty)$. Since each value of x in the domain leads to a single value of y, this relation is a function.

(b) We can solve the equation for y to get

$$y = \frac{6 - 3x}{2}.$$

This shows that the relation is a function.

(c) Solve the equation $x = y^2 + 2$ for y.

$$x - 2 = y^2 \quad \text{Subtract 2 on each side.}$$

Take the square root on each side.

$$y = \sqrt{x - 2}$$

or

$$y = -\sqrt{x - 2}$$

Because we had to write two expressions for y, the relation is not a function.

(d) Because the equation that defines this relation is solved for y, the relation is a function.

● ● ●   **Example 6**   Finding the Domain of a Function from Its Rule

Find the domain of each function.

(a) $y = \dfrac{1}{x - 2}$

Division by 0 is undefined, so x cannot equal 2. (This value of x would make the denominator become $2 - 2$, or 0.) Any other value of x is acceptable, so the domain is all values of x other than 2, that is, $(-\infty, 2) \cup (2, \infty)$.

X	Y1
-1	-.3333
0	-.5
1	-1
2	ERROR
3	1
4	.5
5	.33333

Y1 = 1/(X-2)

Graphing calculators are capable of generating tables of ordered pairs for functions. Here Y_1 is defined as shown in Example 6(a). Notice that an ERROR message is returned for X = 2 since this leads to 0 in the denominator.

(b) $y = \dfrac{8 + x}{(2x - 3)(4x - 1)}$

This denominator is 0 if either

$$2x - 3 = 0 \qquad \text{or} \qquad 4x - 1 = 0.$$

Solve each of these equations.

$$
\begin{aligned}
2x - 3 &= 0 & \text{or} & & 4x - 1 &= 0 \\
2x &= 3 & \text{or} & & 4x &= 1 \\
x &= \frac{3}{2} & \text{or} & & x &= \frac{1}{4}
\end{aligned}
$$

The domain here includes all real numbers x such that $x \neq 3/2$ and $x \neq 1/4$, written using interval notation as $(-\infty, 1/4) \cup (1/4, 3/2) \cup (3/2, \infty)$.

(c) $y = \dfrac{1}{\sqrt{x^2 + 16}}$

For the expression to be defined, the denominator cannot be 0, and the radicand, $x^2 + 16$, must be positive. To find the domain, solve $x^2 + 16 > 0$.

$$x^2 + 16 > 0$$

$$x^2 > -16 \qquad \text{Subtract 16.}$$

Since this last inequality is true for all real numbers, the domain is $(-\infty, \infty)$.

● ● ●

N O T E   In later work, we will encounter trigonometric functions that are defined in such a way that restrictions are necessary since their denominators cannot equal 0. Example 6 illustrates this idea with algebraic functions.

1.1   Exercises

Graph the points on a coordinate system and identify the quadrant or axis for each point.

1. $(3, 2)$ **2.** $(-7, 6)$ **3.** $(-7, -4)$ **4.** $(8, -5)$

5. $(0, 5)$ **6.** $(-8, 0)$ **7.** $(4.5, 7)$ **8.** $(-7.5, 8)$

Give the quadrant in which each point lies.

9. $(-5, \pi)$ **10.** $(\pi, -3)$ **11.** $(-\sqrt{2}, -2\sqrt{2})$ **12.** $\left(1 + \sqrt{3}, \dfrac{\pi}{2}\right)$

13. Suppose the point (a, b) lies in the first quadrant. Describe how you would move from the point (a, b) to the point $(a, -b)$.

14. Suppose the point (a, b) lies in the first quadrant. Describe how you would move from the point (a, b) to the point $(-a, b)$.

15. *Concept Check* If $xy = 1$ is graphed, in which quadrants will the points of the graph lie?

16. *Concept Check* If (a, b) represents a point that lies in quadrant II, in which quadrant will each point lie?
(a) $(-a, b)$ **(b)** $(-a, -b)$ **(c)** $(a, -b)$

Use the distance formula to find the distance between each pair of points. See Example 2.

17. $(2, -1)$ and $(-3, -4)$

18. $(-5, 2)$ and $(3, -7)$

19. $(-1, 0)$ and $(-4, -5)$

20. $(-2, -3)$ and $(-6, 4)$

21. $\left(\sqrt{2}, -\sqrt{5}\right)$ and $\left(3\sqrt{2}, 4\sqrt{5}\right)$

22. $\left(5\sqrt{7}, -\sqrt{3}\right)$ and $\left(-\sqrt{7}, 8\sqrt{3}\right)$

23. Determine the distance between $(5, -6)$ and the x-axis.

24. Determine the distance between $(5, -6)$ and the y-axis.

25. The graphing calculator screen shows two ways to find the distance between two points in the plane. What are the two points?

26. *Concept Check* Suppose the point (a, b) lies in the first quadrant and is 6 units from the origin. In which quadrant is the point $(-a, -b)$ located, and how far is it from the origin?

27. State the Pythagorean theorem in your own words.

28. Suppose the point (a, b) lies in the first quadrant and is 4 units from the origin. Explain why $a < 4$ and $b < 4$.

```
((-2-3)²+(1+π)²)
^.5
          6.492517979
√((-2-3)²+(1+π)²
)
          6.492517979
```

A triple of positive integers (a, b, c) is called a Pythagorean triple if it satisfies the equation of the Pythagorean theorem, $a^2 + b^2 = c^2$. Determine whether each triple is a Pythagorean triple.

29. $(9, 12, 15)$ **30.** $(6, 8, 10)$ **31.** $(5, 10, 15)$ **32.** $(7, 24, 25)$

33. Show by an example that the following statement is true: If, for the positive integers a, b, and c, $a^2 + b^2 = c^2$, then it is not necessarily true that $a + b = c$.

34. Show that $(3, 4, 5)$ is a Pythagorean triple. Then, show that $(3k, 4k, 5k)$ is a Pythagorean triple for $k = 2$, $k = 3$, and $k = 4$. What general conclusion seems likely from this observation? (Although this is not a proof, the conclusion is indeed true and can be proved using algebra.)

The converse of the Pythagorean theorem says that if a triangle has sides of lengths a, b, and c, where c is the longest side, and $a^2 + b^2 = c^2$, then the triangle is a right triangle. Use this result and the distance formula to decide whether the following points are the vertices of right triangles.

35. $(-2, 5), (1, 5), (1, 9)$

36. $(-9, -2), (-1, -2), (-9, 11)$

37. $\left(\sqrt{3}, 2\sqrt{3} + 3\right), \left(\sqrt{3} + 4, -\sqrt{3} + 3\right), \left(2\sqrt{3}, 2\sqrt{3} + 4\right)$

38. $\left(4 - \sqrt{3}, -2\sqrt{3}\right), \left(2 - \sqrt{3}, -\sqrt{3}\right), \left(3 - \sqrt{3}, -2\sqrt{3}\right)$

Find all values of x or y such that the distance between the given points is as indicated.

39. $(x, 7)$ and $(2, 3)$ is 5

40. $(5, y)$ and $(8, -1)$ is 5

41. $(3, y)$ and $(-2, 9)$ is 12

42. $(x, 11)$ and $(5, -4)$ is 17

43. Use the distance formula to write an equation for all points that are 5 units from $(0, 0)$. Sketch a graph showing these points.

44. Write an equation for all points 3 units from $(-5, 6)$. Sketch a graph showing these points.

The following exercises require the Pythagorean theorem. See Example 1.

45. *Slant Height of a Pyramid* The Washington Monument is a marble obelisk with an aluminum pyramid at the top that acts as a lightning rod. See the figure. Find the length x of a slant edge of the pyramid to the nearest hundredth of a foot. (*Source:* National Park Service.)

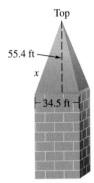

Top

55.4 ft

x

34.5 ft

46. *Square Corners* Carpenters use the Pythagorean theorem to test for square corners. Suppose a rectangular floor has width 5 feet and length 12 feet. What should the diagonal of the rectangle measure to show that the floor has a square corner?

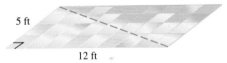

5 ft

12 ft

47. *Railroad Track Expansion* A 1000-foot section of railroad track expands 6 inches because the day is very hot. This causes end C to break off and move to position B, forming right triangle ABC. (See the figure.) Find BC. *Note:* The surprising answer to this simple problem explains why railroad tracks, bridges, and similar structures must be designed to allow for expansion. (*Source: Trigonometry with Calculators* by Lawrence S. Levy. Reprinted by permission of the author.)

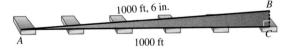

1000 ft, 6 in.

B

A

1000 ft

C

48. *Cutting on the Bias* Clothing manufacturers sometimes cut their material "on the bias" (that is, at 45° to the direction the threads run) to give it more elasticity. A tie maker wants to cut twenty 8-inch strips of silk on the bias from a rectangular piece of material that costs $10 per (linear) yard of material 42 inches wide. (See the figure.) Find the total cost of the material. (*Hint:* First find length AB, using isosceles triangle ABX.) *Note:* This unappealing combination of units—inches,

yards, and dollars—is typical of many practical problems, not just in the clothing industry. (*Source: Trigonometry with Calculators* by Lawrence S. Levy. Reprinted by permission of the author.)

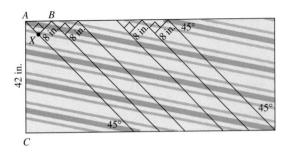

A   B

X   8 in.   8 in.   8 in.   8 in.   45°

42 in.

45°

45°

C

49. *Measurements of the Great Pyramid* The height (h) of the Great Pyramid of Egypt is 144 meters. The apothem (a in the figure) measures 184.7 meters. Assuming the base is a square, find the length ℓ of a side of the base. (*Source:* Hawkes, Jacquetta, *Atlas of Ancient Archaeology,* Barnes & Noble.)

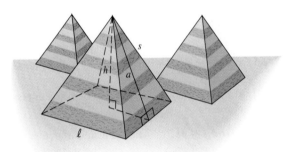

50. *Measurements of the Great Pyramid* Use the result of Exercise 49 to find the length of the edge of the pyramid labeled s in the figure.

51. *Expansion of a Bridge* The longest bridge span in the United States is the 4260-foot Verrazano-Narrows bridge in New York. Suppose that on a hot day it expands 2 feet. If there were no expansion joints to absorb the elongation, the bridge might buckle as shown in the figure. Estimate the height of the bulge, that is, AB. *Note:* The right half of the bridge can be considered to be a relatively small arc of a circle with a very large radius. In such a case, the length of its inscribed chord, BC, is approximately the same as the length of the arc. (*Source: World Almanac and Book of Facts,* 2000.)

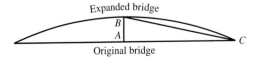

Expanded bridge

B

A

C

Original bridge

Write each set using interval notation.

52. $\{x \mid x > 6\}$ **53.** $\{p \mid p \geq 10\}$ **54.** $\{x \mid -3 < x < 7\}$ **55.** $\{y \mid 8 \leq y \leq 13\}$

56. +——+——+——⊙——+——→
 $\quad$ -1 $\quad$ 0 $\quad$ 1 $\quad$ 2 $\quad$ 3

57. ←——+——+——+——⟩——+——→
 $\quad$ -1 $\quad$ 0 $\quad$ 1 $\quad$ 2 $\quad$ 3

58. +——+——⊙——+——+——⌉——+——→
 $\quad$ -1 $\quad$ 0 $\quad$ 1 $\quad$ 2 $\quad$ 3 $\quad$ 4 $\quad$ 5

59. Explain why the set $\{y \mid |y| \leq 1\}$ is the same as $[-1, 1]$.

60. Explain why the set $\{y \mid |y| \geq 1\}$ is the same as $(-\infty, -1] \cup [1, \infty)$.

Let $f(x) = -2x^2 + 4x + 6$. Find each of the following. See Example 4.

61. $f(0)$ **62.** $f(-2)$ **63.** $f(-1)$ **64.** $f(-m)$ **65.** $f(1 + a)$ **66.** $f(2 - p)$

67. *Concept Check* The table was generated by a graphing calculator for a linear function $Y_1 = f(x)$. Use the table to answer the following questions.

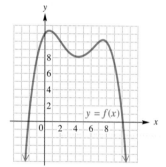

X	Y₁
0	3.6
1	2.4
2	1.2
3	0
4	-1.2
5	-2.4
6	-3.6

X=0

(a) What is $f(2)$?

(b) If $f(x) = -2.4$, what is the value of x?

(c) At what point does the graph of $Y_1 = f(x)$ intersect the y-axis?

(d) At what point does the graph of $Y_1 = f(x)$ intersect the x-axis?

68. *Concept Check* Let $f(x)$ be the function in the graph. Find each of the following.

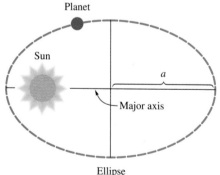

(a) $f(0)$ **(b)** $f(6)$

(c) a negative number a for which $f(a) = 0$

(d) three positive values of x for which $f(x) = 10$

(e) the distance between $(8, f(8))$ and $(10, f(10))$

69. 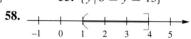 *Orbit of Mars* The German astronomer Johannes Kepler's third law of planetary motion is expressed as the relation $P^2 = a^3$. Here, P is the *sidereal* or *orbital period* of a planet, the time from one orbital point back to that same point in years. The variable a represents half the length of the major axis of the orbital ellipse in *astronomical units*. One astronomical unit (AU) is the average distance between Earth and the sun. The table gives a in astronomical units for selected planets.

Astronomical Units

Planet	Mercury	Venus	Mars	Jupiter	Saturn	Neptune
a (AU)	.387	.723	1.524	5.20	9.54	30.1

Source: Kaler, James B., *Astronomy! A Brief Edition,* Addison Wesley, 1997.

(a) In the relation defined by $P^2 = a^3$, is P a function of a? If the range of P is restricted to $(0, \infty)$, does $P^2 = a^3$ define P as a function of a?

(b) Find P for Mars. Interpret your answer.

(c) For Uranus, $P = 19.18$. Find a.

Find the domain and range of each relation. Identify any relations that define functions. See Example 5.

70. $y = 4x - 3$ **71.** $2x + 5y = 10$ **72.** $y = x^2 + 4$

73. $y = 2x^2 - 5$ **74.** $y = -2(x - 3)^2 + 4$ **75.** $y = 3(x + 1)^2 - 5$

76. $x = y^2$ **77.** $y = \sqrt{4 + x}$ **78.** $y = \sqrt{x^2 + 1}$ **79.** $y = \sqrt{1 - x^2}$

80. 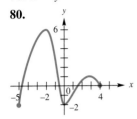   **81.** 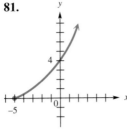   **82.** 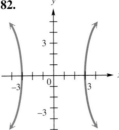   **83.**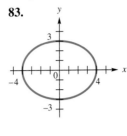

Find the domain of each relation. See Example 6.

84. $y = \dfrac{1}{x}$ **85.** $y = \dfrac{-2}{x + 1}$ **86.** $y = \dfrac{-1}{\sqrt{x^2 + 25}}$

· · · · · · · · · · · · · **Relating Concepts** · · · · · · · · · · · · ·

For individual or collaborative investigation
(Exercises 87–91)

The figure shown is a square made up of four right triangles and a smaller square. By using the method of equal areas, the Pythagorean theorem may be proved. **Work Exercises 87–91 in order,** *filling in the blanks with the missing information.*

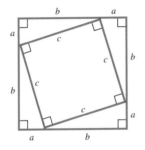

87. The length of a side of the large square is _____, so its area is (_____)^2 or _____.

88. The area of the large square may also be found by obtaining the sum of the areas of the four right triangles and the smaller square. The area of each right triangle is _____, so the sum of the areas of the four right triangles is _____. The area of the smaller square is _____.

89. The sum of the areas of the four right triangles and the smaller square is _____.

90. Since the areas in Exercises 87 and 89 represent the area of the same figure, the expressions there must be equal. Setting them equal to each other we obtain _____ = _____.

91. Subtract $2ab$ from each side of the equation in Exercise 90 to obtain the desired result _____ = _____.

· ·

1.2 Angles

• Basic Terminology • Degree Measure • Standard Position • Coterminal Angles

 Line *AB*

 Segment *AB*

 Ray *AB*

Figure 10

Basic Terminology   Two distinct points A and B determine a line called **line AB.** The portion of the line between A and B, including points A and B themselves, is **segment AB.** The portion of line AB that starts at A and continues through B, and on past B, is called **ray AB.** Point A is the endpoint of the ray. (See Figure 10.)

An **angle** is formed by rotating a ray around its endpoint. The ray in its initial position is called the **initial side** of the angle, while the ray in its location after the rotation is the **terminal side** of the angle. The endpoint of the ray is the **vertex** of the angle. Figure 11 shows the initial and terminal sides of an angle with vertex A.

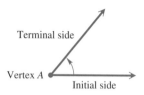

Figure 11

If the rotation of the terminal side is counterclockwise, the angle is **positive.** If the rotation is clockwise, the angle is **negative.** Figure 12 shows two angles, one positive and one negative.

An angle can be named by using the name of its vertex. For example, the angle on the right in Figure 12 can be called angle C. Alternatively, an angle can be named using three letters, with the vertex letter in the middle. Thus, the angle on the right also could be named angle ACB or angle BCA.

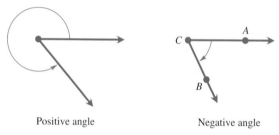

Positive angle   Negative angle

Figure 12

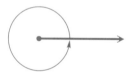

A complete rotation of a ray gives an angle whose measure is 360°.

Figure 13

Degree Measure   There are two systems in common use for measuring the size of angles. The most common unit of measure is the **degree.** (The other common unit of measure, called the *radian,* is discussed in Chapter 3.) Degree measure was developed by the Babylonians, 4000 years ago.* To use degree measure, we assign 360 degrees to a complete rotation of a ray. In Figure 13, notice that the terminal side of the angle corresponds to its initial side when it makes a complete rotation.

One degree, written 1°, represents 1/360 of a rotation. Therefore, 90° represents $90/360 = 1/4$ of a complete rotation, and 180° represents $180/360 = 1/2$ of a complete rotation. Angles of measure 1°, 90°, and 180° are shown in Figure 14.

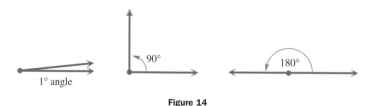

1° angle

Figure 14

---

*The Babylonians were the first to subdivide the circumference of a circle into 360 parts. There are various theories as to why the number 360 was chosen. One is that it is approximately the number of days in a year, and it has many divisors, which makes it convenient to work with. Another involves a roundabout theory dealing with the length of a Babylonian mile.

Special angles are named as shown in the following chart.

Types of Angles

Name	Angle Measure	Example
Acute angle	Between 0° and 90°	60° 82°
Right angle	Exactly 90°	90°
Obtuse angle	Between 90° and 180°	97° 138°
Straight angle	Exactly 180°	180°

If the sum of the measures of two angles is 90°, the angles are called **complementary.** Two angles with measures whose sum is 180° are **supplementary.**

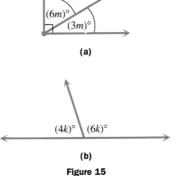

(a)

(b)

Figure 15

● ● ● **Example 1** Finding Measures of Complementary and Supplementary Angles

Find the measure of each angle in Figure 15.

(a) In Figure 15(a), since the two angles form a right angle (as indicated by the ⌐ symbol),

$$6m + 3m = 90$$
$$9m = 90$$
$$m = 10.$$

The two angles have measures of $6 \cdot 10 = 60°$ and $3 \cdot 10 = 30°$.

(b) The angles in Figure 15(b) are supplementary, so

$$4k + 6k = 180$$
$$10k = 180$$
$$k = 18.$$

These angle measures are $4(18) = 72°$ and $6(18) = 108°$.   ● ● ●

Angles can be measured with an instrument called a **protractor.** Figure 16 shows a protractor measuring an angle of 35°.

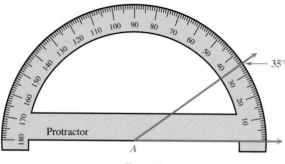

Figure 16

NOTE While much of the study of trigonometry involves finding angle measures, trigonometry does not rely on using protractors to find the measures. For example, if we know the lengths of three sides of a triangle, we can use the law of cosines (Section 7.3) to find the angle measures mathematically, and with more precision than we could get using a protractor.

Do not confuse an angle with its measure. Angle A of Figure 16 is a rotation; the measure of the rotation is 35°. This measure is often expressed by saying that m(angle A) is 35°, where m(angle A) is read "the measure of angle A." It is convenient, however, to abbreviate m(angle A) = 35° as simply angle A = 35°.

Traditionally, portions of a degree have been measured with minutes and seconds. One **minute,** written 1′, is 1/60 of a degree.

$$1' = \frac{1}{60}^{\circ} \qquad \text{or} \qquad 60' = 1^{\circ}$$

One **second,** 1″, is 1/60 of a minute.

$$1'' = \frac{1}{60}^{'} = \frac{1}{3600}^{\circ} \qquad \text{or} \qquad 60'' = 1'$$

The measure 12° 42′ 38″ represents 12 degrees, 42 minutes, 38 seconds.

● ● ● **Example 2** Calculating with Degrees, Minutes, and Seconds

Perform each calculation.

Algebraic Solution

(a) 51° 29′ + 32° 46′

Add the degrees and the minutes separately.

$$\begin{array}{r} 51^{\circ}\ 29' \\ +\ 32^{\circ}\ 46' \\ \hline 83^{\circ}\ 75' \end{array}$$

Graphing Calculator Solution

Some graphing calculators perform calculations with degrees, minutes, and seconds. Your calculator must be in degree (rather than radian) mode. (See your owner's manual for details on

(continued)

Since $75' = 60' + 15' = 1° 15'$, the sum is written

$$83°$$
$$\underline{+ \quad 1° \ 15'}$$
$$84° \ 15'.$$

(b) $90° - 73° \ 12'$
 Write $90°$ as $89° 60'$.

$$89° \ 60'$$
$$\underline{- \ 73° \ 12'}$$
$$16° \ 48'$$

these capabilities.) Figure 17 shows the calculations for parts (a) and (b).

Figure 17

Because calculators are now so prevalent, it is common to measure angles in **decimal degrees.** For example, $12.4238°$ represents

$$12.4238° = 12\frac{4238}{10{,}000}°.$$

● ● ● **Example 3** Converting between Decimal Degrees and Degrees, Minutes, Seconds

Algebraic Solution

(a) Convert $74° \ 8' \ 14''$ to decimal degrees. Round to the nearest thousandth of a degree.

 Since $1' = \dfrac{1}{60}°$ and $1'' = \dfrac{1}{3600}°$,

$$74° \ 8' \ 14'' = 74° + \frac{8}{60}° + \frac{14}{3600}°$$
$$\approx 74° + .1333° + .0039°$$
$$= 74.137° \quad \text{(rounded)}.$$

(b) Convert $34.817°$ to degrees, minutes, and seconds.

$$34.817° = 34° + .817°$$
$$= 34° + .817(60') \qquad 1° = 60'$$
$$= 34° + 49.02'$$
$$= 34° + 49' + .02'$$
$$= 34° + 49' + .02(60'') \quad 1' = 60''$$
$$= 34° + 49' + 1.2''$$
$$= 34° \ 49' \ 1.2''$$

Graphing Calculator Solution

These conversions can be performed on some graphing calculators. A typical screen is shown in Figure 18. The second displayed result was obtained by setting the calculator to show only three decimal places.

```
74°8'14"
          74.13722222
74°8'14"
           74.137
34.817▶DMS
          34°49'1.2"
```

Figure 18

● ● ●

Standard Position An angle is in **standard position** if its vertex is at the origin and its initial side is along the positive x-axis. The angles in Figures 19(a) and 19(b) are in standard position. An angle in standard position is said to lie in the quadrant in which its terminal side lies. For example, an acute angle is in quadrant I (Figure 19(a)) and an obtuse angle is in quadrant II (Figure 19(b)). Figure 19(c) shows ranges of angle measures for each quadrant. Angles in standard position having their terminal sides along the x-axis or y-axis, such as angles with measures 90°, 180°, 270°, and so on, are called **quadrantal angles.**

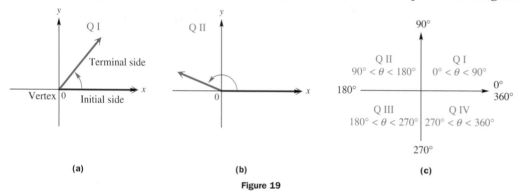

(a)   (b)   (c)

Figure 19

Coterminal Angles A complete rotation of a ray results in an angle measuring 360°. But there is no reason why the rotation has to stop at 360°. By continuing the rotation, angles of measure larger than 360° can be produced. The angles in Figure 20(a) have measures 60° and 420°. These two angles have the same initial side and the same terminal side, but different amounts of rotation. Angles that have the same initial side and the same terminal side are called **coterminal angles,** so the measures of coterminal angles differ by a multiple of 360°. As shown in Figure 20(b), angles with measures 110° and 830° are coterminal.

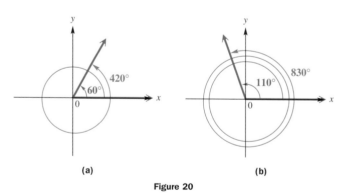

(a)   (b)

Figure 20

● ● ● **Example 4** Finding Measures of Coterminal Angles

Find the angles of smallest possible positive measure coterminal with each angle.

(a) 908°

Add or subtract 360° as many times as needed to get an angle with measure greater than 0° but less than 360°. Since $908° - 2 \cdot 360° = 908° - 720° = 188°$, an angle of 188° is coterminal with an angle of 908°. See Figure 21.

(b) $-75°$

Use a rotation of $360° + (-75°) = 285°$. See Figure 22.   ● ● ●

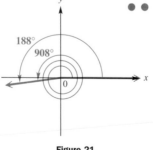

Figure 21

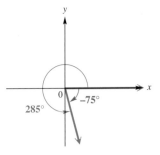

Figure 22

Sometimes it is necessary to find an expression that will generate all angles coterminal with a given angle. For example, any angle coterminal with $60°$ can be obtained by adding an appropriate integer multiple of $360°$ to $60°$. Let n represent any integer; then the expression

$$60° + n \cdot 360°$$

represents all such coterminal angles. The table below shows a few possibilities.

Value of n	Angle Coterminal with $60°$
2	$60° + 2 \cdot 360° = 780°$
1	$60° + 1 \cdot 360° = 420°$
0	$60° + 0 \cdot 360° = 60°$ (the angle itself)
-1	$60° + (-1) \cdot 360° = -300°$

● ● ● **Example 5** Analyzing the Revolutions of a CD Player

There are several types of CD players. CLV (Constant Linear Velocity) players spin faster when reading from the inner tracks of a CD so that the amount of data read per second stays at a fixed level. On the other hand, CAV (Constant Angular Velocity) players always spin at the same speed. Suppose a CAV player makes 480 revolutions per minute. Through how many degrees will a point on the edge of a CD move in 2 seconds?

The player revolves 480 times in 1 minute or $480/60$ times $= 8$ times per second (since 60 seconds $= 1$ minute). In 2 seconds, the player will revolve $2 \cdot 8 = 16$ times. Each revolution is $360°$, so a point on the edge of the CD will revolve $16 \cdot 360° = 5760°$ in 2 seconds. ● ● ●

1.2 Exercises

1. Explain the difference between a segment and a ray.

2. What part of a complete revolution is an angle of $45°$?

3. *Concept Check* What angle is its own complement?

4. *Concept Check* What angle is its own supplement?

Find the measure of the smaller angle formed by the hands of a clock at the following times.

5.

6.

Find the measure of each angle in Exercises 7–12. See Example 1.

7.

$(7x)°$ $(11x)°$

8.

$(2y)°$

$(4y)°$

9.

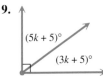

$(5k + 5)°$

$(3k + 5)°$

10. supplementary angles with measures $10m + 7$ and $7m + 3$ degrees

11. supplementary angles with measures $6x - 4$ and $8x - 12$ degrees

12. complementary angles with measures $9z + 6$ and $3z$ degrees

Concept Check   Use the concepts presented in this section to answer each question.

13. If an angle measures $x°$, how can we represent its complement?

14. If an angle measures $x°$, how can we represent its supplement?

15. If a positive angle has measure $x°$ between $0°$ and $60°$, how can we represent the first negative angle coterminal with it?

16. If a negative angle has measure $x°$ between $0°$ and $-60°$, how can we represent the first positive angle coterminal with it?

Perform each calculation. See Example 2.

17. $62° 18' + 21° 41'$

18. $75° 15' + 83° 32'$

19. $71° 18' - 47° 29'$

20. $47° 23' - 73° 48'$

21. $90° - 51° 28'$

22. $180° - 124° 51'$

23. $90° - 72° 58' 11''$

24. $90° - 36° 18' 47''$

Convert each angle measure to decimal degrees. Use a calculator, and round to the nearest thousandth of a degree. See Example 3.

25. $20° 54'$

26. $38° 42'$

27. $91° 35' 54''$

28. $34° 51' 35''$

29. $274° 18' 59''$

30. $165° 51' 9''$

Convert each angle measure to degrees, minutes, and seconds. Use a calculator as necessary. See Example 3.

31. $31.4296°$

32. $59.0854°$

33. $89.9004°$

34. $102.3771°$

35. $178.5994°$

36. $122.6853°$

37. Read about the degree symbol (°) in the manual for your graphing calculator. How is it used?

38. Show that 1.21 hours is the same as 1 hour, 12 minutes, and 36 seconds. Discuss the similarity between converting hours, minutes, and seconds to decimal hours and converting degrees, minutes, and seconds to decimal degrees.

Find the angle of smallest positive measure coterminal with each angle. See Example 4.

39. $-40°$

40. $-98°$

41. $-125°$

42. $-203°$

43. $539°$

44. $699°$

45. $850°$

46. $1000°$

Give an expression that generates all angles coterminal with each angle. Let n represent any integer.

47. $30°$ **48.** $45°$ **49.** $135°$ **50.** $270°$ **51.** $-90°$ **52.** $-135°$

53. Explain why the answers to Exercises 50 and 51 give the same set of angles.

54. *Concept Check*   Which two of the following are not coterminal with $r°$?

  A. $360° + r°$ **B.** $r° - 360°$ **C.** $360° - r°$ **D.** $r° + 180°$

Consider the function $Y_1 = 360((X/360) - \text{int}(X/360))$ *specified on a graphing calculator. (Note: The value of* $\text{int}(x)$ *is the largest integer less than or equal to x. With some calculators, int is found in the MATH menu.) The screen here shows that for* X = 908 *and* X = -75, *the function returns the smallest possible positive measure coterminal with the angle. See Example 4. Use* Y_1 *to do the following.*

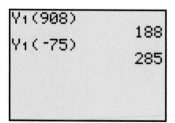

```
Y₁(908)
            188
Y₁(-75)
            285
```

55. Rework Exercise 39 with a graphing calculator.

56. Rework Exercise 40 with a graphing calculator.

Concept Check   *Sketch each angle in standard position. Draw an arrow representing the correct amount of rotation. Find the measure of two other angles, one positive and one negative, that are coterminal with the given angle. Give the quadrant of each angle.*

57. 75°   **58.** 89°   **59.** 174°   **60.** 234°

61. 300°   **62.** 512°   **63.** −61°   **64.** −159°

Concept Check   *Locate each point in a coordinate system. Draw a ray from the origin through the given point. Indicate with an arrow the angle in standard position having smallest positive measure. Then find the distance r from the origin to the point, using the distance formula of Section 1.1.*

65. $(-3, -3)$ **66.** $(-5, 2)$ **67.** $(-3, -5)$

68. $(\sqrt{3}, 1)$ **69.** $(-2, 2\sqrt{3})$ **70.** $(4\sqrt{3}, -4)$

Solve each problem. See Example 5.

71. *Revolutions of a Windmill*   A windmill makes 90 revolutions per minute. How many revolutions does it make per second?

72. *Revolutions of a Turntable*   A turntable in a shop makes 45 revolutions per minute. How many revolutions does it make per second?

73. *Rotating Tire*   A tire is rotating 600 times per minute. Through how many degrees does a point on the edge of the tire move in 1/2 second?

74. *Rotating Airplane Propeller*   An airplane propeller rotates 1000 times per minute. Find the number of de-grees that a point on the edge of the propeller will rotate in 1 second.

75. *Rotating Pulley*   A pulley rotates through 75° in one minute. How many rotations does the pulley make in an hour?

76. *Surveying*   One student in a surveying class measures an angle as 74.25°, while another student measures the same angle as 74° 20′. Find the difference between these measurements, both to the nearest minute and to the nearest hundredth of a degree.

 Angle of a Star   *Refer to the figure and table given in the chapter introduction. For each star in Exercises 77 and 78, find the measure of the other acute angle in the figure, 90° − θ, using the values from the table.*

77. Alpha Centauri   **78.** 61 Cygni

79.  *Viewing Field of a Telescope*   Due to Earth's rotation, celestial objects like the moon and the stars appear to move across the sky, rising in the east and setting in the west. As a result, if a telescope on Earth remains stationary while viewing a celestial object, the object will slowly move outside the viewing field of the telescope. For this reason, a motor is often attached to telescopes so that the telescope rotates at the same rate as Earth. Determine how long it should take the motor to turn the telescope through an angle of 1 minute in a direction perpendicular to Earth's axis.

80. *Angle Measure of a Star on the American Flag*   Determine the measure of the angle in each point of the five-pointed star appearing on the American flag. (*Hint:* Inscribe the star in a circle, and use the following theorem from geometry: *An angle whose vertex lies on the circumference of a circle is equal to half the central angle that cuts off the same arc.* See the figure.)

1.3 Angle Relationships and Similar Triangles

• **Geometric Properties** • **Triangles** • **Applications**

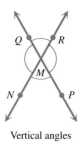

Vertical angles

Figure 23

Geometric Properties In Figure 23, the sides of angle *NMP* have been extended to form another angle, *RMQ*. The pair of angles *NMP* and *RMQ* are called **vertical angles.** Another pair of vertical angles, *NMQ* and *PMR*, are formed at the same time. Vertical angles have the following important property.

> **Vertical Angles**
>
> Vertical angles have equal measures.

Parallel lines are lines that lie in the same plane and do not intersect. Figure 24 shows parallel lines *m* and *n*. When a line *q* intersects two parallel lines, *q* is called a **transversal.** In Figure 24, the transversal intersecting the parallel lines forms eight angles, indicated by numbers.

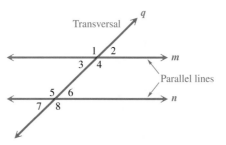

Figure 24

It is shown in geometry that the degree measures of angles 1 through 8 in Figure 24 possess some special properties. The following chart gives the names of these angles and rules about their measures.

Name	Sketch	Rule
Alternate interior angles	(also 3 and 6)	Angle measures are equal.
Alternate exterior angles	(also 2 and 7)	Angle measures are equal.

Name	Sketch	Rule
Interior angles on same side of transversal	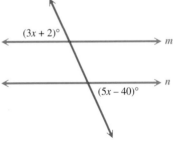 (also 3 and 5)	Angle measures add to 180°.
Corresponding angles	(also 1 and 5, 3 and 7, 4 and 8)	Angle measures are equal.

● ● ●   **Example 1   Finding Angle Measures**

Find the measure of each marked angle in Figure 25, given that lines m and n are parallel.

$(3x + 2)°$

m

n

$(5x - 40)°$

Figure 25

Algebraic Solution

The marked angles are alternate exterior angles, which are equal. Thus,

$$3x + 2 = 5x - 40$$
$$42 = 2x \ ,$$
$$21 = x.$$

One angle has measure

$$3 \cdot 21 + 2 = 65°,$$

and the other has measure

$$5x - 40 = 5 \cdot 21 - 40 = 65°.$$

Graphing Calculator Solution

Current models of graphing calculators feature numerical solve capability. The equation is solved in the screen in Figure 26, leading to the same result as in the algebraic solution. Consult your owner's manual to find the syntax required by your model.

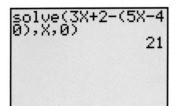

```
solve(3X+2-(5X-4
0),X,0)
                21
```

Figure 26

● ● ●

(a)

(b)

Figure 27

Triangles An important property of triangles that was first proved by Greek geometers deals with the sum of the measures of the angles of any triangle.

> ### Angle Sum of a Triangle
> The sum of the measures of the angles of any triangle is 180°.

While it is not an actual proof, a rather convincing argument for the truth of this statement can be given using any size triangle cut from a piece of paper. Tear each corner from the triangle, as suggested in Figure 27(a). You should be able to rearrange the pieces so that the three angles form a straight angle, as shown in Figure 27(b).

● ● ● **Example 2** Applying the Angle Sum of a Triangle Property

Suppose that the measures of two of the angles of a triangle are 48° and 61°. Find the measure of the third angle, x.

Use the fact that all three angle measures add to 180°.

$$48° + 61° + x = 180°$$
$$109° + x = 180°$$
$$x = 71°$$

The third angle of the triangle measures 71°. ● ● ●

Triangles are classified according to angles and sides, as shown in the following chart.

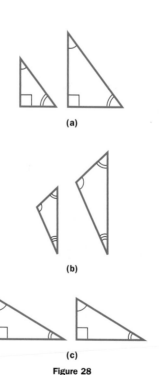

(a)

(b)

(c)

Figure 28

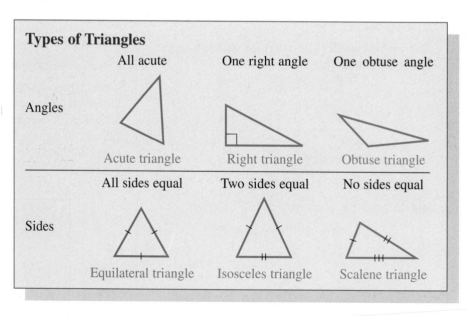

Many key ideas of trigonometry depend on **similar triangles,** which are triangles of exactly the same shape but not necessarily the same size. Figure 28 shows three pairs of similar triangles.

The two triangles in Figure 28(c) not only have the same shape but also the same size. Triangles that are both the same size and the same shape are called **congruent triangles.** If two triangles are congruent, then it is possible to pick one of them up and place it on top of the other so that they coincide. If two triangles are congruent, then they must be similar. However, two similar triangles need not be congruent.

The triangular supports for a child's swing are congruent (and thus similar) triangles, machine-produced with exactly the same dimensions each time. These supports are just one example of similar triangles. The supports of a long bridge, all the same shape but decreasing in size toward the center of the bridge, are examples of similar (but not congruent) triangles.

Suppose a correspondence between two triangles *ABC* and *DEF* is set up as shown in Figure 29.

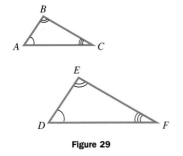

Figure 29

Angle *A* corresponds to angle *D*.	Side *AB* corresponds to side *DE*.
Angle *B* corresponds to angle *E*.	Side *BC* corresponds to side *EF*.
Angle *C* corresponds to angle *F*.	Side *AC* corresponds to side *DF*.

For triangle *ABC* to be similar to triangle *DEF*, the following conditions must hold.

Conditions for Similar Triangles

1. Corresponding angles must have the same measure.
2. Corresponding sides must be proportional. (That is, their ratios must be equal.)

● ● ●   **Example 3**   Finding Angle Measures in Similar Triangles

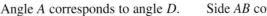

In Figure 30, triangles *ABC* and *NMP* are similar. Find the measures of angles *B* and *C*.

Since the triangles are similar, corresponding angles have the same measure. Since *C* corresponds to *P* and *P* measures 104°, angle *C* also measures 104°. Since angles *B* and *M* correspond, *B* measures 31°.   ● ● ●

Figure 30

N O T E   The small arcs found at the angles in Figures 28–31 are used to denote the corresponding angles in the triangles. This symbolism will be used when appropriate in this book. We will also use △ to denote "triangle."

● ● ●   **Example 4**   Finding Side Lengths in Similar Triangles

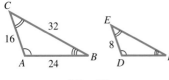

Figure 31

Given that △*ABC* and △*DFE* in Figure 31 are similar, find the lengths of the unknown sides of △*DFE*.

As mentioned before, similar triangles have corresponding sides in proportion. Use this fact to find the unknown side lengths in △*DFE*. Side *DF* of △*DFE* corresponds to side *AB* of △*ABC*, and sides *DE* and *AC* correspond. This leads to the proportion

$$\frac{8}{16} = \frac{DF}{24}.$$

Recall the *cross-multiplication property* of proportions.

$$\text{If } \frac{a}{b} = \frac{c}{d}, \qquad \text{then} \qquad ad = bc.$$

Cross-multiply to solve the equation for *DF*.

$$\frac{8}{16} = \frac{DF}{24}$$

$$8 \cdot 24 = 16 \cdot DF \qquad \text{Cross-multiply.}$$

$$192 = 16 \cdot DF$$

$$12 = DF$$

Side *DF* has length 12.

Side *EF* corresponds to *CB*. This leads to another proportion.

$$\frac{8}{16} = \frac{EF}{32}$$

$$8 \cdot 32 = 16 \cdot EF \qquad \text{Cross-multiply.}$$

$$16 = EF$$

Side *EF* has length 16.   ● ● ●

Applications   Some applied problems can be solved using properties of similar triangles.

● ● ●   **Example 5**   Finding the Height of a Flagpole

Firefighters at the Arcade Fire Station need to measure the height of the station flagpole. They find that at the instant when the shadow of the station is 18 feet long, the shadow of the flagpole is 99 feet long. The station is 10 feet high. Find the height of the flagpole.

Figure 32 shows the information given in the problem. The two triangles shown there are similar, so corresponding sides are in proportion, with

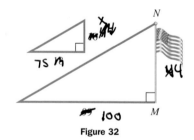

Figure 32

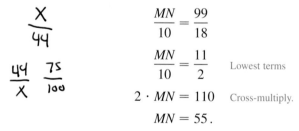

$$\frac{MN}{10} = \frac{99}{18}$$

$$\frac{MN}{10} = \frac{11}{2} \qquad \text{Lowest terms}$$

$$2 \cdot MN = 110 \qquad \text{Cross-multiply.}$$

$$MN = 55.$$

The flagpole is 55 feet high.   ● ● ●

● ● ●   **Example 6**   Determining When a Solar Eclipse Can Occur

The sun has a diameter of about 865,000 miles with a maximum distance from Earth's surface of about 94,500,000 miles. The moon has a smaller diameter of 2159 miles. For a total solar eclipse to occur, the moon must pass between Earth and the sun. The moon must also be close enough to Earth for the moon's *umbra* (shadow) to reach the surface of Earth. See Figure 33.

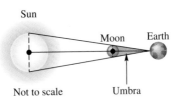

Sun

Moon   Earth

Not to scale   Umbra

Figure 33

(*Source:* Karttunen, H., P. Kröger, H. Oja, M. Putannen, and K. Donners (editors), *Fundamental Astronomy,* Springer-Verlag, 1996.)

(a) Calculate the maximum distance that the moon can be from Earth and still have a total solar eclipse occur.

Let D_s be the Earth–sun distance, d_s the diameter of the sun, D_m the Earth–moon distance, and d_m the diameter of the moon. Then, by similar triangles

$$\frac{D_s}{D_m} = \frac{d_s}{d_m}$$

$$D_m = \frac{D_s d_m}{d_s} = \frac{94{,}500{,}000 \times 2159}{865{,}000} \approx 236{,}000 \text{ miles.}$$

(b) In 1999, the closest approach of the moon to Earth's surface was 225,745 miles and the farthest was 251,978 miles. (*Source: The World Almanac and Book of Facts,* 1999.) Can a total solar eclipse occur every time the moon is between Earth and the sun? Explain.

No. The moon must be less than 236,000 miles away from Earth for an eclipse to occur, and sometimes it is farther than this.   ● ● ●

1.3 Exercises

1. A geometry book states, "When two straight lines intersect, the opposite angles are equal." What term do we use for "opposite angles"?

2. Consider Figure 24. If the measure of one of the angles is known, can the measures of the remaining seven angles be determined? Explain.

Use the properties of angle measures given in this section to find the measure of each marked angle. In Exercises 11–14, m and n are parallel. See Examples 1 and 2.

3.

$(5x - 129)°$ $(2x - 21)°$

4.

$(11x - 37)°$ $(7x + 27)°$

5.

$(x + 20)°$

$x°$ $(210 - 3x)°$

6. $(x + 15)°$

$(x + 5)°$

$(10x - 20)°$

7.

$(2x - 120)°$

$(x - 30)°$

$\left(\frac{1}{2}x + 15\right)°$

8.

$x°$

$x°$ $x°$

9.

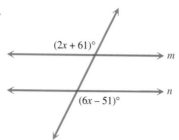

$(3x + 5)°$ $(5x + 15)°$

10.

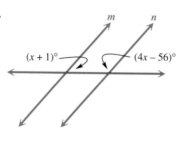

$(5x − 1)°$ $(2x)°$

11.

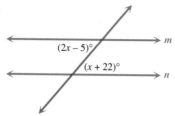

$(2x − 5)°$ *m*

$(x + 22)°$ *n*

12.

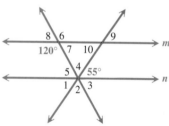

$(2x + 61)°$ *m*

$(6x − 51)°$ *n*

13.

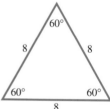

m *n*

$(x + 1)°$ $(4x − 56)°$

14.

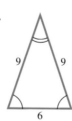

m *n*

$(10x + 11)°$

$(15x − 54)°$

The measures of two angles of a triangle are given. Find the measure of the third angle. See Example 2.

15. $37°, 52°$

16. $29°, 104°$

17. $147° 12', 30° 19'$

18. $136° 50', 41° 38'$

19. $74.2°, 80.4°$

20. $29.6°, 49.7°$

21. Can a triangle have angles of measures $85°$ and $100°$? Explain.

22. Can a triangle have two obtuse angles? Explain.

23. *Concept Check* Use the given figure to find the measures of the numbered angles, given that lines *m* and *n* are parallel.

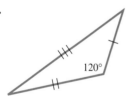

8 6 9 *m*
$120°$ 7 10
4 $55°$
5
1 2 3 *n*

24. *Concept Check* Find the measures of the marked angles, given that $x + y = 40$. (*Hint:* You must solve a system of equations.)

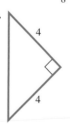

$(x + 2y)°$

$(11x)°$

Classify each triangle in Exercises 25–36 as either acute, right, *or* obtuse. *Also classify each as either* equilateral, isosceles, *or* scalene.

25.

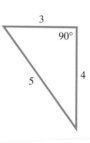

26.

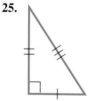

$120°$

27.

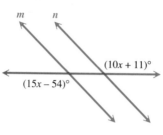

$60°$
8   8
$60°$ $60°$
8

28.

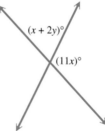

9   9
6

29.

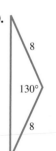

3
$90°$
5   4

30.

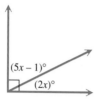

8
$130°$
8

31.

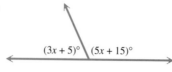

4
4

32.

$60°$

33.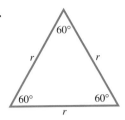

34.

35.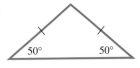

36.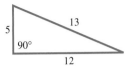

📄 **37.** Write a definition of *isosceles right triangle.*

📄 **38.** Explain why the sum of the lengths of any two sides of a triangle must be greater than the length of the third side.

📄 **39.** Must all equilateral triangles be similar? Explain.

40. *Carpentry Technique*   The following technique is used by carpenters to draw a 60° angle with a straightedge and compass. (*Source:* Hamilton, J. E., and M. S. Hamilton, *Math to Build On,* Construction Trades Press, 1993.)

   Step 1 Draw a straight line segment, and mark a point near the center of the line. See the figure.

   Step 2 Place the compass tip on the marked point and draw a semicircle.

Step 3 Without changing the setting of the compass, place the tip of the compass at the right intersection of the line and the semicircle and then mark a small arc across the semicircle.

Step 4 Draw a line segment from the marked point on the line to the point where the arc crosses the semicircle. This line will make a 60° angle with the original line.

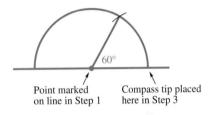

Point marked on line in Step 1   Compass tip placed here in Step 3

Concept Check   Name the corresponding angles and the corresponding sides of each pair of similar triangles.

41.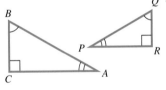

42.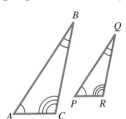

43. (*EA* is parallel to *CD*.)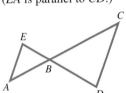

44. (*HK* is parallel to *EF*.)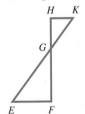

Find all unknown angle measures in each pair of similar triangles. See Example 3.

45.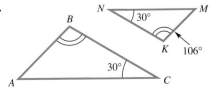

46.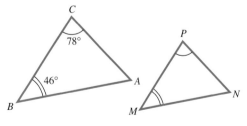

47.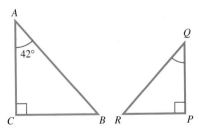

48.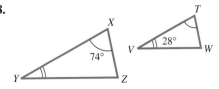

49.

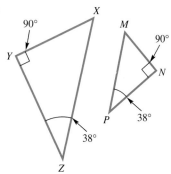

50.

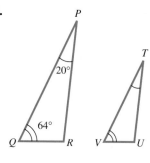

Find the unknown side lengths labeled with a variable in each pair of similar triangles. See Example 4.

51.

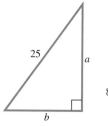

52.

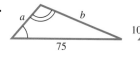

53.

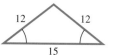

54.

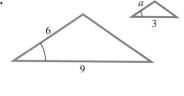

55.

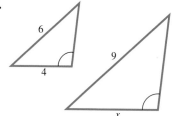

56.

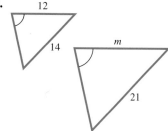

Solve each problem. (Hint: Draw a sketch.) See Example 5.

57. *Height of a Tree*   A tree casts a shadow 45 meters long. At the same time, the shadow cast by a vertical 2-meter stick is 3 meters long. Find the height of the tree.

58. *Height of a Lookout Tower*   A forest fire lookout tower casts a shadow 180 feet long at the same time that the shadow of a 9-foot truck is 15 feet long. Find the height of the tower.

59. *Lengths of Sides of a Triangle*   On a photograph of a triangular piece of land, the lengths of the three sides are 4 cm, 5 cm, and 7 cm, respectively. The shortest side of the actual piece of land is 400 meters long. Find the lengths of the other two sides.

60. *Height of a Lighthouse*   The Santa Cruz lighthouse in the figure casts a shadow 28 meters long at 7 P.M. At the same time, the shadow of the lighthouse keeper, who is 1.75 meters tall, is 3.5 meters long. (See the figure.) How tall is the lighthouse?

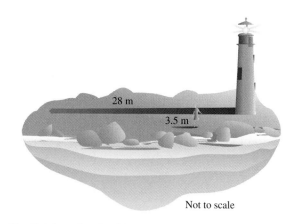

Not to scale

61. *Height of a Building*   A house is 15 feet tall. Its shadow is 40 feet long at the same time the shadow of a nearby building is 300 feet long. Find the height of the building.

62. *Distance between Two Cities* By drawing lines on a map, a triangle can be formed by the cities of Phoenix, Tucson, and Yuma. On the map, the distance between Phoenix and Tucson is 8 cm, the distance between Phoenix and Yuma is 12 cm, and the distance between Tucson and Yuma is 17 cm. The actual straight-line distance from Phoenix to Yuma is 230 km. Find the distances between the other pairs of cities.

In each diagram, there are two similar triangles. Find the unknown measurement. (Hint: In the sketch for Exercise 63, the side of length 100 in the small triangle corresponds to the side of length 100 + 120 = 220 in the large triangle.)

63.

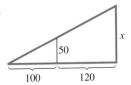

64.

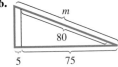

65.

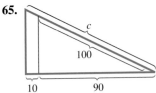

66.

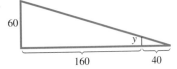

Work each problem.

67. *Lengths of Sides of a Quadrilateral* Two quadrilaterals (four-sided figures) are similar. The lengths of the three shortest sides of the first quadrilateral are 18 cm, 24 cm, and 32 cm. The lengths of the two longest sides of the second quadrilateral are 48 cm and 60 cm. Find the unknown lengths of the sides of these two figures.

68. *Height of a Carving of Lincoln* Assume that Lincoln was 6 1/3 feet tall and his head 3/4 foot long. Knowing that the carved head of Lincoln at Mount Rushmore is 60 feet tall, find how tall his entire body would be if it were carved into the mountain.

In each figure, two similar triangles are present. Find the value of each variable.

69.

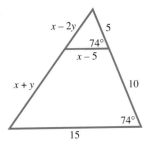

70.

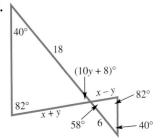

Solve each problem.

71. *Sizes and Distances in the Sky* Astronomers use degrees, minutes, and seconds to measure sizes and distances in the sky along an arc from the horizon to the zenith point directly overhead. An adult observer on Earth can judge distances in the sky using his or her hand at arm's length. An outstretched hand will be about 20 arc degrees wide from the tip of the thumb to the tip of the little finger. A clenched fist at arm's length measures about 10 arc degrees, and a thumb corresponds to about 2 arc degrees. (*Source:* Levy, David H., *Skywatching,* The Nature Company, 1994.)

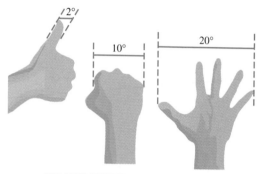

"HANDS ON THE SKY" Your hand can
help you measure distances in the sky.

(a) The apparent size of the moon is about 31 arc minutes. What part of your thumb would cover the moon?

(b) If an outstretched hand plus a fist covers the distance between two bright stars, about how far apart in arc degrees are the stars?

72. *Eclipse on Mars* Refer to Example 6. The sun's distance from the surface of Mars is approximately 142,000,000 miles. One of Mars' two moons, Phobos, has a maximum diameter of 17.4 miles. (*Source:* Zeilik, M., S. Gregory, and E. Smith, *Introductory Astronomy and Astrophysics,* 2nd Edition, Saunders College Publishers, 1998.)

(a) Calculate the maximum distance that the moon Phobos can be from Mars for a total eclipse of the sun to occur on Mars.

(b) Phobos is approximately 5800 miles from Mars. Is it possible for Phobos to cause a total eclipse on Mars?

73. *Eclipse on Jupiter* Refer to Example 6. The sun's distance from the surface of Jupiter is approximately 484,000,000 miles. One of Jupiter's moons, Ganymede, is the largest moon in the solar system with a diameter of 3270 miles. (*Source:* Wright,

John W. (General Editor), *The Universal Almanac,* Andrews and McMeel, 1997.)

(a) Calculate the maximum distance that the moon Ganymede can be from Jupiter for a total eclipse of the sun to occur on Jupiter.

(b) Ganymede is approximately 665,000 miles from Jupiter. Is it possible for Ganymede to cause a total eclipse on Jupiter?

Quantitative Reasoning

74. *Have you ever gazed up at a redwood tree, a skyscraper, a public monument, or perhaps a dinosaur in a museum and wondered how tall it was?* There is a relatively simple way to make a reasonable estimate. All you need is a 1-foot ruler. Hold the ruler vertically at arm's length as you approach the object to be measured. Stop when one end of the ruler lines up with the top of the object and the other end with its base. Now pace off the distance to the object, taking normal strides. The number of paces will be the approximate height of the object in feet.

The reason this method works depends on a concept presented in this section. See whether you can furnish the reasons for each of the following steps, which refer to the figure. (Assume that the length of one pace is *EF*.)

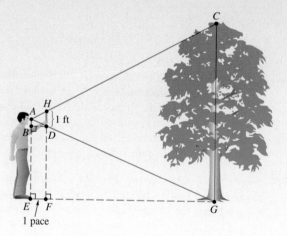

(a) $CG = \dfrac{CG}{1} = \dfrac{AG}{AD}$

(b) $\dfrac{AG}{AD} = \dfrac{EG}{EF}$

(c) $\dfrac{EG}{EF} = \dfrac{EG}{BD} = \dfrac{EG}{1}$

(d) CG feet $= EG$ paces

The height of the tree (in feet) is (approximately) the number of paces.

1.4 Definitions of the Trigonometric Functions

- The Trigonometric Functions • Quadrantal Angles

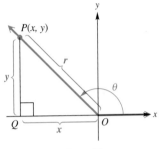

Figure 34

The Trigonometric Functions The study of trigonometry covers the six trigonometric functions defined in this section. Most sections in the remainder of this book involve at least one of these functions. To define these six basic functions, start with an angle θ* in standard position. Choose any point P having coordinates (x, y) on the terminal side of angle θ. (The point P must not be the vertex of the angle.) See Figure 34.

A perpendicular from P to the x-axis at point Q determines a triangle having vertices at O, P, and Q. The distance r from $P(x, y)$ to the origin, $(0, 0)$, can be found using the distance formula.

$$r = \sqrt{(x - 0)^2 + (y - 0)^2} = \sqrt{x^2 + y^2}$$

Notice that $r > 0$ since distance is never negative.

The six trigonometric functions of angle θ are called **sine, cosine, tangent, cotangent, secant,** and **cosecant.** In the following definitions, we use the customary abbreviations for the names of these functions.

Trigonometric Functions

Let (x, y) be a point other than the origin on the terminal side of an angle θ in standard position. The distance from the point to the origin is $r = \sqrt{x^2 + y^2}$. The six trigonometric functions of θ are as follows.

$$\sin \theta = \frac{y}{r} \qquad \cos \theta = \frac{x}{r} \qquad \tan \theta = \frac{y}{x} \ (x \neq 0)$$

$$\csc \theta = \frac{r}{y} \ (y \neq 0) \qquad \sec \theta = \frac{r}{x} \ (x \neq 0) \qquad \cot \theta = \frac{x}{y} \ (y \neq 0)$$

NOTE Although Figure 34 shows a second quadrant angle, these definitions apply to any angle θ. Because of the restrictions on the denominators in the definitions of tangent, cotangent, secant, and cosecant, some angles will have undefined function values.

● ● ● **Example 1** Finding the Function Values of an Angle

The terminal side of an angle α in standard position goes through the point $(8, 15)$. Find the values of the six trigonometric functions of angle α.

Figure 35 on the next page shows angle α and the triangle formed by dropping a perpendicular from the point $(8, 15)$ to the x-axis. The point $(8, 15)$ is 8 units to the right of the y-axis and 15 units above the x-axis, so $x = 8$ and $y = 15$.

---

*Recall that Greek letters are often used to name angles.

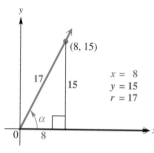

Figure 35

Since $r = \sqrt{x^2 + y^2}$,

$$r = \sqrt{8^2 + 15^2} = \sqrt{64 + 225} = \sqrt{289} = 17.$$

The values of the six trigonometric functions of angle α can now be found with the definitions given above.

$$\sin \alpha = \frac{y}{r} = \frac{15}{17} \qquad \cos \alpha = \frac{x}{r} = \frac{8}{17} \qquad \tan \alpha = \frac{y}{x} = \frac{15}{8}$$

$$\csc \alpha = \frac{r}{y} = \frac{17}{15} \qquad \sec \alpha = \frac{r}{x} = \frac{17}{8} \qquad \cot \alpha = \frac{x}{y} = \frac{8}{15}$$

• • •

• • • **Example 2** Finding the Function Values of an Angle

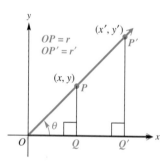

Figure 36

The terminal side of angle β in standard position goes through $(-3, -4)$. Find the values of the six trigonometric functions of β.

As shown in Figure 36, $x = -3$ and $y = -4$. The value of r is

$$r = \sqrt{(-3)^2 + (-4)^2} = \sqrt{25} = 5.$$

(Remember that $r > 0$.) Then by the definitions of the trigonometric functions,

$$\sin \beta = \frac{-4}{5} = -\frac{4}{5} \qquad \cos \beta = \frac{-3}{5} = -\frac{3}{5} \qquad \tan \beta = \frac{-4}{-3} = \frac{4}{3}$$

$$\csc \beta = \frac{5}{-4} = -\frac{5}{4} \qquad \sec \beta = \frac{5}{-3} = -\frac{5}{3} \qquad \cot \beta = \frac{-3}{-4} = \frac{3}{4}.$$

• • •

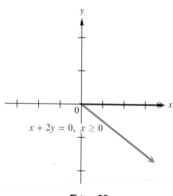

Figure 37

The six trigonometric functions can be found from *any* point other than the origin on the terminal side of an angle. To see why any point may be used, refer to Figure 37, which shows an angle θ and two distinct points on its terminal side. Point P has coordinates (x, y), and point P' (read "P-prime") has coordinates (x', y'). Let r be the length of the hypotenuse of triangle OPQ, and let r' be the length of the hypotenuse of triangle $OP'Q'$. Since corresponding sides of similar triangles are in proportion,

$$\frac{y}{r} = \frac{y'}{r'},$$

so $\sin \theta = y/r$ is the same no matter which point is used to find it. A similar result holds for the other five functions.

We can also find the trigonometric function values of an angle if we know the equation of the line coinciding with the terminal ray. Recall from algebra that the graph of the equation

$$Ax + By = 0$$

is a line that passes through the origin. If we restrict x to have only nonpositive or only nonnegative values, we obtain as the graph a ray with endpoint at the origin. For example, the graph of $x + 2y = 0$, $x \geq 0$, shown in Figure 38, is a ray that can serve as the terminal side of an angle in standard position. By choosing a point on the ray, the trigonometric function values of the angle can be found.

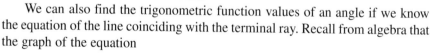

Figure 38

• • • **Example 3** Finding the Function Values of an Angle

Find the six trigonometric function values of the angle θ in standard position, if the terminal side of θ is defined by $x + 2y = 0$, $x \geq 0$.

Algebraic Solution

The angle is shown in Figure 39. We can use *any* point except $(0, 0)$ on the terminal side of θ to find the trigonometric function values, so choosing $x = 2$, we find the corresponding value of y.

$$x + 2y = 0, \, x \geq 0$$
$$2 + 2y = 0 \qquad \text{Arbitrarily choose } x = 2.$$
$$2y = -2$$
$$y = -1$$

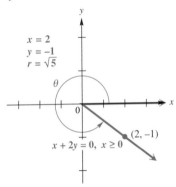

Figure 39

The point $(2, -1)$ lies on the terminal side, and the corresponding value of r is $r = \sqrt{2^2 + (-1)^2} = \sqrt{5}$. Now use the definitions of the trigonometric functions.

$$\sin \theta = \frac{y}{r} = \frac{-1}{\sqrt{5}} = \frac{-1}{\sqrt{5}} \cdot \frac{\sqrt{5}}{\sqrt{5}} = -\frac{\sqrt{5}}{5}$$

$$\cos \theta = \frac{x}{r} = \frac{2}{\sqrt{5}} = \frac{2}{\sqrt{5}} \cdot \frac{\sqrt{5}}{\sqrt{5}} = \frac{2\sqrt{5}}{5}$$

$$\tan \theta = \frac{y}{x} = \frac{-1}{2} = -\frac{1}{2}$$

$$\csc \theta = \frac{r}{y} = \frac{\sqrt{5}}{-1} = -\sqrt{5}$$

$$\sec \theta = \frac{r}{x} = \frac{\sqrt{5}}{2}$$

$$\cot \theta = \frac{x}{y} = \frac{2}{-1} = -2$$

Graphing Calculator Solution

Figure 40 shows the graph of

$$x + 2y = 0, \quad x \geq 0.$$

The ray was graphed by entering the equation as

$$Y_1 = (-1/2)X \quad \text{or} \quad Y_1 = -.5X$$

with the restriction $X \geq 0$. Using the capability of the calculator to give the y-value for $X = 2$ produced the point and the values of x and y shown on the screen.

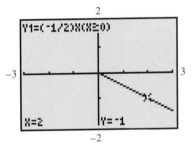

Figure 40

Use these values of x and y and the definitions of the trigonometric functions to find the six function values of θ, as shown in the algebraic solution.

• • •

Recall that when the equation of a line is written in the form $y = mx + b$, the coefficient of x is the slope of the line. In Example 3, $x + 2y = 0$ can be written as $y = (-1/2)x$, so the slope is $-1/2$. Notice that $\tan \theta = -1/2$. In general, it is true that $m = \tan \theta$.

NOTE   The trigonometric function values we found in Examples 1–3 are *exact*. If we were to use a calculator to approximate these values, the decimal results would not be acceptable if exact values were required.

CONNECTIONS   A convenient way to see the sine, cosine, and tangent trigonometric ratios geometrically is shown in Figure 41 for θ in quadrants I and II. The circle, which has a radius of 1, is called a *unit circle*. (We will see the unit circle again in Chapter 3.) By remembering this figure and the segments that represent the sine, cosine, and tangent functions, you can quickly recall the properties of the trigonometric functions. Horizontal line segments to the left of the origin and vertical line segments below the x-axis represent negative values. Note that the tangent line must be tangent to the circle at $(1,0)$, no matter which quadrant θ lies in.

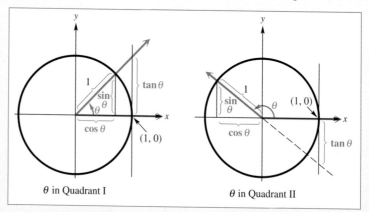

Figure 41

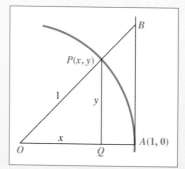

Figure 42

For Discussion or Writing

1. Label the triangles as shown in Figure 42. Use the definitions of the trigonometric functions and similar triangles to show that $PQ = \sin \theta$, $OQ = \cos \theta$, and $AB = \tan \theta$.
2. Sketch similar figures for θ in quadrants III and IV.

Quadrantal Angles   If the terminal side of an angle in standard position lies along the *y*-axis, any point on this terminal side has *x*-coordinate 0. Similarly, an angle with terminal side on the *x*-axis has *y*-coordinate 0 for any point on the terminal side. Since the values of *x* and *y* appear in the denominators of some of the trigonometric functions, and since a fraction is undefined if its denominator is 0, some of the trigonometric function values of quadrantal angles (i.e., those with terminal side on an axis) will be undefined.

● ● ●   **Example 4**   Finding Function Values of a Quadrantal Angle

Find the values of the six trigonometric functions for the following angles.

Algebraic Solution

(a) an angle of 90°

First, select any point on the terminal side of a 90° angle. We select the point $(0, 1)$, as shown in Figure 43(a). Here $x = 0$ and $y = 1$. Verify that $r = 1$. Then, by the definitions of the trigonometric functions,

$$\sin 90° = \frac{1}{1} = 1 \qquad\qquad \csc 90° = \frac{1}{1} = 1$$

$$\cos 90° = \frac{0}{1} = 0 \qquad\qquad \sec 90° = \frac{1}{0}\ \text{(undefined)}$$

$$\tan 90° = \frac{1}{0}\ \text{(undefined)} \qquad \cot 90° = \frac{0}{1} = 0.$$

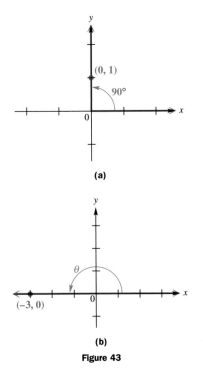

(a)

(b)

Figure 43

(continued)

Graphing Calculator Solution

With the calculator set in degree mode (first screen in Figure 44), it returns the correct values for sin 90° and cos 90°. The last screen shows an ERROR message for tan 90°, because 90° is not in the domain of the tangent function. There are no calculator keys for finding the function values of cotangent, secant, or cosecant, but in Section 1.5 we show how to find these function values with a calculator.

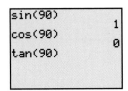

Figure 44

(b) an angle θ in standard position with terminal side through $(-3, 0)$

Figure 43(b) shows the angle. Here, $x = -3$, $y = 0$, and $r = 3$, so the trigonometric functions have the following values.

$$\sin \theta = \frac{0}{3} = 0 \qquad \csc \theta = \frac{3}{0} \text{ (undefined)}$$

$$\cos \theta = \frac{-3}{3} = -1 \qquad \sec \theta = \frac{3}{-3} = -1$$

$$\tan \theta = \frac{0}{-3} = 0 \qquad \cot \theta = \frac{-3}{0} \text{ (undefined)}$$

● ● ●

The conditions under which the trigonometric function values of quadrantal angles are undefined are summarized here.

Undefined Function Values

If the terminal side of a quadrantal angle lies along the y-axis, the tangent and secant functions are undefined. If it lies along the x-axis, the cotangent and cosecant functions are undefined.

Since the most commonly used quadrantal angles are $0°$, $90°$, $180°$, $270°$, and $360°$, the values of the functions of these angles are summarized in the following table. This table is for reference only; you should be able to reproduce it quickly.

Trigonometric Function Values for Quadrantal Angles

θ	$\sin \theta$	$\cos \theta$	$\tan \theta$	$\cot \theta$	$\sec \theta$	$\csc \theta$
$0°$	0	1	0	Undefined	1	Undefined
$90°$	1	0	Undefined	0	Undefined	1
$180°$	0	-1	0	Undefined	-1	Undefined
$270°$	-1	0	Undefined	0	Undefined	-1
$360°$	0	1	0	Undefined	1	Undefined

The values given in this table can also be found with a calculator that has trigonometric function keys. First, make sure the calculator is set in *degree mode*.

CAUTION   One of the most common errors involving calculators in trigonometry occurs when the calculator is set for *radian measure,* rather than *degree measure.* (Radian measure of angles is discussed in Chapter 3.) For this reason, be sure that you know how to set your calculator in *degree mode.*

1.4   Exercises

In Exercises 1–4, sketch an angle θ in standard position such that θ has the smallest possible positive measure, and the given point is on the terminal side of θ.

1. $(-3, 4)$ **2.** $(-4, -3)$ **3.** $(5, -12)$ **4.** $(-12, -5)$

Find the values of the six trigonometric functions for the angles in standard position having the following points on their terminal sides. Rationalize denominators when applicable. Use a calculator in Exercises 11 and 12. See Examples 1, 2, and 4.

5. $(-3, 4)$ **6.** $(-4, -3)$ **7.** $(0, 2)$ **8.** $(-4, 0)$

9. $(1, \sqrt{3})$ **10.** $(-2\sqrt{3}, -2)$ **11.** $(8.7691, -3.2473)$ **12.** $(-5.1021, 7.6132)$

13. For any nonquadrantal angle θ, $\sin \theta$ and $\csc \theta$ will have the same sign. Explain why this is so.

14. How is the value of r interpreted geometrically in the definitions of the sine, cosine, secant, and cosecant functions?

15. *Concept Check*   If $\cot \theta$ is undefined, what is the value of $\tan \theta$?

16. *Concept Check*   If the terminal side of an angle β is in quadrant III, what is the sign of each of the trigonometric function values of β?

Concept Check   *Suppose that the point (x, y) is in the indicated quadrant. Decide whether the given ratio is positive or negative. (Hint: It may be helpful to draw a sketch.)*

17. II, $\dfrac{y}{r}$ **18.** II, $\dfrac{x}{r}$ **19.** III, $\dfrac{y}{r}$ **20.** III, $\dfrac{x}{r}$

21. IV, $\dfrac{x}{r}$ **22.** IV, $\dfrac{y}{r}$ **23.** IV, $\dfrac{y}{x}$ **24.** IV, $\dfrac{x}{y}$

In Exercises 25–30, an equation with a restriction on x is given. This is an equation of the terminal side of an angle θ in standard position. Sketch the smallest positive such angle θ, and find the values of the six trigonometric functions of θ. See Example 3.

25. $2x + y = 0, x \geq 0$ **26.** $3x + 5y = 0, x \geq 0$ **27.** $-4x + 7y = 0, x \leq 0$

28. $-6x - y = 0, x \leq 0$ **29.** $-5x - 3y = 0, x \leq 0$ **30.** $6x - 5y = 0, x \geq 0$

31. Rework Example 3 using a different value for x. Find the corresponding y-value, and then show that the six trigonometric function values you obtain are the same as the ones obtained in Example 3.

32. Rework Example 3 using the values of x and y for which the point (x, y) is on the circle of radius 1 having center at the origin, and then show that the six trigonometric function values you obtain are the same as those obtained in Example 3.

Use the trigonometric function values of quadrantal angles given in this section to evaluate each expression. An expression such as $\cot^2 90°$ means $(\cot 90°)^2$, which is equal to $0^2 = 0$.

33. $\cos 90° + 3 \sin 270°$

34. $\tan 0° - 6 \sin 90°$

35. $3 \sec 180° - 5 \tan 360°$

36. $4 \csc 270° + 3 \cos 180°$

37. $\tan 360° + 4 \sin 180° + 5 \cos^2 180°$

38. $2 \sec 0° + 4 \cot^2 90° + \cos 360°$

39. $\sin^2 180° + \cos^2 180°$

40. $\sin^2 360° + \cos^2 360°$

41. $\sec^2 180° - 3 \sin^2 360° + 2 \cos 180°$

42. $5 \sin^2 90° + 2 \cos^2 270° - 7 \tan^2 360°$

If n is an integer, n · 180° represents an integer multiple of 180°, and (2n + 1) · 90° represents an odd integer multiple of 90°. Decide whether each expression is equal to 0, 1, −1, or is undefined.

43. $\sin[n \cdot 180°]$ **44.** $\cos[(2n + 1) \cdot 90°]$ **45.** $\tan[(2n + 1) \cdot 90°]$ **46.** $\tan[n \cdot 180°]$

Provide conjectures in Exercises 47–50.

47. The angles 15° and 75° are complementary. With your calculator determine sin 15° and cos 75°. Make a conjecture about the sines and cosines of complementary angles, and test your hypothesis with other pairs of complementary angles. (*Note:* This relationship will be discussed in detail in Section 2.1.)

48. The angles 25° and 65° are complementary. With your calculator determine tan 25° and cot 65°. Make a conjecture about the tangents and cotangents of complementary angles, and test your hypothesis with other pairs of complementary angles. (*Note:* This relationship will be discussed in detail in Section 2.1.)

49. With your calculator determine sin 10° and sin (−10°). Make a conjecture about the sines of an angle and its negative, and test your hypothesis with other angles. Also, use a geometry argument with the definition of sin θ to justify your hypothesis. (*Note:* This relationship will be discussed in detail in Section 5.1.)

50. With your calculator determine cos 20° and cos(−20°). Make a conjecture about the cosines of an angle and its negative, and test your hypothesis with other angles. Also, use a geometry argument with the definition of cos θ to justify your hypothesis. (*Note:* This relationship will be discussed in detail in Section 5.1.)

The unit circle has a radius of 1. Figure 41 suggests that the coordinates of the intersection of the unit circle and the terminal side of an angle x is the point (cos x, sin x). *Use this fact for Exercises 51 and 52.*

51. Define the cosine function in terms of the *x*-coordinate of a point on the unit circle.

52. Define the sine function in terms of the *y*-coordinate of a point on the unit circle.

Concept Check   *Use the concepts presented in this section to work each problem.*

53. Give two values of θ, 0° ≤ θ < 360°, for which sin θ = cos θ.

54. Give two values of θ, 0° ≤ θ < 360°, for which tan θ = 1.

55. The equation cos θ = .8 has the approximate solution θ = 36.87°. Find an approximate solution for cos θ = −.8.

56. The equation sin θ = .8 has the approximate solution θ = 53.13°. Find an approximate solution for sin θ = −.8.

In Exercises 57–62, place your graphing calculator in parametric and degree modes. Set the window and functions as shown here, and graph. A circle of radius 1 will appear on the screen. Trace to move a short distance around the circle. In the screen, the point on the circle corresponds to an angle T = 25°, cos 25° *is* .90630779, *and* sin 25° *is* .42261826.

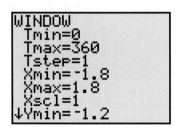
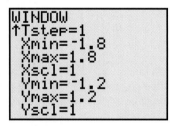

This screen is a continuation of the previous one.

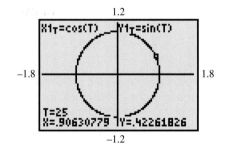

57. Use the right- and left-arrow keys to move to the point corresponding to 20°. What are cos 20° and sin 20°?

In Exercises 58–60 assume 0° ≤ T ≤ 90°.

58. For what angle T is cos T ≈ .766?

59. For what angle T is sin T ≈ .574?

60. For what angle T does cos T = sin T?

61. As T increases from 0° to 90°, does the cosine increase or decrease? What about the sine?

62. As T increases from 90° to 180°, does the cosine increase or decrease? What about the sine?

63. 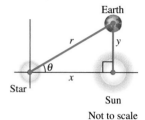 (*Modeling*) *Distance Between the Sun and a Star* Suppose that a star has parallax θ with respect to Earth and the sun. Let the coordinates of Earth be (x, y), the star be $(0, 0)$, and the sun be $(x, 0)$. See the figure.

Earth

r y

θ

x

Star

Sun

Not to scale

Find an equation for x, the distance between the sun and the star, as follows.

(a) Write an equation involving a trigonometric function that relates x, y, and θ.

(b) Solve your equation for x.

64. *Area of a Solar Cell* A solar cell converts the energy of sunlight directly into electrical energy. The amount of energy a cell produces depends on its area. Suppose a solar cell is hexagonal, as shown in the figure. Express its area in terms of $\sin \theta$ and any side x. (*Hint:* Consider one of the six equilateral triangles from the hexagon. See the figure.) (*Source:* Kastner, Bernice, *Space Mathematics,* NASA, 1985.)

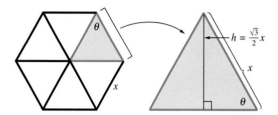

θ

x

$h = \dfrac{\sqrt{3}}{2}x$

x

θ

1.5 Using the Definitions of the Trigonometric Functions

- The Reciprocal Identities
- The Quotient Identities
- Signs and Ranges of Function Values
- The Pythagorean Identities

Identities are equations that are true for all values of the variables for which all expressions are defined. For example, both $(x + y)^2 = x^2 + 2xy + y^2$ and $2(x + 3) = 2x + 6$ are identities. Identities are studied in more detail in Chapter 5.

The Reciprocal Identities Recall the definition of a reciprocal: the *reciprocal* of the nonzero number x is $1/x$. For example, the reciprocal of 2 is $1/2$, and the reciprocal of $8/11$ is $11/8$. There is no reciprocal for 0. Scientific calculators have a reciprocal key, usually labeled $\boxed{1/x}$ or $\boxed{x^{-1}}$. Using this key gives the reciprocal of any nonzero number entered in the display.

The definitions of the trigonometric functions in the previous section were written so that functions in the same column are reciprocals of each other. Since $\sin \theta = y/r$ and $\csc \theta = r/y$,

$$\sin \theta = \frac{1}{\csc \theta} \qquad \text{and} \qquad \csc \theta = \frac{1}{\sin \theta}.$$

Also, $\cos \theta$ and $\sec \theta$ are reciprocals, as are $\tan \theta$ and $\cot \theta$. In summary, we have the **reciprocal identities** that hold for any angle θ that does not lead to a 0 denominator.

Reciprocal Identities

$$\sin \theta = \frac{1}{\csc \theta} \qquad \cos \theta = \frac{1}{\sec \theta} \qquad \tan \theta = \frac{1}{\cot \theta}$$

$$\csc \theta = \frac{1}{\sin \theta} \qquad \sec \theta = \frac{1}{\cos \theta} \qquad \cot \theta = \frac{1}{\tan \theta}$$

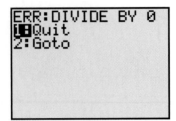

(a)

ERR:DIVIDE BY 0
1▉Quit
2:Goto

(b)

Figure 45

The screen in Figure 45(a) shows how csc 90°, sec 180°, and csc(−270°) are found, using the appropriate reciprocal identities and the reciprocal key of a graphing calculator in degree mode. Be sure *not* to use the *inverse trigonometric function* keys to find the reciprocal function values. Attempting to find sec 90° by entering 1/cos 90° produces an ERROR message, indicating the reciprocal is undefined. See Figure 45(b). Compare these results with the ones found in the chart of quadrantal angle function values in Section 1.4. ∎

N O T E Identities can be written in different forms. For example,

$$\sin \theta = \frac{1}{\csc \theta}$$

can also be written

$$\csc \theta = \frac{1}{\sin \theta} \quad \text{and} \quad (\sin \theta)(\csc \theta) = 1.$$

You should become familiar with all forms of these identities.

● ● ●   **Example 1   Using the Reciprocal Identities**

Find each function value.

(a) cos θ, if sec $\theta = \dfrac{5}{3}$

Since cos θ is the reciprocal of sec θ,

$$\cos \theta = \frac{1}{\sec \theta} = \frac{1}{5/3} = \frac{3}{5}.$$

(b) sin θ, if csc $\theta = -\dfrac{\sqrt{12}}{2}$

$$\sin \theta = \frac{1}{-\sqrt{12}/2}$$

$$= \frac{-2}{\sqrt{12}}$$

$$= \frac{-2}{2\sqrt{3}} \qquad \sqrt{12} = \sqrt{4 \cdot 3} = 2\sqrt{3}$$

$$= \frac{-1}{\sqrt{3}}$$

$$= \frac{-\sqrt{3}}{3} \qquad \text{Multiply by } \frac{\sqrt{3}}{\sqrt{3}} \text{ to rationalize the denominator.}$$

[handwritten: tan if cot is $\dfrac{\sqrt{11}}{4}$]

[handwritten: $\dfrac{1}{\sqrt{11}/4} = \dfrac{4}{\sqrt{11}}$]

● ● ●

Signs and Ranges of Function Values   In the definitions of the trigonometric functions, r is the distance from the origin to the point (x, y). Distance is never negative, so r > 0. If we choose a point (x, y) in quadrant I, then both x and y will be positive. Since r > 0, all six of the fractions used in the definitions of the trigonometric functions will be positive, so the values of all six functions will be positive in quadrant I.

A point (x, y) in quadrant II has $x < 0$ and $y > 0$. This makes the values of sine and cosecant positive for quadrant II angles, while the other four functions take on negative values. Similar results can be obtained for the other quadrants, as summarized next.

Signs of Function Values

θ in Quadrant	$\sin \theta$	$\cos \theta$	$\tan \theta$	$\cot \theta$	$\sec \theta$	$\csc \theta$
I	+	+	+	+	+	+
II	+	−	−	−	−	+
III	−	−	+	+	−	−
IV	−	+	−	−	+	−

NOTE Some students use the sentence "**A**ll **S**tudents **T**ake **C**alculus" to remember which of the three basic functions are positive in each quadrant. **A** indicates "all" in quadrant I, **S** represents "sine" in quadrant II, **T** represents "tangent" in quadrant III, and **C** stands for "cosine" in quadrant IV.

● ● ● **Example 2** Identifying the Quadrant of an Angle

Identify the quadrant (or quadrants) of any angle θ that satisfies $\sin \theta > 0$, $\tan \theta < 0$.

Since $\sin \theta > 0$ in quadrants I and II, while $\tan \theta < 0$ in quadrants II and IV, both conditions are met only in quadrant II. ● ● ●

Figure 46 on the next page shows an angle θ as it increases in measure from near $0°$ toward $90°$. In each case, the value of r is the same. As the measure of the angle increases, y increases but never exceeds r, so $y \leq r$. Dividing both sides by the positive number r gives

$$y \leq r$$

$$\frac{y}{r} \leq 1.$$

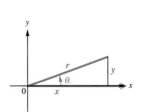 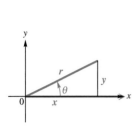 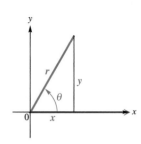 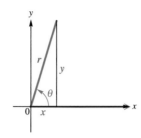

Figure 46

In a similar way, angles in quadrant IV suggest that

$$-1 \le \frac{y}{r},$$

so

$$-1 \le \frac{y}{r} \le 1.$$

Since $y/r = \sin \theta$,

$$-1 \le \sin \theta \le 1$$

for any angle θ. In the same way,

$$-1 \le \cos \theta \le 1.$$

The tangent of an angle is defined as y/x. It is possible that $x < y$, that $x = y$, or that $x > y$. For this reason y/x can take any value at all, so $\tan \theta$ can be any real number, as can $\cot \theta$.

The functions $\sec \theta$ and $\csc \theta$ are reciprocals of the functions $\cos \theta$ and $\sin \theta$, respectively, making

$$\sec \theta \le -1 \text{ or } \sec \theta \ge 1, \qquad \csc \theta \le -1 \text{ or } \csc \theta \ge 1.$$

In summary, the ranges of the trigonometric functions are as follows.

Ranges of Trigonometric Functions

For any angle θ for which the indicated functions exist:

1. $-1 \le \sin \theta \le 1$ and $-1 \le \cos \theta \le 1$;
2. $\tan \theta$ and $\cot \theta$ may be equal to any real number;
3. $\sec \theta \le -1$ or $\sec \theta \ge 1$ and $\csc \theta \le -1$ or $\csc \theta \ge 1$.

(Notice that $\sec \theta$ and $\csc \theta$ are *never* between -1 and 1.)

● ● ● **Example 3** Deciding Whether a Trigonometric Function Value Is in the Range

Decide whether the following statements are *possible* or *impossible*.

(a) $\sin \theta = \sqrt{8}$
For any value of θ, $-1 \le \sin \theta \le 1$. Since $\sqrt{8} > 1$, there is no value of θ with $\sin \theta = \sqrt{8}$.

(b) $\tan \theta = 110.47$
Tangent can take any value. Thus, $\tan \theta = 110.47$ is possible.

(c) $\sec \theta = .6$
Since $\sec \theta \le -1$ or $\sec \theta \ge 1$, the statement $\sec \theta = .6$ is impossible.

● ● ●

The six trigonometric functions are defined in terms of x, y, and r, where the Pythagorean theorem shows that $r^2 = x^2 + y^2$ and $r > 0$. With these relationships, knowing the value of only one function and the quadrant in which the angle lies makes it possible to find the values of all of the other trigonometric functions.

● ● ●   **Example 4**   Finding All Function Values Given One Value and the Quadrant

Suppose that angle α is in quadrant II and $\sin \alpha = 2/3$. Find the values of the other five functions.

Choose any point on the terminal side of angle α. For simplicity, since $\sin \alpha = y/r$, choose the point with $r = 3$. Then

$$\frac{y}{r} = \frac{2}{3}.$$

If $r = 3$, then y will be 2. To find x, use the result $x^2 + y^2 = r^2$.

$$x^2 + y^2 = r^2$$
$$x^2 + 2^2 = 3^2$$
$$x^2 + 4 = 9$$
$$x^2 = 5$$
$$x = \sqrt{5} \quad \text{or} \quad x = -\sqrt{5}$$

Since α is in quadrant II, x must be negative, as shown in Figure 47, so $x = -\sqrt{5}$, and the point $\left(-\sqrt{5}, 2\right)$ is on the terminal side of α.

Now that the values of x, y, and r are known, the values of the remaining trigonometric functions can be found.

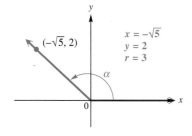

$x = -\sqrt{5}$
$y = 2$
$r = 3$

Figure 47

$$\cos \alpha = \frac{x}{r} = \frac{-\sqrt{5}}{3} = -\frac{\sqrt{5}}{3} \qquad \sec \alpha = \frac{r}{x} = \frac{3}{-\sqrt{5}} = -\frac{3\sqrt{5}}{5}$$

$$\tan \alpha = \frac{y}{x} = \frac{2}{-\sqrt{5}} = -\frac{2\sqrt{5}}{5} \qquad \cot \alpha = \frac{x}{y} = -\frac{\sqrt{5}}{2}$$

$$\csc \alpha = \frac{r}{y} = \frac{3}{2}$$

● ● ●

The Pythagorean Identities   We derive three very useful new identities from the relationship $x^2 + y^2 = r^2$. Dividing both sides by r^2 gives

$$\frac{x^2}{r^2} + \frac{y^2}{r^2} = \frac{r^2}{r^2},$$

or $$\left(\frac{x}{r}\right)^2 + \left(\frac{y}{r}\right)^2 = 1.$$

Since $\cos \theta = x/r$ and $\sin \theta = y/r$, this result becomes

$$(\cos \theta)^2 + (\sin \theta)^2 = 1,$$

or, as it is usually written,

$$\sin^2 \theta + \cos^2 \theta = 1.$$

Starting with $x^2 + y^2 = r^2$ and dividing through by x^2 gives

$$\frac{x^2}{x^2} + \frac{y^2}{x^2} = \frac{r^2}{x^2}$$

$$1 + \left(\frac{y}{x}\right)^2 = \left(\frac{r}{x}\right)^2$$

$$1 + (\tan \theta)^2 = (\sec \theta)^2$$

or $$\tan^2 \theta + 1 = \sec^2 \theta.$$

On the other hand, dividing through by y^2 leads to

$$1 + \cot^2 \theta = \csc^2 \theta.$$

These three identities are called the **Pythagorean identities** since the original equation that led to them, $x^2 + y^2 = r^2$, comes from the Pythagorean theorem.

Pythagorean Identities

$$\sin^2 \theta + \cos^2 \theta = 1 \qquad \tan^2 \theta + 1 = \sec^2 \theta \qquad 1 + \cot^2 \theta = \csc^2 \theta$$

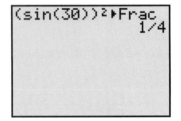

```
(sin(30))²▶Frac
             1/4
```

Figure 48

NOTE   Although we usually write $\sin^2 \theta$, for example, it should be entered as $(\sin \theta)^2$ in your calculator. To test this, verify that $\sin^2 30° = 1/4$. See Figure 48.

As before, we have given only one form of each identity. However, algebraic transformations produce equivalent identities. For example, by subtracting $\sin^2 \theta$ from both sides of $\sin^2 \theta + \cos^2 \theta = 1$, we get the equivalent identity

$$\cos^2 \theta = 1 - \sin^2 \theta.$$

You should be able to transform these identities quickly, and also recognize their equivalent forms.

The Quotient Identities   Recall that $\sin \theta = y/r$ and $\cos \theta = x/r$. Consider the quotient of $\sin \theta$ and $\cos \theta$, where $\cos \theta \neq 0$.

$$\frac{\sin \theta}{\cos \theta} = \frac{y/r}{x/r} = \frac{y}{r} \div \frac{x}{r} = \frac{y}{r} \cdot \frac{r}{x} = \frac{y}{x} = \tan \theta$$

Similarly, $(\cos \theta)/(\sin \theta) = \cot \theta$, for $\sin \theta \neq 0$. Thus we have the **quotient identities.**

Looking Ahead to Calculus

The reciprocal, Pythagorean, and quotient identities are used repeatedly in calculus to find limits, derivatives, and integrals of trigonometric functions. These identities are also used to rewrite expressions in a form that permits simplification of a square root. For example, if $a \geq 0$ and $x = a \sin \theta$,

$$\sqrt{a^2 - x^2} = \sqrt{a^2 - a^2 \sin^2 \theta}$$

$$= \sqrt{a^2(1 - \sin^2 \theta)}$$

$$= \sqrt{a^2 \cos^2 \theta} = a|\cos \theta|.$$

Quotient Identities

$$\frac{\sin \theta}{\cos \theta} = \tan \theta \qquad \frac{\cos \theta}{\sin \theta} = \cot \theta$$

● ● ●   **Example 5**   Finding Other Function Values Given One Value and the Quadrant

Find $\sin \alpha$ and $\tan \alpha$ if $\cos \alpha = -\sqrt{3}/4$ and α is in quadrant II.

Start with $\sin^2 \alpha + \cos^2 \alpha = 1$, and replace $\cos \alpha$ with $-\sqrt{3}/4$ to find $\sin \alpha$.

$$\sin^2 \alpha + \left(-\frac{\sqrt{3}}{4}\right)^2 = 1 \qquad \text{Replace } \cos \alpha \text{ with } -\frac{\sqrt{3}}{4}.$$

$$\sin^2 \alpha + \frac{3}{16} = 1$$

$$\sin^2 \alpha = \frac{13}{16} \qquad \text{Subtract } \frac{3}{16}.$$

$$\sin \alpha = \pm\frac{\sqrt{13}}{4} \qquad \text{Take square roots.}$$

Since α is in quadrant II, $\sin \alpha > 0$, and

$$\sin \alpha = \frac{\sqrt{13}}{4}.$$

To find $\tan \alpha$, use the quotient identity $\tan \alpha = \dfrac{\sin \alpha}{\cos \alpha}$.

$$\tan \alpha = \frac{\sin \alpha}{\cos \alpha} = \frac{\dfrac{\sqrt{13}}{4}}{\dfrac{-\sqrt{3}}{4}} = \frac{\sqrt{13}}{4} \cdot \frac{4}{-\sqrt{3}} = \frac{\sqrt{13}}{-\sqrt{3}}$$

Rationalize the denominator as follows.

$$\frac{\sqrt{13}}{-\sqrt{3}} = \frac{\sqrt{13}}{-\sqrt{3}} \cdot \frac{\sqrt{3}}{\sqrt{3}} = \frac{\sqrt{39}}{-3} = -\frac{\sqrt{39}}{3}$$

Therefore, $\tan \alpha = -\sqrt{39}/3$.   ● ● ●

C A U T I O N   One of the most common errors in problems like those in Examples 4 and 5 involves an incorrect sign choice when square roots are taken. Notice that in Example 5, we chose the positive square root for $\sin \alpha$ since α was in quadrant II and the sine function is positive there. If the problem had specified that α was in quadrant III, then we would have had to choose the negative square root.

● ● ●   **Example 6**   Finding Other Function Values Given One Value and the Quadrant

Find $\sin \theta$ and $\cos \theta$, if $\tan \theta = 4/3$ and θ is in quadrant III.

Since θ is in quadrant III, $\sin \theta$ and $\cos \theta$ will both be negative. It is tempting to say that since $\tan \theta = (\sin \theta)/(\cos \theta)$ and $\tan \theta = 4/3$, then $\sin \theta = -4$ and $\cos \theta = -3$. This is *incorrect,* however, since both $\sin \theta$ and $\cos \theta$ must be in the interval $[-1, 1]$.

Use the Pythagorean identity $\tan^2 \theta + 1 = \sec^2 \theta$ to find $\sec \theta$, and then the reciprocal identity $\cos \theta = 1/\sec \theta$ to find $\cos \theta$.

$$\tan^2 \theta + 1 = \sec^2 \theta$$

$$\left(\frac{4}{3}\right)^2 + 1 = \sec^2 \theta \qquad \tan \theta = \frac{4}{3}$$

$$\frac{16}{9} + 1 = \sec^2 \theta$$

$$\frac{25}{9} = \sec^2 \theta$$

$$-\frac{5}{3} = \sec \theta \qquad \text{Choose the negative square root since } \theta \text{ is in quadrant III.}$$

$$-\frac{3}{5} = \cos \theta \qquad \text{Secant and cosine are reciprocals.}$$

Since $\sin^2 \theta = 1 - \cos^2 \theta$,

$$\sin^2 \theta = 1 - \left(-\frac{3}{5}\right)^2 \qquad \cos \theta = -\frac{3}{5}$$

$$\sin^2 \theta = 1 - \frac{9}{25}$$

$$\sin^2 \theta = \frac{16}{25}$$

$$\sin \theta = -\frac{4}{5}. \qquad \text{Choose the negative square root.}$$

Therefore, we have $\sin \theta = -4/5$ and $\cos \theta = -3/5$. ● ● ●

NOTE Example 6 can also be worked by drawing θ in standard position in quadrant III, finding r to be 5, and then using the definitions of $\sin \theta$ and $\cos \theta$ in terms of x, y, and r.

1.5 Exercises

1. *Concept Check* What positive number a is its own reciprocal? Find a value of θ for which $\sin \theta = \csc \theta = a$.

2. *Concept Check* What negative number a is its own reciprocal? Find a value of θ for which $\cos \theta = \sec \theta = a$.

Use the appropriate reciprocal identity to find each function value. Rationalize denominators when applicable. In Exercises 9 and 10, use a calculator. See Example 1.

3. $\sin \theta$, if $\csc \theta = 3$

4. $\cos \alpha$, if $\sec \alpha = -2.5$

5. $\cot \beta$, if $\tan \beta = -1/5$

6. $\sin \alpha$, if $\csc \alpha = \sqrt{15}$

7. $\sec \beta$, if $\cos \beta = -1/\sqrt{7}$

8. $\tan \theta$, if $\cot \theta = -\sqrt{5}/3$

9. $\sin \theta$, if $\csc \theta = 1.42716321$

10. $\cos \alpha$, if $\sec \alpha = 9.80425133$

11. Can a given angle γ satisfy both $\sin \gamma > 0$ and $\csc \gamma < 0$? Explain.

12. Suppose that the following item appears on a trigonometry test:

Find $\sec \theta$, given that $\cos \theta = 3/2$.

Explain what is wrong with this test item.

13. What is wrong with the following statement?

$$\tan 90° = \frac{1}{\cot 90°}$$

14. *Concept Check* One form of a particular reciprocal identity is $\tan \theta = \frac{1}{\cot \theta}$. Give two other equivalent forms of this identity.

Find the tangent of each angle. See Example 1.

15. $\cot \gamma = 2$

16. $\cot \phi = -3$

17. $\cot \omega = \sqrt{3}/3$

18. $\cot \theta = \sqrt{6}/12$

19. $\cot \alpha = -.01$

20. $\cot \beta = .4$

Find a value of each variable.

21. $\tan(3B - 4°) = \dfrac{1}{\cot(5B - 8°)}$

22. $\cos(6A + 5°) = \dfrac{1}{\sec(4A + 15°)}$

23. $\sec(2\alpha + 6°)\cos(5\alpha + 3°) = 1$

24. $\sin(4\theta + 2°)\csc(3\theta + 5°) = 1$

25. The screen was obtained with the calculator in degree mode. How can we use it to justify that an angle of 14,879° is a quadrant II angle?

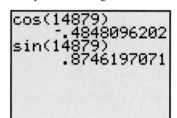

26. The screen was obtained with the calculator in degree mode. An angle of 1294° lies in which quadrant? Explain.

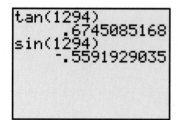

Identify the quadrant or quadrants for the angle satisfying the given conditions. See Example 2.

27. $\sin \alpha > 0, \cos \alpha < 0$

28. $\cos \beta > 0, \tan \beta > 0$

29. $\tan \gamma > 0, \cot \gamma > 0$

30. $\tan \omega < 0, \cot \omega < 0$

31. $\cos \beta < 0$

32. $\tan \theta > 0$

Give the signs of the sine, cosine, and tangent functions for each angle.

33. $74°$

34. $129°$

35. $183°$

36. $298°$

37. $406°$

38. $412°$

39. $-82°$

40. $-121°$

Concept Check In Exercises 41–46, without using a calculator, decide which is greater.

41. $\sin 30°$ or $\tan 30°$

42. $\sin 20°$ or $\sin 21°$

43. $\sin 33°$ or $\sec 33°$

44. $\cos 5°$ or $\cos^2 5°$

45. $\cos 26°$ or $\cos 27°$

46. $\cos 2°$ or $\cot 2°$

Decide whether each statement is possible *or* impossible. *See Example 3.*

47. $\sin \theta = 2$

48. $\cos \alpha = -1.001$

49. $\tan \beta = .92$

50. $\cot \omega = -12.1$

51. $\sec \alpha = 1$

52. $\tan \theta = 1$

53. $\sin \beta + 1 = .6$

54. $\sec \omega + 1 = 1.3$

55. $\sin \alpha = 1/2$ and $\csc \alpha = 2$

56. $\tan \beta = 2$ and $\cot \beta = -2$

Use identities to find each function value. Use a calculator in Exercises 63 and 64. See Examples 4–6.

57. $\tan \alpha$, if $\sec \alpha = 3$, with α in quadrant IV

58. $\sin \alpha$, if $\cos \alpha = -1/4$, with α in quadrant II

59. $\csc \beta$, if $\cot \beta = -1/2$, with β in quadrant IV

60. $\sec \theta$, if $\tan \theta = \sqrt{7}/3$, with θ in quadrant III

61. $\cos \beta$, if $\csc \beta = -4$, with β in quadrant III

62. $\sin \theta$, if $\sec \theta = 2$, with θ in quadrant IV

63. $\cot \alpha$, if $\csc \alpha = -3.5891420$, with α in quadrant III

64. $\tan \beta$, if $\sin \beta = .49268329$, with β in quadrant II

In Exercises 65 and 66, each graphing calculator screen is obtained for a particular stored value of X. *What will the screen display for the value of the expression in the final line of the display?*

65.
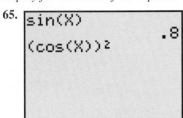

66.
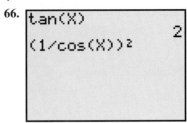

67. *Concept Check*   Does there exist an angle θ with $\cos \theta = -.6$ and $\sin \theta = .8$?

68. *Concept Check*   Does there exist an angle θ with $\cos \theta = 2/3$ and $\sin \theta = 3/4$?

Find all the trigonometric function values for each angle. Use a calculator in Exercises 75 and 76. See Examples 4–6.

69. $\tan \alpha = -15/8$, with α in quadrant II

70. $\cos \alpha = -3/5$, with α in quadrant III

71. $\tan \beta = \sqrt{3}$, with β in quadrant III

72. $\csc \theta = 2$, with θ in quadrant II

73. $\sin \beta = \sqrt{5}/7$, with $\tan \beta > 0$

74. $\cot \alpha = \sqrt{3}/8$, with $\sin \alpha > 0$

75. $\cot \theta = -1.49586$, with θ in quadrant IV

76. $\sin \alpha = .164215$, with α in quadrant II

77. Derive the identity $1 + \cot^2 \theta = \csc^2 \theta$ by dividing $x^2 + y^2 = r^2$ by y^2.

78. Using a method similar to the one given in this section showing that $(\sin \theta)/(\cos \theta) = \tan \theta$, show that

$$\frac{\cos \theta}{\sin \theta} = \cot \theta.$$

79. *Concept Check*   True or false: For all angles θ, $\sin \theta + \cos \theta = 1$. If false, give an example showing why it is false.

80. *Concept Check*   True or false: Since $\cot \theta = \dfrac{\cos \theta}{\sin \theta}$, if $\cot \theta = \dfrac{1}{2}$ with θ in quadrant I, then $\cos \theta = 1$ and $\sin \theta = 2$. If false, explain why.

Use a trigonometric function ratio to solve each problem. (Source: Marla Parker, Editor, She Does Math, Mathematical Association of America, 1995.)

81. *Height of a Tree*   A civil engineer must determine the height of the tree shown in the figure. The given angle was measured with a *clinometer*. She knows that $\sin 70° \approx .9397$, $\cos 70° \approx .3420$, and $\tan 70° \approx 2.747$. Use the pertinent trigonometric function and the measurement given in the figure to find the height of the tree to the nearest whole number.

This is a picture of one type of clinometer, called an Abney hand level and clinometer. The picture is courtesy of Keuffel & Esser Co.

82. *Double Vision*   To correct mild double vision, a small amount of prism is added to a patient's eyeglasses. The amount of light shift this causes is measured in *prism diopters*. A patient needs 12 prism diopters horizontally and 5 prism diopters vertically. A prism that corrects for both requirements should have length r and be set at angle θ. See the figure.

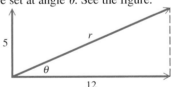

(a) Use the Pythagorean theorem to find r.

(b) Write an equation involving a trigonometric function of θ and the known prism measurements 5 and 12.

83. The straight line in the figure determines both the angle α and the angle β with the positive x-axis. Explain why $\tan \alpha = \tan \beta$.

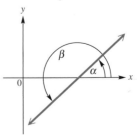

Concept Check *Suppose that* $90° < \theta < 180°$. *Find the sign of each function value.*

84. $\sin 2\theta$

85. $\tan \dfrac{\theta}{2}$

86. $\cot(\theta + 180°)$

87. $\cos(-\theta)$

Chapter 1   Summary

Key Terms & Symbols	Key Ideas
1.1 Basic Concepts	
set-builder notation interval notation open interval closed interval half-open interval relation dependent variable independent variable function domain range	**Pythagorean Theorem** The sum of the squares of the legs of a right triangle equals the square of the hypotenuse, so $a^2 + b^2 = c^2$. **Distance Formula** $$d = \sqrt{(x_2 - x_1)^2 + (y_2 - y_1)^2}$$

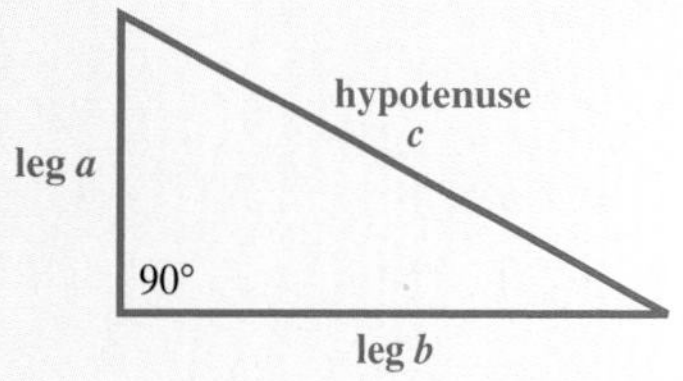

1.2 Angles

line
segment
ray
angle
initial side
terminal side
vertex
positive angle
negative angle
degree (°)
complementary angles
supplementary angles
protractor
minute (′)
second (″)
angle in standard position
quadrantal angle
coterminal angle

Types of Angles

Name	Angle Measure	Example
Acute angle	Between 0° and 90°	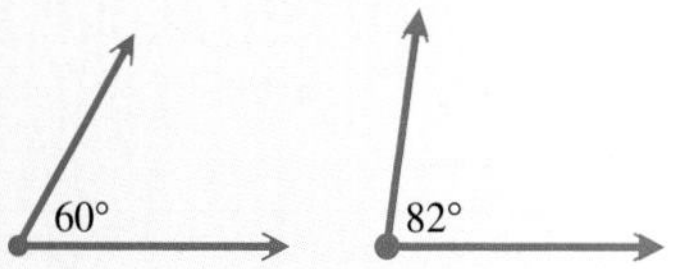
Right angle	Exactly 90°	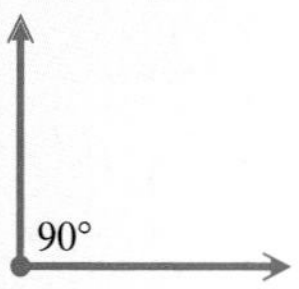
Obtuse angle	Between 90° and 180°	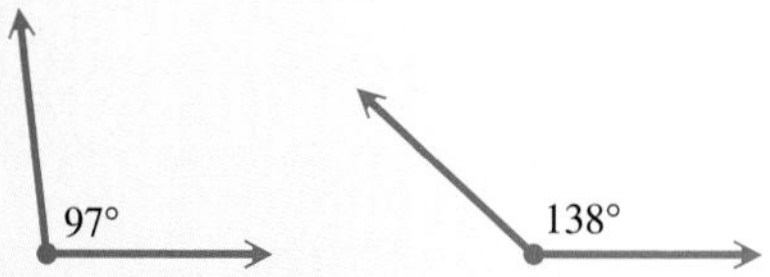
Straight angle	Exactly 180°	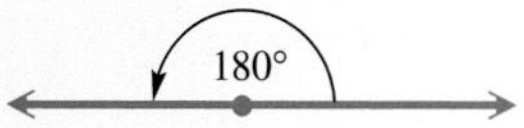

Key Terms & Symbols	Key Ideas

1.3 Angle Relationships and Similar Triangles

vertical angles
parallel lines
transversal

Vertical angles have equal measures.

The sum of the measures of the angles of any triangle is $180°$.

When a transversal intersects parallel lines, the following angles formed have equal measure: alternate interior, alternate exterior, and corresponding. Interior angles on the same side of the transversal are supplementary.

Similar triangles have corresponding angles with the same measures, and corresponding sides proportional.

Congruent triangles are the same size and the same shape.

1.4 Definitions of the Trigonometric Functions

sine
cosine
tangent
cotangent
secant
cosecant

Definitions of the Trigonometric Functions

Let (x, y) be a point other than the origin on the terminal side of an angle θ in standard position. Let $r = \sqrt{x^2 + y^2}$, the distance from the origin to (x, y). Then

$$\sin \theta = \frac{y}{r} \qquad \cos \theta = \frac{x}{r} \qquad \tan \theta = \frac{y}{x} \quad (x \neq 0)$$

$$\csc \theta = \frac{r}{y} \quad (y \neq 0) \quad \sec \theta = \frac{r}{x} \quad (x \neq 0) \quad \cot \theta = \frac{x}{y} \quad (y \neq 0).$$

Trigonometric Function Values for Quadrantal Angles

θ	$\sin \theta$	$\cos \theta$	$\tan \theta$	$\cot \theta$	$\sec \theta$	$\csc \theta$
$0°$	0	1	0	Undefined	1	Undefined
$90°$	1	0	Undefined	0	Undefined	1
$180°$	0	-1	0	Undefined	-1	Undefined
$270°$	-1	0	Undefined	0	Undefined	-1
$360°$	0	1	0	Undefined	1	Undefined

1.5 Using the Definitions of the Trigonometric Functions

identity

Reciprocal Identities

$$\sin \theta = \frac{1}{\csc \theta} \qquad \cos \theta = \frac{1}{\sec \theta} \qquad \tan \theta = \frac{1}{\cot \theta}$$

$$\csc \theta = \frac{1}{\sin \theta} \qquad \sec \theta = \frac{1}{\cos \theta} \qquad \cot \theta = \frac{1}{\tan \theta}$$

Pythagorean Identities

$$\sin^2 \theta + \cos^2 \theta = 1$$

$$\tan^2 \theta + 1 = \sec^2 \theta$$

$$1 + \cot^2 \theta = \csc^2 \theta$$

Key Terms & Symbols	Key Ideas
	Quotient Identities $$\frac{\sin\theta}{\cos\theta}=\tan\theta \qquad \frac{\cos\theta}{\sin\theta}=\cot\theta$$ **Signs of Trigonometric Functions** 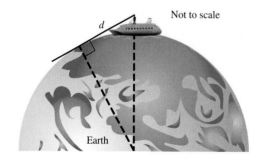

Chapter 1   Review Exercises

 1. Suppose the point (a, b) lies in the first quadrant. Describe how you would move from the point (a, b) to the point $(-a, -b)$.

2. Find the length of the line segment connecting the two points (a, b) and $(0, 0)$.

Write each set using interval notation.

3. $\{x \mid x \le -4\}$

4.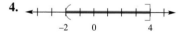

5. *Distance from a Ship to the Horizon*
   (a) How far can you see from the deck of a ship if your eyes are 100 feet above the water line? That is, find the value of d in the figure. (*Note:* The radius of Earth is about 3960 miles, and there are 5280 feet per mile.)
   (b) A rule of thumb is that the distance d (in miles) equals $\sqrt{(3/2)x}$, where x is the height (in feet) of your eyes above sea level. Apply this rule of thumb with $x = 100$. (*Source:* Huff, Darrell, *The Complete How to Figure It,* W. W. Norton, 1996.)

Find the distance between each pair of points.

6. $(4, -2)$ and $(1, -6)$

7. $(-6, 3)$ and $(-2, -5)$

8. Use the distance formula to determine whether the points $(-2, -2)$, $(8, 4)$, and $(2, 14)$ are the vertices of a right triangle.

9. State in your own words the vertical line test for the graph of a function.

Let $f(x) = -x^2 + 3x + 2$. Find each of the following.

10. $f(-2)$

11. $f(x + 1)$

12. The table was generated by a graphing calculator for a function $Y_1 = f(x)$. Use the table to answer the following questions.

(a) What is $f(1)$?

(b) If $f(x) = -14$, what is the value of x?

(c) At what point does the graph of $y = f(x)$ intersect the y-axis?

(d) At what point does the graph of $y = f(x)$ intersect the x-axis?

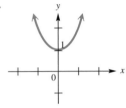

Find the domain and range of each relation. Identify any functions.

13. $y = 9x + 2$ **14.** $y = |x|$ **15.** $y = \sqrt{x}$ **16.** $x + 1 = y^2$

17.

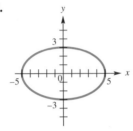

18.

Find the angle of smallest possible positive measure coterminal with each angle.

19. $-51°$ **20.** $-174°$ **21.** $792°$

22. Let n represent any integer, and write an expression for all angles coterminal with an angle of $270°$.

Work each problem.

23. *Rotating Pulley* A pulley is rotating 320 times per minute. Through how many degrees does a point on the edge of the pulley move in 2/3 second?

24. *Rotating Propeller* The propeller of a speedboat rotates 650 times per minute. Through how many degrees will a point on the edge of the propeller rotate in 2.4 seconds?

Convert decimal degrees to degrees, minutes, seconds, and convert degrees, minutes, seconds to decimal degrees. Round to the nearest second or the nearest thousandth of a degree, as appropriate. Use a calculator as necessary.

25. $47° 25' 11''$ **26.** $119° 8' 3''$ **27.** $-61.5034°$ **28.** $275.1005°$

Find the measure of each marked angle.

29.

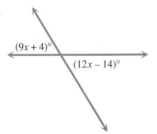

30.

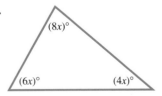

31. Express θ in terms of α and β.

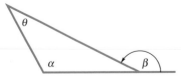

32. *Length of a Road* The flight path CP of a satellite carrying a camera with its lens at C is shown in the figure. Length PC represents the distance from the lens to the film PQ, and BA represents a straight road on the ground. Use the measurements given in the figure to find the length of the road. (*Source:* Kastner, Bernice, *Space Mathematics,* NASA, 1985.)

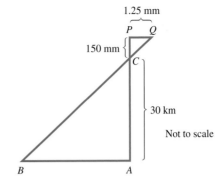

Find all unknown angle measures in each pair of similar triangles.

33.

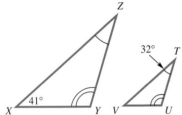

34.

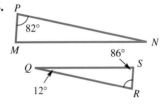

Find the unknown side lengths in each pair of similar triangles.

35.

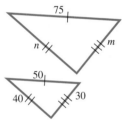
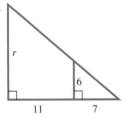

36.

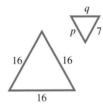

Find the unknown measurement. There are two similar triangles in each figure.

37.

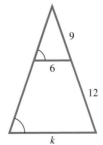

38.

39. *Concept Check* Complete the following statement: If two triangles are similar, their corresponding sides are _____ and their corresponding angles are _____.

40. *Length of a Shadow* If a tree 20 feet tall casts a shadow 8 feet long, how long would the shadow of a 30-foot tree be at the same time and place?

Find the trigonometric function values of each angle. If a value is undefined, say so.

41.

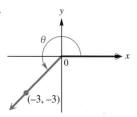

42.

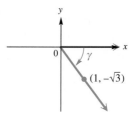

43. 180°

44. 360°

Find the values of all six trigonometric functions for an angle in standard position having each point on its terminal side.

45. $(-8, 15)$

46. $(3, -4)$

47. $(1, -5)$

48. $(9, -2)$

49. $\left(6\sqrt{3}, -6\right)$

50. $\left(-2\sqrt{2}, 2\sqrt{2}\right)$

51. Find the values of all six trigonometric functions for an angle in standard position having its terminal side defined by the equation $5x - 3y = 0$, $x \geq 0$.

52. *Concept Check* If the terminal side of a quadrantal angle lies along the y-axis, which of its trigonometric functions are undefined?

In Exercises 53–55, consider an angle θ in standard position whose terminal side has the equation $y = -5x$, with $x \leq 0$.

53. Sketch θ and use an arrow to show the rotation if $0° \leq \theta < 360°$.

54. Find the exact values of $\sin \theta$ and $\cos \theta$.

55. Find the exact value of $\tan \theta$.

Evaluate each expression.

56. $4 \sec 180° - 2 \sin^2 270°$

57. $-\cot^2 90° + 4 \sin 270° - 3 \tan 180°$

58. Explain why, for any value of θ, the point in the plane with coordinates $(\cos \theta, \sin \theta)$ lies on the unit circle.

59. The graphing calculator screen was obtained for a particular stored value of X. What will the screen display for the value of the expression in the final line of the display?

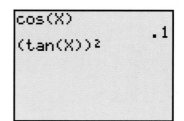

Decide whether each statement is possible *or* impossible.

60. $\sin \theta = 3/4$ and $\csc \theta = 4/3$

61. $\sec \theta = -2/3$

62. $\cos \theta = .25$ and $\sec \theta = -4$

63. $\tan \theta = 1.4$

Concept Check *Without using a calculator, determine which is greater.*

64. $\sin 50°$ or $\tan 50°$

65. $\sin 43°$ or $\sin 44°$

Find all six trigonometric function values for each angle. Rationalize denominators when applicable.

66. $\sin \theta = \sqrt{3}/5$ and $\cos \theta < 0$

67. $\cos \gamma = -5/8$, with γ in quadrant III

68. $\tan \alpha = 2$, with α in quadrant III

69. $\sec \beta = -\sqrt{5}$, with β in quadrant II

70. $\sin \theta = -2/5$, with θ in quadrant III

71. $\sec \alpha = 5/4$, with α in quadrant IV

72. If, for some particular angle θ, $\sin \theta < 0$ and $\cos \theta > 0$, in what quadrant must θ lie? What is the sign of $\tan \theta$?

73. Explain how you would find the cotangent of an angle θ whose tangent is 1.6778490 using a calculator. Then find $\cot \theta$.

Solve each problem.

74.  *Angle the Celestial North Pole Moves* At present, the north star Polaris is located very near the celestial north pole. However, because Earth is inclined 23.5°, the moon's gravitational pull on Earth is uneven. As a result, Earth slowly precesses (moves in) like a spinning top and the direction of the celestial north pole traces out a circular path once every 26,000 years. See the figure. For example, in approximately A.D. 14,000 the star Vega will be located at the celestial north pole—and not the star Polaris. As viewed from the center *C* of this circular path, calculate the angle in seconds that the celestial north pole moves each year. (*Source:* Zeilik, M., S. Gregory, and E. Smith, *Introductory Astronomy and Astrophysics,* 2nd Edition, Saunders College Publishers, 1998.)

75. *Depth of a Crater on the Moon* The depths of unknown craters on the moon can be approximated by comparing the lengths of their shadows to the shadows of nearby craters with known depths. The crater Aristillus is 11,000 feet deep, and its shadow was measured as 1.5 mm on a photograph. Its companion crater, Autolycus, had a shadow of 1.3 mm on the same photograph. Use similar triangles to determine the depth of the crater Autolycus. (*Source:*

Webb, T., *Celestial Objects for Common Telescopes,* Dover Publications, 1962.)

76. 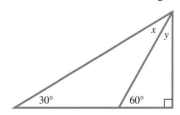 *Height of a Lunar Peak* The lunar mountain peak Huygens has a height of 21,000 feet. The shadow of Huygens on a photograph was 2.8 mm, while the nearby mountain Bradley had a shadow of 1.8 mm on the same photograph. Calculate the height of Bradley. (*Source:* Webb, T., *Celestial Objects for Common Telescopes,* Dover Publications, 1962.)

Chapter 1 Test

1. Find the distance between the points $(2, -5)$ and $(5, -2)$.

2. *Distance from a Satellite to the Horizon* The satellite AE-3 is located at point *P* in the figure. Find the distance from AE-3 to the horizon if it is 150 km above Earth. The radius of Earth is approximately 6400 km. Round your answer to the nearest hundred.

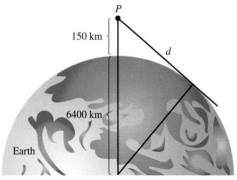

Not to scale

3. If $f(x) = .5x^2 - 2$, find **(a)** the domain, **(b)** the range, **(c)** $f(-2)$.

4. Convert $74° \, 17' \, 54''$ to decimal degrees.

5. Find the angle of smallest positive measure coterminal with $-157°$.

6. *Rotating Tire* A tire rotates 450 times per minute. Through how many degrees does a point on the edge of the tire move in 1 second?

7. Find the measure of each marked angle.

8. *Swimmer in Distress*   A lifeguard located 20 yards from the water spots a swimmer in distress. The swimmer is 30 yards from shore and 100 yards east of the lifeguard. Suppose the lifeguard runs, then swims to the swimmer in a direct line, as shown in the figure. How far east from his original position will he enter the water? (*Hint:* Find the value of *x* in the sketch.)

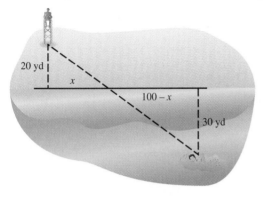

9. If $(2, -5)$ is on the terminal side of an angle θ in standard position, find $\sin \theta$, $\cos \theta$, and $\tan \theta$.

10. If $\cos \theta < 0$ and $\cot \theta > 0$, in what quadrant does θ terminate?

11. If $\cos \theta = 4/5$ and θ is in quadrant IV, find the values of the other trigonometric functions of θ.

12. Is it possible for $\sin \theta$, $\cos \theta$, and $\tan \theta$ to all be negative for the same value of θ? Explain your answer.

Acute Angles and Right Triangles

2

2.1 Trigonometric Functions of Acute Angles

2.2 Trigonometric Functions of Non-Acute Angles

2.3 Finding Trigonometric Function Values Using a Calculator

2.4 Solving Right Triangles

2.5 Further Applications of Right Triangles

Highway transportation is critical to the economy of the United States. In 1970 there were 1150 billion miles traveled, and in the year 2000 this has increased to approximately 2500 billion miles. Designing highways for safety and efficiency saves both lives and time. For example, when an automobile travels around a curve, objects like trees, buildings, and fences situated on the curve, as shown in the figure, may obstruct a driver's vision. Trigonometry is used to determine how far inside the curve land must be cleared to provide visibility for a safe stopping distance.* A problem like this is solved in Section 2.4.

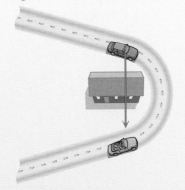

*Source: Mannering, F. and W. Kilareski, *Principles of Highway Engineering and Traffic Analysis,* 2nd Edition, John Wiley & Sons, 1998.

In this chapter, we also see how trigonometry is used to design curves, compute grade resistance, calculate braking distances, and solve other problems related to highway engineering, the theme of this chapter.

• •

2.1 Trigonometric Functions of Acute Angles

• Definitions of the Trigonometric Functions • Cofunctions • Trigonometric Function Values of Special Angles

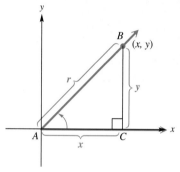

Figure 1

Definitions of the Trigonometric Functions   Figure 1 shows an acute angle A in standard position. The definitions of the trigonometric function values of angle A require x, y, and r. As drawn in Figure 1, x and y are the lengths of the two legs of the right triangle ABC, and r is the length of the hypotenuse.

The side of length y is called the **side opposite** angle A, and the side of length x is called the **side adjacent** to angle A. We use the lengths of these sides to replace x and y in the definition of the trigonometric functions, and the length of the hypotenuse to replace r, to get the following right-triangle-based definitions.

Right-Triangle-Based Definitions of Trigonometric Functions

For any acute angle A in standard position,

$$\sin A = \frac{y}{r} = \frac{\text{side opposite}}{\text{hypotenuse}} \qquad \csc A = \frac{r}{y} = \frac{\text{hypotenuse}}{\text{side opposite}}$$

$$\cos A = \frac{x}{r} = \frac{\text{side adjacent}}{\text{hypotenuse}} \qquad \sec A = \frac{r}{x} = \frac{\text{hypotenuse}}{\text{side adjacent}}$$

$$\tan A = \frac{y}{x} = \frac{\text{side opposite}}{\text{side adjacent}} \qquad \cot A = \frac{x}{y} = \frac{\text{side adjacent}}{\text{side opposite}}.$$

• • • **Example 1** Finding Trigonometric Function Values of an Acute Angle in a Right Triangle

Find the values of the trigonometric functions for angles A and B in the right triangle in Figure 2.

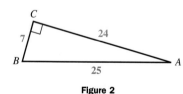

Figure 2

The length of the side opposite angle A is 7, the length of the side adjacent to angle A is 24, and the length of the hypotenuse is 25. Use the relationships given above.

$$\sin A = \frac{\text{side opposite}}{\text{hypotenuse}} = \frac{7}{25} \qquad \csc A = \frac{\text{hypotenuse}}{\text{side opposite}} = \frac{25}{7}$$

$$\cos A = \frac{\text{side adjacent}}{\text{hypotenuse}} = \frac{24}{25} \qquad \sec A = \frac{\text{hypotenuse}}{\text{side adjacent}} = \frac{25}{24}$$

$$\tan A = \frac{\text{side opposite}}{\text{side adjacent}} = \frac{7}{24} \qquad \cot A = \frac{\text{side adjacent}}{\text{side opposite}} = \frac{24}{7}$$

The length of the side opposite angle B is 24, while the length of the side adjacent to B is 7, so

$$\sin B = \frac{24}{25} \qquad \cos B = \frac{7}{25} \qquad \tan B = \frac{24}{7}$$

$$\csc B = \frac{25}{24} \qquad \sec B = \frac{25}{7} \qquad \cot B = \frac{7}{24}.$$

• • •

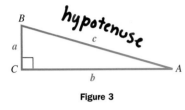

Figure 3

Cofunctions   In Example 1, you may have noticed that $\sin A = \cos B$, $\cos A = \sin B$, and so on. Such relationships are always true for the two acute angles of a right triangle. Figure 3 shows a right triangle with acute angles A and B and a right angle at C. (Whenever we use A, B, and C to name angles in a right triangle, C will be the right angle.) The length of the side opposite A is a, and the length of the side opposite angle B is b. The length of the hypotenuse is c.

By the definitions given above, $\sin A = a/c$. Since $\cos B$ is also equal to a/c,

$$\sin A = \frac{a}{c} = \cos B.$$

In a similar manner,

$$\tan A = \frac{a}{b} = \cot B \qquad \sec A = \frac{c}{b} = \csc B.$$

The sum of the three angles in any triangle is 180°. Since angle C equals 90°, angles A and B must have a sum of $180° - 90° = 90°$. As mentioned in Chapter 1, angles with a sum of 90° are called complementary angles. Since angles A and B are complementary and $\sin A = \cos B$, the functions sine and cosine are called **cofunctions.** Also, tangent and cotangent are cofunctions, as are secant and cosecant. And since the angles A and B are complementary, $A + B = 90°$, or

$$B = 90° - A,$$

giving

$$\sin A = \cos B = \cos(90° - A).$$

Similar results, called the **cofunction identities,** are true for the other trigonometric functions.

Cofunction Identities

For any acute angle A,

$$\sin A = \cos(90° - A) \quad \csc A = \sec(90° - A) \quad \tan A = \cot(90° - A)$$
$$\cos A = \sin(90° - A) \quad \sec A = \csc(90° - A) \quad \cot A = \tan(90° - A).$$

(These identities will be extended to *any* angle A, not just acute angles, in Chapter 5.) You will need to know these identities throughout the book.

● ● ● **Example 2** Writing Functions in Terms of Cofunctions

Write each of the following in terms of its cofunction.

(a) $\cos 52°$

Since $\cos A = \sin(90° - A)$,

$$\cos 52° = \sin(90° - 52°) = \sin 38°.$$

(b) $\tan 71° = \cot 19°$ **(c)** $\sec 24° = \csc 66°$ ● ● ●

● ● ● **Example 3** Solving Equations Using the Cofunction Identities

Find a value of θ satisfying each equation. Assume that all angles involved are acute angles.

(a) $\cos(\theta + 4°) = \sin(3\theta + 2°)$

Since sine and cosine are cofunctions, this equation is true if the sum of the angles is 90°.

$$(\theta + 4°) + (3\theta + 2°) = 90°$$
$$4\theta + 6° = 90°$$
$$4\theta = 84°$$
$$\theta = 21°$$

(b) $\tan(2\theta - 18°) = \cot(\theta + 18°)$

$$(2\theta - 18°) + (\theta + 18°) = 90°$$
$$3\theta = 90°$$
$$\theta = 30°$$ ● ● ●

Figure 4 shows three right triangles. From left to right, the length of each hypotenuse is the same, but angle A increases in measure. As angle A increases in measure from 0° to 90°, the length of the side opposite angle A also increases.

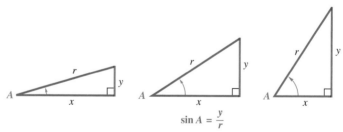

$$\sin A = \frac{y}{r}$$

As A increases, y increases. Since r is fixed, $\sin A$ increases.

Figure 4

Since

$$\sin A = \frac{\text{side opposite}}{\text{hypotenuse}},$$

as angle A increases, the numerator of this fraction also increases, while the denominator is fixed. This means that $\sin A$ *increases* as A increases from 0° to 90°.

As angle A increases from 0° to 90°, the length of the side adjacent to A decreases. Since r is fixed, the ratio x/r will decrease. This ratio gives $\cos A$, showing that the values of cosine *decrease* as the angle measure changes from 0° to

90°. Finally, increasing A from 0° to 90° causes y to increase and x to decrease, making the values of $y/x = \tan A$ increase.

A similar discussion shows that as A increases from 0° to 90°, the values of sec A increase, while the values of cot A and csc A decrease.

● ● ● **Example 4** **Comparing Function Values of Acute Angles**

Tell whether each statement is *true* or *false*.

(a) sin 21° > sin 18°

In the interval from 0° to 90°, as the angle increases, so does the sine of the angle, which makes sin 21° > sin 18° a true statement.

(b) cos 49° ≤ cos 56°

As the angle increases, the cosine of the angle decreases. The given statement cos 49° ≤ cos 56° is false. ● ● ●

Trigonometric Function Values of Special Angles Certain special angles, such as 30°, 45°, and 60°, occur so often in trigonometry and in more advanced mathematics that they deserve special study. We can find the exact trigonometric function values of these angles using properties of geometry and the Pythagorean theorem.

To find the trigonometric function values for 30° and 60°, we start with an equilateral triangle, a triangle with all sides of equal length. Each angle of such a triangle measures 60°. While the results we will obtain are independent of the length, for convenience we choose the length of each side to be 2 units. See Figure 5(a).

Bisecting one angle of this equilateral triangle leads to two right triangles, each of which has angles of 30°, 60°, and 90°, as shown in Figure 5(b). Since the hypotenuse of one of these right triangles has length 2, the shortest side will have length 1. (Why?) If x represents the length of the medium side, then, by the Pythagorean theorem,

$$2^2 = 1^2 + x^2$$
$$4 = 1 + x^2$$
$$3 = x^2$$
$$\sqrt{3} = x.$$

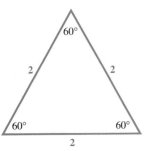

Equilateral triangle

(a)

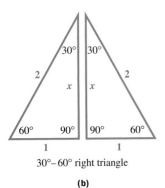

30°–60° right triangle

(b)

Figure 5

Figure 6 summarizes our results, showing a 30°–60° right triangle.

As shown in the figure, the side opposite the 30° angle has length 1; that is, for the 30° angle,

hypotenuse = 2,   side opposite = 1,   side adjacent = $\sqrt{3}$.

Now we use the definitions of the trigonometric functions.

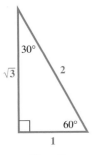

Figure 6

$$\sin 30° = \frac{\text{side opposite}}{\text{hypotenuse}} = \frac{1}{2} \qquad \csc 30° = \frac{2}{1} = 2$$

$$\cos 30° = \frac{\text{side adjacent}}{\text{hypotenuse}} = \frac{\sqrt{3}}{2} \qquad \sec 30° = \frac{2}{\sqrt{3}} = \frac{2\sqrt{3}}{3}$$

$$\tan 30° = \frac{\text{side opposite}}{\text{side adjacent}} = \frac{1}{\sqrt{3}} = \frac{\sqrt{3}}{3} \qquad \cot 30° = \frac{\sqrt{3}}{1} = \sqrt{3}$$

The denominator was rationalized for tan 30° and sec 30°.

In a similar manner,

$$\sin 60° = \frac{\sqrt{3}}{2} \qquad \cos 60° = \frac{1}{2} \qquad \tan 60° = \sqrt{3}$$

$$\csc 60° = \frac{2\sqrt{3}}{3} \qquad \sec 60° = 2 \qquad \cot 60° = \frac{\sqrt{3}}{3}.$$

We find the values of the trigonometric functions for 45° by starting with a 45°–45° right triangle, as shown in Figure 7. This triangle is isosceles, and for convenience, we choose the lengths of the equal sides to be 1 unit. (As before, the results are independent of the length of the equal sides of the right triangle.) Since the shorter sides each have length 1, if r represents the length of the hypotenuse, then

$$1^2 + 1^2 = r^2$$

$$2 = r^2$$

$$\sqrt{2} = r.$$

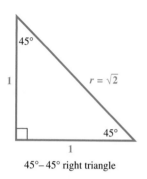

45°–45° right triangle

Figure 7

Now we use the measures indicated on the 45°–45° right triangle in Figure 7.

$$\sin 45° = \frac{1}{\sqrt{2}} = \frac{\sqrt{2}}{2} \qquad \cos 45° = \frac{1}{\sqrt{2}} = \frac{\sqrt{2}}{2} \qquad \tan 45° = \frac{1}{1} = 1$$

$$\csc 45° = \frac{\sqrt{2}}{1} = \sqrt{2} \qquad \sec 45° = \frac{\sqrt{2}}{1} = \sqrt{2} \qquad \cot 45° = \frac{1}{1} = 1$$

The importance of these exact trigonometric function values of 30°, 45°, and 60° angles cannot be overemphasized. It is essential to memorize them. They are summarized in the chart that follows.

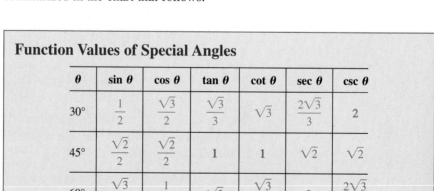

Function Values of Special Angles

θ	$\sin \theta$	$\cos \theta$	$\tan \theta$	$\cot \theta$	$\sec \theta$	$\csc \theta$
30°	$\frac{1}{2}$	$\frac{\sqrt{3}}{2}$	$\frac{\sqrt{3}}{3}$	$\sqrt{3}$	$\frac{2\sqrt{3}}{3}$	2
45°	$\frac{\sqrt{2}}{2}$	$\frac{\sqrt{2}}{2}$	1	1	$\sqrt{2}$	$\sqrt{2}$
60°	$\frac{\sqrt{3}}{2}$	$\frac{1}{2}$	$\sqrt{3}$	$\frac{\sqrt{3}}{3}$	2	$\frac{2\sqrt{3}}{3}$

NOTE You should be able to reproduce this chart quickly. It is not difficult to do if you learn the values of sin 30°, sin 45°, and sin 60°. Then complete the rest of the chart using the reciprocal, cofunction, and quotient identities.

C O N N E C T I O N S A convenient way to quickly produce a chart of the trigonometric function values for the special angles is to produce the chart shown below. Write the angles in the first column. In the second column, each numerator is a radical with the numbers 0, 1, 2, 3, and 4, in order, placed under it. Each denominator is 2. In the third column, each numerator is a radical with the numbers 4, 3, 2, 1, and 0, in order, placed under it. Each denominator is 2. Simplifying these fractions gives the values shown in the chart above for $\sin \theta$ and $\cos \theta$. *Note that this works only for the degree measures shown below and cannot be extended to other values of θ.* The other trigonometric function values are easily found from these basic ones.

θ	$\sin \theta$	$\cos \theta$
0°	$\dfrac{\sqrt{0}}{2}$	$\dfrac{\sqrt{4}}{2}$
30°	$\dfrac{\sqrt{1}}{2}$	$\dfrac{\sqrt{3}}{2}$
45°	$\dfrac{\sqrt{2}}{2}$	$\dfrac{\sqrt{2}}{2}$
60°	$\dfrac{\sqrt{3}}{2}$	$\dfrac{\sqrt{1}}{2}$
90°	$\dfrac{\sqrt{4}}{2}$	$\dfrac{\sqrt{0}}{2}$

For Discussion or Writing

Verify that the simplified forms of the fractions in the table agree with the values shown earlier.

In Exercises 57 and 58, we generalize the relationships among the sides of a 30°–60° right triangle and a 45°–45° right triangle.

Since a calculator finds trigonometric function values at the touch of a key, you may wonder why we spend so much time finding values for special angles. We do this because a calculator gives only *approximate* values in most cases, while we often need *exact* values. For example, tan 30° can be found on a scientific calculator by first setting it in *degree mode,* then entering 30 and pressing the tan key to get

$$\tan 30° \approx .57735027.$$

(The symbol $\approx$ means "is approximately equal to.") Earlier, however, we found the exact value:

$$\tan 30° = \frac{\sqrt{3}}{3}.$$

To use a graphing calculator to approximate sine, cosine, or tangent function values, press the appropriate function key *first,* and then enter the angle measure. (The calculator must be in degree mode to enter the angle measure in degrees.) See Figure 8.

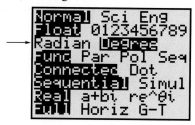

Figure 8

2.1 Exercises

Concept Check *Match each trigonometric function in Column I with its value from Column II.*

I		II	
1. sin 30°		**A.** $\sqrt{3}$	**F.** $\dfrac{\sqrt{3}}{3}$
2. cos 45°		**B.** 1	**G.** 2
3. tan 45°			
4. sec 60°		**C.** $\dfrac{1}{2}$	**H.** $\dfrac{\sqrt{2}}{2}$
5. csc 60°		**D.** $\dfrac{\sqrt{3}}{2}$	**I.** $\sqrt{2}$
6. cot 30°			
		E. $\dfrac{2\sqrt{3}}{3}$	

In each exercise, find the values of the six trigonometric functions for angle A. Leave answers as fractions. See Example 1.

7.
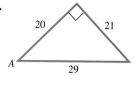

8.

9.
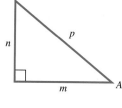

10.
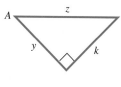

Suppose ABC is a right triangle with sides of lengths a, b, and c with right angle at C. (See Figure 3.) Find the unknown side length using the Pythagorean theorem, and then find the values of the six trigonometric functions for angle B. Rationalize denominators when applicable.

11. $a = 5, b = 12$ **12.** $a = 3, b = 5$ **13.** $a = 6, c = 7$ **14.** $b = 7, c = 12$

15. *Concept Check* Give a summary of the relationships between cofunctions of complementary angles.

Write each of the following in terms of its cofunction. Assume that all angles in which an unknown appears are acute angles. See Example 2.

16. cot 73° **17.** sec 39° **18.** $\cos(\alpha + 20°)$

19. $\cot(\beta - 10°)$ **20.** tan 25.4° **21.** sin 38.7°

22. With a calculator, evaluate $\sin(90° - A)$ and $\cos A$ for various values of A. (Include values greater than 90° and less than 0°.) What do you find?

Find a solution for each equation. Assume that all angles in which an unknown appears are acute angles. See Example 3.

23. $\tan \alpha = \cot(\alpha + 10°)$

24. $\cos \theta = \sin 2\theta$

25. $\sin(2\gamma + 10°) = \cos(3\gamma - 20°)$

26. $\sec(\beta + 10°) = \csc(2\beta + 20°)$

27. $\tan(3B + 4°) = \cot(5B - 10°)$

28. $\cot(5\theta + 2°) = \tan(2\theta + 4°)$

Tell whether each statement is true *or* false. *See Example 4.*

29. $\tan 28° \leq \tan 40°$

30. $\sin 50° > \sin 40°$

31. $\sin 46° < \cos 46°$
(*Hint:* $\cos 46° = \sin 44°$)

32. $\cos 28° < \sin 28°$

33. $\tan 41° < \cot 41°$

34. $\cot 30° < \tan 40°$

Refer to the discussion in this section to give the exact *trigonometric function value. Do not use a calculator.*

35. $\tan 30°$

36. $\cot 30°$

37. $\sin 30°$

38. $\cos 30°$

39. $\csc 45°$

40. $\sec 45°$

41. $\sin 60°$

42. $\cos 60°$

· · · · · · · · · · · · · **Relating Concepts** · · · · · · · · · · · · · · · · ·

For individual or collaborative investigation
(Exercises 43–46)

The figure shows a 45° central angle in a circle with radius 4 units. To find the coordinates of point P on the circle, **work Exercises 43–46 in order.**

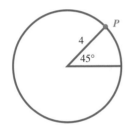

43. Add coordinate axes to the figure so the central angle is in standard position. Add a line from point P perpendicular to the x-axis.

44. Use the trigonometric ratios for a 45° angle to label the sides of the right triangle you sketched in Exercise 43.

45. Which sides of the right triangle give the coordinates of point P? What are the coordinates of P?

46. Follow the same procedure to find the coordinates of P in the figure given here.

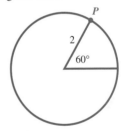

· ·

47. Refer to the table. What trigonometric functions are Y_1 and Y_2?

X	Y₁	Y₂
0	0	0
15	.25882	.26795
30	.5	.57735
45	.70711	1
60	.86603	1.7321
75	.96593	3.7321
90	1	ERROR

X=0

48. Refer to the table. What trigonometric functions are Y_1 and Y_2?

X	Y₁	Y₂
0	1	ERROR
15	.96593	3.8637
30	.86603	2
45	.70711	1.4142
60	.5	1.1547
75	.25882	1.0353
90	0	1

X=0

49. What value of A between 0° and 90° will produce the output for the graphing calculator screen?

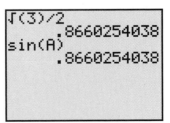

50. A student was asked to give the exact value of sin 45°. Using a calculator, he gave the answer .7071067812. The teacher did not give him credit. What was the teacher's reason for this?

51. With a graphing calculator, find the coordinates of the point of intersection of $y = x$ and $y = \sqrt{1 - x^2}$. These coordinates are the cosine and sine of what angle between 0° and 90°?

Concept Check *Use the concepts of this section to work Exercises 52–56.*

52. Find the equation of the line passing through the origin and making a 60° angle with the x-axis.

53. Find the equation of the line passing through the origin and making a 30° angle with the x-axis.

54. What angle does the line $y = \dfrac{\sqrt{3}}{3}x$ make with the positive x-axis?

55. What angle does the line $y = \sqrt{3}x$ make with the positive x-axis?

56. Which pair of trigonometric functions are both reciprocals and cofunctions?

57. Construct an equilateral triangle with each side having length 2k.
  (a) What is the measure of each angle?
  (b) Label one angle A. Drop a perpendicular from A to the side opposite A. Two 30° angles are formed at A, and two right triangles are formed. What is the length of each side opposite each 30° angle?
  (c) What is the length of the perpendicular constructed in part (b)?
  (d) From the results of parts (a)–(c), complete the following statement: In a 30°–60° right triangle, the hypotenuse is always _____ times as long as the shorter leg, and the longer leg has a length that is _____ times as long as that of the shorter leg. Also, the shorter leg is opposite the _____ angle, and the longer leg is opposite the _____ angle.

58. Construct a square with each side of length k.
  (a) Draw a diagonal of the square. What is the measure of each angle formed by a side of the square and this diagonal?
  (b) What is the length of the diagonal?
  (c) From the results of parts (a) and (b), complete the following statement: In a 45°–45° right triangle, the hypotenuse has a length that is _____ times as long as either leg.

Use the results of Exercises 57 and 58 to find the exact value of each part labeled with a variable in each figure.

59.

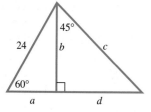

60.

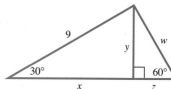

61.

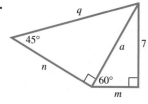

62.

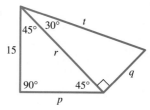

Find a formula for the area of each figure in terms of s.

63.

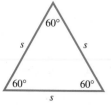

64.

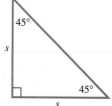

65. Suppose you know the length of one side and one acute angle of a right triangle. Can you determine the measures of all the sides and angles of the triangle?

66. Refer to the table in the Connections box in this section. Explain why this pattern cannot possibly continue past 90°. (*Hint:* What is the maximum value of the sine ratio?)

67. Why is it important to be able to find trigonometric values for the special angles without using a calculator?

Work each problem.

68. 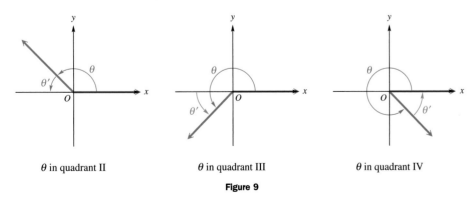 *(Modeling) Braking Distance* If aerodynamic resistance is ignored, the braking distance D (in feet) for an automobile to change its velocity from V_1 to V_2 (feet per second) can be modeled using the equation

$$D = \frac{1.05(V_1^2 - V_2^2)}{64.4(K_1 + K_2 + \sin \theta)}.$$

K_1 is a constant determined by the efficiency of the brakes and tires, K_2 is a constant determined by the

rolling resistance of the automobile, and θ is the grade of the highway. (*Source:* Mannering, F. and W. Kilareski, *Principles of Highway Engineering and Traffic Analysis,* 2nd Edition, John Wiley & Sons, 1998.)

(a) Compute the number of feet required to slow a car from 55 to 30 mph while traveling uphill with a grade of $\theta = 3.5°$. Let $K_1 = .4$ and $K_2 = .02$. (*Hint:* Change miles per hour to feet per second.)

(b) Repeat part (a) with $\theta = -2°$.

(c) How is braking distance affected by the grade θ? Does this agree with your driving experience?

69. *(Modeling) Car's Speed at Collision* Refer to Exercise 68. An automobile is traveling at 90 mph on a highway with a downhill grade of $\theta = -3.5°$. The driver sees a stalled truck in the road 200 feet away and immediately applies the brakes. Assuming that a collision cannot be avoided, how fast (in miles per hour) is the car traveling when it hits the truck? (Use the same values for K_1 and K_2 as in Exercise 68.)

2.2 Trigonometric Functions of Non-Acute Angles

• **Reference Angles** • **Special Angles as Reference Angles**

Reference Angles Associated with every nonquadrantal angle in standard position is a positive acute angle called its *reference angle.* A **reference angle** for an angle θ, written θ', is the positive acute angle made by the terminal side of angle θ and the x-axis. Figure 9 shows several angles θ (each less than one complete counterclockwise revolution) in quadrants II, III, and IV, respectively, with the reference angle θ' also shown. In quadrant I, θ and θ' are the same. If an angle θ is negative or has measure greater than 360°, its reference angle is found by first finding its coterminal angle that is between 0° and 360°, and then using the diagrams in Figure 9.

θ in quadrant II θ in quadrant III θ in quadrant IV

Figure 9

CAUTION   A common error is to find the reference angle by using the terminal side of θ and the y-axis. *The reference angle is always found with reference to the x-axis.*

● ● ● **Example 1** Finding Reference Angles

Find the reference angle for each angle.

(a) 218°

As shown in Figure 10, the positive acute angle made by the terminal side of this angle and the x-axis is $218° - 180° = 38°$. For $\theta = 218°$, the reference angle $\theta' = 38°$.

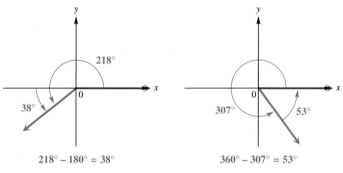

218° − 180° = 38°
Figure 10

360° − 307° = 53°
Figure 11

(b) 1387°

First find a coterminal angle between 0° and 360°. Divide 1387° by 360° to get a quotient of about 3.9. Begin by subtracting 360° three times (because of the 3 in 3.9):

$$1387° - 3 \cdot 360° = 307°.$$

The reference angle for 307° (and thus for 1387°) is $360° - 307° = 53°$. See Figure 11. ● ● ●

The preceding example suggests the following table for finding the reference angle θ' for any angle θ between 0° and 360°.

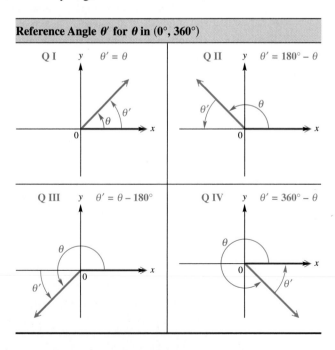

Reference Angle θ' for θ in (0°, 360°)

Special Angles as Reference Angles We can now find exact trigonometric function values of angles with reference angles of 30°, 60°, or 45°.

● ● ● **Example 2** Finding Trigonometric Function Values of a Quadrant III Angle

Find the values of the trigonometric functions for 210°.

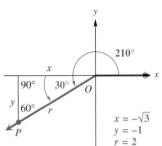

An angle of 210° is shown in Figure 12. The reference angle is 210° − 180° = 30°. To find the trigonometric function values of 210°, choose point P on the terminal side of the angle so that the distance from the origin O to P is 2. By the results from 30°–60° right triangles, the coordinates of point P become $\left(-\sqrt{3}, -1\right)$, with $x = -\sqrt{3}$, $y = -1$, and $r = 2$. Then, by the definitions of the trigonometric functions,

Figure 12

$$\sin 210° = -\frac{1}{2} \qquad \cos 210° = -\frac{\sqrt{3}}{2} \qquad \tan 210° = \frac{\sqrt{3}}{3}$$

$$\csc 210° = -2 \qquad \sec 210° = -\frac{2\sqrt{3}}{3} \qquad \cot 210° = \sqrt{3}.$$

● ● ●

Notice in Example 2 that the trigonometric function values of 210° correspond in absolute value to those of its reference angle 30°. The signs are different for the sine, cosine, secant, and cosecant functions because 210° is a quadrant III angle. These results suggest a shortcut for finding the trigonometric function values of a non-acute angle, using the reference angle. In Example 2, the reference angle for 210° is 30°. Using the trigonometric function values of 30°, and choosing the correct signs for a quadrant III angle, we obtain the same results found in Example 2.

Similarly, the values of the trigonometric functions for any nonquadrantal angle θ can be determined by finding the function values for an angle between 0° and 90°.

Finding Trigonometric Function Values for Any Nonquadrantal Angle

Step 1 If $\theta > 360°$, or if $\theta < 0°$, find a coterminal angle by adding or subtracting 360° as many times as needed to get an angle greater than 0° but less than 360°.

Step 2 Find the reference angle θ'.

Step 3 Find the necessary values of the trigonometric functions for the reference angle θ'.

Step 4 Determine the correct signs for the values found in Step 3. (Use the table of signs in Section 1.5.) This gives the values of the trigonometric functions for angle θ.

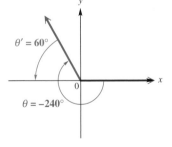

Figure 13

● ● ● **Example 3** Finding Trigonometric Function Values Using Reference Angles

Find the exact value of each of the following.

(a) $\cos(-240°)$

Since an angle of −240° is coterminal with an angle of 360° − 240° = 120°, the reference angle is 180° − 120° = 60°, as shown in Figure 13. Since the cosine is negative in quadrant II,

$$\cos(-240°) = -\cos 60° = -\frac{1}{2}.$$

(b) tan 675°

Begin by subtracting 360° to get a coterminal angle between 0° and 360°.

$$675° - 360° = 315°$$

As shown in Figure 14, the reference angle is 360° − 315° = 45°. An angle of 315° is in quadrant IV, so the tangent will be negative, and

$$\tan 675° = \tan 315° = -\tan 45° = -1.$$

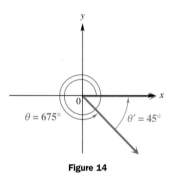

Figure 14

The ideas discussed in this section can be reversed to find the measures of certain angles, given a trigonometric function value and an interval in which the angle must lie. We are most often interested in the interval $[0°, 360°)$.

Example 4 Finding Angle Measures Given an Interval and a Function Value

Find all values of θ, if θ is in the interval $[0°, 360°)$ and $\cos \theta = -\sqrt{2}/2$.

Since cosine here is negative, θ must lie in either quadrant II or III. Also, the absolute value of $\cos \theta$ is $\sqrt{2}/2$, so the reference angle θ' must be 45°. The two possible angles θ are sketched in Figure 15. The quadrant II angle θ must equal 180° − 45° = 135°, and the quadrant III angle θ must equal 180° + 45° = 225°.

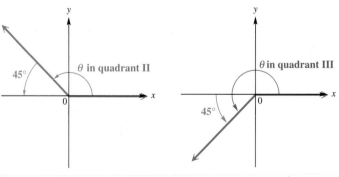

Figure 15

Exact values can be used in evaluating expressions, as shown in the next example.

● ● ● **Example 5** Evaluating an Expression with Function Values of Special Angles

Evaluate $\cos 120° + 2 \sin^2 60° - \tan^2 30°$.

Since $\cos 120° = -1/2$, $\sin 60° = \sqrt{3}/2$, and $\tan 30° = \sqrt{3}/3$,

$$\cos 120° + 2 \sin^2 60° - \tan^2 30° = -\frac{1}{2} + 2\left(\frac{\sqrt{3}}{2}\right)^2 - \left(\frac{\sqrt{3}}{3}\right)^2$$

$$= -\frac{1}{2} + 2\left(\frac{3}{4}\right) - \frac{3}{9}$$

$$= \frac{2}{3}.$$

● ● ●

As mentioned earlier, the values of the trigonometric functions of coterminal angles are the same.

● ● ● **Example 6** Using Coterminal Angles to Find Function Values

Evaluate each of the following by first expressing the function in terms of an angle between 0° and 360°.

(a) $\cos 780°$

Add or subtract 360° as many times as necessary to get an angle between 0° and 360°. Subtracting 720°, which is $2 \cdot 360°$, gives

$$\cos 780° = \cos(60° + 2 \cdot 360°)$$

$$= \cos 60°$$

$$= \frac{1}{2}.$$

(b) $\tan(-405°)$

Add 360° to get $-405° + 360° = -45°$. Following the method of Example 2, we find that $\tan(-45°) = -\tan 45°$ because the reference angle is 45°, and $-45°$ is in quadrant IV where the tangent function is negative. Thus,

$$\tan(-405°) = \tan(-45°) = -\tan 45° = -1.$$

● ● ●

2.2 Exercises

Concept Check   *Match each angle in Column I with its reference angle in Column II. Some reference angles may be used more than once or not at all.*

I	II
1. 98°	**A.** 45°
2. 212°	**B.** 60°
3. −135°	**C.** 82°
4. −60°	**D.** 30°
5. 750°	**E.** 38°
6. 480°	**F.** 32°

Give a short explanation in Exercises 7–10.

7. In Example 2, why was 2 a good choice for r? Could any other positive number have been used?

8. Explain how the reference angle is used to find values of the trigonometric functions for an angle in quadrant III.

9. Explain why two coterminal angles have the same values for their trigonometric functions.

10. If two angles have the same values for each of the six trigonometric functions, must the angles be coterminal? Explain your reasoning.

Note: The remaining exercises in this set are not *to be worked with a calculator.*

Complete the following table with exact trigonometric function values using the methods of this section. See Examples 2 and 3.

θ	$\sin\theta$	$\cos\theta$	$\tan\theta$	$\cot\theta$	$\sec\theta$	$\csc\theta$
11. 30°	1/2	$\sqrt{3}/2$			$2\sqrt{3}/3$	2
12. 45°			1	1		
13. 60°		1/2	$\sqrt{3}$		2	
14. 120°	$\sqrt{3}/2$		$-\sqrt{3}$			$2\sqrt{3}/3$
15. 135°	$\sqrt{2}/2$	$-\sqrt{2}/2$			$-\sqrt{2}$	$\sqrt{2}$
16. 150°		$-\sqrt{3}/2$	$-\sqrt{3}/3$			2
17. 210°	$-1/2$		$\sqrt{3}/3$	$\sqrt{3}$		-2
18. 240°	$-\sqrt{3}/2$	$-1/2$			-2	$-2\sqrt{3}/3$

Use the methods of this section to find the exact values of the six trigonometric functions for each angle. Rationalize denominators when applicable. See Examples 2, 3, and 6.

19. 225° **20.** 300° **21.** 315° **22.** 405°

23. 420° **24.** 480° **25.** 495° **26.** 570°

27. 750° **28.** 1305° **29.** 1500° **30.** 2670°

31. $-390°$ **32.** $-510°$ **33.** $-1020°$ **34.** $-1290°$

Concept Check Find the coordinates of the point P on the circumference of each circle, assuming that the angle is in standard position.

35.

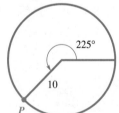

36.

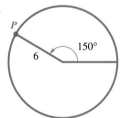

37. *Concept Check* Does there exist an angle θ with the function values $\cos\theta = .6$ and $\sin\theta = -.8$?

38. *Concept Check* Does there exist an angle θ with the function values $\cos\theta = 2/3$ and $\sin\theta = 3/4$?

Suppose θ is in the interval $(90°, 180°)$. Find the sign of each of the following.

39. $\sin\dfrac{\theta}{2}$

40. $\cos\dfrac{\theta}{2}$

41. $\cot(\theta + 180°)$

42. $\sec(\theta + 180°)$

43. $\cos(-\theta)$

44. $\sin(-\theta)$

45. Explain why $\sin\theta = \sin(\theta + n \cdot 360°)$ for any angle θ and any integer n.

46. Explain why $\cos\theta = \cos(\theta + n \cdot 360°)$ for any angle θ and any integer n.

Concept Check Use the concepts of this section to work Exercises 47–50.

47. Without using a calculator, determine which of the following numbers is closest to $\cos 115°$: .4, .6, 0, $-.4$, or $-.6$.

48. Without using a calculator, determine which of the following numbers is closest to $\sin 115°$: .9, .1, 0, $-.9$, or $-.1$.

49. For what angles θ between $0°$ and $360°$ does $\cos \theta = -\sin \theta$?

50. For what angles θ between $0°$ and $360°$ does $\cos \theta = \sin \theta$?

Tell whether each statement is true *or* false. *If false, tell why. See Example 5.*

51. $\sin 30° + \sin 60° = \sin(30° + 60°)$

52. $\sin(30° + 60°) = \sin 30° \cdot \cos 60° + \sin 60° \cdot \cos 30°$

53. $\cos 60° = 2 \cos^2 30° - 1$

54. $\cos 60° = 2 \cos 30°$

55. $\sin 120° = \sin 150° - \sin 30°$

56. $\sin 210° = \sin 180° + \sin 30°$

57. $\sin 120° = \sin 180° \cdot \cos 60° - \sin 60° \cdot \cos 180°$

58. $\cos 300° = \cos 240° \cdot \cos 60° - \sin 240° \cdot \sin 60°$

59. *(Modeling) Length of a Sag Curve* When a highway goes downhill and then uphill it is said to have a *sag curve.* Sag curves are designed so that at night, headlights shine sufficiently far down the road to allow a safe stopping distance. See the figure. The minimum length L of a sag curve is determined by the height h of the car's headlights above the pavement, the downhill grade $\theta_1 < 0°$, the uphill grade $\theta_2 > 0°$, and the safe stopping distance S for a given speed limit. In addition, L is dependent on the vertical alignment of the headlights. Headlights are usually pointed upward at a slight angle α above the horizontal of the car. Using these quantities, L can then be modeled by the formula

$$L = \frac{(\theta_2 - \theta_1)S^2}{200(h + S \tan \alpha)},$$

where $S < L$.

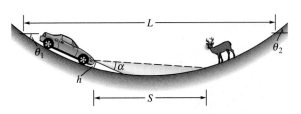

(*Source:* Mannering, F. and W. Kilareski, *Principles of Highway Engineering and Traffic Analysis,* 2nd Edition, John Wiley & Sons, 1998.)

(a) Compute L for a 55 mile per hour speed limit where $h = 1.9$ feet, $\alpha = .9°$, $\theta_1 = -3°$, $\theta_2 = 4°$, and $S = 336$ feet.

(b) Repeat part (a) with $\alpha = 1.5°$.

(c) How does the alignment of the headlights affect the value of L?

2.3 Finding Trigonometric Function Values Using a Calculator

• **Approximating Function Values with a Calculator** • **Finding Angle Measures**

Approximating Function Values with a Calculator   The examples and exercises in this text assume that you have access to a scientific calculator. However, since calculators differ among makes and models, always consult your owner's manual for specific information if questions arise concerning its use.

CAUTION   We have studied only one type of measure for angles—degree measure; another type of measure, radians, will be studied in Chapter 3. When evaluating trigonometric functions of angles given in degrees, remember that the calculator must be set in *degree mode.* Get in the habit of always starting work by finding sin 90°. If the displayed answer is 1, the calculator is set for degree measure; otherwise it is not.

Remember, almost all calculator values of trigonometric functions are *approximations.*

● ● ●   **Example 1**   Finding Function Values with a Calculator

Approximate the value of each trigonometric expression.

Scientific Calculator Solution

(a) sin 49° 12′

Convert 49° 12′ to decimal degrees, as explained in Chapter 1.

$$49° \ 12' = 49\frac{12°}{60} = 49.2°$$

To eight decimal places,

$$\sin 49° \ 12' = \sin 49.2° \approx .75699506.$$

(b) sec 97.977°

Calculators do not have secant keys. However,

$$\sec \theta = \frac{1}{\cos \theta}$$

for all angles θ where $\cos \theta \neq 0$. So find sec 97.977° by first finding cos 97.977° and then taking the reciprocal to get

$$\sec 97.977° \approx -7.205879213.$$

(c) cot 51.4283°

Use the identity $\cot \theta = 1/\tan \theta$.

$$\cot 51.4283° \approx .79748114$$

(d) $\sin(-246°) \approx .91354546$

(e) sin 130° 48′

130° 48′ is equal to 130.8°.

$$\sin 130° \ 48' = \sin 130.8° \approx .75699506$$

Notice that the values in parts (a) and (e) are the same because 49° 12′ is the reference angle for 130° 48′ and the sine function is positive for a quadrant II angle.

Graphing Calculator Solution

The three screens in Figure 16 show the results for parts (a)–(e). Notice that the calculator permits entering the angle measure in degrees and minutes in parts (a) and (e). In the fifth line of the first screen, Ans^{-1} tells the calculator to find the reciprocal of the answer given in the previous line.

Figure 16

Finding Angle Measures   So far in this section we have used a calculator to find trigonometric function values of angles. This process can be reversed; that is, we can use a trigonometric function value to find the measure of an angle. For now, we restrict our attention to angles in the interval [0°, 90°].

● ● ●   **Example 2**   Finding Angle Measures with a Calculator

Find a value of θ in the interval $[0°, 90°]$ satisfying each of the following. Leave answers in decimal degrees.

Scientific Calculator Solution

(a) $\sin \theta = .81815000$

We find θ using a key labeled $\boxed{\text{arc}}$ or $\boxed{\text{INV}}$ together with the $\boxed{\sin}$ key. Some calculators may require a key labeled $\boxed{\sin^{-1}}$ instead. Check your owner's manual to see how your calculator handles this. Again, make sure the calculator is set in degree mode. You should get $\theta \approx 54.900028°$.

(b) $\sec \theta = 1.0545829$

Use the identity $\cos \theta = 1/\sec \theta$. Enter 1.0545829 and find the reciprocal. This gives $\cos \theta \approx .9482421913$. Now find θ as shown in part (a). The result is $\theta \approx 18.514704°$.

Graphing Calculator Solution

As the screen in Figure 17 shows, the procedure is the same with a graphing calculator as with a scientific calculator.

```
sin⁻¹(.81815000)
          54.90002816
1.0545829⁻¹
          .9482421913
cos⁻¹(Ans)
          18.51470432
```

Figure 17

CAUTION   Compare Examples 1(b) and 2(b). Note that the reciprocal is used *before* the inverse cosine key when finding the angle, but *after* the cosine key when finding the trigonometric function.

● ● ●   **Example 3**   Finding Grade Resistance

When an automobile travels uphill or downhill on a highway, it experiences a force due to gravity. This force F in pounds is called *grade resistance* and is modeled by the equation $F = W \sin \theta$, where θ is the grade and W is the weight of the automobile. If the automobile is moving uphill $\theta > 0°$; if downhill $\theta < 0°$. See Figure 18. (*Source:* Mannering, F. and W. Kilareski, *Principles of Highway Engineering and Traffic Analysis,* 2nd Edition, John Wiley & Sons, 1998.)

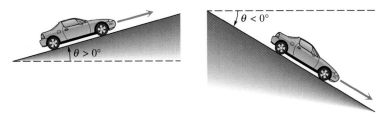

Figure 18

(a) Calculate F to the nearest ten pounds for a 2500-pound car traveling an uphill grade with $\theta = 2.5°$.

$$F = W \sin \theta = 2500 \sin 2.5° \approx 110 \text{ pounds}$$

(b) Calculate F to the nearest ten pounds for a 5000-pound truck traveling a downhill grade with $\theta = -6.1°$.

$$F = W \sin \theta = 5000 \sin(-6.1°) \approx -530 \text{ pounds}$$

F is negative because the truck is moving downhill.

(c) Calculate F for $\theta = 0°$ and $\theta = 90°$. Do these answers agree with your intuition?

$$F = W \sin \theta = W \sin 0° = W(0) = 0 \text{ pounds}$$

$$F = W \sin \theta = W \sin 90° = W(1) = W \text{ pounds}$$

This agrees with intuition because if $\theta = 0°$ then there is level ground and gravity does not cause the vehicle to roll. If $\theta = 90°$, the road would be vertical and the full weight of the vehicle would be pulled downward by gravity, so $F = W$. ● ● ●

2.3 Exercises

Concept Check   Fill in the blanks to complete each statement.

1. The caution at the beginning of this section suggests verifying that a calculator is in degree mode by finding _____ 90°. If the calculator is in degree (sin/cos/tan) mode, the display should be _____.

2. A calculator gives _____ values of (exact/approximate) almost all trigonometric functions.

3. To find trigonometric values of cotangent, secant, and cosecant with a calculator, it is necessary to find the _____ of the _____ function value.

4. The reciprocal is used _____ the inverse func- (before/after) tion key when finding the angle, but _____ (before/after) the function key when finding the trigonometric function value.

Use a calculator to find a decimal approximation for each value. Give as many digits as your calculator displays. In Exercises 18–26, simplify the expression before using the calculator. See Example 1.

5. $\tan 29° 30'$
6. $\sin 38° 42'$
7. $\cot 41° 24'$
8. $\sec 13° 15'$

9. $\csc 44° 30'$
10. $\csc 145° 45'$
11. $\cot 183° 48'$
12. $\cos 421° 30'$

13. $\sec 312° 12'$
14. $\tan(-80° 6')$
15. $\sin(-317° 36')$
16. $\cot(-512° 20')$

17. $\cos(-15')$
18. $\dfrac{1}{\sec 14.8°}$
19. $\dfrac{1}{\csc 514° 24'}$
20. $\dfrac{1}{\cot 23.4°}$

21. $\dfrac{\sin 33°}{\cos 33°}$
22. $\dfrac{\cos 77°}{\sin 77°}$
23. $\cos(90° - 3.69°)$
24. $\cot(90° - 4.72°)$

25. $\sec^2 47.8° - 1$
26. $\sin^2 17.7° + \cos^2 17.7°$

27. A student, wishing to use a calculator to verify the value of $\sin 30°$, enters the information correctly but gets a display of $-.98803162$. He knows that the display should be .5, and he also knows that his calculator is in good working order. What do you think is the problem?

28. A certain make of calculator does not allow the input of angles outside of a particular interval when finding trigonometric function values. For example, trying to find $\cos 2000°$ using the methods of this section would give an error message, despite the fact that $\cos 2000°$ can be evaluated. Explain how you would find $\cos 2000°$ using this calculator.

Find a value of θ in $[0°, 90°]$ that satisfies each statement. Leave answers in decimal degrees. See Example 2.

29. $\sin \theta = .84802194$
30. $\tan \theta = 1.4739716$
31. $\tan \theta = 6.4358841$
32. $\sin \theta = .27843196$

33. $\sec \theta = 1.1606249$
34. $\cot \theta = 1.2575516$
35. $\csc \theta = 1.3861147$
36. $\sec \theta = 2.7496222$

37. What value of A between $0°$ and $90°$ will produce the output in the graphing calculator screen?

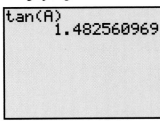

38. What value of A will produce the output in the graphing calculator screen?

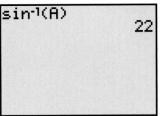

Use a calculator to evaluate each expression. (As shown in Chapter 5, all these answers should be integers.)

39. $\sin 35° \cos 55° + \cos 35° \sin 55°$

40. $\cos 100° \cos 80° - \sin 100° \sin 80°$

41. $\cos 75° \, 29' \cos 14° \, 31' - \sin 75° \, 29' \sin 14° \, 31'$

42. $\sin 28° \, 14' \cos 61° \, 46' + \cos 28° \, 14' \sin 61° \, 46'$

(Modeling) Speed of Light   When a light ray travels from one medium, such as air, to another medium, such as water or glass, the speed of the light changes, and the direction in which the ray is traveling changes. (This is why a fish under water is in a different position than it appears to be.) These changes are given by Snell's law

$$\frac{c_1}{c_2} = \frac{\sin \theta_1}{\sin \theta_2},$$

where c_1 is the speed of light in the first medium, c_2 is the speed of light in the second medium, and θ_1 and θ_2 are the angles shown in the figure. (Source: The Physics Classroom, www.glenbrook.k12.il.us.) In the following exercises, assume that $c_1 = 3 \times 10^8$ m per sec.

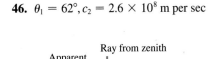

Find the speed of light in the second medium.

43. $\theta_1 = 46°, \theta_2 = 31°$

44. $\theta_1 = 39°, \theta_2 = 28°$

Find θ_2 for the following values of θ_1 and c_2. Round to the nearest degree.

45. $\theta_1 = 40°, c_2 = 1.5 \times 10^8$ m per sec

46. $\theta_1 = 62°, c_2 = 2.6 \times 10^8$ m per sec

(Modeling) Fish's View of the World   The figure here shows a fish's view of the world above the surface of the water. (Source: Walker, Jearl, "The Amateur Scientist," Scientific American, March 1984.) Suppose that a light ray comes from the horizon, enters the water, and strikes the fish's eye.

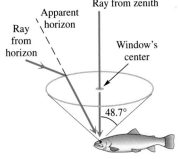

47. Let us assume that this ray gives a value of $90°$ for angle θ_1 in the formula for Snell's law. (In a practical situation, this angle would probably be a little less than $90°$.) The speed of light in water is about 2.254×10^8 m per sec. Find angle θ_2.

(Your result should have been about $48.7°$. This means that a fish sees the world above the water as a cone, making an angle of $48.7°$ with the vertical.)

48. Suppose an object is located at a true angle of $29.6°$ above the horizon. Find the apparent angle above the horizon to a fish.

Use a calculator to decide whether each statement is true *or* false. *It may be that a true statement will lead to results that differ in the last decimal place due to rounding error.*

49. $\cos 40° = 2 \cos 20°$

50. $\sin 10° + \sin 10° = \sin 20°$

51. $\cos 70° = 2 \cos^2 35° - 1$

52. $\sin 50° = 2 \sin 25° \cdot \cos 25°$

53. $2 \cos 38° \, 22' = \cos 76° \, 44'$

54. $\cos 40° = 1 - 2 \sin^2 80°$

55. $\dfrac{1}{2} \sin 40° = \sin \dfrac{1}{2}(40°)$

56. $\sin 39° \, 48' + \cos 39° \, 48' = 1$

 (Modeling) Grade Resistance   See Example 3 to work Exercises 57–64.

57. What is the grade resistance of a 2400-pound car traveling on a $-2.4°$ downhill grade?

58. What is the grade resistance of a 2100-pound car traveling on a $1.8°$ uphill grade?

59. A 3000-pound car traveling uphill has a grade resistance of 150 pounds. What is the angle of the grade?

60. A car traveling on a $1.5°$ uphill grade has a grade resistance of 120 pounds. What is the weight of the car?

61. A car traveling on a $-3°$ downhill grade has a grade resistance of -145 pounds. What is the weight of the car?

62. A 2600-pound car traveling downhill has a grade resistance of -130 pounds. What is the angle of the grade?

63. Which has the greater grade resistance: a 2200-pound car on a $2°$ uphill grade or a 2000-pound car on a $2.2°$ uphill grade?

64. *Highway Grades*   Complete the table for the values of $\sin \theta$, $\tan \theta$, and $\dfrac{\pi \theta}{180}$ to four decimal places.

θ	$\sin \theta$	$\tan \theta$	$\dfrac{\pi \theta}{180}$
0°			
.5°			
1°			
1.5°			
2°			
2.5°			
3°			
3.5°			
4°			

(a) How do $\sin \theta$, $\tan \theta$, and $\dfrac{\pi \theta}{180}$ compare for small grades θ?

(b) Highway grades are usually small. Give two approximations of the grade resistance $F = W \sin \theta$ that do not use the sine function.

(c) A stretch of highway has a 4-foot vertical rise for every 100 feet of horizontal run. Use an approximation from part (a) to estimate the grade resistance for a 2000-pound car on this stretch of highway.

(d) A stretch of highway has a $3.75°$ grade. Without evaluating a trigonometric function, estimate the grade resistance for an 1800-pound car on this stretch of highway.

65.  *(Modeling) Design of Highway Curves*   When highway curves are designed, the outside of the curve is often slightly elevated or inclined above the inside of the curve. See the figure. This inclination is called *superelevation*. For safety reasons, it is important that both the curve's radius and superelevation are correct for a given speed limit. If an automobile is traveling at velocity V (in feet per second), the safe radius R for a curve with superelevation α is modeled by the formula

$$R = \frac{V^2}{g(f + \tan \alpha)},$$

where f and g are constants. (*Source:* Mannering, F. and W. Kilareski, *Principles of Highway Engineering and Traffic Analysis*, 2nd Edition, John Wiley & Sons, 1998.)

(a) A roadway is being designed for automobiles traveling at 45 mph. If $\alpha = 3°$, $g = 32.2$, and $f = .14$, calculate R.

(b) What should the radius of the curve be if the speed in part (a) is increased to 70 mph?

(c) How would increasing the angle α affect the results? Verify your answer by repeating parts (a) and (b) with $\alpha = 4°$.

66. *Speed Limit on a Curve*   Refer to Exercise 65. A highway curve has a radius of $R = 1150$ feet and a superelevation of $\alpha = 2.1°$. What should the speed limit (in miles per hour) be for this curve?

Quantitative Reasoning

67. *Can trigonometry be used to win an Olympic medal?* A shotputter trying to improve performance may wonder: Is there an optimal angle to aim for, or is the velocity (speed) at which the ball is thrown more important? The figure shows the path of a steel ball thrown by a shotputter. The distance D depends on initial velocity v, height h, and angle θ.

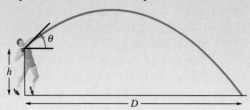

One model developed for this situation gives D as

$$D = \frac{v^2 \sin \theta \cos \theta + v \cos \theta \sqrt{(v \sin \theta)^2 + 64h}}{32}.$$

Typical ranges for the variables are v: 33–46 feet per second, h: 6–8 feet, and θ: 40°–45°. (*Source:* Kreighbaum, E. and K. Barthels, *Biomechanics,* Allyn & Bacon, 1996.)

(a) To see how angle θ affects distance D, let $v = 44$ feet per second and $h = 7$ feet. Calculate D for $\theta = 40°$, $42°$, and $45°$. How does distance D change as θ increases?

(b) To see how velocity v affects distance D, let $h = 7$ and $\theta = 42°$. Calculate D for $v = 43$, 44, and 45 feet per second. How does distance D change as v increases?

(c) Which affects distance D more, v or θ? What should the shotputter do to improve performance?

2.4 Solving Right Triangles

• **Significant Digits** • **Solving Triangles** • **Angles of Elevation or Depression**

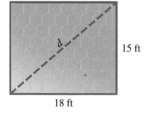

Figure 19

Significant Digits Suppose we quickly measure a room as 15 feet by 18 feet. See Figure 19. To calculate the length of a diagonal of the room, we can use the Pythagorean theorem.

$$d^2 = 15^2 + 18^2$$
$$d^2 = 549$$
$$d = \sqrt{549} \approx 23.430749$$

Should this answer be given as the length of the diagonal of the room? Of course not. The number 23.430749 contains 6 decimal places, while the original data of 15 feet and 18 feet are only accurate to the nearest foot. Since the results of a problem can be no more accurate than the least accurate number in any calculation, we really should say that the diagonal of the 15- by 18-foot room is 23 feet.

If a wall measured to the nearest foot is 18 feet long, this actually means that the wall has length between 17.5 feet and 18.5 feet. If the wall is measured more accurately as 18.3 feet long, then its length is really between 18.25 feet and 18.35 feet. A measurement of 18.00 feet would indicate that the length of the wall is between 17.995 feet and 18.005 feet. The measurement 18 feet is said to have two *significant digits* of accuracy; 18.0 has three significant digits, and 18.00 has four.

What about the measurement 900 meters? We cannot tell whether this represents a measurement to the nearest meter, ten meters, or hundred meters. To avoid this problem, we write the number in scientific notation as 9.00×10^2 to the nearest meter, 9.0×10^2 to the nearest ten meters, or 9×10^2 to the nearest hundred meters. These three cases have three, two, and one significant digits, respectively.

A **significant digit** is a digit obtained by actual measurement. A number that represents the result of counting, or a number that results from theoretical work and is not the result of a measurement, is an **exact number.** There are 50 states in the United States, so 50 is an exact number. The number of states is not 49 3/4 or 50 1/4, nor is the number 50 used here to represent "some number between 45 and 55." In the formula for perimeter of a rectangle, $P = 2L + 2W$, the 2s are obtained from the definition of perimeter and are exact numbers.

Most values of trigonometric functions are approximations, and virtually all measurements are approximations. To perform calculations on such approximate numbers, follow the rules given below.

Calculation with Significant Digits

For *adding* and *subtracting,* round the answer so that the last digit you keep is in the right-most column in which all the numbers have significant digits.

For *multiplying* or *dividing,* round the answer to the least number of significant digits found in any of the given numbers.

For *powers* and *roots,* round the answer so that it has the same number of significant digits as the number whose power or root you are finding.

To **solve a triangle** means to find the measures of all the angles and sides of the triangle. When solving triangles, use the following table to determine the significant digits in angle measure.

Significant Digits for Angles

Number of Significant Digits	Angle Measure to Nearest:
2	Degree
3	Ten minutes, or nearest tenth of a degree
4	Minute, or nearest hundredth of a degree
5	Tenth of a minute, or nearest thousandth of a degree

For example, an angle measuring 52° 30′ has three significant digits (assuming that 30′ is measured to the nearest ten minutes).

Solving Triangles   When solving triangles, a labeled sketch is an important aid. As shown in Figure 20, we use *a* to represent the length of the side opposite angle *A*, *b* for the length of the side opposite angle *B*, and so on. As mentioned earlier, in a right triangle the letter *c* is reserved for the hypotenuse.

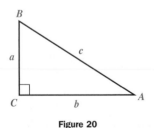

Figure 20

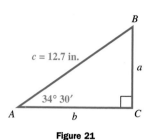

Figure 21

Looking Ahead to Calculus

The derivatives of the *parametric equations* $x = f(t)$ and $y = g(t)$ often represent the rate of change of physical quantities, such as velocities. In such cases, the derivatives are called *related rates* because a change in one causes a related change in the other. Determining these rates in calculus often requires solving a right triangle. Many problems that require the maximum or minimum value of some quantity also involve solving a right triangle.

● ● ● **Example 1** Solving a Right Triangle Given an Angle and a Side

Solve right triangle ABC, with $A = 34° \, 30'$ and $c = 12.7$ inches. See Figure 21.

To solve the triangle, find the measures of the remaining sides and angles. To find the value of a, use a trigonometric function involving the known values of angle A and side c. Since the sine of angle A is given by the quotient of the side opposite A and the hypotenuse, use $\sin A$.

$$\sin A = \frac{a}{c}$$

$$\sin 34° \, 30' = \frac{a}{12.7} \qquad \text{\scriptsize $A = 34° \, 30', c = 12.7$}$$

$$a = 12.7 \sin 34° \, 30' \qquad \text{\scriptsize Multiply by 12.7.}$$

$$a = 12.7(.56640624) \qquad \text{\scriptsize Use a calculator.}$$

$$a = 7.19 \text{ inches} \qquad \text{\scriptsize Three significant digits}$$

We could find the value of b with the Pythagorean theorem. It is better, however, to use the information given in the problem rather than a result just calculated. If a mistake were made in finding a, then b also would be incorrect. Also, rounding more than once may cause the result to be less accurate. Use $\cos A$.

$$\cos A = \frac{\text{side adjacent}}{\text{hypotenuse}} = \frac{b}{c}$$

$$\cos 34° \, 30' = \frac{b}{12.7}$$

$$b = 12.7 \cos 34° \, 30'$$

$$b = 10.5 \text{ inches}$$

Once b is found, the Pythagorean theorem can be used as a check. All that remains to solve triangle ABC is to find the measure of angle B. Since $A = 34° \, 30'$ and $A + B = 90°$,

$$B = 90° - A$$

$$B = 89° \, 60' - 34° \, 30'$$

$$B = 55° \, 30'.$$

● ● ●

NOTE In Example 1, we could have found the measure of angle B first, and then used the trigonometric function values of B to find the unknown sides. The process of solving a right triangle (like many problems in mathematics) can usually be done in several ways, each producing the correct answer. To maintain accuracy, always use given information as much as possible, and avoid rounding off in intermediate steps.

● ● ● **Example 2** Solving a Right Triangle Given Two Sides

Solve right triangle ABC if $a = 29.43$ cm and $c = 53.58$ cm.

Draw a sketch showing the given information, as in Figure 22. One way to begin is to find angle A using sine.

$$\sin A = \frac{\text{side opposite}}{\text{hypotenuse}} = \frac{29.43}{53.58}$$

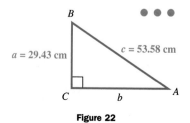

Figure 22

Using ⌊INV⌋ ⌊sin⌋ or ⌊sin⁻¹⌋ on a calculator, we find that $A = 33.32°$. The measure of B is $90° - 33.32° = 56.68°$.

Now find b from the Pythagorean theorem, $a^2 + b^2 = c^2$, with $c = 53.58$ and $a = 29.43$.

$$29.43^2 + b^2 = 53.58^2$$
$$b^2 = 53.58^2 - 29.43^2$$
$$b = 44.77 \text{ cm}$$

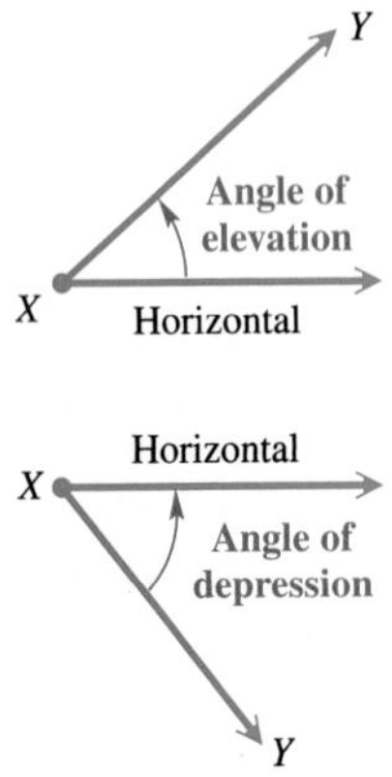

● ● ●

Angles of Elevation or Depression

Many applications of right triangles involve angles of elevation or depression. The **angle of elevation** from point X to point Y (above X) is the acute angle formed by ray XY and a horizontal ray with endpoint at X. See Figure 23. The **angle of depression** from point X to point Y (below X) is the acute angle formed by ray XY and a horizontal ray with endpoint X. Again, see Figure 23.

Angle of elevation

X — Horizontal

Horizontal

X — Angle of depression

Y

Figure 23

CAUTION Be careful when interpreting the angle of depression. Both the angle of elevation *and* the angle of depression are measured between the line of sight and the *horizontal.*

PROBLEM SOLVING To solve applied trigonometry problems, follow the same procedure as solving a triangle. A crucial step is sketching a triangle and labeling it carefully. Then follow the remaining steps.

Solving Applied Trigonometry Problems

Step 1 Draw a sketch, and label it with the given information. Label the quantity to be found with a variable.

Step 2 Use the sketch to write an equation relating the given quantities to the variable.

Step 3 Solve the equation, and check that your answer makes sense.

● ● ● **Example 3** Finding a Length When the Angle of Elevation Is Known

Shelly McCarthy knows that when she stands 123 feet from the base of a flagpole, the angle of elevation to the top is $26° \, 40'$. If her eyes are 5.30 feet above the ground, find the height of the flagpole.

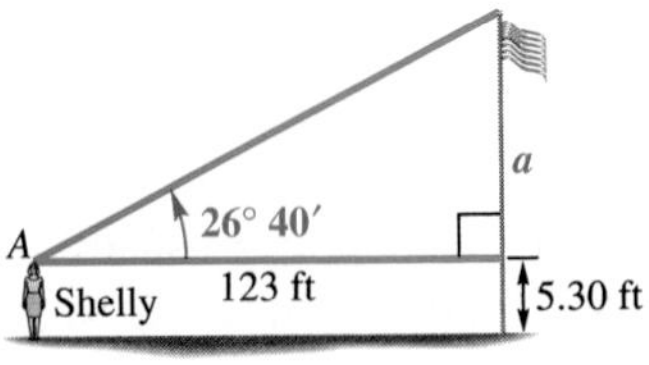

The length of the side adjacent to Shelly is known, and the length of the side opposite her must be found. See Figure 24. The tangent ratio involves these two values.

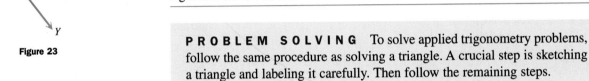

A $26° \, 40'$
Shelly   123 ft   5.30 ft

Figure 24

$$\tan A = \frac{\text{side opposite}}{\text{side adjacent}}$$

$$\tan 26° \, 40' = \frac{a}{123}$$

$$a = 123 \tan 26° \, 40'$$

$$a = 61.8 \text{ feet}$$

Since Shelly's eyes are 5.30 feet above the ground, the height of the flagpole is

$$61.8 + 5.30 = 67.1 \text{ feet.}$$ ● ● ●

● ● ● **Example 4** Finding the Angle of Elevation When Lengths Are Known

The length of the shadow of a building 34.09 meters tall is 37.62 meters. Find the angle of elevation of the sun.

As shown in Figure 25, the angle of elevation of the sun is angle B. Since the side opposite B and the side adjacent to B are known, use the tangent ratio to find B.

$$\tan B = \frac{34.09}{37.62} \quad \text{so } B = 42.18°$$

The angle of elevation of the sun is 42.18°.

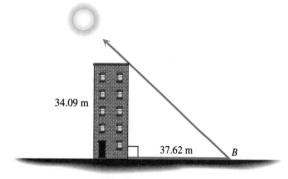

Figure 25   ● ● ●

C O N N E C T I O N S   Probably the most famous study of problem-solving techniques was developed by George Polya (1888–1985). Among his many publications is the modern classic *How to Solve It.* In this book, Polya proposed a four-step process for problem solving.

Polya's Four-Step Problem-Solving Process

Step 1 Understand the problem. You must first decide what you are to find.

Step 2 Devise a plan. Some possible strategies are to use a formula, write an equation and solve it, draw a sketch, make a table or chart.

Step 3 Carry out the plan. This is where the trigonometric techniques you are learning are helpful.

Step 4 Look back and check. Is your answer reasonable? Does it answer the question?

For Discussion or Writing

Compare Polya's four steps with our steps for problem solving given earlier. Which of our steps correspond with each of Polya's steps?

2.4 Exercises

Concept Check Refer to the discussion of accuracy and significant digits in this section to work Exercises 1–10.

  1. *Leading NFL Receiver* At the end of the 1997 National Football League season, San Francisco 49er Jerry Rice was the leading career receiver with 16,455 yards. State the range represented by this number. (*Source: The World Almanac and Book of Facts,* 1999.)

  2. *Height of Mt. Everest* When Mt. Everest was first surveyed, the surveyors obtained a height of 29,000 feet to the nearest foot. State the range represented by this number. (The surveyors thought no one would believe a measurement of 29,000 feet, so they reported it as 29,002.) (*Source:* Dunham, W., *The Mathematical Universe,* John Wiley & Sons, 1994.)

3. *Calories in a Recipe*   A recipe for Chicken Diable is said to contain 547 calories per serving. What is the range of calories represented by this number?

4. 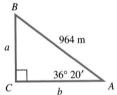 *Longest Vehicular Tunnel*   The E. Johnson Memorial tunnel in Colorado, which measures 8959 feet, is the longest land vehicular tunnel in the United States. What is the range of this number? (*Source: The World Almanac and Book of Facts,* 2000.)

5. *Top WNBA Scorer*   Women's National Basketball Association player Cynthia Cooper of the Houston Comets received the 1999 award for most points scored, 686. Is it appropriate to consider this number as between 685.5 and 686.5? Why or why not? (*Source: The World Almanac and Book of Facts,* 2000.)

6. *Circumference of a Circle*   The formula for the circumference of a circle is $C = 2\pi r$. Suppose you use the $\boxed{\pi}$ key on your calculator to find the circumference

of a circle with radius 54.98 cm, getting 345.44953. Since 2 has only one significant digit, the answer should be given as 3×10^2, or 300 cm. Is this conclusion correct? If not, explain how the answer should be given.

7. Explain the difference between a measurement of 23.0 feet and a measurement of 23.00 feet.

8. What number indicates a measurement between 25.95 and 26.05 pounds?

Fill in the blanks in Exercises 9 and 10.

9. If h is the actual height of a building and the height is measured as 58.6 feet, then $|h - 58.6| \leq$ _____.

10. If w is the actual weight of a car and the weight is measured as 15.00×10^2 pounds, then $|w - 1500| \leq$ _____.

In the remaining exercises in this set, use a calculator as necessary.

Solve each right triangle. See Examples 1 and 2.

11.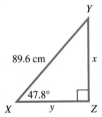

12.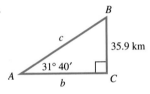

13.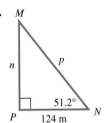

14.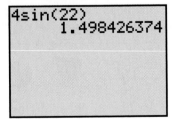

15.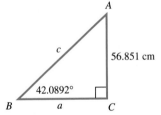

16.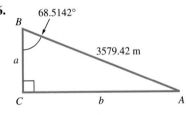

In Exercises 17 and 18, assume the calculator is in degree mode.

17. Sketch and label a right triangle whose solution is obtained from the graphing calculator screen.

```
4sin(22)
          1.498426374
```

18. Sketch and label a right triangle whose solution is obtained from the graphing calculator screen.

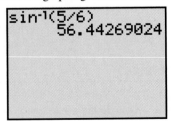

19. Can a right triangle be solved if we are given measures of its two acute angles and no side lengths? Explain.

20. *Concept Check*   If we are given an acute angle and a side in a right triangle, what unknown part of the triangle requires the least work to find?

21. Explain why you can always solve a right triangle if you know the measures of one side and one acute angle.

22. Explain why you can always solve a right triangle if you know the lengths of two sides.

Solve each right triangle. In each case, C = 90°. If angle information is given in degrees and minutes, give answers in the same way. If given in decimal degrees, do likewise in answers. When two sides are given, give angles in degrees and minutes. See Examples 1 and 2.

23. $A = 28.00°, c = 17.4$ ft

24. $B = 46.00°, c = 29.7$ m

25. $B = 73.00°, b = 128$ in.

26. $A = 61° 00', b = 39.2$ cm

27. $a = 76.4$ yd, $b = 39.3$ yd

28. $a = 958$ m, $b = 489$ m

29. $a = 18.9$ cm, $c = 46.3$ cm

30. $b = 219$ m, $c = 647$ m

31. $A = 53° 24', c = 387.1$ ft

32. $A = 13° 47', c = 1285$ m

33. $B = 39° 9', c = .6231$ m

34. $B = 82° 51', c = 4.825$ cm

35. *Concept Check*   When is an angle of elevation equal to 90°?

36. *Concept Check*   Can an angle of elevation be more than 90°?

37. Use the ideas found in Section 1.3 involving a transversal intersecting parallel lines to explain why the angle of depression *DAB* has the same measure as the angle of elevation *ABC* in the figure.

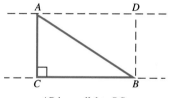

AD is parallel to BC.

38. Why is angle *CAB not* an angle of depression in the figure?

Solve each problem. See Examples 1–3.

39. *Ladder Leaning Against a Wall*   A 13.5-meter fire-truck ladder is leaning against a wall. Find the distance the ladder goes up the wall if it makes an angle of 43° 50′ with the ground.

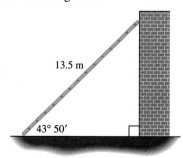

13.5 m

43° 50′

40. *Antenna Mast Guy Wire*   A guy wire 77.4 meters long is attached to the top of an antenna mast that is 71.3 meters high. Find the angle that the wire makes with the ground.

41. *Guy Wire to a Tower*   Find the length of a guy wire that makes an angle of 45° 30′ with the ground if the wire is attached to the top of a tower 63.0 meters high.

42. *Distance Across a Lake*   To find the distance *RS* across a lake, a surveyor lays off *RT* = 53.1 meters,

with angle $T = 32° 10'$ and angle $S = 57° 50'$. Find length *RS*.

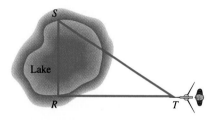

43. *Side Lengths of a Triangle*   The length of the base of an isosceles triangle is 42.36 inches. Each base angle is 38.12°. Find the length of each of the two equal sides of the triangle. (*Hint:* Divide the triangle into two right triangles.)

44. *Altitude of a Triangle*   Find the altitude of an isosceles triangle having a base of 184.2 cm if the angle opposite the base is 68° 44′.

Work each problem involving an angle of elevation or depression. See Examples 3 and 4.

45. *Cloud Ceiling*   The U.S. Weather Bureau defines a *cloud ceiling* as the altitude of the lowest clouds that cover more than half the sky. To determine a cloud ceiling, a powerful searchlight projects a circle of light vertically on the bottom of the cloud. An observer sights the circle of light in the crosshairs of a tube called a *clinometer*. A pendant hanging vertically from the tube and resting on a protractor gives the angle of elevation. Find the cloud ceiling if the searchlight is located 1000 feet from the observer and the angle of elevation is 30.0° as measured with a clinometer at eye-height 6 feet. (Assume three significant digits.)

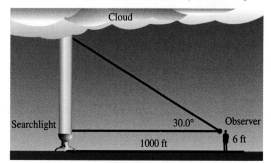

46. *Length of a Shadow* Suppose the angle of elevation of the sun is 23.4°. Find the length of the shadow cast by Cindy Newman, who is 5.75 feet tall.

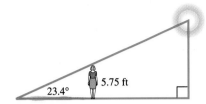

47. *Height of a Tower* The shadow of a vertical tower is 40.6 meters long when the angle of elevation of the sun is 34.6°. Find the height of the tower.

48. *Angle of Elevation of the Sun* Find the angle of elevation of the sun if a 48.6-foot flagpole casts a shadow 63.1 feet long.

49. *Distance from the Ground to the Top of a Building* The angle of depression from the top of a building to a point on the ground is 32° 30′. How far is the point on the ground from the top of the building if the building is 252 meters high?

50. *Airplane Distance* An airplane is flying 10,500 feet above the level ground. The angle of depression from the plane to the base of a tree is 13° 50′. How far horizontally must the plane fly to be directly over the tree?

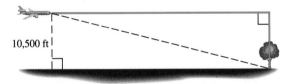

51. *Height of a Building* The angle of elevation from the top of a small building to the top of a nearby taller building is 46° 40′, while the angle of depression to the bottom is 14° 10′. If the smaller building is 28.0 meters high, find the height of the taller building.

52. *Mounting a Video Camera* A video camera is to be mounted on a bank wall so as to have a good view of the head teller. Find the angle of depression that the lens should make with the horizontal.

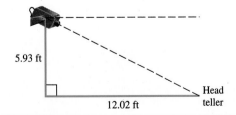

53. *Angle of Depression of a Light* A company safety committee has recommended that a floodlight be mounted in a parking lot so as to illuminate the employee exit. Find the angle of depression of the light.

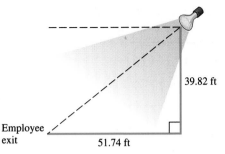

54. *Diameter of the Sun* To determine the diameter of the sun, an astronomer might sight with a *transit* (a device used by surveyors for measuring angles) first to one edge of the sun and then to the other, finding that the included angle equals 1° 4′. Assuming that the distance from Earth to the sun is 92,919,800 miles, calculate the diameter of the sun.

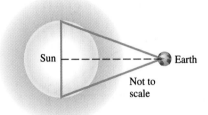

55. *Size of the Threads of a Bolt* The figure shows a magnified view of the threads of a bolt. Find x if d is 2.894 mm.

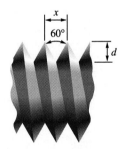

56. *Error in Measurement* A degree may seem like a very small unit, but an error of one degree in measuring an angle may be very significant. For example, suppose a laser beam directed toward the visible center of the moon misses its assigned target by 30 seconds. How far is it (in miles) from its assigned target? Take the distance from the surface of Earth to that of the moon to be 234,000 miles. (*Source: A Sourcebook of Applications of School Mathematics* by Donald Bushaw et al. Copyright © 1980 by The Mathematical Association of America.)

57. *Height of Mt. Everest* The highest mountain peak in the world is Mt. Everest, located in the Himalayas. The height of this enormous mountain was determined in 1856 by surveyors using trigonometry long before it was first climbed in 1953. This difficult measurement had to be done from a great distance. At an altitude of 14,545 feet on a different mountain, the straight line distance to the peak of Mt. Everest is 27.0134 miles and its angle of elevation is $\theta = 5.82°$. (*Source:* Dunham, W., *The Mathematical Universe,* John Wiley & Sons, 1994.)

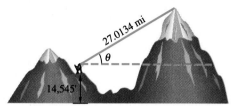

(a) Approximate the height (in feet) of Mt. Everest.
(b) In the actual measurement, Mt. Everest was over 100 miles away and the curvature of Earth had to be taken into account. Would the curvature of Earth make the peak appear taller or shorter than it actually is?

58. *Distance Between an Arc and a Chord* A basic highway curve connecting two straight sections of road is often circular. In the figure, the points P and S mark the beginning and end of the curve. Let Q be the point of intersection where the two straight sections of highway leading into the curve would meet if extended. The radius of the curve is R, and the central angle θ denotes how many degrees the curve turns. (*Source:* Mannering, F. and W. Kilareski, *Principles of Highway Engineering and Traffic Analysis,* 2nd Edition, John Wiley & Sons, 1998.)

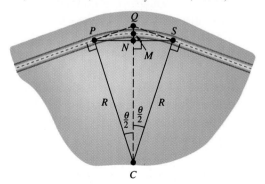

(a) If $R = 965$ feet and $\theta = 37°$, find the distance d between P and Q.
(b) Find an expression in terms of R and θ for the distance between points M and N.

59. *(Modeling) Stopping Distance on a Curve* Refer to Exercise 58. When an automobile travels along a circular curve, objects like trees and buildings situated on the inside of the curve can obstruct a driver's vision. These obstructions prevent the driver from seeing sufficiently far down the highway to ensure a safe stopping distance. In the figure, the *minimum* distance d that should be cleared on the inside of the highway is modeled by the equation

$$d = R\left(1 - \cos\frac{\beta}{2}\right).$$

(*Source:* Mannering, F. and W. Kilareski, *Principles of Highway Engineering and Traffic Analysis,* 2nd Edition, John Wiley & Sons, 1998.)

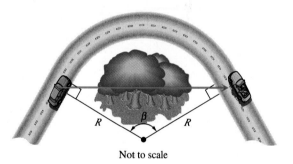

Not to scale

(a) It can be shown that if β is measured in degrees, then $\beta \approx \dfrac{57.3S}{R}$, where S is safe stopping distance for the given speed limit. Compute d for a 55 mph speed limit if $S = 336$ feet and $R = 600$ feet.
(b) Compute d for a 65 mph speed limit if $S = 485$ feet and $R = 600$ feet.
(c) How does the speed limit affect the amount of land that should be cleared on the inside of the curve?

2.5 Further Applications of Right Triangles

● **Bearing** ● **Further Applications**

Bearing Other applications of right triangles involve **bearing,** an important idea in navigation. There are two methods for expressing bearing. When a single angle is given, such as 164°, it is understood that the bearing is measured in a clockwise direction from due north. Several sample bearings using this first method are shown in Figure 26.

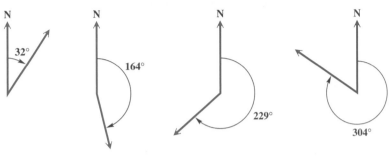

Figure 26

● ● ● **Example 1** Solving a Problem Involving Bearing (First Method)

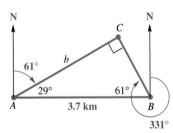

Figure 27

Radar stations *A* and *B* are on an east-west line, 3.7 km apart. Station *A* detects a plane at *C*, on a bearing of 61°. Station *B* simultaneously detects the same plane, on a bearing of 331°. Find the distance from *A* to *C*.

Draw a sketch showing the given information, as in Figure 27. Since a line drawn due north is perpendicular to an east-west line, right angles are formed at *A* and *B*, so angles *CAB* and *CBA* can be found. Angle *C* is a right angle because angles *CAB* and *CBA* are complementary. Find distance *b* by using the cosine function.

$$\cos 29° = \frac{b}{3.7}$$

$$3.7 \cos 29° = b$$

$$b = 3.2 \text{ km} \quad \text{Use a calculator and round to the nearest tenth.} \quad ● ● ●$$

PROBLEM SOLVING The importance of a correctly labeled sketch when solving applications like that in Example 1 cannot be overemphasized. Some of the necessary information is often not directly stated in the problem and can only be determined from the sketch.

The second method for expressing bearing starts with a north-south line and uses an acute angle to show the direction, either east or west, from this line. Figure 28 shows several sample bearings using this system. Either N or S always comes first, followed by an acute angle, and then E or W.

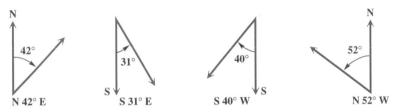

Figure 28

● ● ● **Example 2** Solving a Problem Involving Bearing (Second Method)

The bearing from *A* to *C* is S 52° E. The bearing from *A* to *B* is N 84° E. The bearing from *B* to *C* is S 38° W. A plane flying at 250 miles per hour takes 2.4 hours to go from *A* to *B*. Find the distance from *A* to *C*.

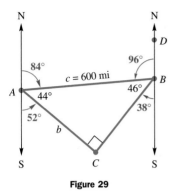

Figure 29

Make a sketch of the situation. First draw the two bearings from point *A*. Choose a point *B* on the bearing N 84° E from *A*, and draw the bearing to *C*. Point *C* will be located where the bearing lines from *A* and *B* intersect, as shown in Figure 29.

Since the bearing from *A* to *B* is N 84° E, angle *ABD* is 180° − 84° = 96°. Thus, angle *ABC* is 46°. Also, angle *BAC* is 180° − (84° + 52°) = 44°. Angle *C* is 180° − (44° + 46°) = 90°. From the statement of the problem, a plane flying at 250 miles per hour takes 2.4 hours to go from *A* to *B*. The distance from *A* to *B* is the product of rate and time, or

$$c = \text{rate} \times \text{time} = 250(2.4) = 600 \text{ miles.}$$

To find *b*, the distance from *A* to *C*, use the sine. (The cosine could also have been used.)

$$\sin 46° = \frac{b}{c}$$

$$\sin 46° = \frac{b}{600}$$

$$600 \sin 46° = b$$

$$b = 430 \text{ miles}$$ ● ● ●

Further Applications

● ● ● **Example 3** Solving a Problem Involving Angle of Elevation

Francisco needs to know the height of a tree. From a given point on the ground, he finds that the angle of elevation to the top of the tree is 36.7°. He then moves back 50 feet. From the second point, the angle of elevation to the top of the tree is 22.2°. See Figure 30. Find the height of the tree.

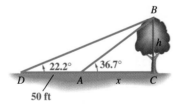

Figure 30

Algebraic Solution

Figure 30 shows two unknowns: *x*, the distance from the center of the trunk of the tree to the point where the first observation was made, and *h*, the height of the tree. Since nothing is given about the length of the hypotenuse of either triangle *ABC* or

Graphing Calculator Solution*

In Figure 31(a) on the next page, we have super-imposed Figure 30 on

(continued)

*Source: Adapted with permission from "Letter to the Editor," by Robert Ruzich (*Mathematics Teacher,* Volume 88, Number 1). Copyright © 1995 by the National Council of Teachers of Mathematics.

triangle BCD, use a ratio that does not involve the hypotenuse—tangent. Refer to Figure 31(a).

In triangle ABC, $\quad \tan 36.7° = \dfrac{h}{x} \quad$ or $\quad h = x \tan 36.7°.$

In triangle BCD, $\quad \tan 22.2° = \dfrac{h}{50 + x}$

$$h = (50 + x) \tan 22.2°.$$

Since each of these expressions equals h, the expressions must be equal, so

$$x \tan 36.7° = (50 + x) \tan 22.2°.$$

Now solve for x.

$$x \tan 36.7° = 50 \tan 22.2° + x \tan 22.2°$$

Distributive property

$$x \tan 36.7° - x \tan 22.2° = 50 \tan 22.2°$$

Get x terms on one side.

$$x(\tan 36.7° - \tan 22.2°) = 50 \tan 22.2°$$

Factor out x on the left.

$$x = \dfrac{50 \tan 22.2°}{\tan 36.7° - \tan 22.2°}$$

Divide by the coefficient of x.

We saw above that $h = x \tan 36.7°$. Substituting for x,

$$h = \left(\dfrac{50 \tan 22.2°}{\tan 36.7° - \tan 22.2°} \right) \tan 36.7°.$$

From a calculator,

$$\tan 36.7° = .74537703$$

$$\tan 22.2° = .40809244$$

so

$$\tan 36.7° - \tan 22.2° = .74537703 - .40809244 = .33728459$$

and

$$h = \left(\dfrac{50(.40809244)}{.33728459} \right) .74537703 = 45 \text{ (rounded)}.$$

The height of the tree is approximately 45 feet.

coordinate axes with the origin at D. By definition, the tangent of the angle between the x-axis and the graph of a line with equation $y = mx + b$ is the slope of the line, m. So for line DB, $m = \tan 22.2°$. Since the y-intercept b is 0 here, the equation of line DB is $Y_1 = (\tan 22.2°)x$. Similarly, the equation of line AB is $Y_2 = (\tan 36.7°)x + b$. However, here $b \neq 0$, so we use the point $A(50, 0)$ and the point-slope form to find the equation.

$$Y_2 - y_1 = m(x - x_1)$$

$$Y_2 - 0 = m(x - 50)$$

Let $x_1 = 50$ and $y_1 = 0$.

$$Y_2 = [\tan(36.7°)](x - 50)$$

Lines Y_1 and Y_2 are graphed in Figure 31(b). The y-coordinate of the point of intersection of the graphs of these two lines gives the length of BC, or h. From the information at the bottom of the screen, we see that $h = 45$ (rounded), which agrees with our algebraic result.

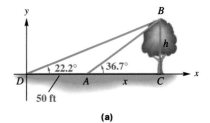

(a)

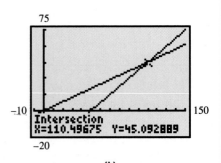

(b)

Figure 31

NOTE In practice, we usually do not write down intermediate calculator approximation steps. We did in Example 3 so you could follow the steps more easily.

● ● ● **Example 4** Using Trigonometry to Measure a Distance

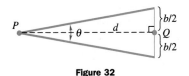

Figure 32

A method that surveyors use to determine a small distance d between two points P and Q is called the *subtense bar method*. The subtense bar with length b is centered at Q and situated perpendicular to the line of sight between P and Q. See Figure 32. Angle θ is measured, and then the distance d can be determined.

(a) Find d when $\theta = 1° \, 23' \, 12''$ and $b = 2$ meters.

From Figure 32, we see that

$$\cot \frac{\theta}{2} = \frac{d}{b/2}$$

$$d = \frac{b}{2} \cot \frac{\theta}{2}.$$

To evaluate $\theta/2$, we change θ to decimal degrees: $1° \, 23' \, 12'' = 1.386667°$, so

$$d = \frac{2}{2} \cot \frac{1.386667°}{2} \approx 82.6341 \text{ meters.}$$

(b) The angle θ usually cannot be measured more accurately than to the nearest $1''$. How much change would there be in the value of d if θ were measured $1''$ larger?

Use $\theta = 1° \, 23' \, 13'' \approx 1.386944°$.

$$d = \frac{2}{2} \cot \frac{1.386944°}{2} \approx 82.6176 \text{ meters.}$$

The difference is $82.6341 - 82.6176 \approx .017$ meter. ● ● ●

2.5 Exercises

Concept Check *Give a short written answer to each question.*

1. When bearing is given as a single angle measure, how is the angle represented in a sketch?

2. When bearing is given as N (or S), then the angle measure, then E (or W), how is the angle represented in a sketch?

3. Why is it important to draw a sketch before solving trigonometric problems like those in the last two sections of this chapter?

4. How should the angle of elevation (or depression) from a point X to a point Y be represented?

An observer for a radar station is located at the origin of a coordinate system. For each of the points in Exercises 5–8, find the bearing of an airplane located at that point. Express the bearing using both methods.

5. $(-4, 0)$ 6. $(-3, -3)$ 7. $(-5, 5)$ 8. $(0, -2)$

9. The ray $y = x$, $x \geq 0$ contains the origin and all points in the coordinate system whose bearing from the origin is $45°$. Determine the equation of a ray consisting of the origin and all points whose bearing from the origin is $240°$.

10. Repeat Exercise 9 for a bearing of $150°$.

Work each problem. In these exercises, assume the course of a plane or ship is on the indicated bearing. See Examples 1 and 2.

11. *Distance Flown by a Plane* A plane flies 1.3 hours at 110 mph on a bearing of 40°. It then turns and flies 1.5 hours at the same speed on a bearing of 130°. How far is the plane from its starting point?

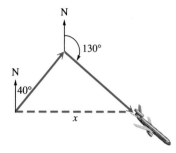

12. *Distance Traveled by a Ship* A ship travels 50 km on a bearing of 27°, then travels on a bearing of 117° for 140 km. Find the distance traveled from the starting point to the ending point.

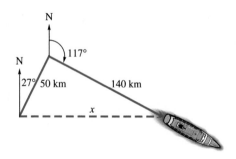

13. *Distance Between Two Ships* Two ships leave a port at the same time. The first ship sails on a bearing of 40° at 18 knots (nautical miles per hour) and the second at a bearing of 130° at 26 knots. How far apart are they after 1.5 hours?

14. *Distance Between Two Lighthouses* Two lighthouses are located on a north-south line. From lighthouse A, the bearing of a ship 3742 meters away is 129° 43′. From lighthouse B, the bearing of the ship is 39° 43′. Find the distance between the lighthouses.

15. *Distance Between Two Cities* The bearing from Winston-Salem, North Carolina, to Danville, Virginia, is N 42° E. The bearing from Danville to Goldsboro, North Carolina, is S 48° E. A car driven by Mark Ferrari, traveling at 60 mph, takes 1 hour to go from Winston-Salem to Danville and 1.8 hours to go from Danville to Goldsboro. Find the distance from Winston-Salem to Goldsboro.

16. *Distance Between Two Cities* The bearing from Atlanta to Macon is S 27° E, and the bearing from Macon to Augusta is N 63° E. An automobile traveling at 60 mph needs 1 1/4 hours to go from Atlanta to Macon and 1 3/4 hours to go from Macon to Augusta. Find the distance from Atlanta to Augusta.

17. *Distance Between Two Ships* A ship leaves its home port and sails on a bearing of N 28° 10′ E. Another ship leaves the same port at the same time and sails on a bearing of S 61° 50′ E. If the first ship sails at 24.0 mph and the second sails at 28.0 mph, find the distance between the two ships after 4 hours.

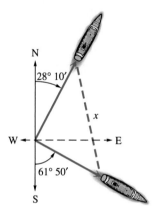

18. *Distance Between Transmitters* Radio direction finders are set up at two points A and B, which are 2.50 miles apart on an east-west line. From A, it is found that the bearing of a signal from a radio transmitter is N 36° 20′ E, while from B the bearing of the same signal is N 53° 40′ W. Find the distance of the transmitter from B.

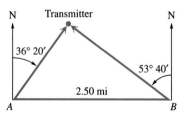

19. Solve the equation $ax = b + cx$ for x in terms of a, b, and c. (*Note:* This is in essence the calculation carried out in Example 3.)

20. Explain why the line $y = (\tan\theta)(x - a)$ passes through the point $(a, 0)$ and makes an angle θ with the x-axis.

21. Find the equation of the line passing through the point $(25, 0)$ that makes an angle of 35° with the x-axis.

22. Find the equation of the line passing through the point $(5, 0)$ that makes an angle of 15° with the x-axis.

In Exercises 23–28, use the method of Example 3. Drawing a sketch for the problems where one is not given may be helpful.

23. Find *h* as indicated in the figure.

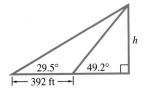

24. Find *h* as indicated in the figure.

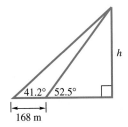

25. *Height of a Pyramid*   The angle of elevation from a point on the ground to the top of a pyramid is 35° 30′. The angle of elevation from a point 135 feet farther back to the top of the pyramid is 21° 10′. Find the height of the pyramid.

26. *Distance Traveled by a Whale*   Debbie Maybury, a whale researcher standing at the top of a tower, is watching a whale approach the tower directly. When she first begins watching the whale, the angle of depression to the whale is 15° 50′. Just as the whale turns away from the tower, the angle of depression is 35° 40′. If the height of the tower is 68.7 meters, find the distance traveled by the whale as it approaches the tower.

27. *Height of an Antenna*   A scanner antenna is on top of the center of a house. The angle of elevation from a point 28.0 meters from the center of the house to the top of the antenna is 27° 10′, and the angle of elevation to the bottom of the antenna is 18° 10′. Find the height of the antenna.

28. *Height of Mt. Whitney*   The angle of elevation from Lone Pine to the top of Mt. Whitney is 10° 50′. Van Dong Le, traveling 7.00 km from Lone Pine along a straight, level road toward Mt. Whitney, finds the angle of elevation to be 22° 40′. Find the height of the top of Mt. Whitney above the level of the road.

29. *Distance Between Two Points*   Refer to Example 4. A variation of the subtense bar method that surveyors use to determine larger distances *d* between two points *P* and *Q* is shown in the figure. In this case the subtense bar with length *b* is placed between the points *P* and *Q* so that the bar is centered on and perpendicular

to the line of sight connecting *P* and *Q*. The angles α and β are measured from points *P* and *Q*, respectively. (*Source:* Mueller, I. and K. Ramsayer, *Introduction to Surveying,* Frederick Ungar Publishing Co., 1979.)

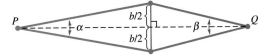

(a) Find a formula for *d* involving α, β, and *b*.

(b) Use your formula to determine *d* if $\alpha = 37'\ 48''$, $\beta = 42'\ 3''$, and *b* = 2.000 meters.

Solve each exercise using the techniques of Section 2.4.

30. *Height of a Plane Above Earth*   Find the minimum height *h* above the surface of Earth so that a pilot at point *A* in the figure can see an object on the horizon at *C*, 125 miles away. Assume that the radius of Earth is 4.00×10^3 miles.

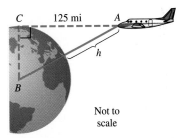

31. *Distance of a Plant from a Fence*   In one area, the lowest angle of elevation of the sun in winter is 23° 20′. Find the minimum distance *x* that a plant needing full sun can be placed from a fence 4.65 feet high.

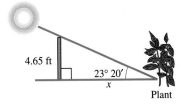

32. *Distance Through a Tunnel*   A tunnel is to be dug from *A* to *B*. Both *A* and *B* are visible from *C*. If *AC* is 1.4923 miles and *BC* is 1.0837 miles, and if *C* is 90°, find the measures of angles *A* and *B*.

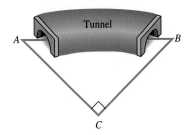

33. *Length of a Side of a Piece of Land*   A piece of land has the shape shown in the figure. Find *x*.

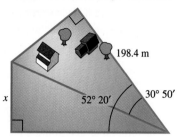

198.4 m

x   52° 20′   30° 50′

34. In Exercise 32, suppose the tunnel is being built by tunneling from both points *A* and *B* to make the straight line *AB*. How can the engineers ensure the two ends will meet? (*Hint:* Consider the angles in the figure.)

Chapter 2   Summary

Key Terms & Symbols	Key Ideas

2.1 Trigonometric Functions of Acute Angles

side opposite
side adjacent
cofunction

Right-Triangle-Based Definitions of the Trigonometric Functions

For any acute angle *A* in standard position,

$$\sin A = \frac{y}{r} = \frac{\text{side opposite}}{\text{hypotenuse}} \qquad \csc A = \frac{r}{y} = \frac{\text{hypotenuse}}{\text{side opposite}}$$

$$\cos A = \frac{x}{r} = \frac{\text{side adjacent}}{\text{hypotenuse}} \qquad \sec A = \frac{r}{x} = \frac{\text{hypotenuse}}{\text{side adjacent}}$$

$$\tan A = \frac{y}{x} = \frac{\text{side opposite}}{\text{side adjacent}} \qquad \cot A = \frac{x}{y} = \frac{\text{side adjacent}}{\text{side opposite}}.$$

Cofunction Identities

For any acute angle *A*,

$$\sin A = \cos(90° - A) \qquad \cos A = \sin(90° - A)$$

$$\tan A = \cot(90° - A) \qquad \cot A = \tan(90° - A)$$

$$\csc A = \sec(90° - A) \qquad \sec A = \csc(90° - A).$$

Function Values of Special Angles

θ	$\sin \theta$	$\cos \theta$	$\tan \theta$	$\cot \theta$	$\sec \theta$	$\csc \theta$
30°	$\frac{1}{2}$	$\frac{\sqrt{3}}{2}$	$\frac{\sqrt{3}}{3}$	$\sqrt{3}$	$\frac{2\sqrt{3}}{3}$	2
45°	$\frac{\sqrt{2}}{2}$	$\frac{\sqrt{2}}{2}$	1	1	$\sqrt{2}$	$\sqrt{2}$
60°	$\frac{\sqrt{3}}{2}$	$\frac{1}{2}$	$\sqrt{3}$	$\frac{\sqrt{3}}{3}$	2	$\frac{2\sqrt{3}}{3}$

Key Terms & Symbols	Key Ideas
2.2 Trigonometric Functions of Non-Acute Angles reference angle	**Reference Angle θ' for θ in $(0°, 360°)$**

	θ in Quadrant	θ' Is
	I	θ
	II	$180° - \theta$
	III	$\theta - 180°$
	IV	$360° - \theta$

Finding Trigonometric Function Values for Any Angle

Step 1 Add or subtract 360° as many times as needed to get an angle of at least 0° but less than 360°.

Step 2 Find the reference angle θ'.

Step 3 Find the trigonometric function values for θ'.

Step 4 Determine the correct signs for the values found in Step 3.

Key Terms & Symbols	Key Ideas
2.4 Solving Right Triangles significant digit exact number solving a triangle angle of elevation angle of depression	**Solving Applied Trigonometry Problems** **Step 1** Draw a sketch, and label it with the given information. Label the quantity to be found with a variable. **Step 2** Use the sketch to write an equation relating the given quantities to the variable. **Step 3** Solve the equation, and check that your answer makes sense.
2.5 Further Applications of Right Triangles bearing	Figures 26 and 28 on page 92 illustrate the two methods for expressing bearing.

Chapter 2   Review Exercises

Find the values of the six trigonometric functions for each angle A.

1.

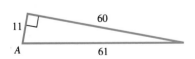

2.

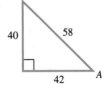

Solve each equation. Assume that all angles are acute angles.

3. $\sin 4\beta = \cos 5\beta$

4. $\sec(2\gamma + 10°) = \csc(4\gamma + 20°)$

5. $\tan(5x + 11°) = \cot(6x + 2°)$

6. $\cos\left(\dfrac{3\theta}{5} + 11°\right) = \sin\left(\dfrac{7\theta}{10} + 40°\right)$

Tell whether each statement is true *or* false. *If false, tell why.*

7. $\sin 46° < \sin 58°$ **8.** $\cos 47° < \cos 58°$ **9.** $\sec 48° \geq \cos 42°$ **10.** $\sin 22° \geq \csc 68°$

11. Explain why, in the figure, the cosine of angle A is equal to the sine of angle B.

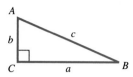

Find the values of the six trigonometric functions for each angle. Give exact values. Do not use a calculator. Rationalize denominators when applicable.

12. 120°

13. 300°

14. −225°

15. −390°

Find all values of θ, if θ is in the interval $[0°, 360°)$ and θ has the given function value.

16. $\sin \theta = -\dfrac{1}{2}$

17. $\cos \theta = -\dfrac{1}{2}$

18. $\cot \theta = -1$

19. $\sec \theta = -\dfrac{2\sqrt{3}}{3}$

Evaluate each expression. Give exact values.

20. $\cos 60° + 2 \sin^2 30°$

21. $\tan^2 120° - 2 \cot 240°$

22. $\sec^2 300° - 2 \cos^2 150° + \tan 45°$

23. If A, B, and C are the three angles of a triangle, then

$$\tan A + \tan B + \tan C = \tan A \tan B \tan C \quad \text{(where } A, B, C \neq 90°\text{)}$$

and

$$\sin^2 A + \sin^2 B + \sin^2 C = 2(1 + \cos A \cos B \cos C).$$

Verify these equations for a 30°–30°–120° triangle.

Use a calculator to find each value.

24. $\sin 72° \, 30'$

25. $\sec 222° \, 30'$

26. $\cot 305.6°$

27. $\csc 78° \, 21'$

28. $\sec 58.9041°$

29. $\tan 11.7689°$

30. *Concept Check* Which one of the following cannot be *exactly* determined using the methods of this chapter?

A. $\cos 135°$ **B.** $\cot(-45°)$ **C.** $\sin 300°$ **D.** $\tan 140°$

Use a calculator to find each value of θ, where θ is in the interval $[0°, 90°)$. Give answers in decimal degrees.

31. $\sin \theta = .82584121$

32. $\cot \theta = 1.1249386$

33. $\cos \theta = .97540415$

34. $\sec \theta = 1.2637891$

35. $\tan \theta = 1.9633124$

36. $\csc \theta = 9.5670466$

Find two angles in the interval $[0°, 360°)$ that satisfy each of the following. Leave answers in decimal degrees.

37. $\sin \theta = .73254290$

38. $\tan \theta = 1.3865342$

Tell whether each statement is true or false. If false, tell why. Use a calculator for Exercises 39 and 42.

39. $\sin 50° + \sin 40° = \sin 90°$

40. $\cos 210° = \cos 180° \cdot \cos 30° - \sin 180° \cdot \sin 30°$

41. $\sin 240° = 2 \sin 120° \cdot \cos 120°$

42. $\sin 42° + \sin 42° = \sin 84°$

43. A student wants to use a calculator to find the value of $\cot 25°$. However, instead of entering $\dfrac{1}{\tan 25}$, he enters $\tan^{-1} 25$. Assuming the calculator is in degree mode, will this produce the correct answer? Explain.

For each angle θ, use a calculator to find $\cos \theta$ and $\sin \theta$. Use your results to decide what quadrant the angle lies in.

44. $\theta = 2976°$

45. $\theta = 1997°$

46. $\theta = 4000°$

Solve each right triangle. In Exercise 48, give angles to the nearest minute. In Exercises 49 and 50, label the triangle as shown in Figure 20 in Section 2.4.

47.

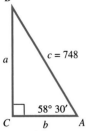

48.

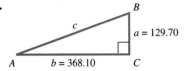

49. $A = 39.72°$, $b = 38.97$ m

50. $B = 47° 53'$, $b = 298.6$ m

Solve each problem.

51. *Height of a Tower*   The angle of elevation from a point 93.2 feet from the base of a tower to the top of the tower is $38° 20'$. Find the height of the tower.

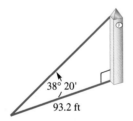

52. *Height of a Tower*   The angle of depression of a television tower to a point on the ground 36.0 meters from the bottom of the tower is $29.5°$. Find the height of the tower.

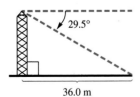

53. *Length of a Diagonal*   A rectangle has adjacent sides measuring 10.93 cm and 15.24 cm. The angle between the diagonal and the longer side is $35.65°$. Find the length of the diagonal.

54. *Length of Sides of an Isosceles Triangle*   An isosceles triangle has a base of length 49.28 meters. The angle opposite the base is $58.746°$. Find the length of each of the two equal sides.

55. *Distance Between Two Points*   The bearing of B from C is $254°$. The bearing of A from C is $344°$. The bearing of A from B is $32°$. The distance from A to C is 780 meters. Find the distance from A to B.

56. *Distance a Ship Sails*   The bearing from point A to point B is S $55°$ E and from point B to point C is N $35°$ E. If a ship sails from A to B, a distance of 80 km, and then from B to C, a distance of 74 km, how far is it from A to C?

57. *Distance Between Two Points*   Two cars leave an intersection at the same time. One heads due south at 55 mph. The other travels due west. After two hours, the bearing of the car headed west from the car headed south is $324°$. How far apart are they at that time?

58. Find a formula for h in terms of k, A, and B. Assume $A < B$.

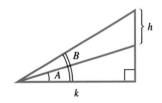

59. Make up a right triangle problem whose solution is $3 \tan 25°$.

60. Make up a right triangle problem whose solution is found from $\sin \theta = \dfrac{3}{4}$.

In Exercises 61–63, find a line segment in the figure whose length is equal to the function value. (Hint: Use right triangles.)

61. $\cot \theta$

62. $\sec \theta$

63. $\csc \theta$

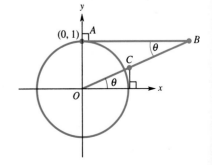

64. *(Modeling) Height of a Satellite* Artificial satellites that orbit Earth often use VHF signals to communicate with the ground. VHF signals travel in straight lines. The height h of the satellite above Earth and the time T that the satellite can communicate with a fixed location on the ground are related by the model

$$h = R\left(\frac{1}{\cos(180T/P)} - 1\right),$$

where $R = 3955$ miles is the radius of Earth and P is the period for the satellite to orbit Earth. (*Source:* Schlosser, W., T. Schmidt-Kaler, and E. Milone, *Challenges of Astronomy,* Springer-Verlag, 1991.)

(a) Find h when $T = 25$ minutes and $P = 140$ minutes. (Evaluate the cosine function in degree mode.)

(b) What is the value of h if T is increased to 30?

65. *Fundamental Surveying Problem* The first fundamental problem of surveying is to determine the coordinates of a point Q given the coordinates of a point P, the distance between P and Q, and the bearing θ from P to Q. See the figure. (*Source:* Mueller, I. and K. Ramsayer, *Introduction to Surveying,* Frederick Ungar Publishing Co., 1979.)

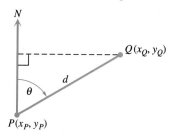

(a) Find a formula for the coordinates (x_Q, y_Q) of the point Q given θ, the coordinates (x_P, y_P) of P, and the distance d between P and Q.

(b) Use your formula to determine (x_Q, y_Q) if $(x_P, y_P) = (123.62, 337.95)$, $\theta = 17° \, 19' \, 22''$, and $d = 193.86$ feet.

Chapter 2   Test

1. Give the six trigonometric function values of angle A.

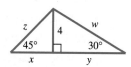

2. Find the exact values of each part labeled with a letter.

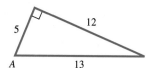

3. Find a solution for $\sin(B + 15°) = \cos(2B + 30°)$.

4. Find a value of θ in $[0°, 90°)$ in decimal degrees, if $\sin \theta = .27843196$.

5. Give two angles in $[0°, 360°)$ that satisfy

$$\cos \theta = -\frac{\sqrt{2}}{2}.$$

6. Explain how you would find $\cot \theta$ using a calculator, if $\tan \theta$ is 1.6778490. Then give $\cot \theta$.

7. Tell whether each statement is *true* or *false*. If false, tell why.
(a) $\sin 24° < \sin 48°$ **(b)** $\cos 24° < \cos 48°$
(c) $\tan 24° < \tan 48°$

8. Find the exact value of $\cot(-750°)$.

9. Use a calculator to approximate the following.
(a) $\sin 78° \, 21'$ **(b)** $\tan 11.7689°$
(c) $\sec 58.9041°$

10. Solve the triangle.

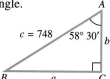

11. *Height of a Flagpole* To measure the height of a flagpole, Amado Carillo found that the angle of elevation from a point 24.7 feet from the base to the top is $32° \, 10'$. What is the height of the flagpole?

12. *Distance of a Ship from a Pier* A ship leaves a pier on a bearing of S 55° E and travels for 80 km. It then turns and continues on a bearing of N 35° E for 74 km. How far is the ship from the pier?

Ball Rolling Down a Plane   When a ball rolls down an inclined plane, as shown in the figure, in the absence of friction and air resistance, velocity v and distance d in feet traveled after t seconds are given as

$$v = 32t \sin \theta \quad \text{and} \quad d = 16t^2 \sin \theta.$$

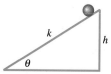

13. Find the time required for the ball to travel the entire distance k when starting at height h.

14. Use your answer from Exercise 13 to find the velocity when the ball reaches the ground. What does the result tell us about how the velocity is related to the quantities h and k?

Radian Measure and the Circular Functions

3

3.1 **Radian Measure**

3.2 **Applications of Radian Measure**

3.3 **Circular Functions of Real Numbers**

3.4 **Linear and Angular Velocity**

Over the past thirty years, production of solar energy has evolved from a mere kilowatt of electricity to hundreds of megawatts. Solar energy has many advantages over traditional energy sources in that it does not pollute and is potentially an unlimited, cheap source of energy. Its use and production is readily available throughout the United States and the world. The North American southwest has some of the brightest sunlight in the world with a potential to provide up to 2500 kilowatt-hours per square meter.*

When designing solar-power plants, engineers must position solar panels perpendicular to the sun's rays so that maximum energy can be collected. Understanding the movement and position of the sun at any time and date are fundamental concepts for solar energy collection. How high in the sky was the sun in Sacramento or in New Orleans on February 29, 2000 at 3 P.M.? How many hours of daylight were there in Hartford, Connecticut, on August 12, 2001? Answers to questions like these are necessary to generate electricity from sunlight. (See Section 3.3.) Because Earth moves in a nearly circular orbit around the sun, the circular functions will be essential for answering these questions.

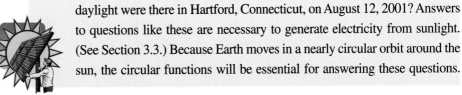

Source: Winter, C., R. Sizmann, and Vant-Hunt (Editors), *Solar Power Plants,* Springer-Verlag, 1991.

3.1 Radian Measure

• **Radian Measure** • **Converting Between Degrees and Radians** • **Finding Function Values for Angles in Radians**

In most work involving applications of trigonometry, angles are measured in degrees. In more advanced work in mathematics, the use of *radian measure* of angles is preferred. Radian measure allows us to treat the trigonometric functions as functions with domains of *real numbers,* rather than angles.

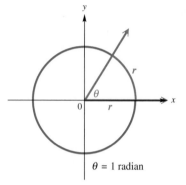

$\theta = 1$ radian

Figure 1

Radian Measure   Figure 1 shows an angle θ in standard position along with a circle of radius r. The vertex of θ is at the center of the circle. Angle θ intercepts an arc on the circle equal in length to the radius of the circle. So, angle θ is said to have a measure of 1 radian.

> **Radian**
>
> An angle with its vertex at the center of a circle that intercepts an arc on the circle equal in length to the radius of the circle has a measure of **1 radian.**

It follows that an angle of measure 2 radians intercepts an arc equal in length to twice the radius of the circle, an angle of measure $1/2$ radian intercepts an arc equal in length to half the radius of the circle, and so on.

Converting Between Degrees and Radians   The circumference of a circle—the distance around the circle—is given by $C = 2\pi r$, where r is the radius of the circle. The formula $C = 2\pi r$ shows that the radius can be laid off 2π times around a circle. Therefore, an angle of 360°, which corresponds to a complete circle, intercepts an arc equal in length to 2π times the radius of the circle. Thus, an angle of 360° has a measure of 2π radians:

$$360° = 2\pi \text{ radians.}$$

An angle of 180° is half the size of an angle of 360°, so an angle of 180° has half the radian measure of an angle of 360°.

$$180° = \frac{1}{2}(2\pi) \text{ radians} = \pi \text{ radians} \quad \text{Degree/radian relationship}$$

We can use the relationship $180° = \pi$ radians to develop a method for converting between degrees and radians as follows.

$$180° = \pi \text{ radians}$$

$$1° = \frac{\pi}{180} \text{ radian} \quad \text{Divide by 180.} \qquad \text{or} \qquad 1 \text{ radian} = \frac{180°}{\pi} \quad \text{Divide by } \pi.$$

Therefore, to change from degrees to radians, multiply by $\pi/180$ radian, and to change from radians to degrees, multiply by $180°/\pi$.

●●● **Example 1** Converting Degrees to Radians

Convert each degree measure to radians.

Algebraic Solution

(a) 45°

$$45° = 45\left(\frac{\pi}{180}\text{ radian}\right) \quad \text{Multiply by } \frac{\pi}{180}\text{ radian.}$$

$$= \frac{\pi}{4}\text{ radian}$$

(b) 249.8°

$$249.8° = 249.8\left(\frac{\pi}{180}\text{ radian}\right)$$

$$= 4.360\text{ radians} \quad \text{Nearest thousandth}$$

Graphing Calculator Solution

The real number π is used extensively in trigonometry to express angle measures. Learn where the $\boxed{\pi}$ key is located on your calculator. Some calculators (in radian mode) have the capability to convert directly between decimal degrees and radians. Figure 2 shows the conversions for this example. Note that when *exact* values involving π are required, such as $\pi/4$ in part (a), calculator approximations are not acceptable.

```
45°
            .7853981634
π/4
            .7853981634
249.8°
           4.359832471
```

Figure 2

●●●

●●● **Example 2** Converting Radians to Degrees

Convert each radian measure to degrees.

Algebraic Solution

(a) $\dfrac{9\pi}{4}$

$$\frac{9\pi}{4} = \frac{9\pi}{4}\left(\frac{180°}{\pi}\right) = 405° \quad \text{Multiply by } \frac{180°}{\pi}.$$

(b) 4.25 (Give the answer in decimal degrees.)

$$4.25 = 4.25\left(\frac{180°}{\pi}\right) \approx 243.5°$$

In the last step we used the $\boxed{\pi}$ key on a calculator to complete the computation.

Graphing Calculator Solution

Figure 3 shows how a calculator in degree mode converts the radian measures in this example to decimal degrees.

```
(9π/4)ʳ
                    405
4.25ʳ
            243.5070629
```

Figure 3

●●●

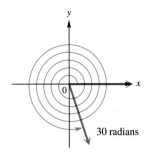

30 radians

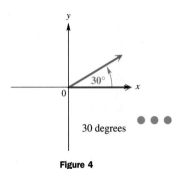

30 degrees

Figure 4

Converting Between Degrees and Radians

1. Multiply a radian measure by $180°/\pi$ and simplify to convert to degrees.
2. Multiply a degree measure by $\pi/180$ radian and simplify to convert to radians.

If no unit of measure is specified for an angle, radian measure is understood.

CAUTION Figure 4 shows angles measuring 30 radians and 30°. These angle measures are not at all close, so be careful not to confuse them.

Finding Function Values for Angles in Radians Trigonometric function values for angles measured in radians can be found by first converting radian measure to degrees. (You should try to skip this intermediate step as soon as possible, and find the function values directly from radian measure.)*

Example 3 Finding a Function Value of an Angle in Radian Measure

Find $\tan \dfrac{2\pi}{3}$.

First convert $\dfrac{2\pi}{3}$ radians to degrees.

$$\tan \frac{2\pi}{3} = \tan\left(\frac{2}{3} \cdot 180°\right) \qquad \text{Substitute } 180° \text{ for } \pi.$$

$$= \tan 120°$$

$$= -\sqrt{3} \qquad \text{Use the methods of Chapter 2.}$$

The following table and Figure 5 give some equivalent angles measured in degrees and radians. It will be useful to remember these equivalent values. Keep in mind that $180° = \pi$ radians. Then it will be easy to reproduce the rest of the table.

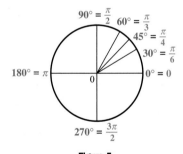

Figure 5

Equivalent Angle Measures in Degrees and Radians

Degrees	Radians		Degrees	Radians	
	Exact	Approximate		Exact	Approximate
0°	0	0	90°	$\dfrac{\pi}{2}$	1.57
30°	$\dfrac{\pi}{6}$	.52	180°	π	3.14
45°	$\dfrac{\pi}{4}$	.79	270°	$\dfrac{3\pi}{2}$	4.71
60°	$\dfrac{\pi}{3}$	1.05	360°	2π	6.28

---

* A table giving the values of the trigonometric functions for common radian and degree measures is given inside the front cover of this book.

● ● ●   **Example 4**   Finding a Function Value of an Angle in Radian Measure

Find $\sin \dfrac{3\pi}{2}$.

From the table, $\dfrac{3\pi}{2}$ radians $= 270°$, so

$$\sin \frac{3\pi}{2} = \sin 270° = -1.$$

● ● ●

3.1   Exercises

Concept Check   In Exercises 1–4, each angle θ is an integer when measured in radians. Give the radian measure of the angle.

1.   **2.**   **3.**   **4.**

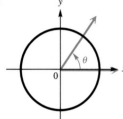   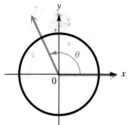   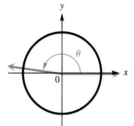   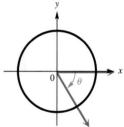

Convert each degree measure to radians. Leave answers as multiples of π. See Example 1.

5. 60°	**6.** 30°	**7.** 90°	**8.** 120°	**9.** 150°
10. 270°	**11.** 300°	**12.** 315°	**13.** 450°	**14.** 480°

Give a short explanation in Exercises 15–20.

15. In your own words, explain how to convert degree measure to radian measure.

16. In your own words, explain how to convert radian measure to degree measure.

17. In your own words, explain the meaning of radian measure.

18. Explain the difference between degree measure and radian measure.

19. Use an example to show that you can convert from radian measure to degree measure by multiplying by $180°/\pi$.

20. Explain why an angle of radian measure t in standard position intercepts an arc of length t on the circle of radius 1.

Convert each radian measure to degrees. See Example 2.

21. $\dfrac{\pi}{3}$	**22.** $\dfrac{8\pi}{3}$	**23.** $\dfrac{7\pi}{4}$	**24.** $\dfrac{2\pi}{3}$	**25.** $\dfrac{11\pi}{6}$	**26.** $\dfrac{15\pi}{4}$	**27.** $-\dfrac{\pi}{6}$
28. $\dfrac{8\pi}{5}$	**29.** $\dfrac{7\pi}{10}$	**30.** $\dfrac{11\pi}{15}$	**31.** $\dfrac{4\pi}{15}$	**32.** $\dfrac{7\pi}{20}$	**33.** $\dfrac{17\pi}{20}$	**34.** $\dfrac{11\pi}{30}$

Convert each degree measure to radians. See Example 1.

35. 39°	**36.** 74°	**37.** 42° 30′	**38.** 53° 40′
39. 139° 10′	**40.** 174° 50′	**41.** 64.29°	**42.** 85.04°
43. 56° 25′	**44.** 122° 37′	**45.** 47.6925°	**46.** 23.0143°

Convert each radian measure to degrees. Write answers to the nearest minute. See Example 2.

47. 2   **48.** 5   **49.** 1.74   **50.** 3.06

51. .3417   **52.** 9.84763   **53.** 5.01095   **54.** −3.47189

55. The value of sin 30 is not 1/2. Explain why.

56. Explain in your own words what is meant by an angle of one radian.

Find the exact value of each expression without using a calculator. See Examples 3 and 4.

57. $\sin \dfrac{\pi}{3}$ **58.** $\cos \dfrac{\pi}{6}$ **59.** $\tan \dfrac{\pi}{4}$ **60.** $\cot \dfrac{\pi}{3}$ **61.** $\sec \dfrac{\pi}{6}$ **62.** $\csc \dfrac{\pi}{4}$

63. $\sin \dfrac{\pi}{2}$ **64.** $\csc \dfrac{\pi}{2}$ **65.** $\tan \dfrac{2\pi}{3}$ **66.** $\cot \dfrac{2\pi}{3}$ **67.** $\sin \dfrac{5\pi}{6}$ **68.** $\tan \dfrac{5\pi}{6}$

69. $\cos 3\pi$ **70.** $\sec \pi$ **71.** $\sin \dfrac{4\pi}{3}$ **72.** $\cot\left(-\dfrac{2\pi}{3}\right)$ **73.** $\sin\left(-\dfrac{7\pi}{6}\right)$ **74.** $\cos\left(-\dfrac{\pi}{6}\right)$

75. *Concept Check* The figure shows the same angles measured in both degrees and radians. Complete the missing measures.

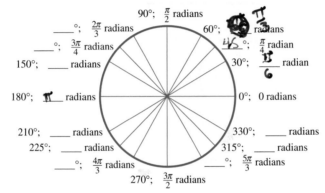

76. Find the measure (in both degrees and radians) of the angle θ formed in the graphing calculator screen by the line passing through the origin and the positive part of the *x*-axis. Use the displayed values of *x* and *y* at the bottom of the screen.

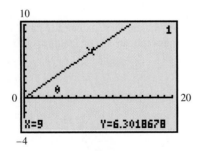

77. *Railroad Engineering* The term *grade* has several different meanings in construction work. Some engineers use the term *grade* to represent 1/100 of a right angle and express grade as a percent. For instance, an angle of .9° would be referred to as a 1% grade. (*Source:* Hay, W., *Railroad Engineering,* John Wiley & Sons, 1982.)

(a) By what number should you multiply a grade to convert it to radians?

(b) In a rapid-transit rail system, the maximum grade allowed between two stations is 3.5%. Express this angle in degrees and radians.

78. In the graphing calculator screen, was the calculator in degree or radian mode?

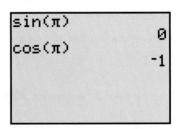

79. *Rotating Pulley* A circular pulley is rotating about its center. Through how many radians would it turn in (a) 8 rotations, and (b) 30 rotations?

80. *Rotating Hour Hand on a Clock* Through how many radians will the hour hand on a clock rotate in (a) 24 hours, and (b) 4 hours?

81. *Orbits of a Space Vehicle* A space vehicle is orbiting Earth in a circular orbit. What radian measure corresponds to (a) 2.5 orbits, and (b) 4/3 orbit?

82. (*Modeling*) *Fluctuation in the Solar Constant* The *solar constant S* is the amount of energy per unit area that reaches Earth's atmosphere from the sun. It is equal to 1367 watts per square meter but varies slightly throughout the seasons. This fluctuation ΔS in S can be calculated using the formula

$$\Delta S = .034S \sin\left[\frac{2\pi(82.5 - N)}{365.25}\right].$$

In this formula, N is the day number covering a four-year period, where $N = 1$ corresponds to January 1 of a leap year and $N = 1461$ corresponds to December 31 of the fourth year. (*Source:* Winter, C., R. Sizmann, and Vant-Hunt (Editors), *Solar Power Plants,* Springer-Verlag, 1991.)

(a) Calculate ΔS for $N = 80$, which is the spring equinox in the first year.
(b) Calculate ΔS for $N = 1268$, which is the summer solstice in the fourth year.
(c) What is the maximum value of ΔS?
(d) Find a value for N where ΔS is equal to 0.

3.2 Applications of Radian Measure

• Arc Length of a Circle • Sector of a Circle

Radian measure is used to simplify certain formulas, two of which are discussed in this section. Both would be more complicated if expressed in degrees.

Arc Length of a Circle The first formula is used to find the length of an arc of a circle. It comes from the fact (proven in plane geometry) that the length of an arc is proportional to the measure of its central angle.

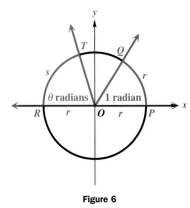

Figure 6

In Figure 6, angle QOP has measure 1 radian and intercepts an arc of length r on the circle. Angle ROT has measure θ radians and intercepts an arc of length s on the circle. Since the lengths of the arcs are proportional to the measures of their central angles,

$$\frac{s}{r} = \frac{\theta}{1}.$$

Multiplying both sides by r gives the following result.

Arc Length

The length s of the arc intercepted on a circle of radius r by a central angle of measure θ radians is given by the product of the radius and the radian measure of the angle, or

$$s = r\theta, \qquad \theta \text{ in radians}.$$

This formula is a good example of the usefulness of radian measure. (See Exercises 18 and 51.)

CAUTION When applying the formula $s = r\theta$, the value of θ *must be expressed in radians.*

● ● ● **Example 1**   Finding Arc Length Using $s = r\theta$

A circle has radius 18.2 cm. Find the length of the arc intercepted by a central angle having each of the following measures.

(a) $\dfrac{3\pi}{8}$ radians

Here $r = 18.2$ cm and $\theta = \dfrac{3\pi}{8}$. Since $s = r\theta$,

$$s = 18.2\left(\frac{3\pi}{8}\right) \text{ cm}$$

$$s = \frac{54.6\pi}{8} \text{ cm} \qquad \text{Exact answer}$$

$$s \approx 21.4 \text{ cm}. \qquad \text{Calculator approximation}$$

(b) $144°$

The formula $s = r\theta$ requires that θ be measured in radians. First, convert θ to radians by multiplying $144°$ by $\pi/180$ radian.

$$144° = 144\left(\frac{\pi}{180}\right) \text{ radians} \qquad \text{Change from degrees to radians.}$$

$$144° = \frac{4\pi}{5} \text{ radians}$$

Now $\qquad s = 18.2\left(\dfrac{4\pi}{5}\right) \text{ cm} \qquad \text{Use } s = r\theta.$

$$s = \frac{72.8\pi}{5} \text{ cm}$$

$$s \approx 45.7 \text{ cm}. \qquad\qquad\qquad\qquad\qquad ● ● ●$$

● ● ● **Example 2**   Using Latitudes to Find the Distance Between Two Cities

Reno, Nevada, is approximately due north of Los Angeles. The latitude of Reno is 40° N, while that of Los Angeles is 34° N. (The N in 34° N means *north* of the equator.) If the radius of Earth is 6400 km, find the north-south distance between the two cities.

Latitude gives the measure of a central angle with vertex at Earth's center whose initial side goes through the equator and whose terminal side goes through the given location. As shown in Figure 7, the central angle between Reno and Los Angeles is 6°. The distance between the two cities can be found by the formula $s = r\theta$, after 6° is first converted to radians.

$$6° = 6\left(\frac{\pi}{180}\right) = \frac{\pi}{30} \text{ radian}$$

The distance between the two cities is

$$s = r\theta$$

$$s = 6400\left(\frac{\pi}{30}\right) \text{ km} \qquad r = 6400, \ \theta = \frac{\pi}{30}$$

$$s \approx 670 \text{ km}. \qquad\qquad\qquad\qquad\qquad ● ● ●$$

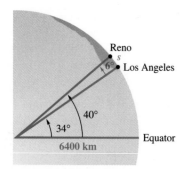

Figure 7

● ● ● **Example 3** Finding a Length Using $s = r\theta$

A rope is being wound around a drum with radius .8725 foot. (See Figure 8.) How much rope will be wound around the drum if the drum is rotated through an angle of 39.72°?

The length of rope wound around the drum is the arc length for a circle of radius .8725 foot and a central angle of 39.72°. Use the formula $s = r\theta$, with the angle converted to radian measure.

$$s = r\theta$$

$$s = .8725\left[39.72\left(\frac{\pi}{180}\right)\right] \quad \text{Convert to radians.}$$

$$s \approx .6049$$

The length of the rope wound around the drum is approximately .6049 foot.

● ● ●

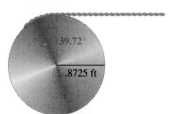

Figure 8

$800 = 14(0$

● ● ● **Example 4** Finding an Angle Measure Using $s = r\theta$

Two gears are adjusted so that the smaller gear drives the larger one, as shown in Figure 9. If the smaller gear rotates through 225°, through how many degrees will the larger gear rotate?

First find the radian measure of the angle, which will give the arc length on the smaller gear that determines the motion of the larger gear. Since $225° = 5\pi/4$ radians, for the smaller gear,

$$s = r\theta = 2.5\left(\frac{5\pi}{4}\right) \text{ cm}.$$

This arc length on the larger gear corresponds to an angle measure θ, in radians, where

$$s = r\theta$$

$$2.4\left(\frac{5\pi}{4}\right) = 4.8\theta$$

$$\theta \approx 2.045307717.$$

Converting back to degrees shows that the larger gear rotates through

$$2.04530771\left(\frac{180°}{\pi}\right) \approx 120°. \quad \text{Two significant digits}$$

● ● ●

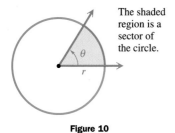

Figure 9

Sector of a Circle A **sector of a circle** is the portion of the interior of a circle intercepted by a central angle. Think of it as a "piece of pie." See Figure 10. A complete circle can be thought of as an angle with measure 2π radians. If a central angle for a sector has measure θ radians, then the sector makes up the fraction $\theta/(2\pi)$ of a complete circle. The area of a complete circle with radius r is $A = \pi r^2$. Therefore, the area of the sector is given by the product of the fraction $\theta/(2\pi)$ and the total area, πr^2.

The shaded region is a sector of the circle.

Figure 10

$$\text{area of sector} = \frac{\theta}{2\pi}(\pi r^2) = \frac{1}{2}r^2\theta, \quad \theta \text{ in radians}$$

This discussion is summarized as follows.

Area of a Sector
The area of a sector of a circle of radius r and central angle θ is given by

$$A = \frac{1}{2}r^2\theta, \qquad \theta \text{ in radians.}$$

CAUTION As in the formula for arc length, the value of θ must be in radians when using this formula for the area of a sector.

Example 5 Finding the Area of a Sector-Shaped Field

Figure 11 shows a field in the shape of a sector of a circle. Find the area of the field.

First, convert 15° to radians.

$$15° = 15\left(\frac{\pi}{180}\right) = \frac{\pi}{12} \text{ radian}$$

Now use the formula for the area of a sector.

$$A = \frac{1}{2}r^2\theta$$

$$A = \frac{1}{2}(321)^2\left(\frac{\pi}{12}\right)$$

$$A \approx 13,500 \text{ m}^2$$

Figure 11

C O N N E C T I O N S *Longitude* is the angular distance (expressed in degrees) East or West of the prime meridian, which goes from the North Pole to the South Pole through Greenwich, England. Arcs of longitude are 110 km apart at the equator. As the figure shows, these sections are similar to those of an orange.

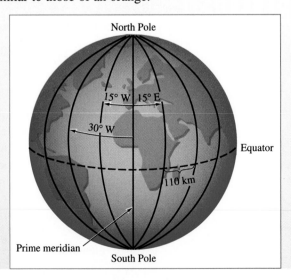

(continued)

Because Earth revolves 15° per hour, longitude is found by taking the difference between time zones multiplied by 15°. For example, if it is 12 noon where you are (in the United States) and 5 P.M. in Greenwich, you are located at longitude $5(15°) =$ longitude 75° W. Thus, determining longitude requires only an accurate measure of time. Before 1772, sailors were unable to determine their position at sea because there were no clocks capable of precise measure of time at sea. In 1772, a clock invented by John Harrison, a carpenter's son with no formal education, solved the problem. (*Sources:* www.members.tripod.com/longitude/html/math.htm. Ola, P. and E. D'Aulaire, "Taking the Measure of Time," *Smithsonian,* December 1999.)

For Discussion or Writing

Use time zones to determine the longitude where you live. What would the longitude be at Greenwich, England? Visit the Internet at the source given above to learn more about longitude.

3.2 Exercises

1. *Rotation of Gas Gauge Arrow*   The arrow on a car's gasoline gauge is one-half inch long. See the figure. Through what angle does the arrow rotate when it moves one inch on the gauge?

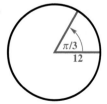

2. *Rotation of a Seesaw*   The seesaw at a playground is 12 feet long. Through what angle does the board rotate when a child rises three feet along the circular arc?

Concept Check   In Exercises 3 and 4, find the exact length of each arc intercepted by the given central angle.

3.

4.

Concept Check   In Exercises 5 and 6, find the radius of each circle.

5.

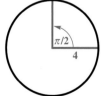

6.

Concept Check In Exercises 7 and 8, find the measure of each central angle (in radians).

7.

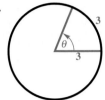

8.

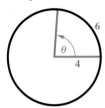

Unless otherwise directed, give calculator approximations in your answers in this exercise set.

Find the length of each arc intercepted by a central angle θ in a circle of radius r. See Example 1.

9. $r = 12.3$ cm, $\theta = \dfrac{2\pi}{3}$ radians

10. $r = .892$ cm, $\theta = \dfrac{11\pi}{10}$ radians

11. $r = 4.82$ m, $\theta = 60°$

12. $r = 71.9$ cm, $\theta = 135°$

13. Find the measure (in radians) of a central angle that intercepts an arc of length 5 inches in a circle of radius 2 inches.

14. Find the radius of a circle in which a central angle of 2 radians intercepts an arc of length 3 feet.

15. Find the radius of a circle in which a central angle of $\pi/5$ radian intercepts an arc of length 4 inches.

16. Find the measure (in radians) of a central angle that intercepts an arc of length 30 cm in a circle of radius 5 cm.

17. If the radius of a circle is doubled, how is the length of the arc intercepted by a fixed central angle changed?

18. The main reason for using radian measure is that it simplifies many formulas, such as the formula for arc length, $s = r\theta$. Give the corresponding formula when θ is measured in degrees instead of radians.

Distance Between Cities Find the distance in kilometers between each pair of cities, assuming they lie on the same north-south line. See Example 2.

19. Panama City, Panama, 9° N, and Pittsburgh, Pennsylvania, 40° N

20. Farmersville, California, 36° N, and Penticton, British Columbia, 49° N

21. New York City, New York, 41° N, and Lima, Peru, 12° S

22. Halifax, Nova Scotia, 45° N, and Buenos Aires, Argentina, 34° S

23. *Latitude of Madison* Madison, South Dakota, and Dallas, Texas, are 1200 km apart and lie on the same north-south line. The latitude of Dallas is 33° N. What is the latitude of Madison?

24. *Latitude of Toronto* Charleston, South Carolina, and Toronto, Canada, are 1100 km apart and lie on the same north-south line. The latitude of Charleston is 33° N. What is the latitude of Toronto?

Work each applied problem. See Examples 3 and 4.

25. *Pulley Raising a Weight*
  (a) How many inches will the weight in the figure rise if the pulley is rotated through an angle of 71° 50′?
  (b) Through what angle, to the nearest minute, must the pulley be rotated to raise the weight 6 inches?

9.27 in.

26. *Pulley Raising a Weight* Find the radius of the pulley in the figure if a rotation of 51.6° raises the weight 11.4 cm.

r

27. *Rotating Wheels*   The rotation of the smaller wheel in the figure causes the larger wheel to rotate. Through how many degrees will the larger wheel rotate if the smaller one rotates through 60.0°?

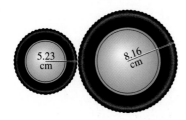

28. *Rotating Wheels*   Find the radius of the larger wheel in the figure if the smaller wheel rotates 80.0° when the larger wheel rotates 50.0°.

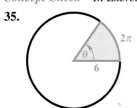

29. *Bicycle Chain Drive*   The figure shows the chain drive of a bicycle. How far will the bicycle move if the pedals are rotated through 180°? Assume the radius of the bicycle wheel is 13.6 inches.

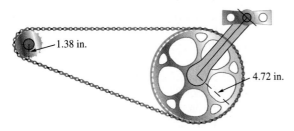

30. *Pickup Truck Speedometer*   The speedometer of a small pickup truck is designed to be accurate with tires of radius 14 inches.
  (a) Find the number of rotations of a tire in 1 hour if the truck is driven at 55 mph.
  (b) Suppose that oversize tires of radius 16 inches are placed on the truck. If the truck is now driven for 1 hour with the speedometer reading 55 mph, how far has the truck gone? If the speed limit is 55 mph, does the driver deserve a speeding ticket?

If a central angle is very small, there is little difference in length between an arc and the inscribed chord. See the figure. Approximate each of the following lengths by finding the necessary arc length. (Note: When a central angle intercepts an arc, the arc is said to **subtend** *the angle.)*

Arc length ≈ length of inscribed chord

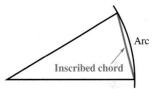

31. *Length of a Train*   A railroad track in the desert is 3.5 km away. A train on the track subtends (horizontally) an angle of 3° 20′. Find the length of the train.

32. *Length of an Oil Tanker*   An oil tanker 2.3 km at sea subtends a 1° 30′ angle horizontally. Find the length of the ship.

33. *Diameter of the Moon*   The full moon subtends an angle of 1/2°. The moon is 240,000 miles away. Find the diameter of the moon.

34. *Distance to a Boat*   The mast of Brent Simon's boat is 32 feet high. If it subtends an angle of 2° 10′, how far away is it?

Concept Check   In Exercises 35 and 36, find the area of each sector.

35.

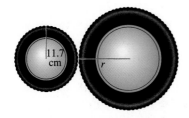

36.

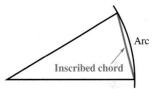

Concept Check   In Exercises 37 and 38, find the measure (in radians) of each central angle. The number inside the sector is the area.

37.

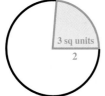

38.

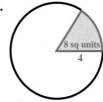

Find the area of a sector of a circle having radius r and central angle θ. See Example 5.

39. $r = 29.2$ m, $\theta = \dfrac{5\pi}{6}$ radians

40. $r = 59.8$ km, $\theta = \dfrac{2\pi}{3}$ radians

41. $r = 52$ cm, $\theta = \dfrac{3\pi}{10}$ radian

42. $r = 25$ mm, $\theta = \dfrac{\pi}{15}$ radian

43. $r = 12.7$ cm, $\theta = 81°$

44. $r = 18.3$ m, $\theta = 125°$

Work each problem.

45. Find the measure (in radians) of a central angle of a sector of area 16 square inches in a circle of radius 3.0 inches.

46. Find the radius of a circle in which a central angle of $\pi/4$ radian determines a sector of area 36 square feet.

47. Find the radius of a circle in which a central angle of $\pi/6$ radian determines a sector of area 64 square meters.

48. Find the measure (in radians) of a central angle of a sector of area 25 square inches in a circle of radius 10 inches.

49. Consider the area-of-a-sector formula $A = (1/2)r^2\theta$. What well-known formula corresponds to the special case $\theta = 2\pi$?

50. If the radius of a circle is doubled and the central angle of a sector is unchanged, how is the area of the sector changed?

51. Give the corresponding formula for the area of a sector when the angle is measured in degrees.

52. The sector in the graphing calculator screen is bounded above by the line $y = \left(\sqrt{3}/3\right)x$, below by the x-axis, and on the right by the circle $x^2 + y^2 = 4$. What is the area of the sector?

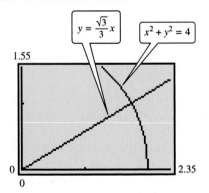

53. *Measures of a Structure*   The figure shows Medicine Wheel, a Native American structure in northern Wyoming. This circular structure is perhaps 2500 years old. There are 27 aboriginal spokes in the wheel, all equally spaced.

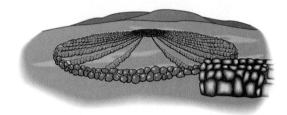

(a) Find the measure of each central angle in degrees and in radians.

(b) If the radius of the wheel is 76 feet, find the circumference.

(c) Find the length of each arc intercepted by consecutive pairs of spokes.

(d) Find the area of each sector formed by consecutive spokes.

54. *Circular Railroad Curves*   In the United States, circular railroad curves are designated by the *degree of curvature,* the central angle subtended by a chord of 100 feet. Suppose a portion of track has curvature 42°. (*Source:* Hay, W., *Railroad Engineering,* John Wiley & Sons, 1982.)

(a) What is the radius of the curve?

(b) What is the length of the arc determined by the 100-foot chord?

(c) What is the area of the segment of the circle bounded by the arc and the 100-foot chord?

55. *Area Cleaned by a Windshield Wiper*   The Ford Model A, built from 1928 to 1931, had a single windshield wiper on the driver's side. The total arm and blade was 10 inches long and rotated back and forth through an angle of 95°.

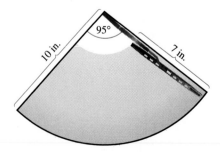

The shaded region in the figure is the portion of the windshield cleaned by the 7-inch wiper blade. What is the area of the region cleaned?

56. *Area of a Lot* A frequent problem in surveying city lots and rural lands adjacent to curves of highways and railways is that of finding the area when one or more of the boundary lines is the arc of a circle. Find the area of the lot shown in the figure. (*Source:* Anderson, J. and E. Michael, *Introduction to Surveying*, McGraw-Hill, 1985.)

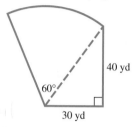

57. *Earth's Diameter* Eratosthenes (*ca.* 230 B.C.) made a famous measurement of Earth. He observed at Syene (the modern Aswan) at noon and at the summer solstice that a vertical stick had no shadow, while at Alexandria (on the same meridian as Syene) the sun's rays were inclined 1/50 of a complete circle to the vertical. See the figure.

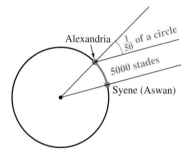

He then calculated the circumference of Earth from the known distance of 5000 stades between Alexandria and Syene. Obtain Eratosthenes' result of 250,000 stades for Earth's circumference. There is reason to suppose that a stade is about equal to 516.7 feet. Assuming this, use Eratosthenes' result to calculate the polar diameter of Earth in miles. (The actual polar diameter of Earth, to the nearest mile, is 7900 miles.) (*Source:* Eves, Howard, *A Survey of Geometry*, Vol. 1. Reprinted by permission of the author.)

Volume of a Solid *Multiply the area of the base by the height to find the volume of each solid.*

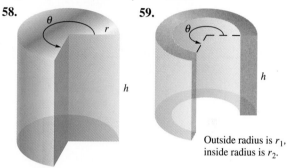

58.

59.

Outside radius is r_1, inside radius is r_2.

60. *Land Required for a Solar-Power Plant* A 300-megawatt solar-power plant requires approximately 950,000 square meters of land area in order to collect the required amount of energy from sunlight.

(a) If this land area is circular, what is its radius?

(b) If this land area is a 35° sector of a circle, what is its radius?

Relating Concepts

For individual or collaborative investigation
(Exercises 61–64)

(Modeling) Measuring Paper Curl *Manufacturers of paper determine its quality by its* curl. *The curl of a sheet of paper is measured by holding it at the center of one edge and comparing the arc formed by the free end to arcs on a chart lying flat on a table. Each arc in the chart corresponds to a number d that gives the* depth *of the arc. See the figure.* (*Source:* Tabakovic, H., J. Paullet, and R. Bertram, "Measuring the Curl of Paper," *The College Mathematics Journal*, Vol. 30 No. 4, September 1999.)

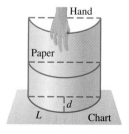
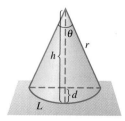

To produce the chart, it is necessary to find a function that relates d to the length of arc L. **Work Exercises 61–64 in order,** *to determine that function. Refer to the figure on the right on the previous page.*

61. Express L in terms of r and θ, and then solve for r.

62. Use a right triangle to relate r, h, and θ. Solve for h.

63. Express d in terms of r and h, then substitute your answer from Exercise 62 for h. Factor out r.

64. Use your answer from Exercise 61 to substitute for r in the result of Exercise 63. This result is a formula that gives d for specific values of θ.

- -

Quantitative Reasoning

65. *Can you find the radius of an Indian artifact given an arc of a circle?* Suppose you find an interesting Indian pottery fragment. The archaeology professor at your college believes it is a piece of the edge of a ceremonial plate and shows you a formula that will give the radius of the original plate using measurements from your fragment, shown in Figure A.

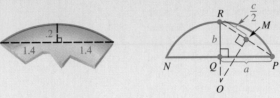

Figure A   Figure B

Other situations also require finding the radius of a circle from an arc of the circle. For instance, a woodworker might be restoring a carving in which a circular arc must be extended. Or, a mechanical engineer might want to know the radius of a flywheel where most of the wheel is below floor level.

In Figure B, a is $1/2$ the length of chord NP and b is the distance from the midpoint of the chord NP to the circle. According to the professor's formula, the radius of the circle, OR, is given by

$$r = \frac{a^2 + b^2}{2b}.$$

Why does this formula work? See if you can use the trigonometry you've studied so far to explain it.

(a) Extend line segment RQ to a radius of the arc, draw line segment RP and line segment MO, which is the perpendicular bisector of line segment RP. See Figure B. Find a pair of similar triangles in the figure. How do you know they are similar?

(b) Find two pairs of corresponding sides, and set their ratios equal. Solve the equation for r.

(c) How are a, b, and c related? Use this relationship to substitute for c in terms of a and b.

(d) What is the radius of the original plate from which your fragment came?

3.3 Circular Functions of Real Numbers

- The Circular Functions • Finding Values of Circular Functions • Determining a Number with a Given Circular Function Value • Applying Circular Functions

We defined the six trigonometric functions for *angles.* The angles can be measured either in degrees or in radians. While the domain of the trigonometric functions is a set of angles, the range is a set of real numbers. In advanced work, such as calculus, it is necessary to modify the trigonometric functions so that the domain contains not angles, but real numbers. We do this by using the relationship between an angle θ and an arc of length s on a circle.

The Circular Functions In Figure 12, starting at the point $(1,0)$, we have marked an arc of length $|s|$ along the circle, with endpoint (x, y). We count counterclockwise if s is positive and clockwise if s is negative. Figure 12 shows a

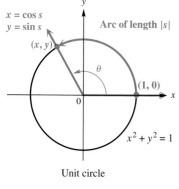

$x = \cos s$
$y = \sin s$

Arc of length $|s|$

(x, y)

θ

$(1, 0)$

$x^2 + y^2 = 1$

Unit circle

Figure 12

Looking Ahead to Calculus

If you plan to go on to calculus, you must become very familiar with radian measure. In calculus, the trigonometric or circular functions are always understood to have real number domains.

unit circle, with center at the origin and radius 1 unit (hence the name *unit circle*). Recall from algebra that the equation of this circle is

$$x^2 + y^2 = 1.$$

We saw earlier that the radian measure of θ is related to the arc length s. In fact, for θ measured in radians, we know that $s = r\theta$. Here, $r = 1$, so s, which is measured in linear units such as inches or centimeters, is numerically equal to θ, measured in radians. Thus, the trigonometric functions of angle θ in radians found by choosing a point (x, y) on the unit circle can be rewritten as functions of the arc length s, a real number. To distinguish these from the trigonometric functions of angles, they are called **circular functions.**

Circular Functions

$$\sin s = y \qquad \cos s = x \qquad \tan s = \frac{y}{x}, \quad x \neq 0$$

$$\csc s = \frac{1}{y}, \quad y \neq 0 \qquad \sec s = \frac{1}{x}, \quad x \neq 0 \qquad \cot s = \frac{x}{y}, \quad y \neq 0$$

NOTE Since $\sin s = y$ and $\cos s = x$, we can replace x and y in the equation $x^2 + y^2 = 1$ and obtain the Pythagorean identity

$$\cos^2 s + \sin^2 s = 1.$$

Since the ordered pair (x, y) represents a point on the unit circle,

$$-1 \leq x \leq 1 \qquad \text{and} \qquad -1 \leq y \leq 1,$$

so

$$-1 \leq \cos s \leq 1 \qquad \text{and} \qquad -1 \leq \sin s \leq 1.$$

For any value of s, both $\sin s$ and $\cos s$ exist, so the domain of these functions is the set of all real numbers. For $\tan s$, defined as y/x, $x \neq 0$. The only way x can equal 0 is when the arc length s is $\pi/2$, $-\pi/2$, $3\pi/2$, $-3\pi/2$, and so on. To avoid a 0 denominator, the domain of the tangent function must be restricted to those values of s satisfying

$$s \neq \frac{\pi}{2} + n\pi, \qquad n \text{ any integer.}$$

The definition of secant also has x in the denominator, so the domain of secant is the same as the domain of tangent. Both cotangent and cosecant are defined with a denominator of y. To guarantee that $y \neq 0$, the domain of these functions must be the set of all values of s satisfying

$$s \neq n\pi, \qquad n \text{ any integer.}$$

> ### Domains of the Circular Functions
>
> The domains of the circular functions are as follows. Assume that n is any integer and s is a real number.
>
> **Sine and Cosine Functions:** $(-\infty, \infty)$
>
> **Tangent and Secant Functions:** $\left\{ s \,\middle|\, s \neq \dfrac{\pi}{2} + n\pi \right\}$
>
> **Cotangent and Cosecant Functions:** $\{ s \mid s \neq n\pi \}$

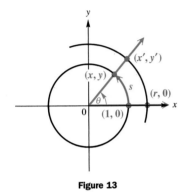

Figure 13

The trigonometric functions were defined by referring to a point (x', y') on the terminal side of an angle θ in standard position, measured in *degrees,* where the distance from the origin to (x', y') is r. On the other hand, the circular function definitions refer to an *arc s* from $(1, 0)$ to the point (x, y) on the *unit* circle (with radius 1). Figure 13 shows the connections between these definitions. Note the radian measure of angle θ is s.

The circular functions are the special case of the trigonometric functions where $\theta = s$ (radians) and $r = 1$. Thus, from the definitions of the trigonometric functions,

$$\sin \theta = \frac{y}{r} = \frac{y}{1} = y = \sin s \qquad \text{and} \qquad \cos \theta = \frac{x}{r} = \frac{x}{1} = x = \cos s.$$

Similar results hold for the other four functions.

Finding Values of Circular Functions As shown above, the trigonometric functions and the circular functions lead to the same function values. Because of this, a value such as $\sin \pi/2$ can be found without worrying about whether $\pi/2$ is a real number or the radian measure of an angle. In either case, $\sin \pi/2 = 1$. All the formulas developed in this book are valid for either angles or real numbers. For example, $\sin \theta = 1/\csc \theta$ is equally valid for θ as the measure of an angle in degrees or radians or for θ as a real number.

We also defined the trigonometric functions as lengths of the sides of a right triangle. Those definitions apply only to acute angles and are appropriate only for applications that involve right triangles.

We can use the ideas of the preceding discussion to find exact circular function values of certain real numbers expressed as rational multiples of π.

● ● ● **Example 1** Finding a Circular Function Value of a Multiple of π

Find the exact circular function value for each of the following.

(a) $\cos \dfrac{2\pi}{3}$

We can consider $2\pi/3$ as the radian measure of an angle. Since

$$\frac{2\pi}{3} = \frac{2}{3} \cdot 180° = 120°,$$

$$\cos \frac{2\pi}{3} = \cos 120° = -\frac{1}{2}.$$

(b) $\tan\left(-\dfrac{\pi}{4}\right) = \tan(-45°) = -1$

(c) $\csc\dfrac{10\pi}{3}$

Because $10\pi/3$ is not between 0 and 2π, we must first find a number between 0 and 2π with which it is coterminal. To do this, we subtract 2π as many times as needed. Here, 2π must be subtracted only once.

$$\frac{10\pi}{3} - 2\pi = \frac{10\pi}{3} - \frac{6\pi}{3} = \frac{4\pi}{3} = 240°$$

$$\csc\frac{10\pi}{3} = \csc\frac{4\pi}{3} = \csc 240° = -\frac{2\sqrt{3}}{3}$$

● ● ●

NOTE The values found in Example 1 can also be determined without converting to degrees, using reference angles measured in radians.

Recall that unless an angle measure is shown in degrees (30°, for example), it is understood to be in radians. Thus, sin 30 indicates the angle has measure 30 radians.

To use a calculator to find an approximation of a circular function of a real number, first set the calculator to *radian mode*.

● ● ● **Example 2** **Approximating Circular Function Values**

Use a calculator to approximate each circular function value.

(a) $\cos .5149 \approx .87034197$

(b) $\cot 1.3209$

As before, to find cotangent, secant, and cosecant function values, we must use the appropriate reciprocal function. To find cot 1.3209, first find tan 1.3209 and then find the reciprocal.

$$\cot 1.3209 = \frac{1}{\tan 1.3209} \approx .25523149$$

(c) $\sec(-2.9234) = \dfrac{1}{\cos(-2.9234)} \approx -1.0242855$

● ● ●

CAUTION A common error in trigonometry is using calculators in degree mode when radian mode should be used. Remember, if you are finding a circular function value of a real number, the calculator *must* be in *radian mode*.

Determining a Number with a Given Circular Function Value Recall from Section 2.3 how we used a calculator to determine an angle measure, given a trigonometric function value of the angle.

● ● ● **Example 3** Finding a Number Given Its Circular Function Value

Approximate the value of s in the interval $[0, \pi/2]$, if $\cos s = .96854556$.

Scientific Calculator Solution

With the calculator set in radian mode, use the [arc], [INV], or [cos⁻¹] key to find $s = .25147856$. Verify that

$$0 < .25147856 < \pi/2.$$

To check, verify that

$$\cos .25147856 = .96854556.$$

Graphing Calculator Solution

See Figure 14. The second line verifies that $.2514785647 < \pi/2$ is true. Recall that 1 indicates a statement is true; 0 would mean it is false.

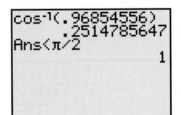

```
cos⁻¹(.96854556)
          .2514785647
Ans<π/2
                     1
```

Figure 14

● ● ●

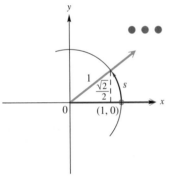

Figure 15

● ● ● **Example 4** Finding a Number Given Its Sine Ratio

Find the exact value of s in the interval $[0, \pi/2]$, if $\sin s = \sqrt{2}/2$.

Sketch a triangle in quadrant I, and use the definition of $\sin s$ to label two sides as shown in Figure 15. To relate it to the definition of the trigonometric function $\sin \theta$, multiply the lengths of each side by 2. We recognize this as a right triangle with two 45° angles. To find s, convert 45° to radians, to get $s = \pi/4$. ● ● ●

Applying Circular Functions The next example answers the first question posed in the chapter introduction.

● ● ● **Example 5** Modeling the Angle of Elevation of the Sun

Knowing the position of the sun in the sky is essential for solar-power plants. Solar panels need to be positioned perpendicular to the sun's rays for maximum efficiency. The angle of elevation θ of the sun in the sky at any latitude L is calculated with the formula

$$\sin \theta = \cos D \cos L \cos \omega + \sin D \sin L,$$

where $\theta = 0$ corresponds to sunrise and $\theta = \pi/2$ occurs if the sun is directly overhead. ω (the Greek letter *omega*) is the number of radians that Earth has rotated through since noon when $\omega = 0$. D is the declination of the sun, which varies because Earth is tilted on its axis. (*Source:* Winter, C., R. Sizmann, and Vant-Hunt (Editors), *Solar Power Plants,* Springer-Verlag, 1991.)

Sacramento, California, has latitude $L = 38.5°$ or .6720 radian. Find the angle of elevation θ of the sun at 3 P.M. on February 29, 2000, where at that time $D \approx -.1425$ and $\omega \approx .7854$.

Use the formula for sin θ.

$$\sin \theta = \cos D \cos L \cos \omega + \sin D \sin L$$
$$= \cos(-.1425) \cos(.6720) \cos(.7854) + \sin(-.1425) \sin(.6720)$$
$$\approx .4593$$

Thus, $\theta \approx .4773$ radian or $27.3°$.   ● ● ●

At the beginning of this section, we defined the cosine and sine of a real number s as the x- and y-coordinates, respectively, of a point on the unit circle $x^2 + y^2 = 1$. We can illustrate this concept using a graphing calculator to graph the circle. We must first solve the equation for y, getting the two functions

$$Y_1 = \sqrt{1 - x^2} \qquad \text{and} \qquad Y_2 = -\sqrt{1 - x^2}.$$

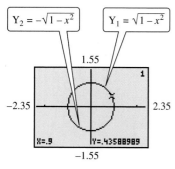

Figure 16

With your calculator in function mode, graph these two equations in the same window to get the graph of the circle. See Figure 16. For an undistorted figure, a *square window* must be used. (See your instruction manual for details.)

The TRACE feature of the calculator allows us to find coordinates of points on the graph. Experiment with this feature, and notice that the x- and y-coordinates are displayed below the graph. The x-coordinate represents the cosine of the length of the arc from the point $(1, 0)$ to the point indicated by the cursor. For example, one such point is

$$x = .9, \qquad y = .43588989.$$

This point is shown in the figure.

While the calculator does not give the arc length, it can be found by setting the calculator in radian mode and finding either $\cos^{-1} .9$ or $\sin^{-1} .43588989$. By doing this, we find the arc length is approximately $.4510268$. ■

Figure 17 shows the unit circle $x^2 + y^2 = 1$ with a great deal of important information. Degree and radian measures are given for the first counterclockwise revolution, and the coordinates of the points on the circle are also given. Remember that the first coordinate is the cosine of the angle (or number) and the second coordinate is the sine. This figure should be helpful as you continue your study of trigonometry.

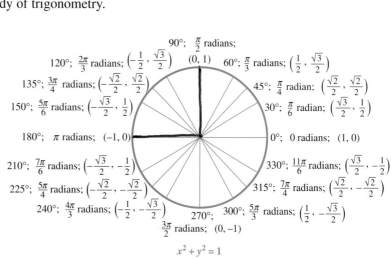

Figure 17

C O N N E C T I O N S   The origin of the term *sinus*— our *sine*—is Indian. The Hindu mathematician and astronomer Aryabhata the Elder (476–ca. 550) called it *ardha-jya* (half-chord), later abbreviated to *jya.* Figure 18 shows why. In the figure,

$$\sin \theta = \frac{PA}{1} = PA,$$

which is half of the chord *PB*. As this term was translated to Arabic and then to Latin, mistranslations changed *jya* to *sinus*.

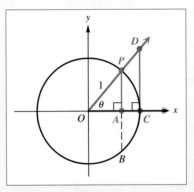

Figure 18

The other trigonometric function names have less complicated histories. *Tangent* comes from the Latin *tangere,* which means to touch. As Figure 18 shows, the tangent segment *CD* just touches (is tangent to) the circle at *C*. Because *secant θ* is represented in the figure by the segment *OD*, which "cuts off" the tangent segment, its name comes from the Latin *secare,* to cut. These two names were introduced by Thomas Fincke (1561–1656), a Dane. The names *cosinus* and *cotangens* were suggested in 1620 by the English mathematician and astronomer Edmund Gunter (1581–1626). They replaced the earlier terms *sinus complementi* and *tangens complementi.* (*Source:* Gullberg, J., *Mathematics from the Birth of Numbers,* W. W. Norton & Company, 1997.)

For Discussion or Writing

Search the Web for a site on the history of trigonometric functions for more information.

3.3 Exercises

Concept Check   *Decide whether each of the following is a circular function or a trigonometric function.*

1. $\sin \dfrac{\pi}{2}$

2. $\cos(-150°)$

3. $\tan 45°$

4. $\cot \dfrac{3\pi}{4}$

5. $\sec\left(-\dfrac{5\pi}{3}\right)$

6. $\csc 240°$

7. $\tan 3$

8. $\cos(-5)$

Concept Check *The figure below displays a unit circle and an angle of* 1 *radian. The tick marks on the circle are spaced at every two-tenths radian. Use the figure to estimate each value.*

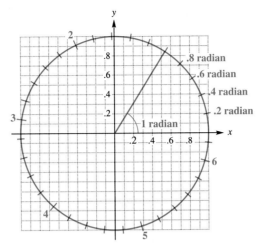

9. cos .8 **10.** cos 4 **11.** sin .2 **12.** sin 4

13. an angle whose cosine is −.65 **14.** an angle whose sine is −.95

15. *Concept Check* Without using a calculator, decide whether each function value is positive or negative. (*Hint:* Consider the radian measures of the quadrantal angles.)

(a) cos 2 (b) tan(−1) (c) sin 5 (d) cos 6

Find the exact circular function value for each of the following. See Example 1.

16. $\sin \dfrac{7\pi}{6}$ **17.** $\cos \dfrac{5\pi}{3}$ **18.** $\tan \dfrac{3\pi}{4}$ **19.** $\sin \dfrac{5\pi}{3}$ **20.** $\tan \dfrac{4\pi}{3}$

21. $\sec \dfrac{2\pi}{3}$ **22.** $\csc \dfrac{11\pi}{6}$ **23.** $\cot \dfrac{5\pi}{6}$ **24.** $\cos\left(-\dfrac{4\pi}{3}\right)$ **25.** $\sin\left(-\dfrac{5\pi}{6}\right)$

26. $\tan \dfrac{17\pi}{3}$ **27.** $\sec \dfrac{23\pi}{6}$ **28.** $\csc \dfrac{13\pi}{3}$ **29.** $\cos \dfrac{13\pi}{4}$

Use a calculator to find an approximation for each circular function value. Be sure that your calculator is set in radian mode. See Example 2.

30. sin .8203 **31.** cos .6429 **32.** tan .9047 **33.** sin 1.5097

34. cot .0465 **35.** csc 1.3875 **36.** sec 7.4526 **37.** cos 6.6701

38. tan 4.0230 **39.** cos 4.2528 **40.** sin 3.4645 **41.** sin(−2.2864)

42. cot(−2.4871) **43.** cos(−3.0602)

Find the value of s in the interval $[0, \pi/2]$ *that makes each statement true. See Example 3.*

44. tan s = .21264138 **45.** cos s = .78269876 **46.** sin s = .99184065 **47.** cot s = .29949853

48. cot s = .62084613 **49.** tan s = 2.6058440 **50.** cos s = .57834328 **51.** sin s = .98771924

52. cot s = .09637041 **53.** csc s = 1.0219553

In Exercises 54–59, find the exact value of s in the given interval that has the given circular function value. Do not use a calculator. See Example 4.

54. $\left[\dfrac{\pi}{2}, \pi\right]$; $\sin s = \dfrac{1}{2}$ **55.** $\left[\dfrac{\pi}{2}, \pi\right]$; $\cos s = -\dfrac{1}{2}$ **56.** $\left[\pi, \dfrac{3\pi}{2}\right]$; $\tan s = \sqrt{3}$

57. $\left[\pi, \dfrac{3\pi}{2}\right]$; $\sin s = -\dfrac{1}{2}$ **58.** $\left[\dfrac{3\pi}{2}, 2\pi\right]$; $\tan s = -1$ **59.** $\left[\dfrac{3\pi}{2}, 2\pi\right]$; $\cos s = \dfrac{\sqrt{3}}{2}$

60. What makes radian measure so important in calculus is the fact that $\sin x/x$ gets closer and closer to 1 as x gets closer and closer to 0. Verify this fact using $x = .1, .01, .001$. Then show that this does not happen when degree measure is used.

Concept Check In Exercises 61 and 62, each graphing calculator screen shows a point on the unit circle. What is the length of the shortest arc of the circle from $(1, 0)$ to the point?

61.

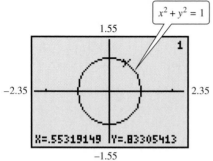

62.

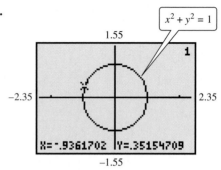

Suppose an arc of length s lies on the unit circle $x^2 + y^2 = 1$, *starting at the point* $(1, 0)$ *and terminating at the point* (x, y). *(See Figure 12.) Use a calculator to find the approximate coordinates for* (x, y). *(Hint:* $x = \cos s$ *and* $y = \sin s$.)

63. $s = 2.5$ **64.** $s = 3.4$ **65.** $s = -7.4$ **66.** $s = -3.9$

Concept Check For each value of s, use a calculator to find $\sin s$ and $\cos s$ and then use the results to decide in which quadrant an angle of s radians lies.

67. $s = 51$ **68.** $s = 49$ **69.** $s = 65$ **70.** $s = 79$

71. *(Modeling) Maximum Temperatures* Because the values of the circular functions repeat every 2π, they are used to describe things that repeat periodically. For example, the maximum afternoon temperature in a given city might be modeled by

$$t = 60 - 30 \cos \frac{x\pi}{6},$$

where t represents the maximum afternoon temperature in month x, with $x = 0$ representing January, $x = 1$ representing February, and so on. Find the maximum afternoon temperature for each of the following months.

(a) January   (b) April   (c) May
(d) June   (e) August   (f) October

72. *(Modeling) Temperature in Fairbanks* The temperature in Fairbanks is modeled by

$$T(x) = 37 \sin \left[\frac{2\pi}{365}(x - 101) \right] + 25,$$

where $T(x)$ is the temperature in degrees Fahrenheit on day x, with $x = 1$ corresponding to January 1 and $x = 365$ corresponding to December 31. Use a calculator to estimate the temperature on the following days. *(Source:* Lando, B. and C. Lando, "Is the Graph of Temperature Variation a Sine Curve?" *The Mathematics Teacher,* 70, September 1977.)

(a) March 1 (day 60)   (b) April 1 (day 91)
(c) Day 150   (d) June 15
(e) September 1   (f) October 31

73. Solve the equation $\sin x = \sin(x + 2)$ for $0 \le x \le 2\pi$. *(Hint:* Use the unit circle definition of sine discussed in this section.)

 Solve each problem related to solar power.

74. *(Modeling) Elevation of the Sun* Refer to Example 5.
(a) Repeat the example for New Orleans, which has latitude $L = 30°$.
(b) Compare your answers. Do they agree with your intuition?

75. *(Modeling) Length of a Day* The ability to calculate the number of daylight hours H at any location is important in estimating potential solar energy production. H can be calculated using the formula

$$\cos(.1309H) = -\tan D \tan L,$$

where D and L are defined in Example 5. Use this trigonometric equation to calculate the shortest and longest days in Minneapolis, Minnesota, if its latitude $L = 44.88°$, the shortest day occurs when $D = -23.44°$, and the longest day occurs when $D = 23.44°$. Remember to convert degrees to radians. *(Source:* Winter, C., R. Sizmann, and Vant-Hunt (Editors), *Solar Power Plants,* Springer-Verlag, 1991.)

76. *(Modeling) Length of a Day* Refer to Exercise 75. Calculate the number of daylight hours at Hartford, Connecticut, on August 12, 2001. *(Hint:* $D \approx .26724$ and latitude $L = 41.93°$.)

3.4 Linear and Angular Velocity

● **Linear Velocity** ● **Angular Velocity** ● **Applications of Linear and Angular Velocity**

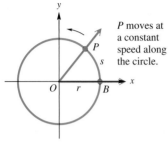

Figure 19

Linear Velocity   In many situations we need to know how fast a point on a circular disk is moving or how fast the central angle of such a disk is changing. Some examples occur with machinery involving gears or pulleys or the speed of a car around a curved portion of highway. Suppose that point P moves at a constant speed along a circle of radius r and center O. See Figure 19. The measure of how fast the position of P is changing is called **linear velocity.** If v represents linear velocity, then

$$\text{velocity} = \frac{\text{distance}}{\text{time}}$$

$$v = \frac{s}{t},$$

where s is the length of the arc traced by point P at time t. (This formula is just a restatement of the familiar result $d = rt$ with s as distance, v as rate, and t as time.)

Angular Velocity   As point P in Figure 19 moves along the circle, ray OP rotates around the origin. Since ray OP is the terminal side of angle POB, the measure of the angle changes as P moves along the circle. The measure of how fast angle POB is changing is called **angular velocity.** Angular velocity, written ω, is given as

$$\omega = \frac{\theta}{t}, \qquad \theta \text{ in radians,}$$

where θ is the measure of angle POB at time t. As with earlier formulas in this chapter, θ must be measured in radians, with ω expressed as radians per unit of time. Angular velocity is used in physics and engineering, among other applications.

In Section 3.2, the length s of the arc intercepted on a circle of radius r by a central angle of measure θ radians was found to be $s = r\theta$. Using this formula, the formula for linear velocity, $v = s/t$, becomes

$$v = \frac{r\theta}{t}$$

$$= r \cdot \frac{\theta}{t}$$

$$= r\omega. \qquad \omega = \frac{\theta}{t}$$

The formula $v = r\omega$ relates linear and angular velocities.

A radian is a "pure number," with no units associated with it. This is why the product of length r, measured in units such as centimeters, and ω, measured in units such as radians per second, is velocity, v, measured in units such as centimeters per second.

The formulas given in this section are summarized below.

Angular and Linear Velocity

Angular Velocity	Linear Velocity
$\omega = \dfrac{\theta}{t}$	$v = \dfrac{s}{t}$
(ω in radians per unit time, θ in radians)	$v = \dfrac{r\theta}{t}$
	$v = r\omega$

Applications of Linear and Angular Velocity

• • • **Example 1** Using Linear and Angular Velocity Formulas

Suppose that point P is on a circle with radius 10 cm, and ray OP is rotating with angular velocity $\pi/18$ radian per second.

(a) Find the angle generated by P in 6 seconds.

The velocity of ray OP is $\omega = \pi/18$ radian per second. Since $\omega = \theta/t$, then in 6 seconds

$$\frac{\pi}{18} = \frac{\theta}{6}$$

$$\theta = \frac{6\pi}{18}$$

$$\theta = \frac{\pi}{3} \text{ radians.}$$

(b) Find the distance traveled by P along the circle in 6 seconds.

In 6 seconds P generates an angle of $\pi/3$ radians. Since $s = r\theta$,

$$s = 10\left(\frac{\pi}{3}\right) = \frac{10\pi}{3} \text{ cm.}$$

(c) Find the linear velocity of P.

Since $v = s/t$, in 6 seconds

$$v = \frac{\dfrac{10\pi}{3}}{6} = \frac{5\pi}{9} \text{ cm per second.}$$

• • •

P R O B L E M   S O L V I N G In practical applications, angular velocity is often given as revolutions per unit of time, which must be converted to radians per unit of time before using the formulas given in this section.

● ● ● **Example 2** Finding Angular Velocity of a Pulley and Linear Velocity of a Belt

A belt runs a pulley of radius 6 cm at 80 revolutions per minute.

(a) Find the angular velocity of the pulley in radians per second.

In 1 minute, the pulley makes 80 revolutions. Each revolution is 2π radians, for a total of

$$80(2\pi) = 160\pi \text{ radians per minute.}$$

Since there are 60 seconds in one minute, ω, the angular velocity in radians per second, is found by dividing 160π by 60.

$$\omega = \frac{160\pi}{60} = \frac{8\pi}{3} \text{ radians per second}$$

(handwritten: $506(2\pi) = 1012\pi$)

(handwritten: 1012π per minute)

(b) Find the linear velocity of the belt in centimeters per second.

The linear velocity of the belt will be the same as that of a point on the circumference of the pulley. Thus,

$$v = r\omega$$

(handwritten: 8π per second)

$$v = 6\left(\frac{8\pi}{3}\right)$$

$$v = 16\pi \text{ cm per second}$$

$$v \approx 50.3 \text{ cm per second.}$$ ● ● ●

● ● ● **Example 3** Finding Linear Velocity and Distance Traveled by a Satellite

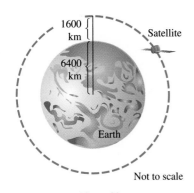

Figure 20

A satellite traveling in a circular orbit 1600 km above the surface of Earth takes two hours to make an orbit. Assume that the radius of Earth is 6400 km. See Figure 20.

(a) Find the linear velocity of the satellite.

The distance of the satellite from the center of Earth is

$$r = 1600 + 6400 = 8000 \text{ km.}$$

For one orbit $\theta = 2\pi$, and

$$s = r\theta = 8000(2\pi) \text{ km.}$$

Since it takes 2 hours to complete an orbit, the linear velocity is

$$v = \frac{s}{t} = \frac{8000(2\pi)}{2} = 8000\pi \approx 25,000 \text{ km per hour.}$$

(b) Find the distance traveled in 4.5 hours.

$$s = vt = 8000\pi(4.5) = 36,000\pi \approx 110,000 \text{ km}$$ ● ● ●

3.4   Exercises

1. *Concept Check*   If a point moves around the circumference of the unit circle at an angular velocity of 1 radian per second, how long will it take for the point to move around the entire circle?

Use the formula $\omega = \theta/t$ to find the value of the missing variable.

2. $\omega = \dfrac{2\pi}{3}$ radians per sec, $t = 3$ sec

3. $\omega = \dfrac{\pi}{4}$ radian per min, $t = 5$ min

4. $\theta = \dfrac{3\pi}{4}$ radians, $t = 8$ sec

5. $\theta = \dfrac{2\pi}{5}$ radians, $t = 10$ sec

6. $\theta = \dfrac{2\pi}{9}$ radian, $\omega = \dfrac{5\pi}{27}$ radian per min

7. $\theta = \dfrac{3\pi}{8}$ radians, $\omega = \dfrac{\pi}{24}$ radian per min

8. $\theta = 3.871142$ radians, $t = 21.4693$ sec

9. $\omega = .90674$ radian per min, $t = 11.876$ min

10. *Concept Check*   If a point moves around the circumference of the unit circle at the speed of 1 unit per second, how long will it take for the point to move around the entire circle?

11. What is the difference between linear velocity and angular velocity?

12. Explain why linear velocity is affected by the radius of the circle, whereas angular velocity is not.

Use the formula $v = r\omega$ to find the value of the missing variable.

13. $r = 12$ m, $\omega = \dfrac{2\pi}{3}$ radians per sec

14. $r = 8$ cm, $\omega = \dfrac{9\pi}{5}$ radians per sec

15. $v = 9$ m per sec, $r = 5$ m

16. $v = 18$ ft per sec, $r = 3$ ft

17. $v = 107.692$ m per sec, $r = 58.7413$ m

18. $r = 24.93215$ cm, $\omega = .372914$ radian per sec

The formula $\omega = \theta/t$ can be rewritten as $\theta = \omega t$. Using ωt for θ changes $s = r\theta$ to $s = r\omega t$. Use the formula $s = r\omega t$ to find the value of the missing variable.

19. $r = 6$ cm, $\omega = \dfrac{\pi}{3}$ radians per sec, $t = 9$ sec

20. $r = 9$ yd, $\omega = \dfrac{2\pi}{5}$ radians per sec, $t = 12$ sec

21. $s = 6\pi$ cm, $r = 2$ cm, $\omega = \dfrac{\pi}{4}$ radian per sec

22. $s = \dfrac{12\pi}{5}$ m, $r = \dfrac{3}{2}$ m, $\omega = \dfrac{2\pi}{5}$ radians per sec

23. $s = \dfrac{3\pi}{4}$ km, $r = 2$ km, $t = 4$ sec

24. $s = \dfrac{8\pi}{9}$ m, $r = \dfrac{4}{3}$ m, $t = 12$ sec

25. Explain the similarities between the familiar $d = rt$ formula and the formula $s = vt$.

26. Suppose you must convert k radians per second to degrees per minute. Explain how you would do this.

Find ω for each of the following.

27. the hour hand of a clock

28. the minute hand of a clock

29. the second hand of a clock

30. a line from the center to the edge of a CD revolving 300 times per minute

Find v for each of the following.

31. the tip of the minute hand of a clock, if the hand is 7 cm long

32. the tip of the second hand of a clock, if the hand is 28 mm long

33. a point on the edge of a flywheel of radius 2 m, rotating 42 times per minute

34. a point on the tread of a tire of radius 18 cm, rotating 35 times per minute

35. the tip of an airplane propeller 3 m long, rotating 500 times per minute (*Hint: r = 1.5 m*)

36. a point on the edge of a gyroscope of radius 83 cm, rotating 680 times per minute

Solve the following problems. See Examples 1–3.

37. *Speed of a Bicycle* The tires of a bicycle have radius 13 inches and are turning at the rate of 200 revolutions per minute. How fast is the bicycle traveling in miles per hour? (*Hint:* 5280 feet = 1 mile.)

13 in.

38. *Hours in a Martian Day* Mars rotates on its axis at the rate of about .2552 radian per hour. Approximately how many hours are in a Martian day? (*Source:* Wright, John W. (General Editor), *The Universal Almanac,* Andrews and McMeel, 1997.)

39. *Angular and Linear Velocities of Earth* Earth travels about the sun in an orbit that is almost circular. Assume that the orbit is a circle, with radius 93,000,000 miles. (See the figure.) Its angular and linear velocities are used in designing solar power facilities.

93,000,000 mi   Earth

θ

Sun

Not to scale

(a) Assume that a year is 365 days, and find θ, the angle formed by Earth's movement in one day.
(b) Give the angular velocity in radians per hour.
(c) Find the linear velocity of Earth in miles per hour.

40. *Angular Velocity of a Pulley* The pulley shown has a radius of 12.96 cm. Suppose it takes 18 seconds for 56 cm of belt to go around the pulley. Find the angular velocity of the pulley in radians per second.

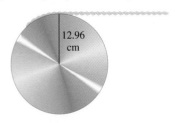

12.96 cm

41. *Angular Velocity of Pulleys* The two pulleys in the figure have radii of 15 cm and 8 cm, respectively. The larger pulley rotates 25 times in 36 seconds. Find the angular velocity of each pulley in radians per second.

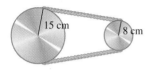

15 cm   8 cm

42. *Radius of a Gear* A gear is driven by a chain that travels 1.46 meters per second. Find the radius of the gear if it makes 46 revolutions per minute.

43. *Radius of a Spool of Thread* A thread is being pulled off a spool at the rate of 59.4 cm per second. Find the radius of the spool if it makes 152 revolutions per minute.

44. 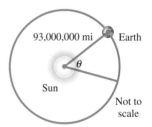 *Angular and Linear Velocities of Earth* Earth revolves on its axis once every 24 hours. Assuming that Earth's radius is 6400 km, find the following.
(a) angular velocity of Earth in radians per day and radians per hour
(b) linear velocity at the North Pole or South Pole
(c) linear velocity at Quito, Ecuador, a city on the equator
(d) linear velocity at Salem, Oregon (halfway from the equator to the North Pole)

45. *Time to Move Along a Railroad Track* A railroad track is laid along the arc of a circle of radius 1800 feet. The circular part of the track subtends a central angle of 40°. How long (in seconds) will it take a point on the front of a train traveling 30 mph to go around this portion of the track?

46. *Angular Velocity of a Motor Propeller* A 90-horsepower outboard motor at full throttle will rotate its propeller at 5000 revolutions per minute. Find the angular velocity of the propeller in radians per second.

Chapter 3 Summary

Key Terms & Symbols	Key Ideas
3.1 Radian Measure radian	An angle that has its vertex at the center of a circle and that intercepts an arc on the circle equal in length to the radius of the circle has a measure of **1 radian.** **Degree/Radian Relationship** $180° = \pi$ radians **Converting Between Degrees and Radians** 1. Multiply a radian measure by $180°/\pi$ and simplify to convert to degrees. 2. Multiply a degree measure by $\pi/180$ radian and simplify to convert to radians.
3.2 Applications of Radian Measure sector of a circle	**Arc Length** The length s of the arc intercepted on a circle of radius r by a central angle of measure θ radians is given by the product of the radius and the radian measure of the angle, or $$s = r\theta, \qquad \theta \text{ in radians.}$$ **Area of Sector** The area of a sector of a circle of radius r and central angle θ is given by $$A = \frac{1}{2}r^2\theta, \qquad \theta \text{ in radians.}$$
3.3 Circular Functions of Real Numbers unit circle	**Circular Functions** Start at the point $(1,0)$ on the unit circle $x^2 + y^2 = 1$ and lay off an arc of length $|s|$ along the circle, going counterclockwise if s is positive, and clockwise if s is negative. Let the endpoint of the arc be at the point (x, y). The six circular functions of s are defined as follows. (Assume that no denominators are 0.) $$\sin s = y \qquad \cos s = x \qquad \tan s = \frac{y}{x}$$ $$\csc s = \frac{1}{y} \qquad \sec s = \frac{1}{x} \qquad \cot s = \frac{x}{y}$$ **Domains of the Circular Functions** The domains of the circular functions are as follows. Assume that n is any integer. Sine and Cosine Functions: $(-\infty, \infty)$ Tangent and Secant Functions: $\left\{ s \mid s \neq \dfrac{\pi}{2} + n\pi \right\}$ Cotangent and Cosecant Functions: $\{ s \mid s \neq n\pi \}$

Key Terms & Symbols	Key Ideas

3.4 Linear and Angular Velocity
linear velocity v
angular velocity ω

Angular Velocity	Linear Velocity
$\omega = \dfrac{\theta}{t}$	$v = \dfrac{s}{t}$
(ω in radians per unit time, θ in radians)	$v = \dfrac{r\theta}{t}$
	$v = r\omega$

Chapter 3   Review Exercises

1. Which is larger—an angle of $1°$ or an angle of 1 radian? Discuss and justify your answer.

2. Consider each angle in standard position having the given radian measure. In what quadrant does the terminal side lie?
   (a) 3   **(b)** 4   **(c)** -2 **(d)** 7

3. Find three angles coterminal to an angle of 1 radian.

4. Give an expression that generates all angles coterminal with an angle of $\pi/6$ radian. Let n represent any integer.

Convert each degree measure to radians. Leave answers as multiples of π.

5. $45°$ **6.** $120°$ **7.** $175°$

8. $330°$ **9.** $800°$ **10.** $1020°$

Convert each radian measure to degrees.

11. $\dfrac{5\pi}{4}$ **12.** $\dfrac{9\pi}{10}$ **13.** $\dfrac{8\pi}{3}$ **14.** $-\dfrac{6\pi}{5}$ **15.** $-\dfrac{11\pi}{18}$ **16.** $\dfrac{21\pi}{5}$

Suppose the tip of the minute hand of a clock is two inches from the center of the clock. For each of the following durations, determine the distance traveled by the tip of the minute hand.

17. 15 minutes   **18.** 20 minutes

19. 3 hours   **20.** $10\frac{1}{2}$ hours

Solve each problem. Use a calculator as necessary.

21. *Arc Length*   The radius of a circle is 15.2 cm. Find the length of an arc of the circle intercepted by a central angle of $3\pi/4$ radians.

22. *Arc Length*   Find the length of an arc intercepted by a central angle of .769 radian on a circle with radius 11.4 cm.

23. *Arc Length*   A circle has radius 8.973 cm. Find the length of an arc on this circle intercepted by a central angle of $49.06°$.

24. *Area of a Sector*   A central angle of $7\pi/4$ radians forms a sector of a circle. Find the area of the sector if the radius of the circle is 28.69 inches.

25. *Area of a Sector*   Find the area of a sector of a circle having a central angle of $21°\,40'$ in a circle of radius 38.0 meters.

26. *Height of a Tree*   A tree 2000 yards away subtends an angle of $1°\,10'$. Find the height of the tree to two significant digits.

Distance Between Cities   Assume that the radius of Earth is 6400 km in the next two exercises.

27. Find the distance in kilometers between cities on a north-south line that are on latitudes 28° N and 12° S, respectively.

28. Two cities on the equator have longitudes of 72° E and 35° W, respectively. Find the distance between the cities.

Concept Check   In Exercises 29 and 30, find the measure of the central angle θ (in radians) and the area of the sector.

29.

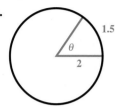

1.5
θ
2

30.

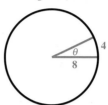

θ   4
8

31. *Concept Check*   The hour hand of a wall clock measures six inches from its tip to the center of the clock.
   (a) Through what angle (in radians) does the hour hand pass between 1 o'clock and 3 o'clock?
   (b) What distance does the tip of the hour hand travel during the time period from 1 o'clock to 3 o'clock?

32. Describe what would happen to the central angle for a given arc length of a circle if the circle's radius were doubled. (Assume everything else is unchanged.)

Convert each radian measure to degrees in order to find the exact function value. Do not use a calculator.

33. $\tan \dfrac{\pi}{3}$

34. $\cos \dfrac{2\pi}{3}$

35. $\sin\left(-\dfrac{5\pi}{6}\right)$

36. $\tan\left(-\dfrac{7\pi}{3}\right)$

37. $\csc\left(-\dfrac{11\pi}{6}\right)$

38. $\cot\left(-\dfrac{17\pi}{3}\right)$

Without using a calculator, determine which of the following is greater.

39. tan 1 or tan 2

40. sin 1 or tan 1

41. cos 2 or sin 2

42. cos(sin(0)) or sin(cos(0))

Use a calculator to find an approximation for each circular function value. Be sure your calculator is set in radian mode.

43. sin 1.0472

44. tan 1.2275

45. cos(−.2443)

46. cot 3.0543

47. tan 7.3159

48. sin 4.8386

Find the value of s in the interval $[0, \pi/2]$ that makes each statement true.

49. cos s = .92500448

50. tan s = 4.0112357

51. sin s = .49244294

52. csc s = 1.2361343

53. cot s = .50221761

54. sec s = 4.5600039

Find the exact value of s in the given interval that has the given circular function value. Do not use a calculator.

55. $\left[0, \dfrac{\pi}{2}\right]$; $\cos s = \dfrac{\sqrt{2}}{2}$

56. $\left[\dfrac{\pi}{2}, \pi\right]$; $\tan s = -\sqrt{3}$

57. $\left[\pi, \dfrac{3\pi}{2}\right]$; $\sec s = -\dfrac{2\sqrt{3}}{3}$

58. $\left[\dfrac{3\pi}{2}, 2\pi\right]$; $\sin s = -\dfrac{1}{2}$

59. Without using a calculator, determine which of the following numbers is closest to cos 2: .4, .6, 0, −.4, or −.6.

60. Without using a calculator, determine which of the following numbers is closest to sin 2: .9, .1, 0, −.9, or −.1.

61. Find the measure (in both degrees and radians) of the angle θ formed in the graphing calculator screen by the line passing through the origin and the positive part of the x-axis. Use the displayed values of x and y at the bottom of the screen.

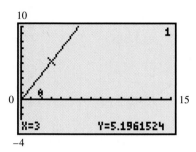

62. *Concept Check* The graphing calculator screen shows a point on the unit circle. What is the length of the shortest arc of the circle from $(1, 0)$ to the point?

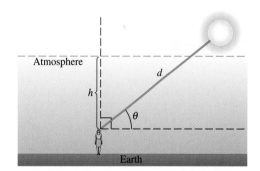

Solve each problem.

63. Find t if $\theta = \dfrac{5\pi}{12}$ radians and $\omega = \dfrac{8\pi}{9}$ radians per sec.

64. Find θ if $t = 12$ sec and $\omega = 9$ radians per sec.

65. Find ω if $t = 8$ sec and $\theta = \dfrac{2\pi}{5}$ radians.

66. Find ω if $s = \dfrac{12\pi}{25}$ ft, $r = \dfrac{3}{5}$ ft, and $t = 15$ sec.

67. Find s if $r = 11.46$ cm, $\omega = 4.283$ radians per sec, and $t = 5.813$ sec.

68. *Linear Velocity of a Flywheel* Find the linear velocity of a point on the edge of a flywheel of radius 7 meters if the flywheel is rotating 90 times per second.

69. *Atmospheric Effect on Sunlight* The shortest path for the sun's rays through Earth's atmosphere occurs when the sun is directly overhead. Disregarding the curvature of Earth, as the sun moves lower on the horizon, the distance that sunlight passes through the atmosphere increases by a factor of $\csc \theta$, where θ is the angle of elevation of the sun. This increased distance reduces both the intensity of the sun and the amount of ultraviolet light that reaches Earth's surface. See the figure. (*Source:* Winter, C., R. Sizmann, and Vant-Hunt (Editors), *Solar Power Plants,* Springer-Verlag, 1991.)

(a) Verify that $d = h \csc \theta$.

(b) Determine θ when $d = 2h$.

(c) The atmosphere filters out the ultraviolet light that causes skin to burn. Compare the difference between sunbathing when $\theta = \pi/2$ and $\pi/3$. Which measure gives less ultraviolet light?

Chapter 3 Test

1. Define what is meant by a measure of 1 radian.

2. Convert $120°$ to radians.

3. Convert $\dfrac{9\pi}{10}$ to degrees.

A central angle of a circle with radius 150 cm cuts off an arc of 200 cm. Find each measure in Exercises 4 and 5.

4. the radian measure of the angle

5. the area of a sector with that central angle

6. Give the sine, cosine, and tangent of s.

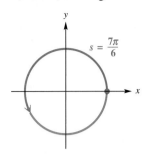

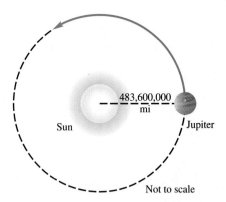

Not to scale

7. What are the domains of the tangent and secant circular functions?

8. Use a calculator to approximate s in the interval $[0, \pi/2]$, if $\sin s = .82584121$.

9. *Orbital Velocity of Jupiter* It takes Jupiter 11.64 years to complete one orbit around the sun. See the figure. If Jupiter's average distance from the sun is 483,600,000 miles, find its orbital velocity (velocity along its orbital path) in miles per second. (*Source:* Wright, John W. (General Editor), *The Universal Almanac*, Andrews and McMeel, 1997.)

10. *Person in a Ferris Wheel* A Ferris wheel has radius 25 feet. A person takes a seat, and then the wheel turns $5\pi/6$ radians. How far is the person above the ground?

11. *Angular Velocity of a Ferris Wheel* In Exercise 10, if it takes 30 seconds for the wheel to turn $5\pi/6$ radians, find the angular velocity of the wheel.

12. *Engine Specifications* The figure shows a rotating wheel (crankshaft) with radius r and a connecting rod AP with length L. The piston slides back and forth along the x-axis as the wheel rotates counterclockwise at a rate of R revolutions per minute (RPM). The 1937 John Deere B engine, used in all models from 1935–1938, had the following specifications.

Maximum RPM: 1340 (no load) $\approx$ 505,168.1 radians per hour

Connecting rod: 10.500000 inches

Stroke: 5.250000 inches

(*Source:* Drost, J. P. and R. H. Kunferman, "Related Rates Challenge Problem: Calculate the Velocity of a Piston," *The AMATYC Review*, Vol. 21 No. 1, Fall 1999.)

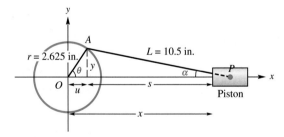

Find an expression for each measure.

(a) y **(b)** u **(c)** s **(d)** x

(e) The velocity v of the piston at the maximum RPM is given by

$$v = -2.625 \sin \theta \left(1 + \frac{\cos \theta}{\sqrt{15 + \cos^2 \theta}} \right) (505,168.1) \left(\frac{1}{12 \cdot 5280} \right),$$

where the last quantity changes inches to miles, so the velocity will be in miles per hour. Use the maximum-finding feature of a graphing calculator, with the window $[0, 6]$ by $[-25, 25]$, to find the value of θ that maximizes velocity. Give this maximum velocity.

Graphs of the Circular Functions

4.1 Graphs of the Sine and Cosine Functions

4.2 Translations of the Graphs of the Sine and Cosine Functions

4.3 Graphs of the Other Circular Functions

Many cities throughout the world experience seasons. Although each city has its own unique weather patterns, there are also similarities between cities as to when seasons occur and how corresponding temperatures vary. The table lists the average monthly temperatures (in °F) in Vancouver, Canada.

Temperatures in Vancouver

Month	°F	Month	°F
Jan	36	July	64
Feb	39	Aug	63
Mar	43	Sept	57
Apr	48	Oct	50
May	55	Nov	43
June	59	Dec	39

Source: Miller, A. and J. Thompson, *Elements of Meteorology,* Charles E. Merrill Publishing Co., 1975.

The average temperatures in Vancouver, coldest in January and warmest in July, cycle yearly and may change only slightly over many years. Seasonal temperature changes occur periodically because Earth's axis is tilted, and its orbit around the sun is nearly circular. When a phenomenon such as temperature results from circular periodic motion, the circular functions are often used to mathematically model the data. The graphs of these functions are important when describing things like world temperatures and seasonal carbon dioxide levels. Their graphs will provide us with both a picture and a better understanding of periodic phenomena, particularly weather, which is the theme of this chapter.

· ·

4.1 Graphs of the Sine and Cosine Functions

• Periodic Functions • Graph of the Sine Function • Graph of the Cosine Function • Graphing Techniques, Amplitude, and Period • Using a Trigonometric Model

Periodic Functions Many things in daily life repeat with a predictable pattern: in warm areas electricity use goes up in summer and down in winter, the price of fresh fruit goes down in summer and up in winter, and attendance at amusement parks increases in spring and declines in autumn. Because the sine and cosine functions repeat their values over and over in a regular pattern, they are examples of *periodic functions*. Figure 1 shows a *sinusoid* (sine graph) that represents a normal heartbeat.

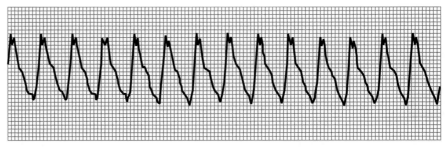

Figure 1

Looking Ahead to Calculus

The periodic functions presented in this chapter are used frequently throughout calculus. To be successful in calculus you will need to know their characteristics. One use of these functions is to describe the location of a point in the plane using *polar coordinates,* an alternative to rectangular coordinates. (See Chapter 8.)

Periodic Function
A **periodic function** is a function f such that
$$f(x) = f(x + np),$$
for every real number x in the domain of f, every integer n, and some positive real number p. The smallest possible positive value of p is the **period** of the function.

The circumference of the unit circle is 2π, so the smallest value of p for which the sine and cosine functions repeat is 2π. Therefore, the sine and cosine functions are periodic functions with period 2π.

Graph of the Sine Function In Section 3.3, we saw that if an arc of length s is traced along the unit circle $x^2 + y^2 = 1$ starting at point $(1, 0)$, the terminal point of the arc has coordinates $(\cos s, \sin s)$. Look at Figure 2, and trace along the circle to verify the results shown in the chart.

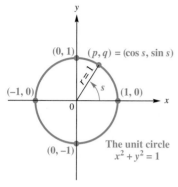

Figure 2

As s Increases from	$\sin s$	$\cos s$
0 to $\pi/2$	Increases from 0 to 1	Decreases from 1 to 0
$\pi/2$ to π	Decreases from 1 to 0	Decreases from 0 to -1
π to $3\pi/2$	Decreases from 0 to -1	Increases from -1 to 0
$3\pi/2$ to 2π	Increases from -1 to 0	Increases from 0 to 1

Any letter can be used instead of s for the arc length, so to avoid confusion when graphing the sine function, we will use x rather than s; this will correspond to our usual choice of letters in the xy coordinate system. Selecting key values of x and finding the corresponding values of $\sin x$ leads to the following table. Note that $\sin x$ values are rounded to the nearest tenth in the horizontal table, and the increment in the graphing calculator table is $\pi/4$.

x	0	$\pi/4$	$\pi/2$	$3\pi/4$	π	$5\pi/4$	$3\pi/2$	$7\pi/4$	2π
$\sin x$	0	.7	1	.7	0	$-.7$	-1	$-.7$	0

X	Y1
0	0
.7854	.70711
1.5708	1
2.3562	.70711
3.1416	0
3.927	-.7071
4.7124	-1

Y1=sin(X)

The calculator must be in radian mode.

To obtain a traditional graph of a portion of the sine function, we plot the points from the table of values and join them with a smooth curve. This results in the graph shown in Figure 3. Since $y = \sin x$ is periodic and has $(-\infty, \infty)$ as its domain, the graph continues in the same pattern in both directions. Figure 3 shows the graph over the interval $[0, 2\pi]$. This graph is called a **sine wave** or **sinusoid.** You should learn this shape and be able to sketch it quickly. Figure 4 shows a graphing calculator version of the graph over the interval $[-2\pi, 2\pi]$.

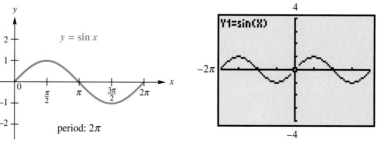

Figure 3

Figure 4

The calculator graph of the sine function in Figure 4 is shown in the *trig window*. In this text, we will refer to the trig window when the *x*-values are in the interval $[-2\pi, 2\pi]$, the *y*-values are in $[-4, 4]$, the scale on the *x*-axis is $\pi/2$, and the scale on the *y*-axis is 1. ∎

The sine function is an example of an **odd function.** For all *x*, $\sin(-x) = -\sin x$.

Characteristics of the Sine Function

Domain: $(-\infty, \infty)$ Range: $[-1, 1]$

Over the interval $[0, 2\pi]$, the sine function exhibits the following behavior.

From 0 to $\pi/2$, sin *x* increases from 0 to 1.
From $\pi/2$ to π, sin *x* decreases from 1 to 0.
From π to $3\pi/2$, sin *x* decreases from 0 to -1.
From $3\pi/2$ to 2π, sin *x* increases from -1 to 0.

The graph is continuous over its entire domain and symmetric with respect to the origin.

x-intercepts: $n\pi$, where *n* is an integer   Period: 2π

Sine graphs occur in many different practical applications. For one application, look back at Figure 2 and assume that the line from the origin to the point (p, q) is part of the pedal of a bicycle, with a foot placed at (p, q). As mentioned earlier, *q* is equal to sin *x*, showing that the height of the pedal from the horizontal axis in Figure 2 is given by sin *x*. By choosing various angles for the pedal and calculating *q* for each angle, the height of the pedal leads to the sine curve shown in Figure 5. Two sample points are also shown.

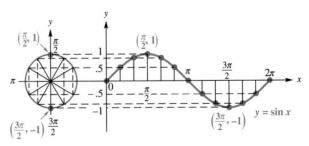

Figure 5

Graph of the Cosine Function
The graph of $y = \cos x$ can be found in much the same way as the graph of $y = \sin x$. Tables of values are shown for $y = \cos x$, using the same values for *x* as before.

x	0	$\pi/4$	$\pi/2$	$3\pi/4$	π	$5\pi/4$	$3\pi/2$	$7\pi/4$	2π
cos *x*	1	.7	0	$-.7$	-1	$-.7$	0	.7	1

Looking Ahead to Calculus

The discussion of the derivative of a function in calculus shows that for the sine function, the slope of the tangent line at any point x is given by $\cos x$. For example, look at the graph of $y = \sin x$ and notice that a tangent line at $x = \pm\dfrac{\pi}{2}, \pm\dfrac{3\pi}{2}, \pm\dfrac{5\pi}{2}, \ldots$ will be horizontal and thus have slope 0. Now look at the graph of $y = \cos x$ and see that for these values, $\cos x = 0$.

Figure 6 shows a traditional graph of $y = \cos x$. Notice that it has the same shape as the graph of $y = \sin x$. It is, in fact, the graph of the sine function shifted, or translated, $\pi/2$ units to the left. Figure 6 shows the graph over the interval $[0, 2\pi]$. Figure 7, generated by a graphing calculator, shows the graph in the trig window.

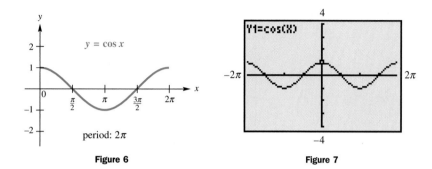

Figure 6   Figure 7

The cosine function is an example of an **even function.** For all x, $\cos(-x) = \cos x$.

Characteristics of the Cosine Function

Domain: $(-\infty, \infty)$ Range: $[-1, 1]$

Over the interval $[0, 2\pi]$, the cosine function exhibits the following behavior.

From 0 to $\pi/2$, $\cos x$ decreases from 1 to 0.
From $\pi/2$ to π, $\cos x$ decreases from 0 to -1.
From π to $3\pi/2$, $\cos x$ increases from -1 to 0.
From $3\pi/2$ to 2π, $\cos x$ increases from 0 to 1.

The graph is continuous over its entire domain and symmetric with respect to the y-axis.

x-intercepts: $\dfrac{\pi}{2} + n\pi$, where n is an integer   Period: 2π

N O T E   Both the sine and cosine functions have domain $(-\infty, \infty)$. In practice, we only show a small portion of the graph, usually over one or two periods. This portion allows us to predict what the rest of the graph looks like.

C O N N E C T I O N S   An even function has the property that $f(-x) = f(x)$ for all x. Because of this property, the graph of an even function is symmetric with respect to the y-axis—that is, if folded along the y-axis, the two halves would match. Also, if (x, y) belongs to the function, then $(-x, y)$ also belongs to the function. The graph of $y = \cos x$ is symmetric with respect to the y-axis, so cosine is an even function and therefore

$$\cos(-x) = \cos x.$$

(continued)

A function is an odd function if $f(-x) = -f(x)$ for all x. The graph of an odd function is symmetric with respect to the origin, which means that if (x, y) belongs to the function, then $(-x, -y)$ also belongs to the function. The sine graph exhibits this property; for example, $(\pi/2, 1)$ and $(-\pi/2, -1)$ are points on the graph of $y = \sin x$. Thus,

$$\sin(-x) = -\sin x.$$

For Discussion or Writing

1. Give an example of a simple polynomial function from algebra that is an even function.
2. Give an example of a simple polynomial function from algebra that is an odd function.
3. Classify each function as even, odd, or neither.
   (a) $f(x) = |x|$ **(b)** $g(x) = \sqrt[3]{x}$ **(c)** $h(x) = x^2 + x$

Graphing Techniques, Amplitude, and Period The examples that follow show graphs that are "stretched" either vertically, horizontally, or both when compared with the graphs of $y = \sin x$ or $y = \cos x$.

● ● ●   **Example 1**   Graphing $y = a \sin x$

Graph $y = 2 \sin x$.

Traditional Approach

For a given value of x, the value of y is twice as large as it would be for $y = \sin x$, as shown in the table of values. The only change in the graph is the range, which becomes $[-2, 2]$. See Figure 8, which also shows a graph of $y = \sin x$ for comparison.

x	0	$\pi/2$	π	$3\pi/2$	2π
$\sin x$	0	1	0	-1	0
$2 \sin x$	0	2	0	-2	0

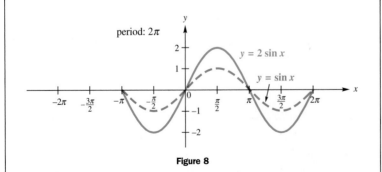

Figure 8

Graphing Calculator Approach

Define Y_1 as $2 \sin x$, and direct the calculator to use a thick line to graph it. In Figure 9, a thin-line graph is shown for $Y_2 = \sin x$, for comparison.

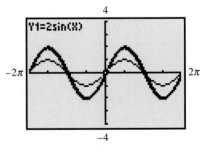

Figure 9

N O T E When graphing periodic functions using a traditional approach, it is customary to first graph over one period. Then more of the graph can be sketched by repeating the cycle over and over.

Generalizing from Example 1 gives the following.

Amplitude of the Sine and Cosine Functions

The graph of $y = a \sin x$ or $y = a \cos x$, with $a \neq 0$, will have the same shape as the graph of $y = \sin x$ or $y = \cos x$, respectively, except with range $[-|a|, |a|]$. The number $|a|$ is called the **amplitude.** (The amplitude of a periodic function can be interpreted as half the difference between its maximum and minimum values.)

No matter what the value of the amplitude, the periods of $y = a \sin x$ and $y = a \cos x$ are still 2π. Now suppose $y = \sin 2x$. We can complete a table of values for the interval $[0, 2\pi]$.

x	0	$\pi/4$	$\pi/2$	$3\pi/4$	π	$5\pi/4$	$3\pi/2$	$7\pi/4$	2π
$\sin 2x$	0	1	0	-1	0	1	0	-1	0

The period here is π, which equals $2\pi/2$. What about $y = \sin 4x$? Look at the table below.

x	0	$\pi/8$	$\pi/4$	$3\pi/8$	$\pi/2$	$5\pi/8$	$3\pi/4$	$7\pi/8$	π
$\sin 4x$	0	1	0	-1	0	1	0	-1	0

These values suggest that a complete cycle is achieved in $\pi/2$ units, which is reasonable since

$$\sin\left(4 \cdot \frac{\pi}{2}\right) = \sin 2\pi = 0.$$

In general, the graph of a function of the form $y = \sin bx$ or $y = \cos bx$, for $b > 0$, will have a period different from 2π when $b \neq 1$. To see why this is so, remember that the values of $\sin bx$ or $\cos bx$ will take on all possible values as bx ranges from 0 to 2π. Therefore, to find the period of either of these functions, we must solve the compound inequality

$$0 \leq bx \leq 2\pi$$

$$0 \leq x \leq \frac{2\pi}{b}. \quad \text{Divide by the positive number } b.$$

Thus, the period is $2\pi/b$. By dividing the interval $[0, 2\pi/b]$ into four equal parts, we obtain the values for which $\sin bx$ or $\cos bx$ is -1, 0, or 1. These values will give minimum points, x-intercepts, and maximum points on the graph. Once these points are determined, the graph can be sketched by joining the points with a smooth sinusoidal curve. (If a function has $b < 0$, then the identities of the next chapter can be used to write the function as one in which $b > 0$.)

NOTE The ability to divide an interval of one period into four equal parts is crucial in sketching the graphs of functions covered in this chapter. In general, for $y = \sin bx$ or $y = \cos bx$, $\dfrac{1}{4}\left(\dfrac{2\pi}{|b|}\right)$ will be the necessary increment from 0 to the end of the period. For example, consider $y = \sin 2x$. The increment is

$$\frac{1}{4}\left(\frac{2\pi}{2}\right) = \frac{\pi}{4}.$$

Thus, to graph the function from 0 to π, we use the following x-values.

$$0$$

$$0 + \frac{\pi}{4} = \frac{\pi}{4}$$

$$\frac{\pi}{4} + \frac{\pi}{4} = \frac{\pi}{2}$$

$$\frac{\pi}{2} + \frac{\pi}{4} = \frac{3\pi}{4}$$

$$\frac{3\pi}{4} + \frac{\pi}{4} = \pi$$

Another way to divide an interval into four equal parts is as follows:

Step 1 Find the midpoint of the interval by adding the x-values of the endpoints and dividing by 2.

Step 2 Find the two midpoints of the intervals found in Step 1, using the same procedure.

● ● ● **Example 2   Graphing $y = \sin bx$**

Graph $y = \sin 2x$.

Traditional Approach

For this function, $b = 2$, so the period is $2\pi/2 = \pi$. Therefore, the graph will complete one period over the interval $[0, \pi]$.

The endpoints are 0 and π, and the three middle points are

$$\frac{1}{4}(0 + \pi), \qquad \frac{1}{2}(0 + \pi), \qquad \text{and} \qquad \frac{3}{4}(0 + \pi),$$

which give the following x-values.

0	$\dfrac{\pi}{4}$	$\dfrac{\pi}{2}$	$\dfrac{3\pi}{4}$	π
↑	↑	↑	↑	↑
Left endpoint	First quarter point	Midpoint	Third quarter point	Right endpoint

We now plot the points from the table of values given earlier, and join them with a smooth sinusoidal curve. More of the graph can be sketched by repeating this cycle over and over, as shown

(continued)

Graphing Calculator Approach

Figure 11 shows the graph of $Y_1 = \sin 2x$ as a thick line and $Y_2 = \sin x$ as a thin line.

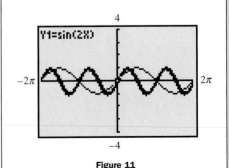

Figure 11

in Figure 10. Notice that the amplitude is not changed. The graph of $y = \sin x$ is included for comparison.

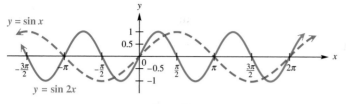

Figure 10

Generalizing from Example 2 leads to the following result.

> ### Period of the Sine and Cosine Functions
> For $b > 0$, the graph of $y = \sin bx$ will look like that of $y = \sin x$, but with a period of $2\pi/b$. Also, the graph of $y = \cos bx$ will look like that of $y = \cos x$, but with a period of $2\pi/b$.

● ● ●   **Example 3   Graphing $y = \cos bx$**

Graph $y = \cos \dfrac{2}{3}x$ over one period.

For this function, the period is $2\pi/(2/3) = 3\pi$. Divide the interval $[0, 3\pi]$ into four equal parts to get the following x-values that yield minimum points, maximum points, and x-intercepts.

$$0 \qquad \frac{3\pi}{4} \qquad \frac{3\pi}{2} \qquad \frac{9\pi}{4} \qquad 3\pi$$

These values are used to get a table of key points for one period.

x	0	$3\pi/4$	$3\pi/2$	$9\pi/4$	3π
$\dfrac{2}{3}x$	0	$\pi/2$	π	$3\pi/2$	2π
$\cos \dfrac{2}{3}x$	1	0	-1	0	1

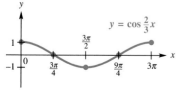

Figure 12

The amplitude is 1 because the maximum value is 1, the minimum value is -1, and half of $1 - (-1) = (1/2)(2) = 1$.

Now plot these points and join them with a smooth curve. The graph is shown in Figure 12.   ● ● ●

N O T E Look at the middle row of the table in Example 3. The method of dividing the interval $[0, 2\pi/b]$ into four equal parts will always give the values 0, $\pi/2$, π, $3\pi/2$, and 2π for this row, resulting in values of -1, 0, or 1 for the circular function. These lead to key points on the graph, which can then be easily sketched.

Guidelines for Sketching Graphs of the Sine and Cosine Functions

To graph $y = a \sin bx$ or $y = a \cos bx$, with $b > 0$, follow these steps.

Step 1 Find the period, $2\pi/b$. Start at 0 on the x-axis, and lay off a distance of $2\pi/b$.

Step 2 Divide the interval into four equal parts. (See the Note preceding Example 2.)

Step 3 Evaluate the function for each of the five x-values resulting from Step 2. The points will be maximum points, minimum points, and x-intercepts.

Step 4 Plot the points found in Step 3, and join them with a sinusoidal curve with amplitude $|a|$.

Step 5 Draw additional cycles of the graph, to the right and to the left, as needed.

The functions in Examples 4 and 5 have both amplitude and period affected by constants.

● ● ● **Example 4** **Graphing** $y = a \sin bx$

Graph $y = -2 \sin 3x$.

Traditional Approach

Step 1 For this function, $b = 3$, so the period is $2\pi/3$. We will first graph the function over the interval $[0, 2\pi/3]$.

Step 2 Dividing the interval $[0, 2\pi/3]$ into four equal parts gives the x-values 0, $\pi/6$, $\pi/3$, $\pi/2$, and $2\pi/3$.

Step 3 Make a table of points determined by the x-values resulting from Step 2.

x	0	$\pi/6$	$\pi/3$	$\pi/2$	$2\pi/3$
$3x$	0	$\pi/2$	π	$3\pi/2$	2π
$\sin 3x$	0	1	0	-1	0
$-2 \sin 3x$	0	-2	0	2	0

Step 4 Plot the points $(0, 0)$, $(\pi/6, -2)$, $(\pi/3, 0)$, $(\pi/2, 2)$, and $(2\pi/3, 0)$, and join them with a sinusoidal curve with amplitude 2. See Figure 13.

Graphing Calculator Approach

Figure 14 shows the graph in the window $[0, 2\pi/3]$ by $[-2.5, 2.5]$. Note that Xscl $= \pi/6$ and Yscl $= 1$.

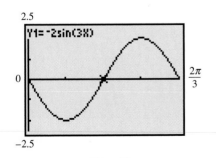

Figure 14

(continued)

Step 5 If necessary, the graph in Figure 13 can be extended by repeating the cycle over and over.

Notice the effect of the negative value of a. When a is negative, the graph of $y = a \sin bx$ will be the reflection across the x-axis of the graph of $y = |a| \sin bx$.

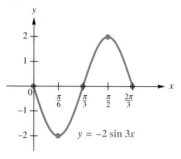

Figure 13

• • • **Example 5** Graphing $y = a \cos bx$

Graph $y = 3 \cos \dfrac{1}{2}x$.

Traditional Approach

The period is $2\pi/(1/2) = 4\pi$. The key points have x-values

$$0, \quad \frac{1}{4}(4\pi) = \pi, \quad \frac{1}{2}(4\pi) = 2\pi, \quad \frac{3}{4}(4\pi) = 3\pi, \quad \text{and} \quad 4\pi.$$

Evaluating the function for these x-values gives the following points.

$$(0, 3) \qquad (\pi, 0) \qquad (2\pi, -3) \qquad (3\pi, 0) \qquad (4\pi, 3)$$

Figure 15 shows these points joined with a smooth curve. The amplitude is 3.

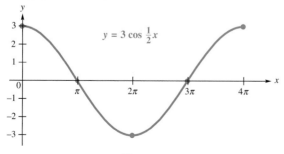

Figure 15

Graphing Calculator Approach

See Figure 16. What are the Xscl and Yscl values?

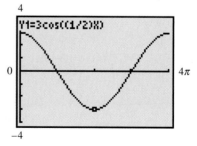

Figure 16

Using a Trigonometric Model

• • •   **Example 6**   Interpreting a Sine Function Model

The average temperature (in °F) at Mould Bay, Canada, can be approximated by the circular function defined by

$$f(x) = 34 \sin\left[\frac{\pi}{6}(x - 4.3)\right],$$

where x is the month and $x = 1$ corresponds to January.

(a) Graph f over the interval $0 \le x \le 25$. Determine the amplitude and period of the graph.

The graph is shown in Figure 17. Its amplitude is 34, and the period is $2\pi/(\pi/6) = 12$. The function f has a period of 12 months or 1 year, which agrees with the changing of the seasons.

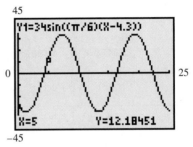

Figure 17

(b) What is the average temperature during the month of May?

May is the fifth month, so the average temperature during May is

$$f(5) = 34 \sin\left[\frac{\pi}{6}(5 - 4.3)\right] \approx 12°F.$$

See the display at the bottom of the screen in Figure 17.

(c) What would be an approximation for the average *yearly* temperature in Mould Bay?

From the graph, it appears that the average yearly temperature is about 0°F since the graph is centered vertically about the line $y = 0$.   • • •

C O N N E C T I O N S   *Note: The discussion that follows is specific to the TI-83 graphing calculator. It may easily be modified to apply to other models of graphing calculators.*

If you have a TI-83 calculator, adjust the settings to correspond to the following screens.

(continued)

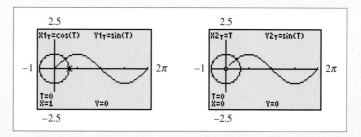

In the final two screens, Tmax is 2π, Tstep is $\pi/40$, Xmax is 2π, and Xscl is $\pi/2$. Now graph these two equations (which are in *parametric form*), and watch as the unit circle and the sine function are graphed simultaneously. Press the $\boxed{\text{TRACE}}$ key once to get the screen shown on the left below, and then press the up-arrow key to get the screen shown on the right below.

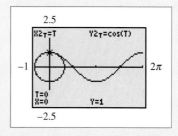

The screen on the left gives a unit circle interpretation of $\cos 0 = 1$ and $\sin 0 = 0$. The screen on the right gives a rectangular coordinate graph interpretation of $\sin 0 = 0$.

Now go back and redefine Y_{2T} as $\cos(T)$. Graph both equations again; the second screen will look like the figure below, after the $\boxed{\text{TRACE}}$ and up-arrow keys are pressed. This screen indicates that $\cos 0 = 1$.

For Discussion or Writing

1. On the unit circle, let $T = 2$. What values of X and Y are displayed? Interpret these values.
2. On the sine graph, trace to the point where $T = 1.9$. What values of X and Y are displayed? Interpret these values with an equation in X and Y.
3. Repeat Exercise 2, but use the cosine graph.

4.1 Exercises

Concept Check In Exercises 1–8, match each defined function with its graph.

1. $y = \sin x$ **G.**
2. $y = \cos x$ **A.**
3. $y = -\sin x$ **E.**
4. $y = -\cos x$ **D.**

5. $y = \sin 2x$ **B.**
6. $y = \cos 2x$ **H.**
7. $y = 2 \sin x$ **F.**
8. $y = 2 \cos x$ **C.**

A.
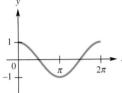

B.
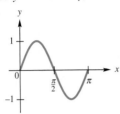

C.
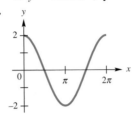

D.
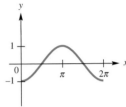

E.
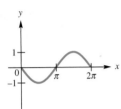

F.
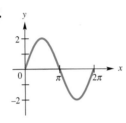

G.
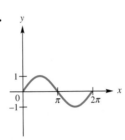

H.
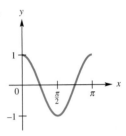

In Exercises 9–28, give a traditional or calculator-generated graph, as directed by your instructor.

Graph each defined function over the interval $[-2\pi, 2\pi]$. Give the amplitude. See Example 1.

9. $y = 2 \cos x$
10. $y = 3 \sin x$
11. $y = \dfrac{2}{3} \sin x$
12. $y = \dfrac{3}{4} \cos x$

13. $y = -\cos x$
14. $y = -\sin x$
15. $y = -2 \sin x$
16. $y = -3 \cos x$

Graph each defined function over a two-period interval. Give the period and the amplitude. See Examples 2–5.

17. $y = \sin \dfrac{1}{2}x$
18. $y = \sin \dfrac{2}{3}x$
19. $y = \cos \dfrac{1}{3}x$
20. $y = \cos \dfrac{3}{4}x$

21. $y = \sin 3x$
22. $y = \cos 2x$
23. $y = 2 \sin \dfrac{1}{4}x$
24. $y = 3 \sin 2x$

25. $y = -2 \cos 3x$
26. $y = -5 \cos 2x$
27. $y = \cos \pi x$
28. $y = -\sin \pi x$

Concept Check In Exercises 29 and 30, give the equation of a sine function having the given graph.

29.
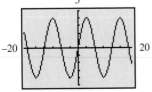

30.
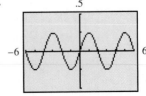

31. *Concept Check*   Use the fact that $\sin x = \cos\left(x - \frac{\pi}{2}\right)$ and the answer to Exercise 29 to give the equation of a cosine function having that graph.

32. *Concept Check*   Use the fact that $\sin x = \cos\left(x - \frac{\pi}{2}\right)$ and the answer to Exercise 30 to give the equation of a cosine function having that graph.

 Tides for Kahului Harbor   *The chart shows the tides for Kahului Harbor (on the island of Maui, Hawaii). To identify high and low tides and times for other Maui areas, the following adjustments must be made.*

Hana: High, +40 minutes, +.1 foot;
          Low, +18 minutes, −.2 foot.

Makena: High, +1:21, −.5 foot;
          Low, +1:09, −.2 foot.

Maalaea: High, +1:52, −.1 foot;
          Low, +1:19, −.2 foot.

Lahaina: High, +1:18, −.2 foot;
          Low, +1:01, −.1 foot.

JANUARY 1997

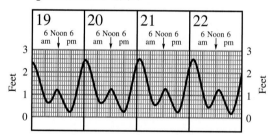

Source: Maui News. Original chart prepared by Edward K. Noda and Associates.

Use the graph to work Exercises 33–38. See Example 6.

33. The graph is an example of a periodic function. What is the period (in hours)?

34. What is the amplitude?

35. At what time on January 20, 1997, was low tide at Kahului? What was the height?

36. Repeat Exercise 35 for Maalaea.

37. At what time on January 22, 1997, was high tide at Kahului? What was the height?

38. Repeat Exercise 37 for Lahaina.

Solve each problem. See Example 6.

39. *Blood Pressure Variation*   The graph gives the variation in blood pressure for a typical person. Systolic and diastolic pressures are the upper and lower limits of the periodic changes in pressure that produce the pulse. The length of time between peaks is called the period of the pulse.

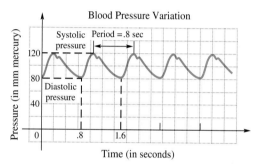

(a) Find the amplitude of the graph.

(b) Find the pulse rate (the number of pulse beats in one minute) for this person.

40. *Activity of a Nocturnal Animal*   Many of the activities of living organisms are periodic. For example, the graph below shows the time that a certain nocturnal animal begins its evening activity.

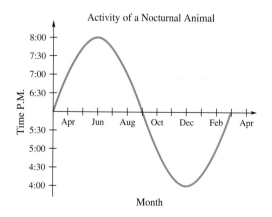

(a) Find the amplitude of this graph.

(b) Find the period.

41. 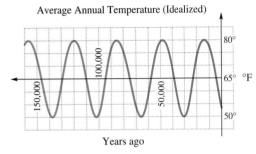 *Average Annual Temperature* Scientists believe that the average annual temperature in a given location is periodic. The average temperature at a given place during a given season fluctuates as time goes on, from colder to warmer, and back to colder. The graph shows an idealized description of the temperature (in °F) for the last few thousand years of a location at the same latitude as Anchorage, AK.

Average Annual Temperature (Idealized)

(a) Find the highest and lowest temperatures recorded.
(b) Use these two numbers to find the amplitude.
(c) Find the period of the function.
(d) What is the trend of the temperature now?

42. *(Modeling) Position of a Moving Arm* The figure shows schematic diagrams of a rhythmically moving arm. The upper arm *RO* rotates back and forth about the point *R*; the position of the arm is measured by the angle *y* between the actual position and the downward vertical position. (*Source:* De Sapio, Rodolfo, *Calculus for the Life Sciences.* Copyright © 1978 by W. H. Freeman and Company. Reprinted by permission.)

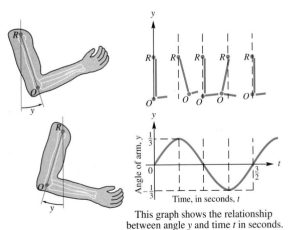

This graph shows the relationship between angle *y* and time *t* in seconds.

(a) Find an equation of the form $y = a \sin kt$ for the graph shown.
(b) How long does it take for a complete movement of the arm?

Musical Sound Waves Pure sounds produce single sine waves on an oscilloscope. Find the amplitude and period of each sine wave graph in Exercises 43 and 44. On the

vertical scale, each square represents .5; on the horizontal scale, each square represents 30° or $\pi/6$.

43.

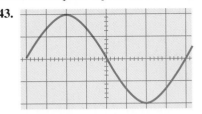

44.

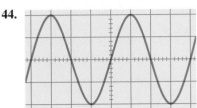

45. *(Modeling) Voltage of an Electrical Circuit* The voltage *E* in an electrical circuit is modeled by

$$E = 5 \cos 120\pi t,$$

where *t* is time measured in seconds.
(a) Find the amplitude and the period.
(b) How many cycles are completed in one second? (The number of cycles (periods) completed in one second is the **frequency** of the function.)
(c) Find *E* when $t = 0, .03, .06, .09, .12$.
(d) Graph *E* for $0 \le t \le 1/30$.

46. *(Modeling) Voltage of an Electrical Circuit* For another electrical circuit, the voltage *E* is modeled by

$$E = 3.8 \cos 40\pi t,$$

where *t* is time measured in seconds.
(a) Find the amplitude and the period.
(b) Find the frequency. See Exercise 45(b).
(c) Find *E* when $t = .02, .04, .08, .12, .14$.
(d) Graph one period of *E*.

47. *(Modeling) Atmospheric Carbon Dioxide* At Mauna Loa, Hawaii, atmospheric carbon dioxide levels in parts per million (ppm) have been measured regularly since 1958. The function defined by

$$L(x) = .022x^2 + .55x + 316 + 3.5 \sin(2\pi x)$$

can be used to model these levels, where *x* is in years and $x = 0$ corresponds to 1960. (*Source:* Nilsson, A., *Greenhouse Earth,* John Wiley & Sons, 1992.)
(a) Graph *L* for $15 \le x \le 35$. (*Hint:* Use $325 \le y \le 365$.)
(b) When do the seasonal maximum and minimum carbon dioxide levels occur?
(c) *L* is the sum of a quadratic function and a sine function. What is the significance of each of these functions? Discuss what physical phenomena may be responsible for each function.

48. *(Modeling) Atmospheric Carbon Dioxide* Refer to Exercise 47. The carbon dioxide content in the atmosphere at Barrow, Alaska, in parts per million (ppm) can be modeled using the function defined by

$$C(x) = .04x^2 + .6x + 330 + 7.5 \sin(2\pi x),$$

where $x = 0$ corresponds to 1970. (*Source:* Zeilik, M., S. Gregory, and E. Smith, *Introductory Astronomy and Astrophysics,* Fourth Edition, Saunders College Publishing, 1998.)

(a) Graph C for $5 \le x \le 25$. (*Hint:* Use $320 \le y \le 380$.)

(b) Discuss possible reasons why the amplitude of the oscillations in the graph of C is larger than the amplitude of the oscillations in the graph of L in Exercise 47, which models Hawaii.

(c) Define a new function C that is valid if x represents the actual year where $1970 \le x \le 1995$.

49. Compare the graphs of $y = \sin 2x$ and $y = 2 \sin x$ over the interval $[0, 2\pi]$. Can we say that, in general, $\sin bx = b \sin x$? Explain.

50. Compare the graphs of $y = \cos 3x$ and $y = 3 \cos x$ over the interval $[0, 2\pi]$. Can we say that, in general, $\cos bx = b \cos x$? Explain.

51. Refer to the graph of $y = \sin x$ in Figure 4. The graph completes one cycle between $x = 0$ and $x = 2\pi$. Consider the statement, "The function $y = \sin(bx)$ completes b cycles between 0 and 2π." Use your graphing calculator to confirm the statement for some positive integer values of b, such as 3, 4, and 5. Interpret and confirm the statement for $b = 1/2$ and $b = 3/2$.

52. Explain how one can observe the graphs of $y = \sin x$ and $y = \cos x$ on the same axes and see that for exactly two x-values in $[0, 2\pi)$, $\sin x = \cos x$. What are the two x-values?

A graphing calculator will return a 1 for a true statement and a 0 for a false statement. Decide which will be returned for any value of x in the given screen. Support your result with your own calculator.

53.
```
cos(X)≤1
```

54.
```
sin(X)≠π
```

Give two values in the interval $[0, 2\pi)$ that can be stored in x for the given screen to be obtained. Support your result with your own calculator.

55.
```
abs(sin(X))
                1
```

56.
```
abs(cos(X))
                1
```

*In part A of the figure, a spring with a weight attached to its free end is in equilibrium (or rest) position. If the weight is pulled down a units and released (part B of the figure), the spring's elastic restoring force causes the weight to rise a units (a > 0) above the equilibrium position, as seen in part C, and then oscillate about the equilibrium position according to an equation of the form $y = a \cos bt$, with t measured in seconds. This behavior is called **simple harmonic motion**. (The system is assumed to be ideal in the sense that it is not subject to friction or air resistance.) The number of oscillations per second is called the* frequency *of the system. The amount of time required for one complete oscillation is called the* period *of the system.*

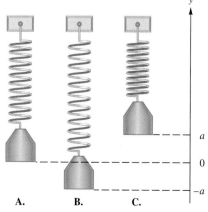

Use these ideas to work Exercises 57–60.

57. *(Modeling) Harmonic Motion* The position of a weight attached to a spring is

$$s(t) = -5 \cos 4\pi t$$

inches after t seconds.

(a) What is the maximum height that the weight rises above the equilibrium position?

(b) What are the frequency and period of the system?

(c) When does the weight first reach its maximum height?

(d) Calculate and interpret $s(1.3)$.

58. *(Modeling) Harmonic Motion* The position of a weight attached to a spring is

$$s(t) = -4 \cos 10t$$

inches after t seconds.

(a) What is the maximum height that the weight rises above the equilibrium position?

(b) What are the frequency and period of the system?

(c) When does the weight first reach its maximum height?

(d) Calculate and interpret $s(1.466)$.

59. *(Modeling) Harmonic Motion* A weight attached to a spring is pulled down 3 inches below the equilibrium position.

(a) Assuming that the frequency of the system is $6/\pi$ cycles per second, determine a trigonometric model that gives the position of the weight at time t seconds.

(b) What is the period of the system?

60. *(Modeling) Harmonic Motion* A weight attached to a spring is pulled down 2 inches below the equilibrium position.

(a) Assuming that the period of the system is $1/3$ second, determine a trigonometric model that gives the position of the weight at time t seconds.

(b) What is the frequency of the system?

4.2   Translations of the Graphs of the Sine and Cosine Functions

> • **Horizontal Translations** • **Vertical Translations** • **Combinations of Translations** • **Determining a Trigonometric Model Using Curve Fitting**

Horizontal Translations   In general, the graph of the function $y = f(x - d)$ is translated *horizontally* when compared to the graph of $y = f(x)$. The translation is d units to the right if $d > 0$ and $|d|$ units to the left if $d < 0$. See Figure 18. With circular functions, a horizontal translation is called a **phase shift.** In the function $y = f(x - d)$, the expression $x - d$ is called the **argument.**

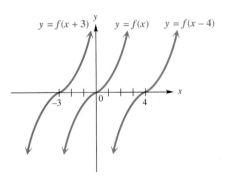

Horizontal Translations of $y = f(x)$

Figure 18

We give two methods that can be used to sketch the graph of a circular function involving a phase shift.

● ● ● **Example 1** Graphing $y = \sin(x - d)$

Graph $y = \sin\left(x - \dfrac{\pi}{3}\right)$.

Method 1 The argument $x - \pi/3$ indicates that the graph will be translated $\pi/3$ units to the *right* (the phase shift) as compared to the graph of $y = \sin x$. Notice that in Figure 19 the graph of $y = \sin x$ is shown as a dashed curve, and the graph of $y = \sin(x - \pi/3)$ is shown as a solid curve. Therefore, to graph a function using this method, first graph the basic circular function, and then graph the desired function by using the appropriate translation.

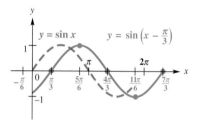

Figure 19

Method 2 For the argument $x - \pi/3$ to result in all possible values throughout one period, it must take on all values between 0 and 2π, inclusive. Therefore, to find an interval of one period, we solve the compound inequality

$$0 \le x - \frac{\pi}{3} \le 2\pi.$$

Add $\pi/3$ to each expression to find the interval

$$\frac{\pi}{3} \le x \le \frac{7\pi}{3} \qquad \text{or} \qquad \left[\frac{\pi}{3}, \frac{7\pi}{3}\right].$$

As first shown in Section 4.1, divide this interval into four equal parts to get the following x-values.

$$\frac{\pi}{3} \quad \frac{5\pi}{6} \quad \frac{4\pi}{3} \quad \frac{11\pi}{6} \quad \frac{7\pi}{3}$$

Make a table of points using the x-values above.

x	$\pi/3$	$5\pi/6$	$4\pi/3$	$11\pi/6$	$7\pi/3$
$x - \dfrac{\pi}{3}$	0	$\pi/2$	π	$3\pi/2$	2π
$\sin\left(x - \dfrac{\pi}{3}\right)$	0	1	0	-1	0

Join these points to get the graph shown in Figure 19. The period is 2π, and the amplitude is 1.

● ● ●

● ● ● **Example 2** Graphing $y = a \cos(x - d)$

Graph $y = 3 \cos\left(x + \dfrac{\pi}{4}\right)$.

Traditional Approach

Method 1 Start by writing $3 \cos(x + \pi/4)$ in the form $a \cos(x - d)$.

$$3 \cos\left(x + \frac{\pi}{4}\right) = 3 \cos\left[x - \left(-\frac{\pi}{4}\right)\right]$$

This result shows that $d = -\pi/4$. Since $-\pi/4$ is negative, the phase shift is $|-\pi/4| = \pi/4$ unit to the left. The period is 2π, and the amplitude is 3. The graph is the same as that of $y = 3 \cos x$, except that it is translated $\pi/4$ unit to the left. See Figure 20.

Graphing Calculator Approach

Figure 21 shows the graph of

$$Y_1 = 3 \cos\left(x + \frac{\pi}{4}\right)$$

as a thick line. The graph of

$$Y_2 = 3 \cos x$$

is shown as a thin line for comparison.

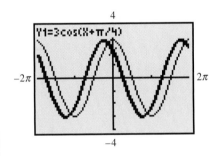

Figure 21

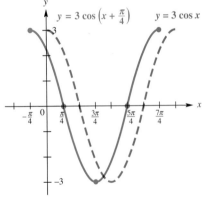

Figure 20

Method 2 The graph can be sketched by first solving the inequality

$$0 \le x + \frac{\pi}{4} \le 2\pi$$

$$-\frac{\pi}{4} \le x \le \frac{7\pi}{4}. \quad \text{Add } -\pi/4 \text{ to each expression.}$$

This gives an interval over which one period of the function can be graphed. Dividing this interval into four equal parts gives x-values of $-\pi/4$, $\pi/4$, $3\pi/4$, $5\pi/4$, and $7\pi/4$. A table of points for these x-values once again leads to maximum points, minimum points, and x-intercepts.

(continued)

x	$-\pi/4$	$\pi/4$	$3\pi/4$	$5\pi/4$	$7\pi/4$
$x + \dfrac{\pi}{4}$	0	$\pi/2$	π	$3\pi/2$	2π
$3\cos\left(x + \dfrac{\pi}{4}\right)$	3	0	-3	0	3

This method produces the same graph as the one shown in Figure 20.

● ● ●

A function of the form $y = a\cos b(x - d)$ has both a phase shift (if $d \neq 0$) and a period different from 2π (if $|b| \neq 1$).

● ● ●   **Example 3   Graphing $y = a\cos b(x - d)$**

Graph $y = -2\cos(3x + \pi)$.

Traditional Approach

Method 1   First factor 3 out of the argument by writing the expression in the form $a\cos b(x - d)$.

$$y = -2\cos(3x + \pi) = -2\cos 3\left(x + \frac{\pi}{3}\right)$$

Then $a = -2$, $b = 3$, and $d = -\pi/3$. The amplitude is $|-2| = 2$, and the period is $2\pi/3$ (since the value of b is 3). The phase shift is $|-\pi/3| = \pi/3$ units to the left as compared to the graph of $y = -2\cos 3x$. See Figure 22.

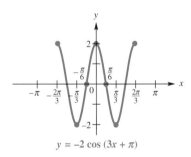

$$y = -2\cos(3x + \pi)$$

Figure 22

Graphing Calculator Approach

Figure 23 shows a table of values that correspond to the points found using Method 2 in the traditional approach. Figure 24 shows the graph as generated by a graphing calculator.

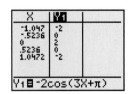

Figure 23

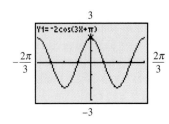

Figure 24

(continued)

Method 2 The function can be sketched over one period by solving the compound inequality

$$0 \le 3\left(x + \frac{\pi}{3}\right) \le 2\pi$$

to get the interval $[-\pi/3, \pi/3]$. Divide this interval into four equal parts to get the following points.

$$\left(-\frac{\pi}{3}, -2\right) \quad \left(-\frac{\pi}{6}, 0\right) \quad (0, 2) \quad \left(\frac{\pi}{6}, 0\right) \quad \left(\frac{\pi}{3}, -2\right)$$

Plot these points and then join them with a smooth curve. By graphing an additional half-period to the left and to the right, we obtain the same graph as the one shown in Figure 22.

● ● ●

Vertical Translations The graph of a function of the form $y = c + f(x)$ is translated *vertically* as compared with the graph of $y = f(x)$. See Figure 25. The translation is c units up if $c > 0$ and $|c|$ units down if $c < 0$.

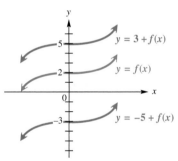

Vertical Translations of $y = f(x)$

Figure 25

● ● ● **Example 4** Graphing $y = c + a \cos bx$

Graph $y = 3 - 2 \cos 3x$.

Traditional Approach

Method 2 (only) The values of y will be 3 greater than the corresponding values of y in $y = -2 \cos 3x$. This means that the graph of $y = 3 - 2 \cos 3x$ is the same as the graph of $y = -2 \cos 3x$, vertically translated 3 units up. Since the period of $y = -2 \cos 3x$ is $2\pi/3$, the key points have x-values

$$0 \quad \frac{\pi}{6} \quad \frac{\pi}{3} \quad \frac{\pi}{2} \quad \frac{2\pi}{3}.$$

Graphing Calculator Approach

Figure 27 shows the graph of

$$Y_1 = 3 - 2 \cos 3x$$

as a thick line. For comparison, the graph of

$$Y_2 = -2 \cos 3x$$

(continued)

The key points are shown on the graph in Figure 26, along with more of the graph, sketched using the fact that the function is periodic.

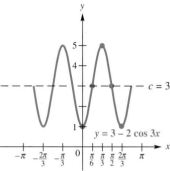

Figure 26

is shown as a thin line. Note the vertical translation.

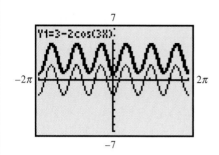

Figure 27

● ● ●

Combinations of Translations In the next example, we graph a function that involves all the types of stretching, compressing, and translating presented in the previous section and this one.

● ● ● **Example 5** Graphing $y = c + a \sin b(x - d)$

Graph $y = -1 + 2 \sin 4\left(x + \dfrac{\pi}{4}\right)$.

Traditional Approach

Method 1 Here, the amplitude is 2, the period is $2\pi/4 = \pi/2$, and the graph is translated down 1 unit and $\pi/4$ unit to the left as compared to the graph of $y = 2 \sin 4x$. Since the graph is translated $\pi/4$ unit to the left, start at the x-value $0 - \pi/4 = -\pi/4$. The first period will end at $-\pi/4 + \pi/2 = \pi/4$. The maximum y-value will be $2 - 1 = 1$, and the minimum y-value will be $-2 - 1 = -3$. Sketch the graph using the typical sine curve. See Figure 28.

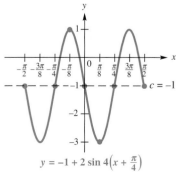

$$y = -1 + 2 \sin 4\left(x + \tfrac{\pi}{4}\right)$$

Figure 28

Graphing Calculator Approach

Figure 29 shows the graph of this function. Notice that the line $Y_2 = -1$ is also graphed; compare this graph with the graph in Figure 28.

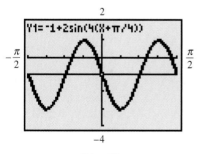

Figure 29

(continued)

Method 2 Start by finding an interval over which the graph will complete one cycle. To do this, use the argument $4(x + \pi/4)$ in a compound inequality, with 0 as one endpoint and 2π as the other. Then solve the inequality for x.

$$0 \le 4\left(x + \frac{\pi}{4}\right) \le 2\pi$$

$$0 \le x + \frac{\pi}{4} < \frac{\pi}{2} \qquad \text{Divide by 4.}$$

$$-\frac{\pi}{4} \le x \le \frac{\pi}{4} \qquad \text{Subtract } \frac{\pi}{4}.$$

Divide the interval $[-\pi/4, \pi/4]$ into four equal parts to find the key points on the graph, as shown in the following table.

x	$-\pi/4$	$-\pi/8$	0	$\pi/8$	$\pi/4$
$x + \pi/4$	0	$\pi/8$	$\pi/4$	$3\pi/8$	$\pi/2$
$4(x + \pi/4)$	0	$\pi/2$	π	$3\pi/2$	2π
$-1 + 2\sin 4(x + \pi/4)$	-1	1	-1	-3	-1

Join the key points with a smooth curve as shown in Figure 28. This function has period $\pi/2$ and amplitude $[1 - (-3)]/2 = 2$. The phase shift is $\pi/4$ unit to the left, and there is a vertical translation 1 unit down.

● ● ●

Guidelines for Sketching Graphs of the Sine and Cosine Functions

Use one of these methods to graph the function

$$y = c + a \sin b(x - d) \quad \text{or} \quad y = c + a \cos b(x - d), \quad b > 0.$$

Method 1 First graph the basic circular function. The amplitude of the function is $|a|$ and the period is $2\pi/b$. Then use translations to graph the desired function. The vertical translation is c units up if $c > 0$ and $|c|$ units down if $c < 0$. The horizontal translation (phase shift) is d units to the right if $d > 0$ and $|d|$ units to the left if $d < 0$.

Method 2 Follow these steps.

Step 1 Find an interval whose length is one period $(2\pi/b)$ by solving the compound inequality $0 \le b(x - d) \le 2\pi$.

Step 2 Divide the interval into four equal parts.

(continued)

> **Step 3** Evaluate the function for each of the five *x*-values resulting from Step 2. The points will be maximum points, minimum points, and points that intersect the line $y = c$ ("middle" points of the wave).
> **Step 4** Plot the points found in Step 3, and join them with a sinusoidal curve.
> **Step 5** Draw the graph over additional periods, to the right and to the left, as needed.

Determining a Trigonometric Model Using Curve Fitting A sinusoidal function is often a good approximation of a set of data points from a real situation. This is an example of a more general technique called **curve fitting.** In Example 6 of Section 4.1, we used a trigonometric model for temperatures. The final example of this section shows how such a model is determined from given data by fitting a sine curve to the data.

● ● ● **Example 6** Modeling Temperature with a Sine Function

 The maximum average monthly temperature in New Orleans is 82°F and the minimum is 54°F. The table shows the average monthly temperatures.

Temperatures in New Orleans

Month	°F	Month	°F
Jan	54	July	82
Feb	55	Aug	81
Mar	61	Sept	77
Apr	69	Oct	71
May	73	Nov	59
June	79	Dec	55

Source: Miller, A. and J. Thompson, *Elements of Meteorology,* Charles E. Merrill Publishing Co., 1975.

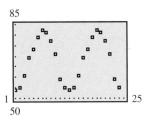

Figure 30

The **scatter diagram** (that is, a graph of ordered pairs) for a two-year interval in Figure 30 strongly suggests that the temperatures can be modeled with a sine curve.

(a) Using only the maximum and minimum temperatures, determine a function of the form $f(x) = a \sin b(x - d) + c$, where a, b, c, and d are constants, that models the average monthly temperature in New Orleans. Let x represent the month, with January corresponding to $x = 1$.

We can use the maximum and minimum average monthly temperatures to find the amplitude a.

$$a = \frac{82 - 54}{2} = 14$$

The average of the maximum and minimum temperatures is a good choice for *c*. The average is

$$\frac{82° + 54°}{2} = 68°F.$$

Since the coldest month is January, when $x = 1$, and the hottest month is July, when $x = 7$, we should choose *d* to be about 4. The table shows that temperatures are actually a little warmer after July than before, so we experiment with values just greater than 4 to find *d*. Trial and error leads to $d = 4.2$. Since temperatures repeat every 12 months, *b* is $2\pi/12 = \pi/6$. Thus,

$$f(x) = a \sin b(x - d) + c = 14 \sin\left[\frac{\pi}{6}(x - 4.2)\right] + 68.$$

(b) On the same coordinate axes, graph *f* for a two-year period together with the actual data values found in the table.

See Figure 31. The figure also shows the graph of $y = 14 \sin(\pi/6)x + 68$ for comparison. The horizontal translation of the model is fairly obvious here.

● ● ●

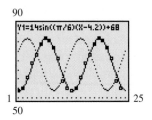

Figure 31

 Some graphing calculators are capable of fitting a sine curve to a set of data points. This is called *sine regression.* Using the data of Example 6, the screen in Figure 32 shows the equation of the model. Figure 33 shows the graph of the model along with the data points.

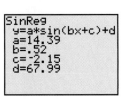

Values are rounded to the nearest hundredth.

Figure 32

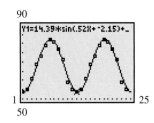

Figure 33

■

4.2 Exercises

Concept Check   Match the functions defined in Column I with the appropriate descriptions in Column II.

I	II
1. $y = 3 \sin(2x - 4)$	**A.** amplitude = 2, period = $\frac{\pi}{2}$, phase shift = $\frac{3}{4}$
2. $y = 2 \sin(3x - 4)$	**B.** amplitude = 3, period = π, phase shift = 2
3. $y = 4 \sin(3x - 2)$	**C.** amplitude = 4, period = $\frac{2\pi}{3}$, phase shift = $\frac{2}{3}$
4. $y = 2 \sin(4x - 3)$	**D.** amplitude = 2, period = $\frac{2\pi}{3}$, phase shift = $\frac{4}{3}$

Concept Check   *In Exercises 5–12, match each defined function with its graph.*

5. $y = \sin\left(x - \dfrac{\pi}{4}\right)$

6. $y = \sin\left(x + \dfrac{\pi}{4}\right)$

7. $y = \cos\left(x - \dfrac{\pi}{4}\right)$

8. $y = \cos\left(x + \dfrac{\pi}{4}\right)$

9. $y = 1 + \sin x$

10. $y = -1 + \sin x$

11. $y = 1 + \cos x$

12. $y = -1 + \cos x$

A.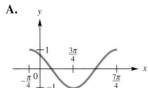

B.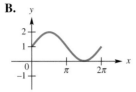

C.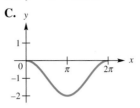

D.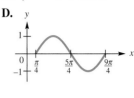

E.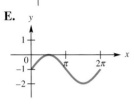

F.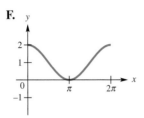

G.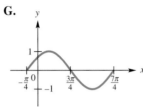

H.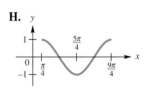

Concept Check   *In Exercises 13 and 14, fill in the blanks with the word* right *or the word* left.

13. If the graph of $y = \cos x$ is translated $\pi/2$ units horizontally to the _____, it will coincide with the graph of $y = \sin x$.

14. If the graph of $y = \sin x$ is translated $\pi/2$ units horizontally to the _____, it will coincide with the graph of $y = \cos x$.

Find the amplitude, the period, any vertical translation, and any phase shift of each graph. See Examples 1–5.

15. $y = 2 \sin(x - \pi)$

16. $y = \dfrac{2}{3} \sin\left(x + \dfrac{\pi}{2}\right)$

17. $y = 4 \cos\left(\dfrac{x}{2} + \dfrac{\pi}{2}\right)$

18. $y = -\cos \dfrac{2}{3}\left(x - \dfrac{\pi}{3}\right)$

19. $y = 3 \cos 2\left(x - \dfrac{\pi}{4}\right)$

20. $y = \dfrac{1}{2} \sin\left(\dfrac{x}{2} + \pi\right)$

21. $y = 2 - \sin\left(3x - \dfrac{\pi}{5}\right)$

22. $y = -1 + \dfrac{1}{2} \cos(2x - 3\pi)$

In Exercises 23–46, give a traditional or calculator-generated graph, as directed by your instructor.

Graph each defined function over a two-period interval. See Examples 1 and 2.

23. $y = \cos\left(x - \dfrac{\pi}{2}\right)$

24. $y = \sin\left(x - \dfrac{\pi}{4}\right)$

25. $y = \sin\left(x + \dfrac{\pi}{4}\right)$

26. $y = \cos\left(x - \dfrac{\pi}{3}\right)$

27. $y = 2 \cos\left(x - \dfrac{\pi}{3}\right)$

28. $y = 3 \sin\left(x - \dfrac{3\pi}{2}\right)$

Graph each defined function over a one-period interval. See Example 3.

29. $y = \dfrac{3}{2} \sin 2\left(x + \dfrac{\pi}{4}\right)$

30. $y = -\dfrac{1}{2} \cos 4\left(x + \dfrac{\pi}{2}\right)$

31. $y = -4 \sin(2x - \pi)$

32. $y = 3 \cos(4x + \pi)$

33. $y = \dfrac{1}{2} \cos\left(\dfrac{1}{2}x - \dfrac{\pi}{4}\right)$

34. $y = -\dfrac{1}{4} \sin\left(\dfrac{3}{4}x + \dfrac{\pi}{8}\right)$

Graph each defined function over a two-period interval. See Example 4.

35. $y = -3 + 2 \sin x$

36. $y = 2 - 3 \cos x$

37. $y = 1 - \dfrac{2}{3} \sin \dfrac{3}{4}x$

38. $y = -1 - 2 \cos 5x$

39. $y = 1 - 2 \cos \dfrac{1}{2}x$

40. $y = -3 + 3 \sin \dfrac{1}{2}x$

41. $y = -2 + \dfrac{1}{2} \sin 3x$

42. $y = 1 + \dfrac{2}{3} \cos \dfrac{1}{2}x$

Graph each defined function over a one-period interval. See Example 5.

43. $y = -3 + 2 \sin\left(x + \dfrac{\pi}{2}\right)$

44. $y = 4 - 3 \cos(x - \pi)$

45. $y = \dfrac{1}{2} + \sin 2\left(x + \dfrac{\pi}{4}\right)$

46. $y = -\dfrac{5}{2} + \cos 3\left(x - \dfrac{\pi}{6}\right)$

 47. Without actually graphing the function defined by $y = -4 - 3 \sin 2(x - \pi/6)$, write an explanation of how the constants -4, -3, 2, and $\pi/6$ affect the graph, using the graph of $y = \sin x$ as a basis for comparison.

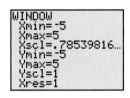 **48.** The graph of $y = \dfrac{1}{2} \sin x + \dfrac{\sqrt{3}}{2} \cos x$ is the same as the graph of a function of the form $y = a \sin(x + \alpha)$. Graph the function with a graphing calculator and then determine the values of a and α.

Concept Check In Exercises 49 and 50, find the equation of a sine function having the given graph.

49. (*Note:* Xscl $= \pi/4$.)

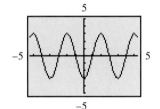

50. (*Note:* Yscl $= \pi$.)

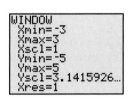

51. *Concept Check* Use the fact that

$$\sin x = \cos\left(x - \dfrac{\pi}{2}\right)$$

and the answer to Exercise 49 to give the equation of a cosine function having that graph.

52. *Concept Check* Use the fact that

$$\sin x = \cos\left(x - \dfrac{\pi}{2}\right)$$

and the answer to Exercise 50 to give the equation of a cosine function having that graph.

Solve each problem. See Example 6.

53. *(Modeling) Average Monthly Temperature* As discussed in the chapter introduction, the average monthly temperature (in °F) in Vancouver, Canada, is shown in the table.

Temperatures in Vancouver

Month	°F	Month	°F
Jan	36	July	64
Feb	39	Aug	63
Mar	43	Sept	57
Apr	48	Oct	50
May	55	Nov	43
June	59	Dec	39

Source: Miller, A. and J. Thompson, *Elements of Meteorology,* Charles E. Merrill Publishing Co., 1975.

(a) Plot the average monthly temperature over a two-year period by letting $x = 1$ correspond to the month of January during the first year. Do the data seem to indicate a translated sine graph?

(b) The highest average monthly temperature is 64°F in July, and the lowest average monthly temperature is 36°F in January. Their average is 50°F. Graph the data together with the line $y = 50$. What does this line represent with regard to temperature in Vancouver?

(c) Approximate the amplitude, period, and phase shift of the translated sine wave indicated by the data.

(d) Determine a function of the form $f(x) = a \sin b(x - d) + c$, where a, b, c, and d are constants, that models the data.

(e) Graph f together with the data on the same coordinate axes. How well does f model the given data?

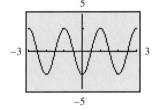 **(f)** Use the sine regression capability of a graphing calculator to find the equation of a sine curve that fits these data.

54. *(Modeling) Average Monthly Temperature*
The average monthly temperature (in °F) in
Phoenix, Arizona, is shown in the table.

Temperatures in Phoenix

Month	°F	Month	°F
Jan	51	July	90
Feb	55	Aug	90
Mar	63	Sept	84
Apr	67	Oct	71
May	77	Nov	59
June	86	Dec	52

Source: Miller, A. and J. Thompson,
Elements of Meteorology, Charles E.
Merrill Publishing Co., 1975.

(a) Predict the average yearly temperature and compare it to the actual value of 70°F.

(b) Plot the average monthly temperature over a two-year period by letting $x = 1$ correspond to January of the first year.

(c) Determine a function of the form $f(x) = a \cos b(x - d) + c$, where a, b, c, and d are constants, that models the data.

(d) Graph f together with the data on the same coordinate axes. How well does f model the data?

(e) Use the sine regression capability of a graphing calculator to find the equation of a sine curve that fits these data.

Quantitative Reasoning

55. *Does the fact that average monthly temperatures are periodic affect your utility bills?* In an article entitled "I Found Sinusoids in My Gas Bill" (*Mathematics Teacher,* January 2000), Cathy G. Schloemer presents the following graph that accompanied her gas bill.

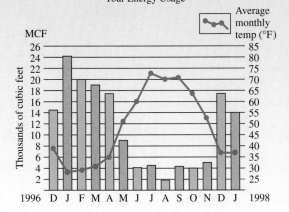

Notice that two sinusoids are suggested here: one for the behavior of the average monthly temperature and another for gas use in MCF (thousands of cubic feet).

(a) If January 1997 is represented by $x = 1$, the data of estimated ordered pairs (month, temperature) is

given in the list shown on the two graphing calculator screens below.

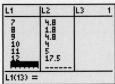

Use the sine regression feature of a graphing calculator to find a sine function that fits these data points. Then make a scatter diagram, and graph the function.

(b) Again, if January 1997 is represented by $x = 1$, the data of estimated ordered pairs (month, gas use in MCF) is given in the list shown on the two graphing calculator screens below.

Use the sine regression feature of a graphing calculator to find a sine function that fits these data points. Then make a scatter diagram, and graph the function.

(c) Answer the question posed at the beginning of this exercise, in the form of a short paragraph.

4.3 Graphs of the Other Circular Functions

- Graphs of the Cosecant and Secant Functions • Graphs of the Tangent and Cotangent Functions
- Addition of Ordinates

Graphs of the Cosecant and Secant Functions Since cosecant values are reciprocals of the corresponding sine values, the period of the function $y = \csc x$ is 2π, the same as for $y = \sin x$. The following table shows several values for $y = \sin x$ and the corresponding values of $y = \csc x$.

x	0	$\pi/4$	$\pi/2$	$3\pi/4$	π	$5\pi/4$	$3\pi/2$	2π
sin x	0	$\sqrt{2}/2$	1	$\sqrt{2}/2$	0	$-\sqrt{2}/2$	-1	0
csc x	undefined	$\sqrt{2}$	1	$\sqrt{2}$	undefined	$-\sqrt{2}$	-1	undefined

When $\sin x = 1$, the value of $\csc x$ is also 1, and when $0 < \sin x < 1$, then $\csc x > 1$. Also, if $-1 < \sin x < 0$, then $\csc x < -1$. As x approaches 0, $\sin x$ approaches 0, and $|\csc x|$ gets larger and larger. The graph of $\csc x$ approaches the vertical line $x = 0$ but never touches it. The line $x = 0$ is called a **vertical asymptote**. In fact, the lines $x = n\pi$, where n is any integer, are all vertical asymptotes. Using this information and plotting a few points shows that the graph takes the shape of the solid curve shown in Figure 34. To show how the two graphs are related, the graph of $y = \sin x$ is also shown, as a dashed curve. The domain of the function $y = \csc x$ is $\{x \mid x \neq n\pi,$ where n is any integer$\}$, and the range is $(-\infty, -1] \cup [1, \infty)$. It is an odd function, and its graph is symmetric with respect to the origin.

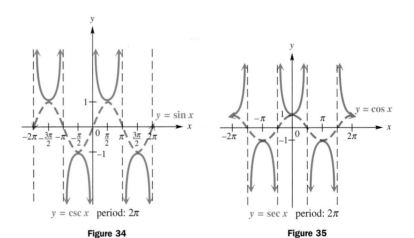

$y = \csc x$ period: 2π
Figure 34

$y = \sec x$ period: 2π
Figure 35

The graph of $y = \sec x$, shown in Figure 35, is related to the cosine graph in the same way that the graph of $y = \csc x$ is related to the sine graph because $\sec x = 1/\cos x$. The domain of the function $y = \sec x$ is $\{x \mid x \neq \pi/2 + n\pi,$ where n is any integer$\}$, and the range is $(-\infty, -1] \cup [1, \infty)$. It is an even function, and its graph is symmetric with respect to the y-axis.

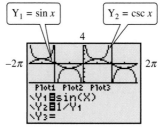

Trig window; connected mode

Figure 36

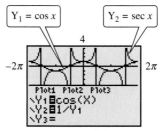

Trig window; connected mode

Figure 37

Typically, calculators do not have keys for the cosecant and secant functions. To graph $y = \csc x$ with a graphing calculator, use the fact that $\csc x = \dfrac{1}{\sin x}$. The graphs of $Y_1 = \sin x$ and $Y_2 = \csc x$ are shown in Figure 36. The calculator is in split screen and connected modes. Similarly, the secant function is graphed by using the identity $\sec x = \dfrac{1}{\cos x}$, as shown in Figure 37.

NOTE Using dot mode for graphing will eliminate the vertical lines that appear in Figures 36 and 37. While they suggest asymptotes and are sometimes called *pseudo-asymptotes,* they are not actually parts of the graphs.

Characteristics of the Cosecant and Secant Functions

Cosecant

Domain: $\{x \mid x \neq n\pi$, where n is an integer$\}$ Range: $(-\infty, -1] \cup [1, \infty)$

Over the interval $(0, 2\pi)$, the cosecant function exhibits the following behavior.

From 0 to $\pi/2$, $\csc x$ decreases from ∞ to 1.

From $\pi/2$ to π, $\csc x$ increases from 1 to ∞.

From π to $3\pi/2$, $\csc x$ increases from $-\infty$ to -1.

From $3\pi/2$ to 2π, $\csc x$ decreases from -1 to $-\infty$.

The graph is discontinuous at values of x of the form $n\pi$, has vertical asymptotes at these values, and is symmetric with respect to the origin.

   x-intercepts: none Amplitude: none Period: 2π

Secant

Domain: $\left\{ x \mid x \neq \dfrac{\pi}{2} + n\pi$, where n is an integer $\right\}$

Range: $(-\infty, -1] \cup [1, \infty)$

Over the interval $[0, 2\pi]$, the secant function exhibits the following behavior.

From 0 to $\pi/2$, $\sec x$ increases from 1 to ∞.

From $\pi/2$ to π, $\sec x$ increases from $-\infty$ to -1.

From π to $3\pi/2$, $\sec x$ decreases from -1 to $-\infty$.

From $3\pi/2$ to 2π, $\sec x$ decreases from ∞ to 1.

The graph is discontinuous at values of x of the form $(2n + 1)\dfrac{\pi}{2}$, has vertical asymptotes at these values, and is symmetric with respect to the y-axis.

   x-intercepts: none Amplitude: none Period: 2π

Guidelines for Sketching Graphs of the Cosecant and Secant Functions

To graph $y = a \csc bx$ or $y = a \sec bx$, with $b > 0$, follow these steps.

Step 1 Graph the corresponding reciprocal function as a guide, using a dashed curve.

To Graph	Use as a Guide
$y = a \csc bx$	$y = a \sin bx$
$y = a \sec bx$	$y = a \cos bx$

Step 2 Sketch the vertical asymptotes. They will have equations of the form $x = k$, where k is an x-intercept of the graph of the guide function.

Step 3 Sketch the graph of the desired function by drawing the typical U-shaped branches between the adjacent asymptotes. The branches will be above the graph of the guide function when the guide function values are positive and below the graph of the guide function when the guide function values are negative. The graph will resemble those in Figures 34 and 35.

Like the sine and cosine functions, the secant and cosecant function graphs may be translated vertically and horizontally. The period of both functions is 2π.

● ● ●   **Example 1   Graphing $y = a \sec bx$**

Graph $y = 2 \sec \dfrac{1}{2}x$.

Step 1 This function involves the secant, so the corresponding reciprocal function will involve the cosine. The guide function we will graph is

$$y = 2 \cos \frac{1}{2}x.$$

Using the guidelines of Section 4.1, we find that one period of the graph lies along the interval that satisfies the inequality

$$0 \leq \frac{1}{2}x \leq 2\pi,$$

or $[0, 4\pi]$. Dividing this interval into four equal parts gives the following key points.

$(0, 2) \qquad (\pi, 0) \qquad (2\pi, -2) \qquad (3\pi, 0) \qquad (4\pi, 2)$

These are joined with a smooth curve; it is dashed to indicate that this graph is only a guide. An additional period is graphed as seen in Figure 38(a).

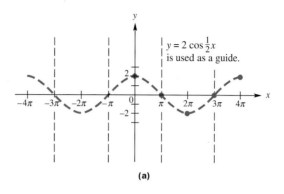

(a)

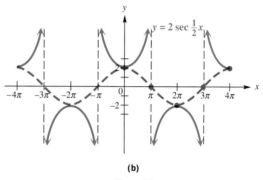

(b)

Figure 38

Step 2 Sketch the vertical asymptotes. These occur at x-values for which the guide function equals 0, such as

$$x = -3\pi \qquad x = -\pi \qquad x = \pi \qquad x = 3\pi.$$

See Figure 38(a).

Step 3 Sketch the graph of $y = 2 \sec(1/2)x$ by drawing the typical U-shaped branches, approaching the asymptotes. See Figure 38(b). ● ● ●

● ● ● **Example 2** Graphing $y = a \csc(x - d)$

Graph $y = \dfrac{3}{2} \csc\left(x - \dfrac{\pi}{2}\right)$.

Traditional Approach

This function can be graphed as in Example 1, by first graphing the corresponding reciprocal function

$$y = \frac{3}{2} \sin\left(x - \frac{\pi}{2}\right).$$

Graphing Calculator Approach

Figures 40 and 41 on the next page show the graph of $y = (3/2)$ $\csc(x - \pi/2)$. Connected mode will

(continued)

We can alternatively analyze the function as follows. Compared with the graph of $y = \csc x$, this graph has phase shift $\pi/2$ units to the right. Thus, the asymptotes are the lines $x = \pi/2, 3\pi/2$, and so on. Also, there are no values of y between $-3/2$ and $3/2$. As shown in Figure 39, this is related to the increased amplitude of $y = (3/2) \sin x$ compared with $y = \sin x$. (Amplitude does not apply to the secant or cosecant functions; it enters only indirectly from the corresponding cosine or sine graphs.) This means that the graph goes through the points $(\pi, 3/2)$, $(2\pi, -3/2)$, and so on. Two periods are shown in Figure 39. (The graph of the guide function, $y = (3/2) \sin(x - \pi/2)$, is shown in red.)

draw vertical lines appearing between the portions of the graph, while dot mode does not. Compare these graphs with Figure 39.

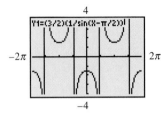

Trig window; connected mode

Figure 40

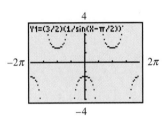

Trig window; dot mode

Figure 41

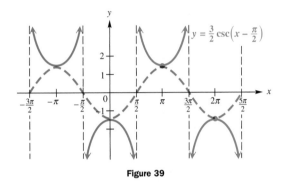

Figure 39

● ● ●

Graphs of the Tangent and Cotangent Functions Since the values of $y = \tan x$ are positive in quadrants I and III and negative in quadrants II and IV,

$$\tan(x + \pi) = \tan x,$$

so the period of $y = \tan x$ is π. Thus, the tangent function can be investigated within an interval of only π units. A convenient interval for this purpose is $(-\pi/2, \pi/2)$ because, although the endpoints $-\pi/2$ and $\pi/2$ are not in the domain of $y = \tan x$ (why?), $\tan x$ exists for all other values in the interval. In the interval $(0, \pi/2)$, $\tan x$ is positive. As x goes from 0 to $\pi/2$, a calculator shows that $\tan x$ gets larger and larger without bound. As x goes from $-\pi/2$ to 0, the values of $\tan x$ approach 0 through negative values. These results are summarized in the following table.

As x Increases from	$\tan x$
0 to $\pi/2$	Increases from 0, without bound
$-\pi/2$ to 0	Increases to 0

Based on these results, the graph of $y = \tan x$ will approach the vertical line $x = \pi/2$ but never touch it, so the line $x = \pi/2$ is a vertical asymptote. The lines $x = \pi/2 + n\pi$, where n is any integer, are all vertical asymptotes. These asymptotes are indicated with dashed lines on the graph in Figure 42. In the interval $(-\pi/2, 0)$, which corresponds to quadrant IV on the unit circle, $\tan x$ is negative, and as x decreases from 0 to $-\pi/2$, $\tan x$ gets smaller and smaller. A table of values for $\tan x$, where $-\pi/2 < x < \pi/2$, follows.

x	$-\pi/3$	$-\pi/4$	$-\pi/6$	0	$\pi/6$	$\pi/4$	$\pi/3$
$\tan x$	-1.7	-1	$-.6$	0	$.6$	1	1.7

Plotting the points from the table and letting the graph approach the asymptotes at $x = \pi/2$ and $x = -\pi/2$ gives the portions of the graph closest to the origin in Figure 42. More of the graph can be sketched by repeating the same curve, as shown in the figure. This graph, like the graphs for the sine and cosine functions, should be learned well enough so that a quick sketch can easily be made. Convenient key points are $(-\pi/4, -1)$, $(0, 0)$, and $(\pi/4, 1)$. These points are shown in Figure 42. The lines $x = \pi/2$ and $x = -\pi/2$ are vertical asymptotes. (The concept of *amplitude,* discussed earlier, applies only to the sine and cosine functions. However, here it means that each y-value of $y = a \tan x$ is a times the corresponding y-value of $y = \tan x$.) The domain of the tangent function is $\{x \mid x \ne \pi/2 + n\pi, \text{ where } n \text{ is any integer}\}$. The range is $(-\infty, \infty)$. The tangent function is an odd function, and its graph is symmetric with respect to the origin.

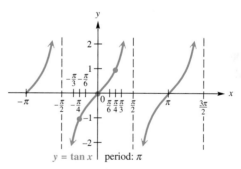

$y = \tan x$ period: π

Figure 42

The definition $\cot x = 1/(\tan x)$ can be used to find the graph of $y = \cot x$. The period of the cotangent, like that of the tangent, is π. The domain of $y = \cot x$ excludes $0 + n\pi$, where n is any integer, since $1/\tan x$ is undefined for these values of x. Thus, the vertical lines $x = n\pi$ are asymptotes. Values of x that lead to asymptotes for $\tan x$ will make $\cot x = 0$, so $\cot(-\pi/2) = 0$, $\cot \pi/2 = 0$, $\cot 3\pi/2 = 0$, and so on. The values of $\tan x$ increase as x goes from $-\pi/2$ to $\pi/2$, so the values of $\cot x$ will *decrease* as x goes from $-\pi/2$ to $\pi/2$. See Figure 43. A table of values for $\cot x$, where $0 < x < \pi$, is shown below.

x	$\pi/6$	$\pi/4$	$\pi/3$	$\pi/2$	$2\pi/3$	$3\pi/4$	$5\pi/6$
$\cot x$	1.7	1	$.6$	0	$-.6$	-1	-1.7

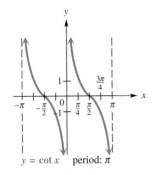

$y = \cot x$ period: π

Figure 43

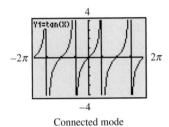

Connected mode

Figure 44

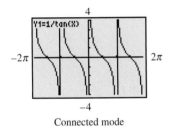

Connected mode

Figure 45

Plotting these points and using the information discussed above gives the graph of $y = \cot x$ shown in Figure 43. (The graph shows two periods.) The domain of the cotangent function is $\{x \mid x \neq n\pi$, where n is any integer$\}$. The range is $(-\infty, \infty)$. The cotangent is also an odd function.

The tangent function can be graphed directly with a graphing calculator, using the tangent key. See Figure 44. To graph the cotangent function, however, we must use one of the identities $\cot x = 1/\tan x$ or $\cot x = \cos x/\sin x$ since, in general, graphing calculators do not have a cotangent key. See Figure 45. ■

Characteristics of the Tangent and Cotangent Functions

Tangent

Domain: $\left\{x \mid x \neq \dfrac{\pi}{2} + n\pi, \text{ where } n \text{ is an integer}\right\}$ Range: $(-\infty, \infty)$

Over the interval $[0, \pi]$, the tangent function exhibits the following behavior.
     From 0 to $\pi/2$, $\tan x$ increases from 0 to ∞.
     From $\pi/2$ to π, $\tan x$ increases from $-\infty$ to 0.

The graph is discontinuous at values of x of the form $(2n + 1)\dfrac{\pi}{2}$, has vertical asymptotes at these values, and is symmetric with respect to the origin.

     x-intercepts: $n\pi$ Amplitude: none   Period: π

Cotangent

Domain: $\{x \mid x \neq n\pi, \text{ where } n \text{ is an integer}\}$ Range: $(-\infty, \infty)$

Over the interval $(0, \pi)$, the cotangent function exhibits the following behavior.
     From 0 to π, $\cot x$ decreases from ∞ to $-\infty$.

The graph is discontinuous at values of x of the form $n\pi$, has vertical asymptotes at these values, and is symmetric with respect to the origin.

     x-intercepts: $\dfrac{\pi}{2} + n\pi$ Amplitude: none   Period: π

Guidelines for Sketching Graphs of the Tangent and Cotangent Functions

To graph $y = a \tan bx$ or $y = a \cot bx$, with $b > 0$, follow these steps.

Step 1 The period is π/b. To locate two adjacent vertical asymptotes, solve the following equations for x:

     For $y = a \tan bx$: $bx = -\dfrac{\pi}{2}$ and $bx = \dfrac{\pi}{2}$

     For $y = a \cot bx$: $bx = 0$ and $bx = \pi$.

(continued)

Step 2 Sketch the two vertical asymptotes found in Step 1.

Step 3 Divide the interval formed by the vertical asymptotes into four equal parts.

Step 4 Evaluate the function for the first-quarter point, midpoint, and third-quarter point, using the x-values found in Step 3.

Step 5 Join the points with a smooth curve, approaching the vertical asymptotes. Indicate additional asymptotes and periods of the graph as necessary.

Like the other circular functions, the graphs of the tangent and cotangent functions may be translated horizontally as well as vertically.

● ● ● **Example 3 Graphing $y = \tan bx$**

Graph $y = \tan 2x$.

Step 1 The period of this function is $\pi/2$. To locate two adjacent vertical asymptotes, solve $2x = -\pi/2$ and $2x = \pi/2$ (since this is a tangent function). The two asymptotes have equations

$$x = -\frac{\pi}{4} \quad \text{and} \quad x = \frac{\pi}{4}.$$

Step 2 Sketch the two vertical asymptotes $x = \pm\pi/4$, as shown in Figure 46.

Step 3 Divide the interval $(-\pi/4, \pi/4)$ into four equal parts. This gives the following key x-values.

first-quarter value: $-\dfrac{\pi}{8}$ middle value: 0 third-quarter value: $\dfrac{\pi}{8}$

Step 4 Evaluate the function for the x-values found in Step 3.

x	$-\pi/8$	0	$\pi/8$
$2x$	$-\pi/4$	0	$\pi/4$
$\tan 2x$	-1	0	1

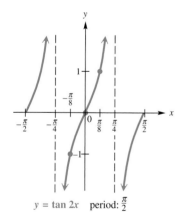

$y = \tan 2x$ period: $\frac{\pi}{2}$

Figure 46

Step 5 Join these points with a smooth curve, approaching the vertical asymptotes. See Figure 46. Another period has been graphed as well, one half-period to the left and one half-period to the right.   ● ● ●

● ● ● **Example 4 Graphing $y = a \tan bx$**

Graph $y = -3 \tan \dfrac{1}{2}x$.

Traditional Approach

The period is $\pi/(1/2) = 2\pi$. Adjacent asymptotes are at $x = -\pi$ and $x = \pi$. Dividing the interval $-\pi < x < \pi$ into four equal parts gives key x-values of $-\pi/2$, 0, and $\pi/2$. Evaluating the function at these x-values gives these key points.

$$\left(-\frac{\pi}{2}, 3\right) \qquad (0,0) \qquad \left(\frac{\pi}{2}, -3\right)$$

Graphing Calculator Approach

On the next page, Figure 48 shows a calculator graph of the same portion of the curve shown in Figure 47.

(continued)

Plotting these points and joining them with a smooth curve gives the graph shown in Figure 47. Notice that, because the coefficient -3 is negative, the graph is reflected across the x-axis compared to the graph of $y = 3 \tan \frac{1}{2}x$.

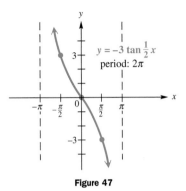

Figure 47

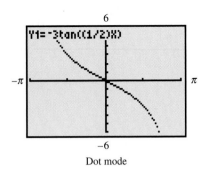

Dot mode

Figure 48

Because dot mode of the calculator is used, the pseudo-asymptotes do not appear here.

• • •

N O T E   The function defined by $y = -3 \tan \frac{1}{2}x$ in Example 4, graphed in Figures 47 and 48, has a graph that compares to the graph of $y = \tan x$ as follows.

1. The period is larger because $b = 1/2$, and $1/2 < 1$.
2. The graph is "stretched" because $a = -3$, and $|-3| > 1$.
3. Each branch of the graph goes down from left to right (that is, the function decreases) between each pair of adjacent asymptotes because $a = -3 < 0$. When $a < 0$, the graph is reflected across the x-axis, compared to the graph of $y = a \tan bx$.

• • •   **Example 5**   Graphing $y = a \cot bx$

Graph $y = \frac{1}{2} \cot 2x$.

Traditional Approach

Because this function involves the cotangent, we can locate two adjacent asymptotes by solving the equations $2x = 0$ and $2x = \pi$. The lines $x = 0$ (the y-axis) and $x = \pi/2$ are two such asymptotes. Divide the interval $0 < x < \pi/2$ into four equal parts, getting key x-values of $\pi/8$, $\pi/4$, and $3\pi/8$. Evaluating the function at these x-values gives the following key points.

$$\left(\frac{\pi}{8}, \frac{1}{2}\right) \qquad \left(\frac{\pi}{4}, 0\right) \qquad \left(\frac{3\pi}{8}, -\frac{1}{2}\right)$$

Graphing Calculator Approach

Figure 50 shows the graph and how the function was entered as Y_1.

(continued)

Joining these points with a smooth curve approaching the asymptotes gives the graph shown in Figure 49.

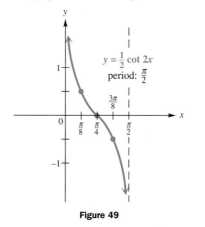

$y = \frac{1}{2} \cot 2x$

period: $\frac{\pi}{2}$

Figure 49

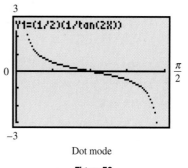

Dot mode

Figure 50

● ● ●   **Example 6**   Graphing a Tangent Function with a Vertical Translation

Graph $y = 2 + \tan x$.

Traditional Approach

Every value of y for this function will be 2 units more than the corresponding value of y in $y = \tan x$, causing the graph of $y = 2 + \tan x$ to be translated 2 units upward as compared with the graph of $y = \tan x$. See Figure 51.

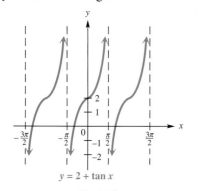

$y = 2 + \tan x$

Figure 51

Graphing Calculator Approach

To illustrate the vertical translation, observe the coordinates displayed at the bottoms of the screens in Figures 52 and 53. For $x = \pi/4 \approx .78539816$, $Y_1 = \tan x = 1$, while for the same x-value, $Y_2 = 2 + \tan x = 2 + 1 = 3$.

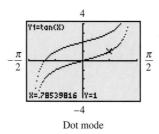

Dot mode

Figure 52

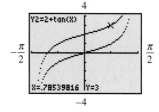

Dot mode

Figure 53

● ● ● **Example 7** Graphing a Cotangent Function with Vertical and Horizontal Translations

Graph $y = -2 - \cot\left(x - \dfrac{\pi}{4}\right)$.

Traditional Approach

Here $b = 1$, so the period is π. The graph will be translated down 2 units (because $c = -2$), reflected across the x-axis (because of the negative sign in front of the cotangent), and will have a phase shift (horizontal translation) $\pi/4$ unit to the right (because of the argument $(x - \pi/4)$). To locate adjacent asymptotes, since this function involves the cotangent, we solve the following equations:

$$x - \frac{\pi}{4} = 0, \quad \text{so } x = \frac{\pi}{4}.$$

$$x - \frac{\pi}{4} = \pi, \quad \text{so } x = \frac{5\pi}{4}.$$

Dividing the interval $\pi/4 < x < 5\pi/4$ into four equal parts and evaluating the function at the three key x-values within the interval gives these points.

$$\left(\frac{\pi}{2}, -3\right) \qquad \left(\frac{3\pi}{4}, -2\right) \qquad (\pi, -1)$$

Join these points with a smooth curve. This period of the graph, along with the one in the interval $-3\pi/4 < x < \pi/4$, is shown in Figure 54.

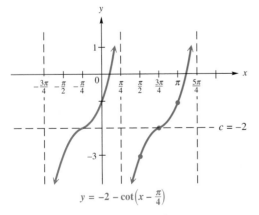

$$y = -2 - \cot\left(x - \frac{\pi}{4}\right)$$

Figure 54

Graphing Calculator Approach

Figure 55 shows the same portion of the graph as Figure 54. The pseudo-asymptotes are visible in the connected mode.

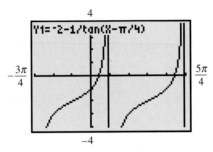

Connected mode

Figure 55

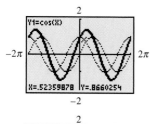

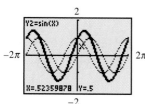

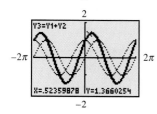

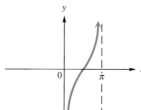

Figure 56

Addition of Ordinates New functions are often formed by adding or subtracting other functions. A function formed by combining two other functions, such as

$$y = \cos x + \sin x,$$

has historically been graphed using a method known as *addition of ordinates*. (The *x*-value of a point is sometimes called its *abscissa,* while its *y*-value is called its *ordinate.*) To apply this method to this function, we would graph the functions $y = \cos x$ and $y = \sin x$. Then, for selected values of *x*, we would add $\cos x$ and $\sin x$, and plot the points $(x, \cos x + \sin x)$. Connecting the selected points with a typical circular function-type curve would give the graph of the desired function. While this method illustrates some valuable concepts involving the arithmetic of functions, it is very time-consuming.

 With graphing calculators, this technique is more easily illustrated. Let $Y_1 = \cos x$, $Y_2 = \sin x$, and $Y_3 = Y_1 + Y_2$. Figure 56 shows the result when Y_1 and Y_2 are graphed with thin lines, and $Y_3 = \cos x + \sin x$ is graphed with a thick line. Notice that for $x = \pi/6 \approx .52359878$, $Y_1 + Y_2 = Y_3$. ■

4.3 Exercises

Concept Check Tell whether each statement is true or false. If false, tell why.

1. The smallest positive number *k* for which $x = k$ is an asymptote for the tangent function is $\pi/2$.

2. The smallest positive number *k* for which $x = k$ is an asymptote for the cotangent function is $\pi/2$.

3. The tangent and secant functions are undefined for the same values.

4. The secant and cosecant functions are undefined for the same values.

5. The graph of $y = \tan x$ in Figure 42 suggests that $\tan(-x) = \tan x$ for all *x* in the domain of tan *x*.

6. The graph of $y = \sec x$ in Figure 35 suggests that $\sec(-x) = \sec x$ for all *x* in the domain of sec *x*.

Concept Check In Exercises 7–12, match each defined function with its graph.

7. $y = -\csc x$

8. $y = -\sec x$

9. $y = -\tan x$

10. $y = -\cot x$

11. $y = \tan\left(x - \dfrac{\pi}{4}\right)$

12. $y = \cot\left(x - \dfrac{\pi}{4}\right)$

A.

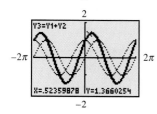

B.

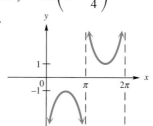

C.

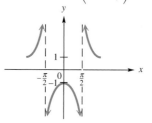

D.

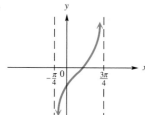

E.

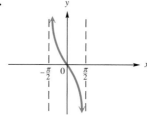

F.

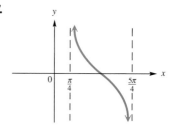

In Exercises 13–42, give a traditional or calculator-generated graph, as directed by your instructor.

Graph each defined function over a one-period interval. See Examples 1 and 2.

13. $y = \csc\left(x - \dfrac{\pi}{4}\right)$

14. $y = \sec\left(x + \dfrac{3\pi}{4}\right)$

15. $y = \sec\left(x + \dfrac{\pi}{4}\right)$

16. $y = \csc\left(x + \dfrac{\pi}{3}\right)$

17. $y = \sec\left(\dfrac{1}{2}x + \dfrac{\pi}{3}\right)$

18. $y = \csc\left(\dfrac{1}{2}x - \dfrac{\pi}{4}\right)$

19. $y = 2 + 3\sec(2x - \pi)$

20. $y = 1 - 2\csc\left(x + \dfrac{\pi}{2}\right)$

21. $y = 1 - \dfrac{1}{2}\csc\left(x - \dfrac{3\pi}{4}\right)$

22. $y = 2 + \dfrac{1}{4}\sec\left(\dfrac{1}{2}x - \pi\right)$

Graph each defined function over a one-period interval. See Examples 3–5.

23. $y = 2\tan x$

24. $y = 2\cot x$

25. $y = \dfrac{1}{2}\cot x$

26. $y = 2\tan \dfrac{1}{4}x$

27. $y = \cot 3x$

28. $y = -\cot \dfrac{1}{2}x$

Graph each defined function over a two-period interval. See Examples 6 and 7.

29. $y = \tan(2x - \pi)$

30. $y = \tan\left(\dfrac{x}{2} + \pi\right)$

31. $y = \cot\left(3x + \dfrac{\pi}{4}\right)$

32. $y = \cot\left(2x - \dfrac{3\pi}{2}\right)$

33. $y = 1 + \tan x$

34. $y = -2 + \tan x$

35. $y = 1 - \cot x$

36. $y = -2 - \cot x$

37. $y = -1 + 2\tan x$

38. $y = 3 + \dfrac{1}{2}\tan x$

39. $y = -1 + \dfrac{1}{2}\cot(2x - 3\pi)$

40. $y = -2 + 3\tan(4x + \pi)$

41. $y = \dfrac{2}{3}\tan\left(\dfrac{3}{4}x - \pi\right) - 2$

42. $y = 1 - 2\cot 2\left(x + \dfrac{\pi}{2}\right)$

43. Consider the function defined by $f(x) = -4\tan(2x + \pi)$. What is the domain of f? What is its range?

44. Consider the function defined by $g(x) = -2\csc(4x + \pi)$. What is the domain of g? What is its range?

45. *Concept Check*   If c is any number, how many solutions does the equation $c = \tan x$ have in the interval $(-2\pi, 2\pi]$?

46. *Concept Check*   If c is any number such that $-1 < c < 1$, how many solutions does the equation $c = \sec x$ have over the entire domain of the secant function?

Solve each problem.

47. *Distance of a Rotating Beacon*   A rotating beacon is located at point A next to a long wall. (See the figure.) The beacon is 4 meters from the wall. The distance d is given by

$$d = 4\tan 2\pi t,$$

where t is time measured in seconds since the beacon started rotating. (When $t = 0$, the beacon is aimed at point R. When the beacon is aimed to the right of R, the value of d is positive; d is negative if the beacon is aimed to the left of R.) Find d for the following times.

(a) $t = 0$
(b) $t = .4$
(c) $t = .8$
(d) $t = 1.2$
(e) Why is .25 a meaningless value for t?

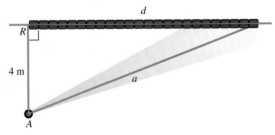

48. *Distance of a Rotating Beacon*   In the figure for Exercise 47, the distance a is given by

$$a = 4|\sec 2\pi t|.$$

Find a for the following times.
(a) $t = 0$
(b) $t = .86$
(c) $t = 1.24$

49. Simultaneously graph $y = \tan x$ and $y = x$ for $-1 \leq x \leq 1$ and $-1 \leq y \leq 1$ with a graphing calculator. Write a sentence or two describing the relationship of $\tan x$ and x for small x-values.

50. Between each pair of successive asymptotes, a portion of the graph of $y = \sec x$ or $y = \csc x$ resembles a parabola. Can each of these portions actually be a parabola? Explain.

Use a graphing calculator to graph Y_1, Y_2, and $Y_1 + Y_2$ on the same screen. Evaluate each of the three functions at $x = \pi/6$, and verify that $Y_1(\pi/6) + Y_2(\pi/6) = (Y_1 + Y_2)(\pi/6)$. See the discussion on addition of ordinates.

51. $Y_1 = \sin x, \qquad Y_2 = \sin 2x$

52. $Y_1 = \cos x, \qquad Y_2 = \sec x$

· · · · · · · · · · · · · **Relating Concepts** · · · · · · · · · ·

For individual or collaborative investigation
(Exercises 53–58)

Consider the function defined by $y = -2 - \cot\left(x - \dfrac{\pi}{4}\right)$ from Example 7. Work these exercises in order.

53. What is the smallest positive number for which $y = \cot x$ is undefined?

54. Let k represent the number you found in Exercise 53. Set $x - \dfrac{\pi}{4}$ equal to k, and solve to find the smallest positive number for which $\cot\left(x - \dfrac{\pi}{4}\right)$ is undefined.

55. Based on your answer in Exercise 54 and the fact that the cotangent function has period π, give the general form of the equations of the asymptotes of the graph of $y = -2 - \cot\left(x - \dfrac{\pi}{4}\right)$. Let n represent any integer.

56. Use the capabilities of your calculator to find the smallest positive x-intercept of the graph of this function.

57. Use the fact that the period of this function is π to find the next positive x-intercept.

58. Give the solution set of the equation

$$-2 - \cot\left(x - \frac{\pi}{4}\right) = 0$$

over all real numbers. Let n represent any integer.

· ·

Chapter 4 Summary

Key Terms & Symbols	Key Ideas
4.1 Graphs of the Sine and Cosine Functions **4.2 Translations of the Graphs of the Sine and Cosine Functions** periodic function period sine wave (sinusoid) odd function even function amplitude phase shift argument curve fitting scatter diagram	**Cosine and Sine Functions** 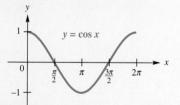 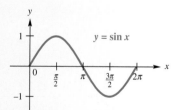 **Domain:** $(-\infty, \infty)$ **Domain:** $(-\infty, \infty)$ **Range:** $[-1, 1]$ **Range:** $[-1, 1]$ **Amplitude:** 1 **Amplitude:** 1 **Period:** 2π **Period:** 2π Assume $b > 0$. The graph of $$y = c + a \sin b(x - d) \text{ or } y = c + a \cos b(x - d)$$ has **1.** amplitude $|a|$, **2.** period $2\pi/b$, **3.** vertical translation c units up if $c > 0$ or $|c|$ units down if $c < 0$, and **4.** phase shift d units to the right if $d > 0$ or $|d|$ units to the left if $d < 0$. See pages 160 and 161 for a summary of graphing techniques.

Key Terms & Symbols	Key Ideas

4.3 Graphs of the Other Circular Functions

vertical asymptote

Secant and Cosecant Functions

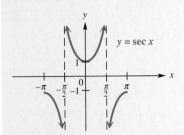

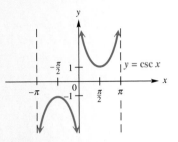

Domain: $\{x \mid x \neq \pi/2 + n\pi,$
  n any integer$\}$
Range: $(-\infty, -1] \cup [1, \infty)$
Period: 2π

Domain: $\{x \mid x \neq n\pi,$
  n any integer$\}$
Range: $(-\infty, -1] \cup [1, \infty)$
Period: 2π

See page 168 for a summary of graphing techniques.

Tangent and Cotangent Functions

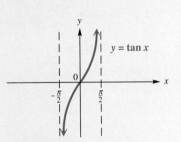

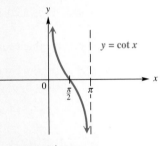

Domain: $\{x \mid x \neq \pi/2 + n\pi,$
  n any integer$\}$
Range: $(-\infty, \infty)$
Period: π

Domain: $\{x \mid x \neq n\pi,$
  n any integer$\}$
Range: $(-\infty, \infty)$
Period: π

See pages 172 and 173 for a summary of graphing techniques.

Chapter 4   Review Exercises

1. *Concept Check*   Which one of the following is true about the graph of $y = 4 \sin 2x$?

   A. It has amplitude 2 and period $\dfrac{\pi}{2}$.

   B. It has amplitude 4 and period π.
   C. Its range is $[0, 4]$.
   D. Its range is $[-4, 0]$.

2. *Concept Check*   Which one of the following is false about the graph of $y = -3 \cos \dfrac{1}{2}x$?
   A. Its range is $[-3, 3]$.
   B. Its domain is $(-\infty, \infty)$.
   C. Its amplitude is 3, and its period is 4π.
   D. Its amplitude is 3, and its period is π.

3. *Concept Check*   Which of the basic trigonometric functions can have y-value $1/2$?

4. *Concept Check*   Which of the basic trigonometric functions can have y-value 2?

For each defined function, give the amplitude, period, vertical translation, and phase shift, as applicable.

5. $y = 2 \sin x$

6. $y = \tan 3x$

7. $y = -\dfrac{1}{2} \cos 3x$

8. $y = 2 \sin 5x$

9. $y = 1 + 2 \sin \dfrac{1}{4} x$

10. $y = 3 - \dfrac{1}{4} \cos \dfrac{2}{3} x$

11. $y = 3 \cos\left(x + \dfrac{\pi}{2}\right)$

12. $y = -\sin\left(x - \dfrac{3\pi}{4}\right)$

13. $y = \dfrac{1}{2} \csc\left(2x - \dfrac{\pi}{4}\right)$

14. $y = 2 \sec(\pi x - 2\pi)$

15. $y = \dfrac{1}{3} \tan\left(3x - \dfrac{\pi}{3}\right)$

16. $y = \cot\left(\dfrac{x}{2} + \dfrac{3\pi}{4}\right)$

Concept Check   Use the concepts presented in this chapter to identify the basic circular function that satisfies the description.

17. period is π, x intercepts are of the form $n\pi$, where n is an integer

18. period is 2π, passes through the origin

19. period is 2π, passes through the point $(\pi/2, 0)$

20. period is 2π, domain is $\{x \mid x \neq n\pi$, where n is an integer$\}$

21. period is π, function is decreasing on the interval $0 < x < \pi$

22. period is 2π, has vertical asymptotes of the form $x = \pi/2 + n\pi$, where n is an integer

23. Suppose that f is a sine function with period 10 and $f(5) = 2$. Explain why $f(25) = 2$.

24. Suppose that f is a sine function with period π and $f(6\pi/5) = 1$. Explain why $f(-4\pi/5) = 1$.

In Exercises 25–42, give a traditional or calculator graph, as directed by your instructor.

Graph each defined function over a one-period interval.

25. $y = 3 \sin x$

26. $y = \dfrac{1}{2} \sec x$

27. $y = -\tan x$

28. $y = -2 \cos x$

29. $y = 2 + \cot x$

30. $y = -1 + \csc x$

31. $y = \sin 2x$

32. $y = \tan 3x$

33. $y = 3 \cos 2x$

34. $y = \dfrac{1}{2} \cot 3x$

35. $y = \cos\left(x - \dfrac{\pi}{4}\right)$

36. $y = \tan\left(x - \dfrac{\pi}{2}\right)$

37. $y = \sec\left(2x + \dfrac{\pi}{3}\right)$

38. $y = \sin\left(3x + \dfrac{\pi}{2}\right)$

39. $y = 1 + 2 \cos 3x$

40. $y = -1 - 3 \sin 2x$

41. $y = 2 \sin \pi x$

42. $y = -\dfrac{1}{2} \cos(\pi x - \pi)$

43. Explain why a function of the form $f(x) = 2 \sin(bx + c)$ has range $[-2, 2]$.

44. Explain why a function of the form $f(x) = 2 \csc(bx + c)$ has range $(-\infty, -2] \cup [2, \infty)$.

Solve each problem.

45. *(Modeling) Average Monthly Temperature*
The average monthly temperature (in °F) in Chicago, Illinois, is shown in the table.

(a) Plot the average monthly temperature over a 2-year period by letting $x = 1$ correspond to January of the first year.

(b) Determine a model function of the form $f(x) = a \sin b(x - d) + c$, where a, b, c, and d are constants.

(c) Explain the significance of each constant.

(d) Graph f together with the data on the same coordinate axes. How well does f model the data?

(e) Use the sine regression capability of a graphing calculator to find the equation of a sine curve that fits these data.

Temperatures in Chicago

Month	°F	Month	°F
Jan	25	July	74
Feb	28	Aug	75
Mar	36	Sept	66
Apr	48	Oct	55
May	61	Nov	39
June	72	Dec	28

Source: Miller, A. and J. Thompson, *Elements of Meteorology,* Charles E. Merrill Publishing Co., 1975.

46. *Viewing Angle to an Object* Let a person h_1 feet tall stand d feet from an object h_2 feet tall, where $h_2 > h_1$. Let θ be the angle of elevation to the top of the object.

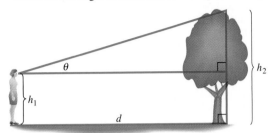

(a) Show that $d = (h_2 - h_1) \cot \theta$.
(b) Let $h_2 = 55$ and $h_1 = 5$. Graph d for the interval $0 < \theta \le \pi/2$.

 47. *(Modeling) Pollution Trends* The amount of pollution in the air fluctuates with the seasons. It is lower after heavy spring rains and higher after periods of little rain. In addition to this seasonal fluctuation, the long-term trend is upward. An idealized graph of this situation is shown in the figure.

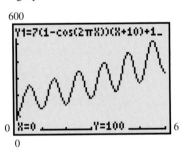

Circular functions can be used to model the fluctuating part of the pollution levels. Powers of the number e (e is the base of the natural logarithm; to six decimal places, $e = 2.718282$) can be used to model long-term growth. In fact, the pollution level in a certain area might be given by

$$y = 7(1 - \cos 2\pi x)(x + 10) + 100e^{.2x},$$

where x is time in years, with $x = 0$ representing January 1 of the base year. July 1 of the same year would be represented by $x = .5$, October 1 of the following year would be represented by $x = 1.75$, and so on. Find the pollution levels on the following dates.
(a) January 1, base year
(b) July 1, base year
(c) January 1, following year
(d) July 1, following year

48. *Lynx and Hare Populations* The figure shows the populations of lynx and hares in Canada for the years 1847–1903. The hares are food for the lynx. An increase in hare population causes an increase in lynx population some time later. The increasing lynx population then causes a decline in hare population. The two graphs have the same period.

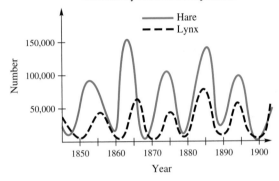

(a) Estimate the length of one period.
(b) Estimate maximum and minimum hare populations.

49. Explain how the graph of $y = 2\cos(3x + 1)$ differs from the graph of $y = 2\cos 3x + 1$.

50. If $f(x) = \sin 2x$ and $g(x) = \cos 2x$, what is $(f + g)(\pi)$? Can $(f + g)(x)$ ever exceed 2?

Chapter 4 Test

1. Consider the function defined by $y = 3 - 6\sin\left(2x + \dfrac{\pi}{2}\right)$.
   (a) What is its period?
   (b) What is the amplitude of its graph?
   (c) What is its range?
   (d) What is the y-intercept of its graph?
   (e) What is its phase shift?

Graph each defined function over a two-period interval. Identify asymptotes when applicable. (Give a traditional or calculator-generated graph, as directed by your instructor.)

2. $y = -1 + 2\sin(x + \pi)$

3. $y = -\cos 2x$

4. $y = \tan\left(x - \dfrac{\pi}{2}\right)$

5. $y = -\cot\dfrac{1}{2}x$

6. $y = -\sec x$

7. $y = 3\csc \pi x$

8. *(Modeling) Average Monthly Temperature* The average monthly temperature (in °F) in Austin, Texas, can be modeled using the trigonometric function defined by

$$f(x) = 17.5\sin\left[\dfrac{\pi}{6}(x - 4)\right] + 67.5,$$

where x is the month and $x = 1$ corresponds to January. (*Source:* Miller, A. and J. Thompson, *Elements of Meteorology*, Charles E. Merrill Publishing Co., 1975.)
  (a) Graph f over the interval $1 \le x \le 25$.
  (b) Determine the amplitude, period, phase shift, and vertical translation of f.
  (c) What is the average monthly temperature for the month of December?
  (d) Determine the maximum and minimum average monthly temperatures and the months when they occur.
  (e) What would be an approximation for the average *yearly* temperature in Austin? How is this related to the vertical translation of the sine function in the formula for f?

9. Explain why the domains of the tangent and secant functions are the same, and then give a similar explanation for the cotangent and cosecant functions.

10. Which one of the following functions is graphed here? In this screen, $X\text{scl} = \dfrac{\pi}{2}$ and $Y\text{scl} = 1$.

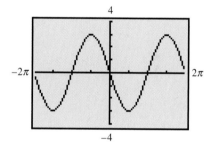

  A. $y = 3\sin x$ **B.** $y = 3\cos x$
  C. $y = -3\sin x$ **D.** $y = -3\cos x$

Chapter 4  Internet Project

Modeling Sunset Times

Sunset time at a location varies depending on the time of year. For example, see the table on the next page for sunset times at a location of 40° N on the longitude of Greenwich, England. (*Note:* Times are for the first day of each month.)

  Since the pattern for sunset time repeats itself every year, the data is periodic (with period 1 year or 12 months). This means that there exists a transformation of a sine function of the form

$$y = a\sin[b(x - d)] + c$$

that fits the data closely. You should be able to find such a function based on our discussion of modeling in this chapter. The Web site for this text at www.awl.com/lhs provides more information on this topic.

Sunset Times

Month	Sunset	Month	Sunset
January	4:46 P.M.	July	7:33 P.M.
February	5:19 P.M.	August	7:14 P.M.
March	5:52 P.M.	September	6:32 P.M.
April	6:24 P.M.	October	5:42 P.M.
May	6:55 P.M.	November	4:58 P.M.
June	7:23 P.M.	December	4:35 P.M.

5 Trigonometric Identities

5.1 **Fundamental Identities**

5.2 **Verifying Trigonometric Identities**

5.3 **Sum and Difference Identities for Cosine**

5.4 **Sum and Difference Identities for Sine and Tangent**

5.5 **Double-Angle Identities**

5.6 **Half-Angle Identities**

In 1831 Michael Faraday discovered that when a wire passes by a magnet, a small electric current is produced in the wire. This phenomenon became known as Faraday's law. Since then, people have used this property to generate massive amounts of electricity by simultaneously rotating thousands of wires near large electromagnets. The electricity supplied to most homes is produced by electric generators that rotate at 60 cycles per second. Because of this rotation, electric current alternates its direction in electrical wires and can be modeled accurately by either the sine or cosine function.*

Understanding electric current requires knowledge of the trigonometric functions themselves and trigonometric identities that relate the trigonometric functions to one another. In this chapter, we will see that these concepts can be applied to phenomena such as sound waves and stress on muscles as well as to electricity, the theme of this chapter.

Source: Weidner, R. and R. Sells, *Elementary Classical Physics,* Vol. 2, Allyn & Bacon, 1973.

5.1   Fundamental Identities

- Review of Basic Identities • Negative-Angle Identities • Fundamental Identities

Recall that an *identity* is an equation that is true for *every* value in the domain of its variable. Examples of identities include

$$5(x + 3) = 5x + 15 \quad \text{and} \quad (a + b)^2 = a^2 + 2ab + b^2.$$

In this chapter, we discuss identities involving trigonometric and circular functions.* The variables in these functions represent either angles or real numbers. The domain of the variable is assumed to be all values for which a given function is defined. For reference, we repeat the basic definitions of the trigonometric functions of an angle θ in standard position. See Figure 1.

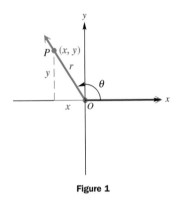

Figure 1

Trigonometric Functions

Let (x, y) be a point other than the origin on the terminal side of an angle θ in standard position. Then $r = \sqrt{x^2 + y^2}$ is the distance from the point to the origin. The six trigonometric functions of θ are defined as follows.

$$\sin \theta = \frac{y}{r} \qquad \cos \theta = \frac{x}{r} \qquad \tan \theta = \frac{y}{x} \ \ (x \neq 0)$$

$$\csc \theta = \frac{r}{y} \ \ (y \neq 0) \qquad \sec \theta = \frac{r}{x} \ \ (x \neq 0) \qquad \cot \theta = \frac{x}{y} \ \ (y \neq 0)$$

Review of Basic Identities   In Chapter 1, we used these definitions to derive the following identities.

Reciprocal Identities $\cot \theta = \dfrac{1}{\tan \theta} \qquad \csc \theta = \dfrac{1}{\sin \theta} \qquad \sec \theta = \dfrac{1}{\cos \theta}$

Quotient Identities $\tan \theta = \dfrac{\sin \theta}{\cos \theta} \qquad \cot \theta = \dfrac{\cos \theta}{\sin \theta}$

Pythagorean Identities $\sin^2 \theta + \cos^2 \theta = 1 \qquad \tan^2 \theta + 1 = \sec^2 \theta$

$$1 + \cot^2 \theta = \csc^2 \theta$$

Each of these identities leads to other forms. For example,

$$\csc \theta = \frac{1}{\sin \theta} \quad \text{gives} \quad \sin \theta = \frac{1}{\csc \theta},$$

$$\tan \theta = \frac{\sin \theta}{\cos \theta} \quad \text{gives} \quad \cos \theta \tan \theta = \sin \theta,$$

and $\qquad \tan^2 \theta + 1 = \sec^2 \theta \quad \text{gives} \quad \tan^2 \theta = \sec^2 \theta - 1.$

---

* All the identities given in this chapter are summarized in the Chapter Summary and inside the back cover of the text.

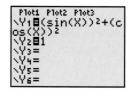

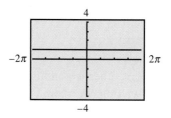

Figure 2

We can use a graphing calculator to decide whether two functions are identical. For example, to support the identity $\sin^2 x + \cos^2 x = 1$, let $Y_1 = \sin^2 x + \cos^2 x$ and let $Y_2 = 1$. See Figure 2. (Be sure your calculator is set in radian mode.) Now, graph the two functions. If it is an identity, you should see no difference in the two graphs. If the equation is not an identity, the graphs of Y_1 and Y_2 will not coincide. As a check, to guard against the possibility that the graphs are different but one of them is not showing in the window being used, graph each function separately. ∎

Negative-Angle Identities   As suggested by the circle shown in Figure 3, an angle θ having the point (x, y) on its terminal side has a corresponding angle $-\theta$ with a point $(x, -y)$ on its terminal side. From the definition of sine,

$$\sin(-\theta) = \frac{-y}{r} \qquad \text{and} \qquad \sin\theta = \frac{y}{r},$$

so $\sin(-\theta)$ and $\sin\theta$ are negatives of each other, or

$$\sin(-\theta) = -\sin\theta.$$

Figure 3 shows an angle θ in quadrant II, but the same result holds for θ in any quadrant. Also, by definition,

$$\cos(-\theta) = \frac{x}{r} \qquad \text{and} \qquad \cos\theta = \frac{x}{r},$$

so $$\cos(-\theta) = \cos\theta.$$

These formulas for $\sin(-\theta)$ and $\cos(-\theta)$ can be used to find $\tan(-\theta)$ in terms of $\tan\theta$:

$$\tan(-\theta) = \frac{\sin(-\theta)}{\cos(-\theta)} = \frac{-\sin\theta}{\cos\theta} = -\frac{\sin\theta}{\cos\theta}$$

or $$\tan(-\theta) = -\tan\theta.$$

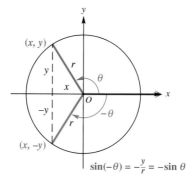

$$\sin(-\theta) = -\frac{y}{r} = -\sin\theta$$

Figure 3

The preceding three identities are **negative-angle identities.**

Fundamental Identities   As a group the identities given in this section are called the **fundamental identities.**

Fundamental Identities

$$\sec\theta = \frac{1}{\cos\theta}$$

Reciprocal Identities

$$\cot\theta = \frac{1}{\tan\theta} \qquad \sec\theta = \frac{1}{\cos\theta} \qquad \csc\theta = \frac{1}{\sin\theta}$$

Quotient Identities

$$\tan\theta = \frac{\sin\theta}{\cos\theta} \qquad \cot\theta = \frac{\cos\theta}{\sin\theta}$$

Pythagorean Identities

$$\sin^2\theta + \cos^2\theta = 1 \qquad \tan^2\theta + 1 = \sec^2\theta \qquad 1 + \cot^2\theta = \csc^2\theta$$

Negative-Angle Identities

$$\sin(-\theta) = -\sin\theta \qquad \cos(-\theta) = \cos\theta \qquad \tan(-\theta) = -\tan\theta$$

NOTE The most commonly recognized forms of the fundamental identities are given above. Throughout this chapter you must also recognize alternative forms of these identities. For example, two other forms of $\sin^2 \theta + \cos^2 \theta = 1$ are

$$\sin^2 \theta = 1 - \cos^2 \theta \quad \text{and} \quad \cos^2 \theta = 1 - \sin^2 \theta.$$

You should be able to transform the basic identities using other algebraic transformations.

One way we use these identities is to find the values of other trigonometric functions from the value of a given trigonometric function. We could find these values by using a right triangle instead, but this is a good way to practice using the fundamental identities. For example, given a value of $\tan \theta$, we can find the value of $\cot \theta$ from the identity $\cot \theta = 1/\tan \theta$. In fact, given any trigonometric function value and the quadrant in which θ lies, we can find the values of all other trigonometric functions by using identities, as in the following example.

● ● ●   **Example 1**   Finding All Trigonometric Function Values, Given One Value and the Quadrant

If $\tan \theta = -5/3$ and θ is in quadrant II, find the values of the other trigonometric functions.

Use the fundamental identities. The identity $\cot \theta = 1/\tan \theta$ leads to $\cot \theta = -3/5$. Next, find $\sec \theta$ from the identity $\tan^2 \theta + 1 = \sec^2 \theta$.

$$\left(-\frac{5}{3}\right)^2 + 1 = \sec^2 \theta$$

$$\frac{25}{9} + 1 = \sec^2 \theta$$

$$\frac{34}{9} = \sec^2 \theta$$

$$-\sqrt{\frac{34}{9}} = \sec \theta$$

$$-\frac{\sqrt{34}}{3} = \sec \theta$$

We choose the negative square root since $\sec \theta$ is negative in quadrant II. Now find $\cos \theta$:

$$\cos \theta = \frac{1}{\sec \theta} = \frac{-3}{\sqrt{34}} = -\frac{3\sqrt{34}}{34},$$

after rationalizing the denominator. Find $\sin \theta$ by using the identity $\sin^2 \theta + \cos^2 \theta = 1$, with $\cos \theta = -3/\sqrt{34}$.

$$\sin^2 \theta + \left(\frac{-3}{\sqrt{34}}\right)^2 = 1$$

$$\sin^2 \theta = 1 - \frac{9}{34}$$

$$\sin^2 \theta = \frac{25}{34}$$

$$\sin \theta = \frac{5}{\sqrt{34}}$$

$$\sin \theta = \frac{5\sqrt{34}}{34} \qquad \text{Rationalize the denominator.}$$

Use the positive square root because $\sin \theta$ is positive in quadrant II. Finally, since $\csc \theta$ is the reciprocal of $\sin \theta$,

$$\csc \theta = \frac{\sqrt{34}}{5}.$$

• • •

C A U T I O N   Several comments must be made concerning Example 1.

1. We are given $\tan \theta = -5/3$. Although $\tan \theta = (\sin \theta)/(\cos \theta)$, do *not* assume that $\sin \theta = -5$ and $\cos \theta = 3$. (Why can these values not possibly be correct?)
2. We can usually work problems of this type in more than one way. For example, after finding $\cot \theta = -3/5$, we could have then found $\csc \theta$ using the identity $1 + \cot^2 \theta = \csc^2 \theta$. The remaining function values could then be found as well.
3. The most common error made in problems like this is an incorrect sign choice for the functions. When taking the square root, be sure to choose the sign based on the quadrant of θ and the function being found.

Since $\tan \theta$, $\cot \theta$, $\sec \theta$, and $\csc \theta$ can easily be expressed in terms of $\sin \theta$ and/or $\cos \theta$, we often make such substitutions in an expression so the expression can be simplified.

• • •   **Example 2**   Simplifying an Expression by Writing in Terms of Sine and Cosine

Write $\tan \theta + \cot \theta$ in terms of $\sin \theta$ and $\cos \theta$, and simplify.

Algebraic Solution

From the fundamental identities,

$$\tan \theta + \cot \theta = \frac{\sin \theta}{\cos \theta} + \frac{\cos \theta}{\sin \theta}.$$

Graphing Calculator Support

To support the algebraic solution, graph $Y_1 = \tan x + \cot x$ and $Y_2 = \dfrac{1}{\cos x \sin x}$ in the same window. See

(continued)

Simplify this expression by adding the two fractions on the right side, using the common denominator $\cos\theta\sin\theta$.

$$\tan\theta + \cot\theta = \frac{\sin\theta}{\cos\theta} + \frac{\cos\theta}{\sin\theta}$$

$$= \frac{\sin^2\theta}{\cos\theta\sin\theta} + \frac{\cos^2\theta}{\cos\theta\sin\theta}$$

$$= \frac{\sin^2\theta + \cos^2\theta}{\cos\theta\sin\theta}$$

$$\tan\theta + \cot\theta = \frac{1}{\cos\theta\sin\theta} \qquad \sin^2\theta + \cos^2\theta = 1$$

Figure 4. The graphs coincide, indicating that the functions are equivalent.

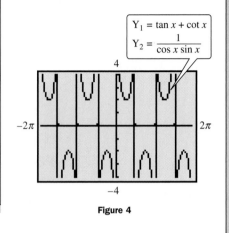

$$Y_1 = \tan x + \cot x$$
$$Y_2 = \frac{1}{\cos x \sin x}$$

Figure 4

Every trigonometric function of an angle θ or a number x can be expressed in terms of every other function.

● ● ● **Example 3** Expressing One Function in Terms of Another

Express $\cos x$ in terms of $\tan x$.

Since $\sec x$ is related to both $\cos x$ and $\tan x$ by identities, start with $\tan^2 x + 1 = \sec^2 x$. Then take reciprocals to get

$$\frac{1}{\tan^2 x + 1} = \frac{1}{\sec^2 x}$$

$$\frac{1}{\tan^2 x + 1} = \cos^2 x$$

$$\pm\sqrt{\frac{1}{\tan^2 x + 1}} = \cos x \qquad \text{Take the square root of both sides.}$$

$$\cos x = \frac{\pm 1}{\sqrt{\tan^2 x + 1}}$$

$$\cos x = \frac{\pm\sqrt{\tan^2 x + 1}}{\tan^2 x + 1}. \qquad \text{Rationalize the denominator.}$$

Choose the $+$ sign or the $-$ sign, depending on the quadrant of x. ● ● ●

CAUTION When working with trigonometric expressions and identities, be sure to write the argument of the function. For example, we would *not* write $\sin^2 + \cos^2 = 1$; an argument such as θ is necessary in this identity.

5.1 Exercises

Concept Check Fill in the blanks.

1. If $\tan x = 2.6$, then $\tan(-x) =$ _____.

2. If $\cos x = -.65$, then $\cos(-x) =$ _____.

3. If $\tan x = 1.6$, then $\cot x =$ _____.

4. If $\cos x = .8$ and $\sin x = .6$, then $\tan(-x) =$ _____.

Find $\sin s$. *See Example 1.*

5. $\cos s = \dfrac{3}{4}$, s in quadrant I

6. $\cot s = -\dfrac{1}{3}$, s in quadrant IV

7. $\cos s = \dfrac{\sqrt{5}}{5}$, $\tan s < 0$

8. $\tan s = -\dfrac{\sqrt{7}}{2}$, $\sec s > 0$

9. $\sec s = \dfrac{11}{4}$, $\tan s < 0$

10. $\csc s = -\dfrac{8}{5}$

11. Why is it unnecessary to give the quadrant of s in Exercise 10?

Concept Check For each graph, determine whether $f(-x) = f(x)$ or $f(-x) = -f(x)$.

12.
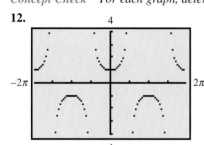

13.
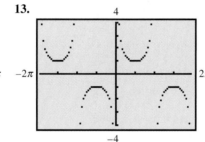

14.
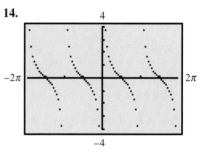

Use the fundamental identities to find the remaining five trigonometric functions of θ. *See Example 1.*

15. $\sin \theta = \dfrac{2}{3}$, θ in quadrant II

16. $\cos \theta = \dfrac{1}{5}$, θ in quadrant I

17. $\tan \theta = -\dfrac{1}{4}$, θ in quadrant IV

18. $\csc \theta = -\dfrac{5}{2}$, θ in quadrant III

19. $\cot \theta = \dfrac{4}{3}$, $\sin \theta > 0$

20. $\sin \theta = -\dfrac{4}{5}$, $\cos \theta < 0$

21. $\sec \theta = \dfrac{4}{3}$, $\sin \theta < 0$

22. $\cos \theta = -\dfrac{1}{4}$, $\sin \theta > 0$

Concept Check For each trigonometric expression in Column I, choose the expression from Column II that completes a fundamental identity.

I

23. $\dfrac{\cos x}{\sin x}$

24. $\tan x$

25. $\cos(-x)$

26. $\tan^2 x + 1$

27. 1

II

A. $\sin^2 x + \cos^2 x$

B. $\cot x$

C. $\sec^2 x$

D. $\dfrac{\sin x}{\cos x}$

E. $\cos x$

Concept Check For each expression in Column I, choose the expression from Column II that completes an identity. You will have to rewrite one or both expressions, using a fundamental identity, to recognize the matches.

I **II**

28. $-\tan x \cos x$

A. $\dfrac{\sin^2 x}{\cos^2 x}$

29. $\sec^2 x - 1$

B. $\dfrac{1}{\sec^2 x}$

30. $\dfrac{\sec x}{\csc x}$

C. $\sin(-x)$

31. $1 + \sin^2 x$

D. $\csc^2 x - \cot^2 x + \sin^2 x$

32. $\cos^2 x$

E. $\tan x$

 33. A student writes "$1 + \cot^2 = \csc^2$." Comment on this student's work.

34. Another student makes the following claim: "Since $\sin^2 \theta + \cos^2 \theta = 1$, I should be able to also say that $\sin \theta + \cos \theta = 1$ if I take the square root of both sides." Comment on this student's statement.

35. *Concept Check* Suppose that $\cos \theta = x/(x + 1)$. Find $\sin \theta$.

36. *Concept Check* Find $\tan \alpha$ if $\sec \alpha = (p + 4)/p$.

Use the fundamental identities to get an equivalent expression involving only sines and cosines, and then simplify it. See Example 2.

37. $\cot \theta \sin \theta$

38. $\sec \theta \cot \theta \sin \theta$

39. $\cos \theta \csc \theta$

40. $\cot^2 \theta (1 + \tan^2 \theta)$

41. $\sin^2 \theta (\csc^2 \theta - 1)$

42. $(\sec \theta - 1)(\sec \theta + 1)$

43. $(1 - \cos \theta)(1 + \sec \theta)$

44. $\dfrac{\cos \theta + \sin \theta}{\sin \theta}$

45. $\dfrac{\cos^2 \theta - \sin^2 \theta}{\sin \theta \cos \theta}$

46. $\dfrac{1 - \sin^2 \theta}{1 + \cot^2 \theta}$

47. $\tan \theta + \cot \theta$

48. $(\sec \theta + \csc \theta)(\cos \theta - \sin \theta)$

49. $\sin \theta (\csc \theta - \sin \theta)$

50. $\dfrac{1 + \tan^2 \theta}{1 + \cot^2 \theta}$

51. $\sin^2 \theta + \tan^2 \theta + \cos^2 \theta$

52. $\dfrac{\tan(-\theta)}{\sec \theta}$

Complete this chart so that each trigonometric function in the column at the left is expressed in terms of the functions given across the top. See Example 3.

	$\sin \theta$	$\cos \theta$	$\tan \theta$	$\cot \theta$	$\sec \theta$	$\csc \theta$
53. $\sin \theta$	$\sin \theta$	$\pm\sqrt{1 - \cos^2 \theta}$	$\dfrac{\pm\tan \theta\sqrt{1 + \tan^2 \theta}}{1 + \tan^2 \theta}$			$\dfrac{1}{\csc \theta}$
54. $\cos \theta$		$\cos \theta$	$\dfrac{\pm\sqrt{\tan^2 \theta + 1}}{\tan^2 \theta + 1}$		$\dfrac{1}{\sec \theta}$	
55. $\tan \theta$			$\tan \theta$	$\dfrac{1}{\cot \theta}$		
56. $\cot \theta$			$\dfrac{1}{\tan \theta}$	$\cot \theta$	$\dfrac{\pm\sqrt{\sec^2 \theta - 1}}{\sec^2 \theta - 1}$	
57. $\sec \theta$		$\dfrac{1}{\cos \theta}$			$\sec \theta$	
58. $\csc \theta$	$\dfrac{1}{\sin \theta}$					$\csc \theta$

59. *Concept Check* Let $\cos x = \dfrac{1}{5}$. Find all possible values for $\dfrac{\sec x - \tan x}{\sin x}$.

60. *Concept Check* Let $\csc x = -3$. Find all possible values for $\dfrac{\sin x + \cos x}{\sec x}$.

· · · · · · · · · · · · · · **Relating Concepts** · · · · · · · · · · · · ·

For individual or collaborative investigation

(Exercises 61–65)

*In Chapter 4 we graphed functions defined by $y = c + a \cdot f[b(x - d)]$ with the assumption that $b > 0$. To see what happens when $b < 0$, **work Exercises 61–65 in order.***

61. Use a negative-angle identity to write $y = \sin(-2x)$ as a function of $2x$.

62. How does your answer to Exercise 61 relate to $y = \sin(2x)$?

63. Use a negative-angle identity to write $y = \cos(-4x)$ as a function of $4x$.

64. How does your answer to Exercise 63 relate to $y = \cos(4x)$?

65. Use your results from Exercises 61–64 to rewrite the following with a positive value of b.
  (a) $y = \sin(-4x)$ **(b)** $y = \cos(-2x)$ **(c)** $y = -5\sin(-3x)$

· ·

Concept Check *Use the concepts of this section to answer each question.*

66. How does the graph of $y = \sec(-x)$ compare to that of $y = \sec x$?

67. How does the graph of $y = \csc(-x)$ compare to that of $y = \csc x$?

68. How does the graph of $y = \tan(-x)$ compare to that of $y = \tan x$?

69. How does the graph of $y = \cot(-x)$ compare to that of $y = \cot x$?

Use a graphing calculator to decide whether each equation is an identity. See Example 2. (Hint: In Exercises 74 and 75, graph the functions of x for a few different values of y (in radians).)

70. $\cos 2x = 1 - 2\sin^2 x$ **71.** $2 \sin s = \sin 2s$ **72.** $\sin x = \sqrt{1 - \cos^2 x}$

73. $\cos 2x = \cos^2 x - \sin^2 x$ **74.** $\cos(x - y) = \cos x - \cos y$ **75.** $\sin(x + y) = \sin x + \sin y$

5.2 Verifying Trigonometric Identities

• **Verify Identities by Working with One Side** • **Verify Identities by Working with Both Sides**

One of the skills required for more advanced work in mathematics, especially in calculus, is the ability to use trigonometric identities to write trigonometric expressions in alternative forms. This skill is developed by using the fundamental identities to verify that a trigonometric equation is an identity (for those values of the variable for which it is defined). Here are some hints to help you get started.

Hints for Verifying Identities

1. Learn the fundamental identities given in the last section. Whenever you see either side of a fundamental identity, the other side should come to mind. Also, be aware of equivalent forms of the fundamental identities. For example, $\sin^2 \theta = 1 - \cos^2 \theta$ is an alternative form of $\sin^2 \theta + \cos^2 \theta = 1$.

2. Try to rewrite the more complicated side of the equation so that it is identical to the simpler side.

3. It is often helpful to express all trigonometric functions in the equation in terms of sine and cosine and then simplify the result.

4. Usually any factoring or indicated algebraic operations should be performed. For example, the expression $\sin^2 x + 2 \sin x + 1$ can be factored as $(\sin x + 1)^2$. The sum or difference of two trigonometric expressions, such as

$$\frac{1}{\sin \theta} + \frac{1}{\cos \theta},$$

can be added or subtracted in the same way as any other rational expression.

$$\frac{1}{\sin \theta} + \frac{1}{\cos \theta} = \frac{\cos \theta}{\sin \theta \cos \theta} + \frac{\sin \theta}{\sin \theta \cos \theta}$$

$$= \frac{\cos \theta + \sin \theta}{\sin \theta \cos \theta}$$

5. As you select substitutions, keep in mind the side you are not changing, because it represents your goal. For example, to verify the identity

$$\tan^2 x + 1 = \frac{1}{\cos^2 x},$$

try to think of an identity that relates $\tan x$ to $\cos x$. Here, since $\sec x = 1/\cos x$ and $\sec^2 x = \tan^2 x + 1$, the secant function is the best link between the two sides.

6. If an expression contains $1 + \sin x$, multiplying both numerator and denominator by $1 - \sin x$ would give $1 - \sin^2 x$, which could be replaced with $\cos^2 x$. Similar results for $1 - \sin x$, $1 + \cos x$, and $1 - \cos x$ may be useful.

CAUTION Verifying identities is not the same as solving equations. Techniques used in solving equations, such as adding the same terms to both sides, or multiplying both sides by the same term, are not valid when working with identities since you are starting with a statement (to be verified) that may not be true.

Verify Identities by Working with One Side   To avoid the temptation to use algebraic properties of equations to verify identities, *work with only one side and rewrite it to match the other side.*

● ● ●   **Example 1**   Verifying an Identity (Working with One Side)

Verify that the following equation is an identity.

$$\cot s + 1 = \csc s(\cos s + \sin s)$$

Algebraic Solution

Use the fundamental identities to rewrite one side of the equation so that it is identical to the other side. Since the right side is more complicated, we work with it. Here we use the third hint, and change all the trigonometric functions to sine or cosine.

Steps   **Reasons**

$$\csc s(\cos s + \sin s) = \frac{1}{\sin s}(\cos s + \sin s) \qquad \csc s = \frac{1}{\sin s}$$

$$= \frac{\cos s}{\sin s} + \frac{\sin s}{\sin s} \qquad \text{Distributive property}$$

$$= \cot s + 1 \qquad \frac{\cos s}{\sin s} = \cot s;$$
$$\frac{\sin s}{\sin s} = 1$$

The given equation is an identity since the right side equals the left side.

Graphing Calculator Support

To support the algebraic solution, graph the two expressions in the same window. Let

$$Y_1 = \cot x + 1 = (1/\tan x) + 1$$

and

$$Y_2 = \csc x(\cos x + \sin x)$$
$$= (\cos x + \sin x)/\sin x.$$

Notice that we wrote each expression in a form that can be entered in the calculator. The two graphs coincide, as shown in Figure 5, which supports the algebraic result.

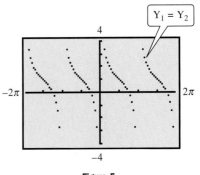

Figure 5

● ● ●

● ● ●   **Example 2**   Verifying an Identity (Working with One Side)

Verify that the following equation is an identity.

$$\tan^2 \alpha (1 + \cot^2 \alpha) = \frac{1}{1 - \sin^2 \alpha}$$

Algebraic Solution

Work with the more complicated left side, as suggested in the second hint.

$$\tan^2 \alpha (1 + \cot^2 \alpha) = \tan^2 \alpha + \tan^2 \alpha \cot^2 \alpha \qquad \text{Distributive property}$$

$$= \tan^2 \alpha + \tan^2 \alpha \cdot \frac{1}{\tan^2 \alpha} \qquad \cot^2 \alpha = \frac{1}{\tan^2 \alpha}$$

$$= \tan^2 \alpha + 1 \qquad \tan^2 \alpha \cdot \frac{1}{\tan^2 \alpha} = 1$$

$$= \sec^2 \alpha \qquad \tan^2 \alpha + 1 = \sec^2 \alpha$$

$$= \frac{1}{\cos^2 \alpha} \qquad \sec^2 \alpha = \frac{1}{\cos^2 \alpha}$$

$$= \frac{1}{1 - \sin^2 \alpha} \qquad \cos^2 \alpha = 1 - \sin^2 \alpha$$

Since the left side is identical to the right side, the given equation is an identity.

cost sec t = 1

Graphing Calculator Support

The table feature also can be used to support an algebraic result, although it can be misleading because not every point in an interval is represented. In the table of values in Figure 6(a),

$$Y_1 = \tan^2 x (1 + \cot^2 x)$$

and $$Y_2 = \frac{1}{1 - \sin^2 x}.$$

The table supports the identity for selected values in $[1, 4]$. The screen in Figure 6(b) further supports the identity.

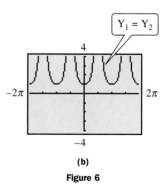

(a)

(b)

Figure 6

● ● ●

● ● ●   **Example 3**   Verifying an Identity (Working with One Side)

Verify that the following equation is an identity.

$$\frac{\tan t - \cot t}{\sin t \cos t} = \sec^2 t - \csc^2 t$$

Since the left side is more complicated, transform it to match the right side.

$$\frac{\tan t - \cot t}{\sin t \cos t} = \frac{\tan t}{\sin t \cos t} - \frac{\cot t}{\sin t \cos t} \qquad \frac{a-b}{c} = \frac{a}{c} - \frac{b}{c}$$

$$= \tan t \cdot \frac{1}{\sin t \cos t} - \cot t \cdot \frac{1}{\sin t \cos t} \qquad \frac{a}{b} = a \cdot \frac{1}{b}$$

$$= \frac{\sin t}{\cos t} \cdot \frac{1}{\sin t \cos t} - \frac{\cos t}{\sin t} \cdot \frac{1}{\sin t \cos t} \qquad \tan t = \frac{\sin t}{\cos t};$$
$$\cot t = \frac{\cos t}{\sin t}$$

$$= \frac{1}{\cos^2 t} - \frac{1}{\sin^2 t}$$

$$= \sec^2 t - \csc^2 t \qquad \frac{1}{\cos^2 t} = \sec^2 t;$$
$$\frac{1}{\sin^2 t} = \csc^2 t$$

Here, the hint about writing all trigonometric functions in terms of sine and cosine was used in the third line of the solution.   ● ● ●

● ● ●   **Example 4**   Verifying an Identity (Working with One Side)

Verify that the following equation is an identity.

$$\frac{\cos x}{1 - \sin x} = \frac{1 + \sin x}{\cos x}$$

Work on the right side. Use the last hint given at the beginning of the section to multiply numerator and denominator on the right by $1 - \sin x$.

$$\frac{1 + \sin x}{\cos x} = \frac{(1 + \sin x)(1 - \sin x)}{\cos x(1 - \sin x)} \qquad \text{Multiply by 1.}$$

$$= \frac{1 - \sin^2 x}{\cos x(1 - \sin x)}$$

$$= \frac{\cos^2 x}{\cos x(1 - \sin x)} \qquad 1 - \sin^2 x = \cos^2 x$$

$$= \frac{\cos x}{1 - \sin x} \qquad \text{Write in lowest terms.} \qquad ● ● ●$$

Verify Identities by Working with Both Sides   If both sides of an identity appear to be equally complex, the identity can be verified by working independently on the left side and on the right side, until each side is changed into some common third result. *Each step, on each side, must be reversible.* With all steps reversible, the procedure is as follows.

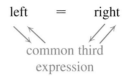

left = right

common third
expression

The left side leads to the third expression, which leads back to the right side. This procedure is just a shortcut for the procedure used in Examples 1–4: the left side is changed into the right side, but by going through an intermediate step.

● ● ● **Example 5** Verifying an Identity (Working with Both Sides)

Verify that the following equation is an identity.

$$\frac{\sec \alpha + \tan \alpha}{\sec \alpha - \tan \alpha} = \frac{1 + 2 \sin \alpha + \sin^2 \alpha}{\cos^2 \alpha}$$

Both sides appear equally complex, so verify the identity by changing each side into a common third expression. Work first on the left, multiplying numerator and denominator by $\cos \alpha$.

$$\frac{\sec \alpha + \tan \alpha}{\sec \alpha - \tan \alpha} = \frac{(\sec \alpha + \tan \alpha)\cos \alpha}{(\sec \alpha - \tan \alpha)\cos \alpha} \qquad \frac{\cos \alpha}{\cos \alpha} = 1;$$
multiplicative identity

$$= \frac{\sec \alpha \cos \alpha + \tan \alpha \cos \alpha}{\sec \alpha \cos \alpha - \tan \alpha \cos \alpha} \qquad \text{Distributive property}$$

$$= \frac{1 + \tan \alpha \cos \alpha}{1 - \tan \alpha \cos \alpha} \qquad \sec \alpha \cos \alpha = 1$$

$$= \frac{1 + \dfrac{\sin \alpha}{\cos \alpha} \cdot \cos \alpha}{1 - \dfrac{\sin \alpha}{\cos \alpha} \cdot \cos \alpha} \qquad \tan \alpha = \frac{\sin \alpha}{\cos \alpha}$$

$$= \frac{1 + \sin \alpha}{1 - \sin \alpha}$$

On the right side of the original equation, begin by factoring.

$$\frac{1 + 2 \sin \alpha + \sin^2 \alpha}{\cos^2 \alpha} = \frac{(1 + \sin \alpha)^2}{\cos^2 \alpha} \qquad a^2 + 2ab + b^2 = (a + b)^2$$

$$= \frac{(1 + \sin \alpha)^2}{1 - \sin^2 \alpha} \qquad \cos^2 \alpha = 1 - \sin^2 \alpha$$

$$= \frac{(1 + \sin \alpha)^2}{(1 + \sin \alpha)(1 - \sin \alpha)} \qquad \text{Factor } 1 - \sin^2 \alpha.$$

$$= \frac{1 + \sin \alpha}{1 - \sin \alpha} \qquad \text{Write in lowest terms.}$$

We now have shown that

$$\frac{\sec \alpha + \tan \alpha}{\sec \alpha - \tan \alpha} = \frac{1 + \sin \alpha}{1 - \sin \alpha} = \frac{1 + 2 \sin \alpha + \sin^2 \alpha}{\cos^2 \alpha},$$

verifying that the original equation is an identity.

● ● ●

C A U T I O N   Use this method *only* if the steps are reversible.

There are usually several ways to verify a given identity. For instance, another way to begin verifying the identity in Example 5 is to work on the left as follows.

$$\frac{\sec \alpha + \tan \alpha}{\sec \alpha - \tan \alpha} = \frac{\dfrac{1}{\cos \alpha} + \dfrac{\sin \alpha}{\cos \alpha}}{\dfrac{1}{\cos \alpha} - \dfrac{\sin \alpha}{\cos \alpha}} \qquad \text{Fundamental identities}$$

$$= \frac{\dfrac{1 + \sin \alpha}{\cos \alpha}}{\dfrac{1 - \sin \alpha}{\cos \alpha}} \qquad \text{Add fractions; subtract fractions.}$$

$$= \frac{1 + \sin \alpha}{1 - \sin \alpha} \qquad \text{Divide fractions.}$$

Compare this with the result shown in Example 5 for the right side to see that the two sides indeed agree.

Looking Ahead to Calculus

Much of our work with identities in this chapter is preparation for calculus, which uses many of the identities we verify here. Some calculus problems are simplified by making an appropriate trigonometric substitution, as shown in the Connections box.

C O N N E C T I O N S   Trigonometric substitutions and identities make it possible to replace an expression such as $\sqrt{9 + x^2}$ with a trigonometric expression without a radical. To do this, we choose $x = 3 \tan \theta$. The reason for this choice will become clear as we continue.

Letting $x = 3 \tan \theta$ gives

$$\sqrt{9 + x^2} = \sqrt{9 + (3 \tan \theta)^2}$$
$$= \sqrt{9 + 9 \tan^2 \theta}$$
$$= \sqrt{9(1 + \tan^2 \theta)}$$
$$= 3\sqrt{1 + \tan^2 \theta}$$
$$= 3\sqrt{\sec^2 \theta}.$$

In the interval $(0, \pi/2)$, the value of $\sec \theta$ is positive, giving

$$\sqrt{9 + x^2} = 3 \sec \theta.$$

For Discussion or Writing

Substitute $\cos \theta$ for x in $\sqrt{(1 - x^2)^3}$ and simplify. Why is $\cos \theta$ an appropriate choice here?

5.2 Exercises

Perform each indicated operation and simplify the result.

1. $\cot \theta + \dfrac{1}{\cot \theta}$

2. $\dfrac{\sec x}{\csc x} + \dfrac{\csc x}{\sec x}$

3. $\tan s(\cot s + \csc s)$

4. $\cos \beta(\sec \beta + \csc \beta)$

5. $\dfrac{1}{\csc^2 \theta} + \dfrac{1}{\sec^2 \theta}$

6. $\dfrac{1}{\sin \alpha - 1} - \dfrac{1}{\sin \alpha + 1}$

7. $\dfrac{\cos x}{\sec x} + \dfrac{\sin x}{\csc x}$

8. $\dfrac{\cos \gamma}{\sin \gamma} + \dfrac{\sin \gamma}{1 + \cos \gamma}$

9. $(1 + \sin t)^2 + \cos^2 t$

10. $(1 + \tan s)^2 - 2 \tan s$

11. $\dfrac{1}{1 + \cos x} - \dfrac{1}{1 - \cos x}$

12. $(\sin \alpha - \cos \alpha)^2$

Factor each trigonometric expression.

13. $\sin^2 \gamma - 1$

14. $\sec^2 \theta - 1$

15. $(\sin x + 1)^2 - (\sin x - 1)^2$

16. $(\tan x + \cot x)^2 - (\tan x - \cot x)^2$

17. $2 \sin^2 x + 3 \sin x + 1$

18. $4 \tan^2 \beta + \tan \beta - 3$

19. $\cos^4 x + 2 \cos^2 x + 1$

20. $\cot^4 x + 3 \cot^2 x + 2$

21. $\sin^3 x - \cos^3 x$

22. $\sin^3 \alpha + \cos^3 \alpha$

Each expression simplifies to a constant, a single circular function, or a power of a circular function. Use fundamental identities to simplify each expression.

23. $\tan \theta \cos \theta$

24. $\cot \alpha \sin \alpha$

25. $\sec r \cos r$

26. $\cot t \tan t$

27. $\dfrac{\sin \beta \tan \beta}{\cos \beta}$

28. $\dfrac{\csc \theta \sec \theta}{\cot \theta}$

29. $\sec^2 x - 1$

30. $\csc^2 t - 1$

31. $\dfrac{\sin^2 x}{\cos^2 x} + \sin x \csc x$

32. $\dfrac{1}{\tan^2 \alpha} + \cot \alpha \tan \alpha$

In Exercises 33–68, verify that each trigonometric equation is an identity. See Examples 1–5.

33. $\dfrac{\cot \theta}{\csc \theta} = \cos \theta$

34. $\dfrac{\tan \alpha}{\sec \alpha} = \sin \alpha$

35. $\dfrac{1 - \sin^2 \beta}{\cos \beta} = \cos \beta$

36. $\dfrac{\tan^2 \gamma + 1}{\sec \gamma} = \sec \gamma$

37. $\cos^2 \theta(\tan^2 \theta + 1) = 1$

38. $\sin^2 \beta(1 + \cot^2 \beta) = 1$

39. $\cot s + \tan s = \sec s \csc s$

40. $\sin^2 \alpha + \tan^2 \alpha + \cos^2 \alpha = \sec^2 \alpha$

41. $\dfrac{\cos \alpha}{\sec \alpha} + \dfrac{\sin \alpha}{\csc \alpha} = \sec^2 \alpha - \tan^2 \alpha$

42. $\dfrac{\sin^2 \gamma}{\cos \gamma} = \sec \gamma - \cos \gamma$

43. $\sin^4 \theta - \cos^4 \theta = 2 \sin^2 \theta - 1$

44. $\dfrac{\cos \theta}{\sin \theta \cot \theta} = 1$

45. $(1 - \cos^2 \alpha)(1 + \cos^2 \alpha) = 2 \sin^2 \alpha - \sin^4 \alpha$

46. $\tan^2 \gamma \sin^2 \gamma = \tan^2 \gamma + \cos^2 \gamma - 1$

47. $\dfrac{\cos \theta + 1}{\tan^2 \theta} = \dfrac{\cos \theta}{\sec \theta - 1}$

48. $\dfrac{(\sec \theta - \tan \theta)^2 + 1}{\sec \theta \csc \theta - \tan \theta \csc \theta} = 2 \tan \theta$

49. $\dfrac{1}{1 - \sin \theta} + \dfrac{1}{1 + \sin \theta} = 2 \sec^2 \theta$

50. $\dfrac{1}{\sec \alpha - \tan \alpha} = \sec \alpha + \tan \alpha$

51. $\dfrac{\tan s}{1 + \cos s} + \dfrac{\sin s}{1 - \cos s} = \cot s + \sec s \csc s$

52. $\dfrac{1 - \cos x}{1 + \cos x} = (\cot x - \csc x)^2$

53. $\dfrac{\cot \alpha + 1}{\cot \alpha - 1} = \dfrac{1 + \tan \alpha}{1 - \tan \alpha}$

54. $\dfrac{1}{\tan \alpha - \sec \alpha} + \dfrac{1}{\tan \alpha + \sec \alpha} = -2 \tan \alpha$

55. $\sin^2 \alpha \sec^2 \alpha + \sin^2 \alpha \csc^2 \alpha = \sec^2 \alpha$

56. $\dfrac{\csc \theta + \cot \theta}{\tan \theta + \sin \theta} = \cot \theta \csc \theta$

57. $\sec^4 x - \sec^2 x = \tan^4 x + \tan^2 x$

58. $\dfrac{1 - \sin \theta}{1 + \sin \theta} = \sec^2 \theta - 2 \sec \theta \tan \theta + \tan^2 \theta$

59. $\sin \theta + \cos \theta = \dfrac{\sin \theta}{1 - \dfrac{\cos \theta}{\sin \theta}} + \dfrac{\cos \theta}{1 - \dfrac{\sin \theta}{\cos \theta}}$

60. $\dfrac{\sin \theta}{1 - \cos \theta} - \dfrac{\sin \theta \cos \theta}{1 + \cos \theta} = \csc \theta(1 + \cos^2 \theta)$

61. $\dfrac{\sec^4 s - \tan^4 s}{\sec^2 s + \tan^2 s} = \sec^2 s - \tan^2 s$

62. $\dfrac{\cot^2 t - 1}{1 + \cot^2 t} = 1 - 2 \sin^2 t$

63. $\dfrac{\tan^2 t - 1}{\sec^2 t} = \dfrac{\tan t - \cot t}{\tan t + \cot t}$

64. $(1 + \sin x + \cos x)^2 = 2(1 + \sin x)(1 + \cos x)$

65. $\dfrac{1 + \cos x}{1 - \cos x} - \dfrac{1 - \cos x}{1 + \cos x} = 4 \cot x \csc x$

66. $(\sec \alpha - \tan \alpha)^2 = \dfrac{1 - \sin \alpha}{1 + \sin \alpha}$

67. $(\sec \alpha + \csc \alpha)(\cos \alpha - \sin \alpha) = \cot \alpha - \tan \alpha$

68. $\dfrac{\sin^4 \alpha - \cos^4 \alpha}{\sin^2 \alpha - \cos^2 \alpha} = 1$

69. A student claims that the equation
$$\cos \theta + \sin \theta = 1$$
is an identity, since by letting $\theta = 90°$ (or $\pi/2$ radians) we get $0 + 1 = 1$, a true statement. Comment on this student's reasoning.

70. The table suggests that $Y_1 = Y_2$ is an identity. Here, $Y_1 = \sin x$ and $Y_2 = \sqrt{1 - \cos^2 x}$. Is $\sin x = \sqrt{1 - \cos^2 x}$ true for all real numbers x? Explain.

X	Y1	Y2
0	0	0
.3927	.38268	.38268
.7854	.70711	.70711
1.1781	.92388	.92388
1.5708	1	1
1.9635	.92388	.92388
2.3562	.70711	.70711

X=0

Concept Check   Graph each expression and conjecture an identity. Then prove your conjecture.

71. $(\sec \theta + \tan \theta)(1 - \sin \theta)$

72. $(\csc \theta + \cot \theta)(\sec \theta - 1)$

73. $\dfrac{\cos \theta + 1}{\sin \theta + \tan \theta}$

74. $\tan \theta \sin \theta + \cos \theta$

Graph the expressions on each side of the equals sign to determine whether the equation might be an identity. (Note: Use a domain whose length is at least 2π.) If the equation looks like an identity, prove it algebraically. See Example 1.

75. $\dfrac{2 + 5 \cos s}{\sin s} = 2 \csc s + 5 \cot s$

76. $1 + \cot^2 s = \dfrac{\sec^2 s}{\sec^2 s - 1}$

77. $\dfrac{\tan s - \cot s}{\tan s + \cot s} = 2 \sin^2 s$

78. $\dfrac{1}{1 + \sin s} + \dfrac{1}{1 - \sin s} = \sec^2 s$

79. $\dfrac{1 - \tan^2 s}{1 + \tan^2 s} = \cos^2 s - \sin s$

80. $\dfrac{\sin^3 s - \cos^3 s}{\sin s - \cos s} = \sin^2 s + 2 \sin s \cos s + \cos^2 s$

Decide whether each equation might be an identity by using a table of values. Scroll through a domain with length at least 2π. See Example 2.

81. $\sin^2 s + \cos^2 s = \dfrac{1}{2}(1 - \cos 4s)$

82. $\cos 3s = 3 \cos s + 4 \cos^3 s$

83. $\tan^2 x - \sin^2 x = (\tan x \sin x)^2$

84. $\dfrac{\cot \theta}{\csc \theta + 1} = \sec \theta - \tan \theta$

By substituting a number for s or t, show that the equation is not an identity for all real numbers s and t.

85. $\sin(\csc s) = 1$ **86.** $\sqrt{\cos^2 s} = \cos s$ **87.** $\csc t = \sqrt{1 + \cot^2 t}$ **88.** $\cos t = \sqrt{1 - \sin^2 t}$

89. Let $\tan \theta = t$ and show that $\sin \theta \cos \theta = \dfrac{t}{t^2 + 1}$.

90. An equation that is an identity has an infinite number of solutions. If an equation has an infinite number of solutions, is it necessarily an identity? Discuss.

91. Explain why the method described in the text involving working on both sides of an identity to show that each side is equal to the same expression is a valid method of verifying an identity. When using this method, what must be true about each step taken? (*Hint:* See the discussion preceding Example 5.)

5.3 Sum and Difference Identities for Cosine

• **Difference Identity for Cosine** • **Sum Identity for Cosine** • **Cofunction Identities** • **Applying the Sum and Difference Identities**

Difference Identity for Cosine Several examples presented throughout this book should have convinced you by now that $\cos(A - B)$ does *not* equal $\cos A - \cos B$. For example, if $A = \pi/2$ and $B = 0$,

$$\cos(A - B) = \cos\left(\frac{\pi}{2} - 0\right) = \cos \frac{\pi}{2} = 0,$$

while

$$\cos A - \cos B = \cos \frac{\pi}{2} - \cos 0 = 0 - 1 = -1.$$

We derive the actual formula for $\cos(A - B)$ in this section. Start by locating angles A and B in standard position on a unit circle, with $B < A$. Let S and Q be the points where the terminal sides of angles A and B, respectively, intersect the circle. Locate point R on the unit circle so that angle POR equals the difference $A - B$. See Figure 7.

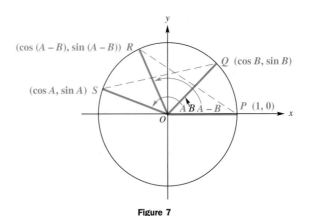

Figure 7

Point Q is on the unit circle, so by the work with circular functions in Chapter 3, the x-coordinate of Q is given by the cosine of angle B. The y-coordinate of Q is given by the sine of angle B.

Q has coordinates $(\cos B, \sin B)$.

In the same way,

$$S \text{ has coordinates } (\cos A, \sin A),$$

and

$$R \text{ has coordinates } (\cos(A - B), \sin(A - B)).$$

Angle SOQ also equals $A - B$. Since the central angles SOQ and POR are equal, chords PR and SQ are equal. By the distance formula, since $PR = SQ$,

$$\sqrt{[\cos(A - B) - 1]^2 + [\sin(A - B) - 0]^2}$$
$$= \sqrt{(\cos A - \cos B)^2 + (\sin A - \sin B)^2}.$$

Squaring both sides and clearing parentheses gives

$$\cos^2(A - B) - 2\cos(A - B) + 1 + \sin^2(A - B)$$
$$= \cos^2 A - 2\cos A \cos B + \cos^2 B + \sin^2 A - 2\sin A \sin B + \sin^2 B.$$

Since $\sin^2 x + \cos^2 x = 1$ for any value of x, we can rewrite the equation as

$$2 - 2\cos(A - B) = 2 - 2\cos A \cos B - 2\sin A \sin B$$

$$\cos(A - B) = \cos A \cos B + \sin A \sin B.$$

This is the identity for $\cos(A - B)$. Although Figure 7 shows angles A and B in the second and first quadrants, respectively, this result is the same for any values of these angles.

Sum Identity for Cosine   To find a similar expression for $\cos(A + B)$, rewrite $A + B$ as $A - (-B)$ and use the identity for $\cos(A - B)$.

$$\cos(A + B) = \cos[A - (-B)]$$

$$= \cos A \cos(-B) + \sin A \sin(-B) \quad \text{Cosine difference identity}$$

$$= \cos A \cos B + \sin A(-\sin B) \quad \text{Negative-angle identities}$$

$$\cos(A + B) = \cos A \cos B - \sin A \sin B$$

Cosine of Sum or Difference

$$\cos(A - B) = \cos A \cos B + \sin A \sin B$$

$$\cos(A + B) = \cos A \cos B - \sin A \sin B$$

These identities are important in calculus and other areas of mathematics and useful in certain applications. Although a calculator can be used to find an approximation for $\cos 15°$, for example, the method shown below can be applied to practice using the sum and difference identities, as well as to get an exact value.

● ● ● **Example 1** Using the Cosine Sum and Difference Identities to Find Exact Values

Find the *exact* value of the following.

Algebraic Solution

(a) $\cos 15°$

To find $\cos 15°$, write $15°$ as the sum or difference of two angles with known function values. Since we know the exact trigonometric function values of both $45°$ and $30°$, we write $15°$ as $45° - 30°$. (We could also use $60° - 45°$.) Then we use the identity for the cosine of the difference of two angles.

$$\cos 15° = \cos(45° - 30°)$$
$$= \cos 45° \cos 30° + \sin 45° \sin 30° \quad \text{Cosine difference identity}$$
$$= \frac{\sqrt{2}}{2} \cdot \frac{\sqrt{3}}{2} + \frac{\sqrt{2}}{2} \cdot \frac{1}{2}$$
$$= \frac{\sqrt{6} + \sqrt{2}}{4}$$

(b) $\cos \frac{5}{12}\pi = \cos\left(\frac{\pi}{6} + \frac{\pi}{4}\right)$

$$= \cos \frac{\pi}{6} \cos \frac{\pi}{4} - \sin \frac{\pi}{6} \sin \frac{\pi}{4} \quad \text{Cosine sum identity}$$
$$= \frac{\sqrt{3}}{2} \cdot \frac{\sqrt{2}}{2} - \frac{1}{2} \cdot \frac{\sqrt{2}}{2}$$
$$= \frac{\sqrt{6} - \sqrt{2}}{4}$$

(c) $\cos 87° \cos 93° - \sin 87° \sin 93° = \cos(87° + 93°)$
$$\text{Cosine sum identity}$$
$$= \cos(180°)$$
$$= -1$$

Graphing Calculator Support

The calculator screen in Figure 8(a) supports the algebraic solution in part (b) by giving the same approximation for both $\cos \frac{5}{12}\pi$ and $\frac{\sqrt{6} - \sqrt{2}}{4}$. Alternatively, in Figure 8(b), we entered $\frac{5}{12}\pi$ for x, and the corresponding y-value is the calculator approximation for $\frac{\sqrt{6} - \sqrt{2}}{4}$ shown in Figure 8(a).

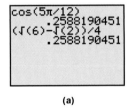

(a)

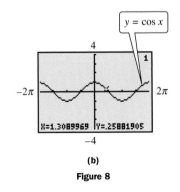

(b)

Figure 8

● ● ●

NOTE In Example 1(b), we used the fact that $5\pi/12 = \pi/6 + \pi/4$. At first glance, this sum may not be obvious. Think of the values $\pi/6$ and $\pi/4$ in terms of fractions with denominator 12: $\pi/6 = 2\pi/12$ and $\pi/4 = 3\pi/12$. The list below may help you with problems of this type.

$$\frac{\pi}{3} = \frac{4\pi}{12}, \quad \frac{\pi}{4} = \frac{3\pi}{12}, \quad \frac{\pi}{6} = \frac{2\pi}{12}$$

Using this list, for example, we see that $\pi/12 = \pi/3 - \pi/4$ (or $\pi/4 - \pi/6$).

Cofunction Identities   We can use the identities for the cosine of the sum and difference of two angles to derive other identities. Recall the *cofunction identities* presented earlier for values of θ in the interval $[0°, 90°]$.

Cofunction Identities

$$\cos(90° - \theta) = \sin\theta \qquad \cot(90° - \theta) = \tan\theta$$
$$\sin(90° - \theta) = \cos\theta \qquad \sec(90° - \theta) = \csc\theta$$
$$\tan(90° - \theta) = \cot\theta \qquad \csc(90° - \theta) = \sec\theta$$

Similar identities can be obtained for a real number domain by replacing $90°$ by $\pi/2$.

These identities now can be generalized for any angle θ, not just those between $0°$ and $90°$. For example, substituting $90°$ for A and θ for B in the identity given above for $\cos(A - B)$ gives

$$\cos(90° - \theta) = \cos 90° \cos\theta + \sin 90° \sin\theta$$
$$= 0 \cdot \cos\theta + 1 \cdot \sin\theta$$
$$= \sin\theta.$$

This result is true for *any* value of θ since the identity for $\cos(A - B)$ is true for any values of A and B. For the derivations of other cofunction identities, see Exercises 77 and 78.

● ● ●   **Example 2**   **Using the Cofunction Identities to Find θ**

Find an angle θ that satisfies each of the following.

(a) $\cot\theta = \tan 25°$
Since tangent and cotangent are cofunctions, $\tan(90° - \theta) = \cot\theta$.

$$\tan(90° - \theta) = \tan 25° \qquad \text{Substitute.}$$
$$90° - \theta = 25°$$
$$\theta = 65°$$

(b) $\sin\theta = \cos(-30°)$
In the same way,

$$\cos(90° - \theta) = \sin\theta = \cos(-30°),$$
$$90° - \theta = -30°$$
$$\theta = 120°.$$

$$\tan\theta = \cot(30 + 5\theta)$$

(c) $\csc \dfrac{3\pi}{4} = \sec \theta$

$$\csc \dfrac{3\pi}{4} = \sec\left(\dfrac{\pi}{2} - \dfrac{3\pi}{4}\right) = \sec \theta \qquad \text{Cofunction identity}$$

$$\sec\left(-\dfrac{\pi}{4}\right) = \sec \theta \qquad \text{Combine terms.}$$

$$-\dfrac{\pi}{4} = \theta$$

● ● ●

N O T E   Because trigonometric (and circular) functions are periodic, the solutions in Example 2 are not unique. In each case, we give only one of infinitely many possibilities.

Applying the Sum and Difference Identities   If one of the angles A or B in the identities for $\cos(A + B)$ and $\cos(A - B)$ is a quadrantal angle, then the identity allows us to write the expression in terms of a single function of A or B.

● ● ●   **Example 3   Reducing $\cos(A - B)$ to a Function of a Single Variable**

Write $\cos(180° - \theta)$ as a trigonometric function of θ.
    Use the difference identity. Replace A with $180°$ and B with θ.

$$\cos(180° - \theta) = \cos 180° \cos \theta + \sin 180° \sin \theta$$
$$= (-1) \cos \theta + (0) \sin \theta$$
$$= -\cos \theta$$

● ● ●

● ● ●   **Example 4   Finding $\cos(s + t)$ Given Information about s and t**

Suppose that $\sin s = 3/5$, $\cos t = -12/13$, and both s and t are in quadrant II. Find $\cos(s + t)$.
    By the identity above, $\cos(s + t) = \cos s \cos t - \sin s \sin t$. The values of $\sin s$ and $\cos t$ are given, so $\cos(s + t)$ can be found if $\cos s$ and $\sin t$ are known. We can find $\cos s$ and $\sin t$ by sketching two angles in the second quadrant, one with $\sin s = 3/5$ and the other with $\cos t = -12/13$. See Figure 9.
    In Figure 9(a), since $\sin s = 3/5 = y/r$, let $y = 3$ and $r = 5$. Substituting in the Pythagorean theorem gives $x^2 + 3^2 = 5^2$; solve to get $x = -4$. Thus, $\cos s = -4/5$. In Figure 9(b), $\cos t = -12/13 = x/r$, so let $x = -12$ and $r = 13$. Then $(-12)^2 + y^2 = 13^2$; solve to get $y = 5$. Thus, $\sin t = 5/13$. Now find $\cos(s + t)$.

$$\cos(s + t) = \cos s \cos t - \sin s \sin t$$
$$= -\dfrac{4}{5}\left(-\dfrac{12}{13}\right) - \dfrac{3}{5} \cdot \dfrac{5}{13}$$
$$= \dfrac{48}{65} - \dfrac{15}{65}$$
$$= \dfrac{33}{65}$$

● ● ●

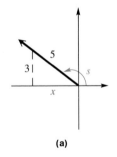

(a)

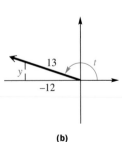

(b)

Figure 9

N O T E In Example 4, the values of cos s and sin t could also be found by using the Pythagorean identities. The problem could then be solved using the identity for $\cos(s + t)$ in the same way as shown in the example.

● ● ● **Example 5** Applying the Cosine Difference Identity to Voltage

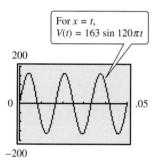

Common household electrical current is called alternating current because the current alternates direction within the wires. The voltage V in a typical 115-volt outlet can be expressed using the equation $V = 163 \sin \omega t$, where ω is the angular velocity (in radians per second) of the rotating generator at the electrical plant, and t is time measured in seconds. (*Source:* Bell, D., *Fundamentals of Electric Circuits,* Fourth Edition, Prentice-Hall, 1988.)

(a) It is essential for electrical generators to rotate at precisely 60 cycles per second so household appliances and computers will function properly. Determine ω for these electrical generators.

Since each cycle is 2π radians, at 60 cycles per second, $\omega = 60(2\pi) = 120\pi$ radians per second.

For $x = t$,
$V(t) = 163 \sin 120\pi t$

(b) Graph V on the interval $0 \le t \le .05$.

$V = 163 \sin \omega t = 163 \sin 120\pi t$. Because the amplitude is 163 here, we choose $-200 \le V \le 200$ for the range, as shown in Figure 10.

(c) For what value of ϕ will the graph of $V = 163 \cos(\omega t - \phi)$ be the same as the graph of $V = 163 \sin \omega t$?

Since $\cos(x - \pi/2) = \cos(\pi/2 - x) = \sin x$, choose $\phi = \pi/2$. In the Chapter Test you will be asked to use the cosine difference identity to show that $V = 163 \sin \omega t$ and $V = 163 \cos(\omega t - \pi/2)$ are equivalent equations.

● ● ●

Figure 10

5.3 Exercises

1. Compare the formulas for $\cos(A - B)$ and $\cos(A + B)$. How do they differ? How are they alike?

2. What does the cofunction identity $\cos(\pi/2 - \theta) = \sin \theta$ imply about the graphs of the cosine and sine functions? (*Hint:* First observe that $\cos(\pi/2 - \theta)$ is the same as $\cos(\theta - \pi/2)$.)

Use the cosine sum and difference identities to find each exact value. (Do not use a calculator.) See Example 1.

3. $\cos 75°$

4. $\cos(-15°)$

5. $\cos 105°$ (*Hint:* $105° = 60° + 45°$)

6. $\cos(-105°)$ (*Hint:* $-105° = -60° + (-45°)$)

7. $\cos \dfrac{7\pi}{12}$

8. $\cos\left(-\dfrac{\pi}{12}\right)$

9. $\cos 40° \cos 50° - \sin 40° \sin 50°$

10. $\cos(-10°) \cos 35° + \sin(-10°) \sin 35°$

11. $\cos \dfrac{2\pi}{5} \cos \dfrac{\pi}{10} - \sin \dfrac{2\pi}{5} \sin \dfrac{\pi}{10}$

12. $\cos \dfrac{7\pi}{9} \cos \dfrac{2\pi}{9} - \sin \dfrac{7\pi}{9} \sin \dfrac{2\pi}{9}$

Use a graphing calculator to support your answer for each of the following. See Example 1.

13. Exercise 9

14. Exercise 10

Write each function value in terms of the cofunction of a complementary angle. See Example 2.

15. $\tan 87°$

16. $\sin 15°$

17. $\cos \dfrac{\pi}{12}$

18. $\sin \dfrac{2\pi}{5}$

19. $\csc(-14° \, 24')$

20. $\sin 142° \, 14'$

21. $\sin \dfrac{5\pi}{8}$

22. $\cot \dfrac{9\pi}{10}$

23. $\sec 146° \, 42'$

24. $\tan 174° \, 3'$

25. $\cot 176.9814°$

26. $\sin 98.0142°$

Use the cofunction identities to fill in each blank with the appropriate trigonometric function name. See Example 2.

27. $\cot \dfrac{\pi}{3} = \underline{\hspace{1cm}} \dfrac{\pi}{6}$

28. $\sin \dfrac{2\pi}{3} = \underline{\hspace{1cm}} \left(-\dfrac{\pi}{6}\right)$

29. $\underline{\hspace{1cm}} 33° = \sin 57°$

30. $\underline{\hspace{1cm}} 72° = \cot 18°$

31. $\cos 70° = \dfrac{1}{\underline{\hspace{1cm}} 20°}$

32. $\tan 24° = \dfrac{1}{\underline{\hspace{1cm}} 66°}$

Use the cofunction identities to find an angle θ that makes each statement true. See Example 2.

33. $\tan \theta = \cot(45° + 2\theta)$

34. $\sin \theta = \cos(2\theta - 10°)$

35. $\sec \theta = \csc\left(\dfrac{\theta}{2} + 20°\right)$

36. $\cos \theta = \sin(3\theta + 10°)$

37. $\sin(3\theta - 15°) = \cos(\theta + 25°)$

38. $\cot(\theta - 10°) = \tan(2\theta + 20°)$

Use the identities for the cosine of a sum or a difference to write each expression as a single function of θ. See Example 3.

39. $\cos(0° - \theta)$

40. $\cos(90° - \theta)$

41. $\cos(180° - \theta)$

42. $\cos(270° - \theta)$

43. $\cos(0° + \theta)$

44. $\cos(90° + \theta)$

45. $\cos(180° + \theta)$

46. $\cos(270° + \theta)$

Find $\cos(s + t)$ and $\cos(s - t)$. See Example 4.

47. $\cos s = -1/5$ and $\sin t = 3/5$, s and t in quadrant II

48. $\sin s = 2/3$ and $\sin t = -1/3$, s in quadrant II and t in quadrant IV

49. $\sin s = 3/5$ and $\sin t = -12/13$, s in quadrant I and t in quadrant III

50. $\cos s = -8/17$ and $\cos t = -3/5$, s and t in quadrant III

51. $\sin s = \sqrt{5}/7$ and $\sin t = \sqrt{6}/8$, s and t in quadrant I

52. $\cos s = \sqrt{2}/4$ and $\sin t = -\sqrt{5}/6$, s and t in quadrant IV

Concept Check Tell whether each statement is true *or* false.

53. $\cos 42° = \cos(30° + 12°)$

54. $\cos(-24°) = \cos 16° - \cos 40°$

55. $\cos 74° = \cos 60° \cos 14° + \sin 60° \sin 14°$

56. $\cos 140° = \cos 60° \cos 80° - \sin 60° \sin 80°$

57. $\cos \dfrac{\pi}{3} = \cos \dfrac{\pi}{12} \cos \dfrac{\pi}{4} - \sin \dfrac{\pi}{12} \sin \dfrac{\pi}{4}$

58. $\cos \dfrac{2\pi}{3} = \cos \dfrac{11\pi}{12} \cos \dfrac{\pi}{4} + \sin \dfrac{11\pi}{12} \sin \dfrac{\pi}{4}$

59. $\cos 70° \cos 20° - \sin 70° \sin 20° = 0$

60. $\cos 85° \cos 40° + \sin 85° \sin 40° = \dfrac{\sqrt{2}}{2}$

61. $\tan\left(\theta - \dfrac{\pi}{2}\right) = \cot \theta$

62. $\sin\left(\theta - \dfrac{\pi}{2}\right) = \cos \theta$

Verify that each equation is an identity.

63. $\cos\left(\dfrac{\pi}{2} + x\right) = -\sin x$

64. $\sec(\pi - x) = -\sec x$

65. $\cos 2x = \cos^2 x - \sin^2 x$ (*Hint:* $\cos 2x = \cos(x + x)$.)

66. $1 + \cos 2x - \cos^2 x = \cos^2 x$ (*Hint:* Use the result from Exercise 65.)

Use a graphing calculator to support each identity.

67. $\cos(\pi + s - t) = -\sin s \sin t - \cos s \cos t$

68. $\cos\left(\dfrac{\pi}{2} + s - t\right) = \sin(t - s)$

69. $\cos(\alpha + \beta)\cos(\alpha - \beta) = 1 - \sin^2\alpha - \sin^2\beta$

70. $\cos 4x \cos 7x - \sin 4x \sin 7x = \cos 11x$

71. Suppose a fellow student tells you that the cosine of the sum of two angles is the sum of their cosines. Write in your own words how you would correct this student's statement.

72. *Concept Check* By a cofunction identity, $\cos 20° = \sin 70°$. What are some values other than 70° that make $\cos 20° = \sin \theta$ a true statement?

· · · · · · · · · · · · **Relating Concepts** · · · · · · · · · · ·

For individual or collaborative investigation

(Exercises 73–76)

The identities for $\cos(A + B)$ and $\cos(A - B)$ can be used to find exact values of expressions like $\cos 195°$ and $\cos 255°$, where the angle is not in the first quadrant. **Work Exercises 73–76 in order,** *to see how this is done.*

73. By writing 195° as 180° + 15°, use the identity for $\cos(A + B)$ to express $\cos 195°$ as $-\cos 15°$.

74. Use the identity for $\cos(A - B)$ to find $-\cos 15°$.

75. By the results of Exercises 73 and 74, $\cos 195° =$ _____.

76. Find the exact value of each of the following using the method shown in Exercises 73–75.

 (a) $\cos 255°$ **(b)** $\cos \dfrac{11\pi}{12}$

· ·

77. Use the identity $\cos(90° - \theta) = \sin \theta$, and replace θ with $90° - A$, to derive the identity $\cos A = \sin(90° - A)$.

78. Use the identities in Exercise 77 to derive the identity $\tan A = \cot(90° - A)$.

79. Let $f(x) = \cos x$. Prove that $\dfrac{f(x + h) - f(x)}{h} = \cos x\left(\dfrac{\cos h - 1}{h}\right) - \sin x\left(\dfrac{\sin h}{h}\right)$.

80. Without using a calculator, show that $\cos 140° + \cos 100° + \cos 20° = 0$.

 Electric Current *Exercises 81 and 82 refer to Example 5.*

81. How many times does the current oscillate in .05 second?

82. What are the maximum and minimum voltages in this outlet? Is the voltage always equal to 115 volts?

83. *(Modeling) Sound Waves* Sound is a result of waves applying pressure to a person's eardrum. For a pure sound wave radiating outward in a spherical shape, the trigonometric function

$$P = \frac{a}{r}\cos\left[\frac{2\pi r}{\lambda} - ct\right]$$

can be used to model the sound pressure at a radius of r feet from the source: t is time in seconds, λ is length of the sound wave in feet, c is speed of sound in feet per second, and a is maximum sound pressure at the source measured in pounds per square foot. (*Source:* Beranek, L., *Noise and Vibration Control,* Institute of Noise Control Engineering, Washington, D.C., 1988.) Let $\lambda = 4.9$ feet and $c = 1026$ feet per second.

 (a) Let $a = .4$ pound per square foot. Graph the sound pressure at distance $r = 10$ feet from its source over the interval $0 \le t \le .05$. Describe P at this distance.

 (b) Now let $a = 3$ and $t = 10$. Graph the sound pressure for $0 \le r \le 20$. What happens to pressure P as radius r increases?

 (c) Suppose a person stands at a radius r so that $r = n\lambda$, where n is a positive integer. Use the difference identity for cosine to simplify P in this situation.

84. *Spacecraft Coordinate Systems*   A conventional three-dimensional spacecraft coordinate system is shown in the figure.

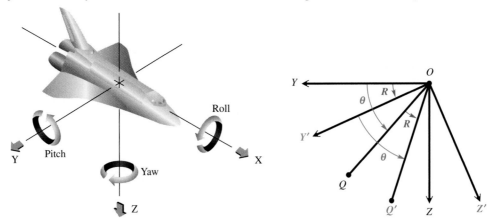

Angle $YOQ = \theta$ and $OQ = r$. The coordinates of Q are (x, y, z), where

$$y = r \cos \theta \quad \text{and} \quad z = r \sin \theta.$$

When the spacecraft performs a rotation, it is necessary to find the coordinates in the spacecraft system after the rotation takes place. For example, suppose the spacecraft undergoes roll through angle R. The coordinates (x, y, z) of point Q become (x', y', z'), the coordinates of the corresponding point Q'. In the new reference system, $OQ' = r$ and, since the roll is around the x-axis and angle $Y'OQ' = YOQ = \theta$,

$$x' = x, \quad y' = r \cos(\theta + R), \quad \text{and} \quad z' = r \sin(\theta + R).$$

Use the cosine sum and difference identities to write expressions for y' and z' in terms of y, z, and R. (*Source:* Kastner, B., *Space Mathematics,* NASA.)

5.4   Sum and Difference Identities for Sine and Tangent

• **Sum and Difference Identities for Sine** **•** **Sum and Difference Identities for Tangent** **•** **Applying the Sum and Difference Identities**

Sum and Difference Identities for Sine   We can develop formulas for both $\sin(A + B)$ and $\sin(A - B)$ from the results in Section 5.3. Starting with the cofunction identity

$$\sin \theta = \cos(90° - \theta),$$

replace θ with $A + B$.

$$\begin{aligned} \sin(A + B) &= \cos[90° - (A + B)] \\ &= \cos[(90° - A) - B] \end{aligned}$$

Using the formula for $\cos(A - B)$ from the previous section gives

$$\sin(A + B) = \cos(90° - A) \cos B + \sin(90° - A) \sin B$$

or $\sin(A + B) = \sin A \cos B + \cos A \sin B$.

(The cofunction identities were used in the last step.)

Now write $\sin(A - B)$ as $\sin[A + (-B)]$, and then use the identity for $\sin(A + B)$.

$$\begin{aligned} \sin(A - B) &= \sin[A + (-B)] \\ &= \sin A \cos(-B) + \cos A \sin(-B) \quad \text{Identity for } \sin(A + B) \\ \sin(A - B) &= \sin A \cos B - \cos A \sin B \quad \text{Negative-angle identities} \end{aligned}$$

Sum and Difference Identities for Tangent Using the identities for $\sin(A + B)$, $\cos(A + B)$, $\sin(A - B)$, and $\cos(A - B)$, and the identity $\tan \theta = \sin \theta/\cos \theta$, gives the following identities.

$$\tan(A + B) = \frac{\tan A + \tan B}{1 - \tan A \tan B} \qquad \tan(A - B) = \frac{\tan A - \tan B}{1 + \tan A \tan B}$$

We show a proof for the first of these two identities. The proof for the other is very similar. Start with

$$\tan(A + B) = \frac{\sin(A + B)}{\cos(A + B)}$$

$$= \frac{\sin A \cos B + \cos A \sin B}{\cos A \cos B - \sin A \sin B}.$$

To express this result in terms of the tangent function, multiply both numerator and denominator by $1/(\cos A \cos B)$.

$$\tan(A + B) = \frac{\dfrac{\sin A \cos B + \cos A \sin B}{1}}{\dfrac{\cos A \cos B - \sin A \sin B}{1}} \cdot \frac{\dfrac{1}{\cos A \cos B}}{\dfrac{1}{\cos A \cos B}}$$

$$= \frac{\dfrac{\sin A \cos B}{\cos A \cos B} + \dfrac{\cos A \sin B}{\cos A \cos B}}{\dfrac{\cos A \cos B}{\cos A \cos B} - \dfrac{\sin A \sin B}{\cos A \cos B}}$$

$$= \frac{\dfrac{\sin A}{\cos A} + \dfrac{\sin B}{\cos B}}{1 - \dfrac{\sin A}{\cos A} \cdot \dfrac{\sin B}{\cos B}}$$

$$\tan(A + B) = \frac{\tan A + \tan B}{1 - \tan A \tan B} \qquad \tan \theta = \frac{\sin \theta}{\cos \theta}$$

Sine and Tangent of Sum or Difference

$$\sin(A + B) = \sin A \cos B + \cos A \sin B$$

$$\sin(A - B) = \sin A \cos B - \cos A \sin B$$

$$\tan(A + B) = \frac{\tan A + \tan B}{1 - \tan A \tan B}$$

$$\tan(A - B) = \frac{\tan A - \tan B}{1 + \tan A \tan B}$$

Applying the Sum and Difference Identities Again, we give the following examples and the corresponding exercises primarily to offer practice in using these new identities.

● ● ● **Example 1** Using the Sine and Tangent Sum and Difference Identities to Find Exact Values

Find the *exact* value of the following.

Algebraic Solution

(a) $\sin 75° = \sin(45° + 30°)$

$= \sin 45° \cos 30° + \cos 45° \sin 30°$ Sine sum identity

$= \dfrac{\sqrt{2}}{2} \cdot \dfrac{\sqrt{3}}{2} + \dfrac{\sqrt{2}}{2} \cdot \dfrac{1}{2}$

$= \dfrac{\sqrt{6} + \sqrt{2}}{4}$

sin

(b) $\tan \dfrac{7\pi}{12} = \tan\left(\dfrac{\pi}{3} + \dfrac{\pi}{4}\right)$

$= \dfrac{\tan \dfrac{\pi}{3} + \tan \dfrac{\pi}{4}}{1 - \tan \dfrac{\pi}{3} \tan \dfrac{\pi}{4}}$ Tangent sum identity

$= \dfrac{\sqrt{3} + 1}{1 - \sqrt{3} \cdot 1}$

$= \dfrac{\sqrt{3} + 1}{1 - \sqrt{3}} \cdot \dfrac{1 + \sqrt{3}}{1 + \sqrt{3}}$ Rationalize the denominator.

$= \dfrac{\sqrt{3} + 3 + 1 + \sqrt{3}}{1 - 3}$

$= \dfrac{4 + 2\sqrt{3}}{-2}$

$= \dfrac{2(2 + \sqrt{3})}{-2}$ Factor out 2.

$= -2 - \sqrt{3}$ Write in lowest terms.

(c) $\sin 40° \cos 160° - \cos 40° \sin 160° = \sin(40° - 160°)$
Sine difference identity

$= \sin(-120°)$

$= -\sin 120°$

$= -\dfrac{\sqrt{3}}{2}$

Graphing Calculator Support

The screen in Figure 11(a) supports the algebraic solution for part (b) by giving the same approximation for $\tan \dfrac{7\pi}{12}$ and $-2 - \sqrt{3}$. Figure 11(b) indicates that the point $(1.8325957, -3.732051)$, which approximates $\left(\dfrac{7\pi}{12}, -2 - \sqrt{3}\right)$, lies on the graph of $Y = \tan x$, further showing that $\tan \dfrac{7\pi}{12} = -2 - \sqrt{3}$.

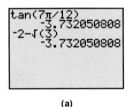

(a)

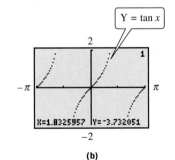

(b)

Figure 11

● ● ●

● ● ● **Example 2** Writing a Function as an Expression Involving Functions of θ

Write each of the following as an expression involving functions of θ.

(a) $\sin(30° + \theta)$

Using the identity for $\sin(A + B)$,

$$\sin(30° + \theta) = \sin 30° \cos \theta + \cos 30° \sin \theta$$
$$= \frac{1}{2} \cos \theta + \frac{\sqrt{3}}{2} \sin \theta.$$

(b) $\tan(45° - \theta) = \dfrac{\tan 45° - \tan \theta}{1 + \tan 45° \tan \theta} = \dfrac{1 - \tan \theta}{1 + \tan \theta}$

(c) $\sin(180° + \theta) = \sin 180° \cos \theta + \cos 180° \sin \theta$
$$= 0 \cdot \cos \theta + (-1) \sin \theta$$
$$= -\sin \theta$$

● ● ●

● ● ● **Example 3** Finding Functions and the Quadrant of $A + B$ Given Information about A and B

If $\sin A = 4/5$ and $\cos B = -5/13$, where A is in quadrant II and B is in quadrant III, find each value.

(a) $\sin(A + B)$

The identity for $\sin(A + B)$ requires $\sin A$, $\cos A$, $\sin B$, and $\cos B$. Two of these values are given. We must find the two missing values, $\cos A$ and $\sin B$, first, using the identity $\sin^2 x + \cos^2 x = 1$. For $\cos A$,

$$\sin^2 A + \cos^2 A = 1$$
$$\frac{16}{25} + \cos^2 A = 1 \qquad \sin A = \frac{4}{5}$$
$$\cos^2 A = \frac{9}{25}$$
$$\cos A = -\frac{3}{5}. \quad \text{Since } A \text{ is in quadrant II, } \cos A < 0.$$

In the same way, $\sin B = -12/13$. Now use the formula for $\sin(A + B)$.

$$\sin(A + B) = \frac{4}{5}\left(-\frac{5}{13}\right) + \left(-\frac{3}{5}\right)\left(-\frac{12}{13}\right)$$
$$= -\frac{20}{65} + \frac{36}{65} = \frac{16}{65}$$

(b) $\tan(A + B)$

Use the values of sine and cosine from part (a) to get $\tan A = -4/3$ and $\tan B = 12/5$. Then

$$\tan(A + B) = \frac{-\dfrac{4}{3} + \dfrac{12}{5}}{1 - \left(-\dfrac{4}{3}\right)\left(\dfrac{12}{5}\right)} = \frac{\dfrac{16}{15}}{1 + \dfrac{48}{15}} = \frac{\dfrac{16}{15}}{\dfrac{63}{15}} = \frac{16}{63}.$$

(c) the quadrant of $A + B$

From the results of parts (a) and (b), $\sin(A + B)$ is positive and $\tan(A + B)$ is also positive. Therefore, $A + B$ must be in quadrant I, since it is the only quadrant in which both sine and tangent are positive. ● ● ●

● ● ● **Example 4** Verifying an Identity Using Sum and Difference Identities

Verify that the equation is an identity.

$$\sin\left(\frac{\pi}{6} + s\right) + \cos\left(\frac{\pi}{3} + s\right) = \cos s$$

Work on the left side, using the identities for $\sin(A + B)$ and $\cos(A + B)$.

$$\sin\left(\frac{\pi}{6} + s\right) + \cos\left(\frac{\pi}{3} + s\right)$$

$$= \left(\sin\frac{\pi}{6}\cos s + \cos\frac{\pi}{6}\sin s\right) \qquad \sin(A + B) = \sin A \cos B + \cos A \sin B$$

$$+ \left(\cos\frac{\pi}{3}\cos s - \sin\frac{\pi}{3}\sin s\right) \qquad \cos(A + B) = \cos A \cos B - \sin A \sin B$$

$$= \left(\frac{1}{2}\cos s + \frac{\sqrt{3}}{2}\sin s\right) \qquad \sin\frac{\pi}{6} = \frac{1}{2}; \quad \cos\frac{\pi}{6} = \frac{\sqrt{3}}{2}$$

$$+ \left(\frac{1}{2}\cos s - \frac{\sqrt{3}}{2}\sin s\right) \qquad \cos\frac{\pi}{3} = \frac{1}{2}; \quad \sin\frac{\pi}{3} = \frac{\sqrt{3}}{2}$$

$$= \frac{1}{2}\cos s + \frac{1}{2}\cos s$$

$$= \cos s \qquad\qquad\qquad\qquad\qquad ● ● ●$$

5.4 Exercises

1. Compare the formulas for $\sin(A - B)$ and $\sin(A + B)$. How do they differ? How are they alike?

2. Compare the formulas for $\tan(A - B)$ and $\tan(A + B)$. How do they differ? How are they alike?

Concept Check Match each expression in Column I with its value in Column II. See Example 1.

I

3. $\sin 15°$

4. $\sin 105°$

5. $\tan 15°$

6. $\tan 105°$

7. $\sin(-105°)$

8. $\tan(-105°)$

II

A. $\dfrac{\sqrt{6} + \sqrt{2}}{4}$

B. $\dfrac{-\sqrt{6} - \sqrt{2}}{4}$

C. $\dfrac{\sqrt{6} - \sqrt{2}}{4}$

D. $2 + \sqrt{3}$

E. $2 - \sqrt{3}$

F. $-2 - \sqrt{3}$

Use the identities of this section to find the exact value of each of the following. See Example 1.

9. $\sin \dfrac{5\pi}{12}$
10. $\tan \dfrac{5\pi}{12}$
11. $\tan \dfrac{\pi}{12}$
12. $\sin \dfrac{\pi}{12}$
13. $\sin\left(-\dfrac{7\pi}{12}\right)$
14. $\tan\left(-\dfrac{7\pi}{12}\right)$

15. $\sin 76° \cos 31° - \cos 76° \sin 31°$

16. $\sin 40° \cos 50° + \cos 40° \sin 50°$

17. $\dfrac{\tan 80° + \tan 55°}{1 - \tan 80° \tan 55°}$

18. $\dfrac{\tan 80° - \tan(-55°)}{1 + \tan 80° \tan(-55°)}$

19. $\dfrac{\tan 100° + \tan 80°}{1 - \tan 100° \tan 80°}$

20. $\sin 100° \cos 10° - \cos 100° \sin 10°$

21. $\sin \dfrac{\pi}{5} \cos \dfrac{3\pi}{10} + \cos \dfrac{\pi}{5} \sin \dfrac{3\pi}{10}$

22. $\dfrac{\tan \dfrac{5\pi}{12} + \tan \dfrac{\pi}{4}}{1 - \tan \dfrac{5\pi}{12} \tan \dfrac{\pi}{4}}$

Use the identities of this section and the previous one to write each of the following as an expression involving functions of x or θ. See Example 2.

23. $\cos(30° + \theta)$
24. $\cos(45° - \theta)$
25. $\cos(60° + \theta)$
26. $\cos(\theta - 30°)$
27. $\cos\left(\dfrac{3\pi}{4} - x\right)$

28. $\sin(45° + \theta)$
29. $\tan(\theta + 30°)$
30. $\tan\left(\dfrac{\pi}{4} + x\right)$
31. $\sin\left(\dfrac{\pi}{4} + x\right)$
32. $\sin(180° - \theta)$

33. $\sin(270° - \theta)$
34. $\tan(180° + \theta)$
35. $\tan(360° - \theta)$
36. $\sin(\pi + \theta)$
37. $\tan(\pi - \theta)$

38. Why is it not possible to follow Example 2 to find a formula for $\tan(270° - \theta)$?

39. What happens when you try to evaluate $\dfrac{\tan 65.902° + \tan 24.098°}{1 - \tan 65.902° \tan 24.098°}$?

40. Show that if A, B, and C are the angles of a triangle, then $\sin(A + B + C) = 0$.

For each of the following, find $\sin(s + t)$, $\sin(s - t)$, $\tan(s + t)$, $\tan(s - t)$, the quadrant of $s + t$, and the quadrant of $s - t$. See Example 3.

41. $\cos s = 3/5$ and $\sin t = 5/13$, s and t in quadrant I

42. $\cos s = -1/5$ and $\sin t = 3/5$, s and t in quadrant II

43. $\sin s = 2/3$ and $\sin t = -1/3$, s in quadrant II and t in quadrant IV

44. $\sin s = 3/5$ and $\sin t = -12/13$, s in quadrant I and t in quadrant III

45. $\cos s = -8/17$ and $\cos t = -3/5$, s and t in quadrant III

46. $\cos s = -15/17$ and $\sin t = 4/5$, s in quadrant II and t in quadrant I

47. $\sin s = -4/5$ and $\cos t = 12/13$, s in quadrant III and t in quadrant IV

48. $\sin s = -5/13$ and $\sin t = 3/5$, s in quadrant III and t in quadrant II

49. $\cos s = -\sqrt{7}/4$ and $\sin t = \sqrt{3}/5$, s and t in quadrant II

50. $\cos s = \sqrt{11}/5$ and $\cos t = \sqrt{2}/6$, s and t in quadrant IV

Graph each expression and use the graph to conjecture an identity. Then verify your conjecture algebraically. See Example 4.

51. $\sin\left(\dfrac{\pi}{2} + x\right)$
52. $\sin\left(\dfrac{3\pi}{2} + x\right)$
53. $\tan\left(\dfrac{\pi}{2} + x\right)$
54. $\dfrac{1 + \tan x}{1 - \tan x}$

Verify that each equation is an identity. See Example 4.

55. $\sin 2x = 2 \sin x \cos x$ (*Hint:* $\sin 2x = \sin(x + x)$)

56. $\sin(x + y) + \sin(x - y) = 2 \sin x \cos y$

57. $\tan(x - y) - \tan(y - x) = \dfrac{2(\tan x - \tan y)}{1 + \tan x \tan y}$

58. $\sin(210° + x) - \cos(120° + x) = 0$

59. $\dfrac{\cos(\alpha - \beta)}{\cos \alpha \sin \beta} = \tan \alpha + \cot \beta$

60. $\dfrac{\sin(s + t)}{\cos s \cos t} = \tan s + \tan t$

61. $\dfrac{\sin(x - y)}{\sin(x + y)} = \dfrac{\tan x - \tan y}{\tan x + \tan y}$

62. $\dfrac{\sin(x + y)}{\cos(x - y)} = \dfrac{\cot x + \cot y}{1 + \cot x \cot y}$

63. $\dfrac{\sin(s - t)}{\sin t} + \dfrac{\cos(s - t)}{\cos t} = \dfrac{\sin s}{\sin t \cos t}$

64. $\dfrac{\tan(\alpha + \beta) - \tan \beta}{1 + \tan(\alpha + \beta) \tan \beta} = \tan \alpha$

Find each exact value. (See the technique developed in Exercises 73–76 of Section 5.3.)

65. $\sin 165°$ **66.** $\tan 165°$ **67.** $\tan 255°$ **68.** $\sin 255°$ **69.** $\sin 285°$

70. $\tan 285°$ **71.** $\sin \dfrac{11\pi}{12}$ **72.** $\tan \dfrac{11\pi}{12}$ **73.** $\tan\left(-\dfrac{13\pi}{12}\right)$ **74.** $\sin\left(-\dfrac{13\pi}{12}\right)$

75. Derive the identity for $\tan(A - B)$ using the identity for $\tan(A + B)$ and the fact that $A - B = A + (-B)$.

76. Derive the identity for $\tan(A - B)$ using the identities for $\sin(A - B)$ and $\cos(A - B)$ and the fact that
$$\tan(A - B) = \frac{\sin(A - B)}{\cos(A - B)}.$$

77. Let $f(x) = \sin x$. Show that $\dfrac{f(x + h) - f(x)}{h} = \sin x\left(\dfrac{\cos h - 1}{h}\right) + \cos x\left(\dfrac{\sin h}{h}\right).$

· · · · · · · · · · · · · · **Relating Concepts** · · · · · · · · · · · · · ·

For individual or collaborative investigation
(Exercises 78–83)

Refer to the figure on the left below. By the definition of $\tan \theta$, *m* $= \tan \theta$, *where m is the slope and* θ *is the angle of inclination of the line. The following exercises, which depend on properties of triangles, refer to triangle ABC in the figure on the right below.* **Work Exercises 78–81 in order.** *Assume that all angles are measured in degrees.*

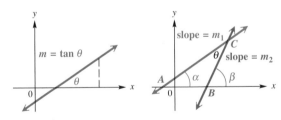

78. In terms of β, what is the measure of angle ABC?

79. Use the fact that the sum of the angles in a triangle is $180°$ to express θ in terms of α and β.

80. Apply the formula for $\tan(A - B)$ to obtain an expression for $\tan \theta$ in terms of $\tan \alpha$ and $\tan \beta$.

81. Replace $\tan \alpha$ with m_1 and $\tan \beta$ with m_2 to obtain $\tan \theta = \dfrac{m_2 - m_1}{1 + m_1 m_2}$.

In Exercises 82 and 83, use the result from Exercise 81 to find the angle between each pair of lines. Use a calculator and round to the nearest tenth of a degree.

82. $x + y = 9$, $2x + y = -1$

83. $5x - 2y + 4 = 0$, $3x + 5y = 6$

84. *Voltage* A coil of wire rotating in a magnetic field induces a voltage

$$e = 20 \sin\left(\frac{\pi t}{4} - \frac{\pi}{2}\right).$$

Use an identity from this section to express this in terms of $\cos \frac{\pi t}{4}$.

85. *Voltage* When the two voltages $V_1 = 30 \sin 120\pi t$ and $V_2 = 40 \cos 120\pi t$ are applied to the same circuit, the resulting voltage V will be equal to their sum. (*Source:* Bell, D., *Fundamentals of Electric Circuits,* Fourth Edition, Prentice-Hall, 1988.)

 (a) Graph $V = V_1 + V_2$ over the interval $0 \le t \le .05$.

(b) Use the graph and our work in Chapter 4 to estimate values for a and ϕ so that the voltage $V = a \sin(120\pi t + \phi)$.

(c) Use identities to verify that your expression for V is valid.

86. (*Modeling*) *Force on Back Muscles* If a person bends at the waist with a straight back making an angle of θ degrees with the horizontal, then the force F exerted on the back muscles can be modeled by the equation

$$F = \frac{.6W \sin(\theta + 90°)}{\sin 12°},$$

where W is the weight of the person. (*Source:* Metcalf, H., *Topics in Classical Biophysics,* Prentice-Hall, 1980.)

(a) Calculate F when $W = 170$ pounds and $\theta = 30°$.

(b) Use an identity to show that F is approximately equal to $2.9W \cos \theta$.

(c) For what value of θ is F maximum?

5.5 Double-Angle Identities

• **Double-Angle Identities** • **Verifying Identities with Double Angles** • **Applying a Double-Angle Identity**

Double-Angle Identities When $A = B$ in the identities for the sum of two angles, these identities are called the **double-angle identities.** For example, to derive an expression for $\cos 2A$, we let $B = A$ in the identity $\cos(A + B) = \cos A \cos B - \sin A \sin B$.

$$\cos 2A = \cos(A + A)$$
$$= \cos A \cos A - \sin A \sin A$$
$$\cos 2A = \cos^2 A - \sin^2 A$$

Two other useful forms of this identity can be obtained by substituting either $\cos^2 A = 1 - \sin^2 A$ or $\sin^2 A = 1 - \cos^2 A$. Replace $\cos^2 A$ with $1 - \sin^2 A$ to get

$$\cos 2A = \cos^2 A - \sin^2 A$$
$$= (1 - \sin^2 A) - \sin^2 A$$
$$\cos 2A = 1 - 2 \sin^2 A,$$

and replace $\sin^2 A$ with $1 - \cos^2 A$ to get

$$\cos 2A = \cos^2 A - (1 - \cos^2 A)$$
$$= \cos^2 A - 1 + \cos^2 A$$
$$\cos 2A = 2 \cos^2 A - 1.$$

We find $\sin 2A$ with the identity $\sin(A + B) = \sin A \cos B + \cos A \sin B$, letting $B = A$.

$$\sin 2A = \sin(A + A)$$
$$= \sin A \cos A + \cos A \sin A$$
$$\sin 2A = 2 \sin A \cos A$$

Using the identity for $\tan(A + B)$, we find $\tan 2A$.

$$\tan 2A = \tan(A + A)$$
$$= \frac{\tan A + \tan A}{1 - \tan A \tan A}$$
$$\tan 2A = \frac{2 \tan A}{1 - \tan^2 A}$$

Double-Angle Identities

$$\cos 2A = \cos^2 A - \sin^2 A \qquad \cos 2A = 1 - 2 \sin^2 A$$
$$\cos 2A = 2 \cos^2 A - 1 \qquad \sin 2A = 2 \sin A \cos A$$
$$\tan 2A = \frac{2 \tan A}{1 - \tan^2 A}$$

● ● ● **Example 1** Using the Double-Angle Identities

Given $\cos \theta = 3/5$ and $\sin \theta < 0$, use identities to find $\sin 2\theta$, $\cos 2\theta$, and $\tan 2\theta$.

To find $\sin 2\theta$, we must first find the value of $\sin \theta$.

$$\sin^2 \theta + \left(\frac{3}{5}\right)^2 = 1 \qquad \sin^2 \theta + \cos^2 \theta = 1 \text{ and } \cos \theta = \frac{3}{5}$$

$$\sin^2 \theta = \frac{16}{25}$$

$$\sin \theta = -\frac{4}{5}. \qquad \text{Choose the negative square root since } \sin \theta < 0.$$

Using the double-angle identity for sine, we get

$$\sin 2\theta = 2 \sin \theta \cos \theta = 2\left(-\frac{4}{5}\right)\left(\frac{3}{5}\right) = -\frac{24}{25}.$$

Now we find $\cos 2\theta$, using the first form of the identity. (Any form may be used.)

$$\cos 2\theta = \cos^2 \theta - \sin^2 \theta = \frac{9}{25} - \frac{16}{25} = -\frac{7}{25}$$

The value of $\tan 2\theta$ can be found in either of two ways. We can use the double-angle identity and the fact that $\tan \theta = (\sin \theta)/(\cos \theta) = (-4/5)/(3/5) = -4/3$.

$$\tan 2\theta = \frac{2 \tan \theta}{1 - \tan^2 \theta} = \frac{2\left(-\frac{4}{3}\right)}{1 - \frac{16}{9}} = \frac{-\frac{8}{3}}{-\frac{7}{9}} = \frac{24}{7}$$

Alternatively, we can find $\tan 2\theta$ by finding the quotient of $\sin 2\theta$ and $\cos 2\theta$.

$$\tan 2\theta = \frac{\sin 2\theta}{\cos 2\theta} = \frac{-24/25}{-7/25} = \frac{24}{7}$$

● ● ●

● ● ● **Example 2** Finding Functions of θ Given Information about 2θ

Find the values of the six trigonometric functions of θ if $\cos 2\theta = 4/5$ and $90° < \theta < 180°$.

Use one of the double-angle identities for cosine to get a trigonometric function value for θ.

$$\cos 2\theta = 1 - 2\sin^2\theta$$

$$\frac{4}{5} = 1 - 2\sin^2\theta$$

$$-\frac{1}{5} = -2\sin^2\theta$$

$$\frac{1}{10} = \sin^2\theta$$

$$\sin\theta = \sqrt{\frac{1}{10}} = \frac{\sqrt{10}}{10}$$

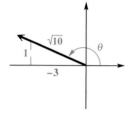

Figure 12

Choose the positive square root since θ terminates in quadrant II. Now find values of $\cos\theta$ and $\tan\theta$ using the fundamental identities or by sketching and labeling a right triangle in quadrant II. Using a triangle as in Figure 12, we have

$$\cos\theta = \frac{-3}{\sqrt{10}} = -\frac{3\sqrt{10}}{10}, \quad \text{and} \quad \tan\theta = \frac{1}{-3} = -\frac{1}{3}.$$

Find the other three functions using reciprocals.

$$\csc\theta = \frac{1}{\sin\theta} = \sqrt{10}, \quad \sec\theta = \frac{1}{\cos\theta} = -\frac{\sqrt{10}}{3}, \quad \cot\theta = \frac{1}{\tan\theta} = -3$$

● ● ●

● ● ● **Example 3** Simplifying Expressions Using Double-Angle Identities

Simplify each expression.

(a) $\cos^2 7x - \sin^2 7x$

This expression suggests one of the identities for $\cos 2A$: $\cos 2A = \cos^2 A - \sin^2 A$. Substituting $7x$ for A gives

$$\cos^2 7x - \sin^2 7x = \cos 2(7x) = \cos 14x.$$

(b) $\sin 15° \cos 15°$

If this expression were $2\sin 15° \cos 15°$, we could apply the identity for $\sin 2A$ directly since $\sin 2A = 2\sin A \cos A$. We can still apply the identity with

$A = 15°$ by writing the multiplicative identity element 1 as $(1/2)(2)$.

$$\sin 15° \cos 15° = \frac{1}{2}(2) \sin 15° \cos 15° \qquad \text{Multiply by 1 in the form } \frac{1}{2}(2).$$

$$= \frac{1}{2}(2 \sin 15° \cos 15°) \qquad \text{Associative property}$$

$$= \frac{1}{2} \sin(2 \cdot 15°) \qquad 2 \sin A \cos A = \sin 2A, \text{ with } A = 15°$$

$$= \frac{1}{2} \sin 30°$$

$$= \frac{1}{2} \cdot \frac{1}{2} \qquad \sin 30° = \frac{1}{2}$$

$$= \frac{1}{4}$$

● ● ●

Verifying Identities with Double Angles

● ● ● **Example 4** Verifying an Identity

Verify that the following equation is an identity.

$$\cot x \sin 2x = 1 + \cos 2x$$

Algebraic Solution

Work on the left side.

$$\cot x \sin 2x = \frac{\cos x}{\sin x} \cdot \sin 2x \qquad \cot x = \frac{\cos x}{\sin x}$$

$$= \frac{\cos x}{\sin x}(2 \sin x \cos x) \qquad \sin 2x = 2 \sin x \cos x$$

$$= 2 \cos^2 x$$

$$= 1 + \cos 2x \qquad 2 \cos^2 x - 1 = \cos 2x$$

Graphing Calculator Support

The graph in Figure 13 supports the algebraic solution.

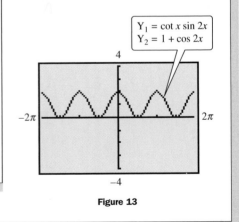

$Y_1 = \cot x \sin 2x$
$Y_2 = 1 + \cos 2x$

Figure 13

● ● ●

The methods used earlier to derive the identities for double angles can also be used to find identities for expressions such as $\sin 3s$.

• • • **Example 5** Deriving a Multiple-Angle Identity

Write $\sin 3s$ in terms of $\sin s$.

Algebraic Solution

$$\sin 3s = \sin(2s + s)$$

$= \sin 2s \cos s + \cos 2s \sin s$ Sine sum identity

$= (2 \sin s \cos s) \cos s + (\cos^2 s - \sin^2 s) \sin s$

                           Double-angle identities

$= 2 \sin s \cos^2 s + \cos^2 s \sin s - \sin^3 s$

$= 2 \sin s(1 - \sin^2 s) + (1 - \sin^2 s) \sin s - \sin^3 s$

                    $\cos^2 s = 1 - \sin^2 s$

$= 2 \sin s - 2 \sin^3 s + \sin s - \sin^3 s - \sin^3 s$

                    Distributive property

$= 3 \sin s - 4 \sin^3 s$

Graphing Calculator Support

The table in Figure 14 supports the algebraic solution. Scroll up and down to see more of the function values for Y_1 and Y_2. Although the table gives strong support to the identity, it does not verify it because it does not list every possible x-value.

X	Y1	Y2
-1.178	.38268	.38268
-.7854	-.7071	-.7071
-.3927	-.9239	-.9239
0	0	0
.3927	.92388	.92388
.7854	.70711	.70711
1.1781	-.3827	-.3827

Y₁■sin(3X)

Figure 14

• • •

C O N N E C T I O N S Because they make it possible to rewrite a product as a sum, the identities for $\cos(A + B)$ and $\cos(A - B)$ are used to derive a group of identities useful in calculus.

Adding the identities for $\cos(A + B)$ and $\cos(A - B)$ gives

$$\cos(A + B) = \cos A \cos B - \sin A \sin B$$
$$\cos(A - B) = \cos A \cos B + \sin A \sin B$$
$$\overline{\cos(A + B) + \cos(A - B) = 2 \cos A \cos B}$$

or $\cos A \cos B = \dfrac{1}{2}[\cos(A + B) + \cos(A - B)].$

Similarly, subtracting $\cos(A + B)$ from $\cos(A - B)$ gives

$$\sin A \sin B = \dfrac{1}{2}[\cos(A - B) - \cos(A + B)].$$

(continued)

Using the identities for $\sin(A + B)$ and $\sin(A - B)$ in the same way, we get two more identities.

$$\sin A \cos B = \frac{1}{2}[\sin(A + B) + \sin(A - B)]$$

$$\cos A \sin B = \frac{1}{2}[\sin(A + B) - \sin(A - B)]$$

These last two identities make it possible to rewrite an expression involving both sine and cosine as an expression with just one of these functions. In solving a conditional equation, this conversion is very useful.

For Discussion or Writing

1. Show that the double-angle identity $\sin 2A = 2 \sin A \cos A$ is a special case of the identity for $\sin A \cos B$ given above.
2. Show that the double-angle identity $\cos 2A = 2 \cos^2 A - 1$ is a special case of the identity for $\cos A \cos B$ given above.

Applying a Double-Angle Identity

● ● ● **Example 6** Using a Model to Determine Wattage Consumption

 If a toaster is plugged into a common household outlet, the wattage consumed is not constant. Instead, it varies at a high frequency according to the model

$$W = \frac{V^2}{R},$$

where V is the voltage and R is a constant that measures the resistance of the toaster in ohms. (*Source:* Bell, D., *Fundamentals of Electric Circuits,* Fourth Edition, Prentice-Hall, 1988.) Graph the wattage W consumed by a typical toaster with $R = 15$ and $V = 163 \sin 120\pi t$ over the interval $0 \le t \le .05$. How many oscillations are there?

By substituting the given values into the wattage equation, we get

$$W = \frac{V^2}{R} = \frac{(163 \sin 120\pi t)^2}{15}.$$

The graph is shown in Figure 15. To determine the range for W, we note that $\sin 120\pi t$ has maximum value 1, so the expression for W has maximum value $163^2/15 \approx 1771$. The minimum value is 0. The graph shows that there are 6 oscillations. (In Exercise 84 you will need to use a double-angle identity to show that the equation for W is equivalent to the equation $W = a \cos(\omega t) + c$, for specific values of a, c, and ω.) ● ● ●

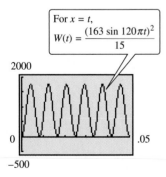

For $x = t$,
$$W(t) = \frac{(163 \sin 120\pi t)^2}{15}$$

Figure 15

5.5 Exercises

Concept Check   *Match each expression in Column I with its value in Column II.*

I	II

1. $2 \cos^2 15° - 1$

A. $\dfrac{1}{2}$

2. $\dfrac{2 \tan 15°}{1 - \tan^2 15°}$

B. $\dfrac{\sqrt{2}}{2}$

3. $2 \sin 22.5° \cos 22.5°$

C. $\dfrac{\sqrt{3}}{2}$

4. $\cos^2 \dfrac{\pi}{6} - \sin^2 \dfrac{\pi}{6}$

D. $-\sqrt{3}$

5. $2 \sin \dfrac{\pi}{3} \cos \dfrac{\pi}{3}$

E. $\dfrac{\sqrt{3}}{3}$

6. $\dfrac{2 \tan \dfrac{\pi}{3}}{1 - \tan^2 \dfrac{\pi}{3}}$

Use the identities in this section to find values of the six trigonometric functions for each of the following. See Examples 1 and 2.

7. θ, given $\cos 2\theta = 3/5$ and θ terminates in quadrant I

8. α, given $\cos 2\alpha = 3/4$ and α terminates in quadrant III

9. x, given $\cos 2x = -5/12$ and $\pi/2 < x < \pi$

10. t, given $\cos 2t = 2/3$ and $\pi/2 < t < \pi$

11. 2θ, given $\sin \theta = 2/5$ and $\cos \theta < 0$

12. 2β, given $\cos \beta = -12/13$ and $\sin \beta > 0$

13. $2x$, given $\tan x = 2$ and $\cos x > 0$

14. $2x$, given $\tan x = 5/3$ and $\sin x < 0$

15. 2α, given $\sin \alpha = -\sqrt{5}/7$ and $\cos \alpha > 0$

16. 2α, given $\cos \alpha = \sqrt{3}/5$ and $\sin \alpha > 0$

Concept Check   *In Exercises 17 and 18, the given graphing calculator screen was obtained for a particular stored value of X. What will the screen display for the value of the expression in the final line of the display?*

17.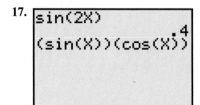

18.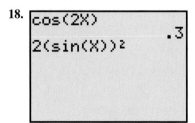

19. State in your own words each of the following identities. (*Example:* The sine of twice an angle is two times the sine of the angle times the cosine of the angle.)
   (a) $\cos 2A$ (first form derived in this section)   **(b)** $\cos 2A$ (second form)
   (c) $\cos 2A$ (third form)   **(d)** $\tan 2A$

20. Specific identities for $\sec 2A$, $\csc 2A$, and $\cot 2A$ are not usually studied in detail. Why do you think this is so? Give these identities in terms of $\cos A$, $\sin A$, and $\tan A$.

Use an identity to write each expression as a single trigonometric function or as a single number. See Example 3.

21. $\cos^2 15° - \sin^2 15°$

22. $\dfrac{2 \tan 15°}{1 - \tan^2 15°}$

23. $1 - 2 \sin^2 15°$

24. $1 - 2 \sin^2 22\dfrac{1}{2}°$

25. $2 \cos^2 67\dfrac{1}{2}° - 1$

26. $\cos^2 \dfrac{\pi}{8} - \dfrac{1}{2}$

27. $\dfrac{\tan 51°}{1 - \tan^2 51°}$

28. $\dfrac{\tan 34°}{2(1 - \tan^2 34°)}$

29. $\dfrac{1}{4} - \dfrac{1}{2} \sin^2 47.1°$

30. $\dfrac{1}{8} \sin 29.5° \cos 29.5°$

31. $\sin^2 \dfrac{2\pi}{5} - \cos^2 \dfrac{2\pi}{5}$

32. $\cos^2 2\alpha - \sin^2 2\alpha$

33. $2 \sin 5x \cos 5x$

34. $2 \cos^2 6\alpha - 1$

Find the exact value of each of the following in two ways: evaluate it directly, and use an appropriate identity. Do not use a calculator.

35. $\sin 2(45°)$

36. $\cos 2(45°)$

37. $\cos 2(60°)$

38. $\tan 2(60°)$

39. $\cos 2\left(\dfrac{5\pi}{3}\right)$

40. $\sin 2\left(\dfrac{\pi}{3}\right)$

41. $\tan 2\left(-\dfrac{\pi}{3}\right)$

42. $\cos 2\left(-\dfrac{9\pi}{4}\right)$

43. $\tan 2\left(-\dfrac{4\pi}{3}\right)$

44. $\tan 2\left(-\dfrac{13\pi}{6}\right)$

45. $\sin 2\left(-\dfrac{11\pi}{2}\right)$

46. $\sin 2\left(-\dfrac{17\pi}{2}\right)$

Graph each expression and use the graph to conjecture an identity. Then verify your conjecture as in Example 4.

47. $\cos^4 x - \sin^4 x$

48. $\dfrac{4 \tan x \cos^2 x - 2 \tan x}{1 - \tan^2 x}$

49. $\dfrac{\cot^2 x - 1}{2 \cot x}$

50. $\dfrac{2 \tan x}{2 - \sec^2 x}$

Verify that each equation is an identity. See Example 4.

51. $(\sin \gamma + \cos \gamma)^2 = \sin 2\gamma + 1$

52. $\sec 2x = \dfrac{\sec^2 x + \sec^4 x}{2 + \sec^2 x - \sec^4 x}$

53. $\tan 8k - \tan 8k \tan^2 4k = 2 \tan 4k$

54. $\sin 2\gamma = \dfrac{2 \tan \gamma}{1 + \tan^2 \gamma}$

55. $\cos 2y = \dfrac{2 - \sec^2 y}{\sec^2 y}$

56. $-\tan 2\theta = \dfrac{2 \tan \theta}{\sec^2 \theta - 2}$

57. $\sin 4\alpha = 4 \sin \alpha \cos \alpha \cos 2\alpha$

58. $\dfrac{1 + \cos 2x}{\sin 2x} = \cot x$

59. $\tan(\theta - 45°) + \tan(\theta + 45°) = 2 \tan 2\theta$

60. $\cot 4\theta = \dfrac{1 - \tan^2 2\theta}{2 \tan 2\theta}$

61. $\dfrac{2 \cos 2\alpha}{\sin 2\alpha} = \cot \alpha - \tan \alpha$

62. $\sin 4\gamma = 4 \sin \gamma \cos \gamma - 8 \sin^3 \gamma \cos \gamma$

63. $\sin 2\alpha \cos 2\alpha = \sin 2\alpha - 4 \sin^3 \alpha \cos \alpha$

64. $\cos 2x = \dfrac{1 - \tan^2 x}{1 + \tan^2 x}$

65. $\tan s + \cot s = 2 \csc 2s$

66. $\dfrac{\cot \alpha - \tan \alpha}{\cot \alpha + \tan \alpha} = \cos 2\alpha$

67. $1 + \tan x \tan 2x = \sec 2x$

68. $\cot \theta \tan(\theta + \pi) - \sin(\pi - \theta) \cos\left(\dfrac{\pi}{2} - \theta\right) = \cos^2 \theta$

Express each function as a trigonometric function of x. See Example 5.

69. $\cos 3x$

70. $\sin 4x$

71. $\tan 3x$

72. $\cos 4x$

Distance a Dropped Object Falls   *If an object is dropped in a vacuum, then the distance, d, that the object falls in t seconds is given by*

$$d = \frac{1}{2}gt^2,$$

where g is the acceleration due to gravity. At any particular point on Earth's surface, the value of g is a constant, roughly 978 cm per sec². A more exact value of g at any point on Earth's surface is given by

$$g = 978.0524(1 + .005297 \sin^2 \phi - .0000059 \sin^2 2\phi) - .000094h$$

in cm per sec², where φ is the latitude of the point and h is the altitude of the point in feet. Find g, rounding to the nearest thousandth, given the following.

73. $\phi = 47°\ 12',\ h = 387.0$ feet

74. $\phi = 68°\ 47',\ h = 1145$ feet

Use the formulas developed in the Connections box to rewrite each of the following as a sum or difference of trigonometric functions.

75. $\cos 45° \sin 25°$

76. $2 \sin 74° \cos 114°$

77. $3 \cos 5x \cos 3x$

78. $2 \sin 2x \sin 4x$

79. $\sin(-\theta) \sin(-3\theta)$

80. $4 \cos 8\alpha \sin(-4\alpha)$

81. $-8 \cos 4y \cos 5y$

82. $2 \sin 3k \sin 14k$

Solve each problem.

83. *(Modeling) Distance Traveled by a Stone*   The distance D of an object thrown (or propelled) from height h (feet) at angle θ with initial velocity v is modeled by the formula

$$D = \frac{v^2 \sin \theta \cos \theta + v \cos \theta \sqrt{(v \sin \theta)^2 + 64h}}{32}.$$

See the figure. (*Source:* Kreighbaum, E. and K. Barthels, *Biomechanics,* Allyn & Bacon, 1996.) Also see Exercise 67 in Section 2.3.

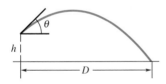

(a) Find D when $h = 0$; that is, when the object is propelled from the ground.

(b) Suppose a car driving over loose gravel kicks up a small stone at a velocity of 36 feet per second (about 25 mph) and an angle $\theta = 30°$. How far will the stone travel?

 84. *Wattage Consumption*   Refer to Example 6. Use an identity to determine values for a, c, and ω so that $W = a \cos(\omega t) + c$. Check your answer by graphing both expressions for W on the same coordinate axes.

 85. *Amperage, Wattage, and Voltage*   Amperage is a measure of the amount of electricity that is moving through a circuit, whereas voltage is a measure of the force pushing the electricity. The wattage W consumed by an electrical device can be determined by calculating the product of the amperage I and voltage V. (*Source:* Wilcox, G. and C. Hesselberth, *Electricity for Engineering Technology,* Allyn & Bacon, 1970.)

(a) A household circuit has voltage

$$V = 163 \sin(120\pi t)$$

when an incandescent light bulb is turned on with amperage

$$I = 1.23 \sin(120\pi t).$$

Graph the wattage $W = VI$ consumed by the light bulb over the interval $0 \le t \le .05$.

(b) Determine the maximum and minimum wattages used by the light bulb.

(c) Use identities to determine values for a, c, and ω so that $W = a \cos(\omega t) + c$.

(d) Check your answer by graphing both expressions for W on the same coordinate axes.

(e) Use the graph to estimate the average wattage used by the light. For how many watts do you think this incandescent light bulb is rated?

86. *Amperage, Wattage, and Voltage*   Refer to Exercise 85. Suppose that for an electric heater, voltage is given by $V = a \sin(2\pi\omega t)$ and amperage by $I = b \sin(2\pi\omega t)$, where t is time in seconds.

(a) Find the period of the graph for the voltage.

(b) Show that the graph of the wattage $W = VI$ will have half the period of the voltage. Interpret this result.

5.6 Half-Angle Identities

• **Half-Angle Identities** • **Using the Half-Angle Identities**

From the alternative forms of the identity for cos $2A$, we derive three additional identities for $\sin A/2$, $\cos A/2$, and $\tan A/2$. These are known as **half-angle identities.**

Half-Angle Identities To derive the identity for $\sin A/2$, start with the following double-angle identity for cosine and solve for $\sin x$.

$$\cos 2x = 1 - 2\sin^2 x$$

$$2\sin^2 x = 1 - \cos 2x$$

$$\sin x = \pm\sqrt{\frac{1 - \cos 2x}{2}}$$

Looking Ahead to Calculus

Half-angle identities for sine and cosine (using radians) are used in calculus when eliminating the xy-term from an equation in the form $Ax^2 + Bxy + Cy^2 + Dx + Ey + F = 0$, so the type of conic section it represents can be determined.

Now let $2x = A$, so $x = A/2$, and substitute into this last expression.

$$\sin\frac{A}{2} = \pm\sqrt{\frac{1 - \cos A}{2}}$$

The $\pm$ sign in this identity indicates that the appropriate sign is chosen depending on the quadrant of $A/2$. For example, if $A/2$ is a quadrant III angle, we choose the negative sign since the sine function is negative there.

The identity for $\cos A/2$ is derived in a similar way, starting with the double-angle identity $\cos 2x = 2\cos^2 x - 1$. Solve for $\cos x$.

$$1 + \cos 2x = 2\cos^2 x$$

$$\cos x = \pm\sqrt{\frac{1 + \cos 2x}{2}}$$

$$\cos\frac{A}{2} = \pm\sqrt{\frac{1 + \cos A}{2}} \qquad \text{Replace } x \text{ with } \frac{A}{2}.$$

The $\pm$ sign is chosen as described above.

Finally, an identity for $\tan A/2$ comes from the half-angle identities for sine and cosine.

$$\tan\frac{A}{2} = \frac{\pm\sqrt{\dfrac{1 - \cos A}{2}}}{\pm\sqrt{\dfrac{1 + \cos A}{2}}}$$

$$\tan\frac{A}{2} = \pm\sqrt{\frac{1 - \cos A}{1 + \cos A}} \qquad \pm \text{ chosen depending on the quadrant of } \frac{A}{2}$$

We derive an alternative identity for $\tan A/2$ using the fact that $\tan A/2 = (\sin A/2)/(\cos A/2)$.

$$\tan \frac{A}{2} = \frac{\sin \dfrac{A}{2}}{\cos \dfrac{A}{2}}$$

$$= \frac{2 \sin \dfrac{A}{2} \cos \dfrac{A}{2}}{2 \cos^2 \dfrac{A}{2}} \qquad \text{Multiply by } 2 \cos \dfrac{A}{2} \text{ in numerator and denominator.}$$

$$= \frac{\sin 2\left(\dfrac{A}{2}\right)}{1 + \cos 2\left(\dfrac{A}{2}\right)} \qquad \text{Double-angle identities}$$

$$\tan \frac{A}{2} = \frac{\sin A}{1 + \cos A}$$

From this identity for $\tan A/2$, we can also derive

$$\tan \frac{A}{2} = \frac{1 - \cos A}{\sin A}.$$

See Exercise 57. These last two identities for $\tan A/2$ do not require a sign choice as required in the others.

Half-Angle Identities

$$\cos \frac{A}{2} = \pm \sqrt{\frac{1 + \cos A}{2}} \qquad \sin \frac{A}{2} = \pm \sqrt{\frac{1 - \cos A}{2}}$$

$$\tan \frac{A}{2} = \pm \sqrt{\frac{1 - \cos A}{1 + \cos A}} \qquad \tan \frac{A}{2} = \frac{\sin A}{1 + \cos A} \qquad \tan \frac{A}{2} = \frac{1 - \cos A}{\sin A}$$

NOTE As mentioned earlier, the plus or minus sign is selected according to the quadrant in which $\dfrac{A}{2}$ terminates. For example, if A represents an angle of $324°$, then $\dfrac{A}{2} = 162°$, which lies in quadrant II. In quadrant II, $\cos \dfrac{A}{2}$ and $\tan \dfrac{A}{2}$ are negative, while $\sin \dfrac{A}{2}$ is positive.

Using the Half-Angle Identities

● ● ● **Example 1** Using a Half-Angle Identity to Find an Exact Value

Find the exact value of cos 15° using the half-angle identity for cosine.

$$\cos 15° = \cos \frac{1}{2}(30°)$$

$$= \sqrt{\frac{1 + \cos 30°}{2}} \qquad \text{Choose the positive square root.}$$

$$= \sqrt{\frac{1 + \dfrac{\sqrt{3}}{2}}{2}}$$

$$= \sqrt{\frac{\left(1 + \dfrac{\sqrt{3}}{2}\right) \cdot 2}{2 \cdot 2}}$$

$$= \frac{\sqrt{2 + \sqrt{3}}}{2}$$

● ● ●

N O T E Compare the value of cos 15° in Example 1 to the value in Example 1 of Section 5.3, where we used the identity for the cosine of the difference of two angles. Although the expressions look completely different, they are equal, as suggested by a calculator approximation for both, .96592583.

● ● ● **Example 2** Using a Half-Angle Identity to Find an Exact Value

Find the exact value of tan 22.5° using the identity $\tan \dfrac{A}{2} = \dfrac{\sin A}{1 + \cos A}$.

Since 22.5° = (1/2)(45°), replace A with 45°.

$$\tan 22.5° = \tan \frac{45°}{2} = \frac{\sin 45°}{1 + \cos 45°} = \frac{\dfrac{\sqrt{2}}{2}}{1 + \dfrac{\sqrt{2}}{2}}$$

Now multiply numerator and denominator by 2. Then rationalize the denominator.

$$\tan 22.5° = \frac{\sqrt{2}}{2 + \sqrt{2}} = \frac{\sqrt{2}}{2 + \sqrt{2}} \cdot \frac{2 - \sqrt{2}}{2 - \sqrt{2}}$$

$$= \frac{2\sqrt{2} - 2}{2}$$

$$= \frac{2(\sqrt{2} - 1)}{2}$$

$$= \sqrt{2} - 1$$

● ● ●

● ● ●   **Example 3**   Finding Functions of $A/2$ Given Information about A

Given $\cos s = 2/3$, with $3\pi/2 < s < 2\pi$, find $\cos s/2$, $\sin s/2$, and $\tan s/2$.
Since

$$\frac{3\pi}{2} < s < 2\pi,$$

and $$\frac{3\pi}{4} < \frac{s}{2} < \pi,$$ Divide by 2.

$s/2$ terminates in quadrant II. See Figure 16. In this quadrant the value of $\cos s/2$ is negative and the value of $\sin s/2$ is positive. Now use the appropriate half-angle identities.

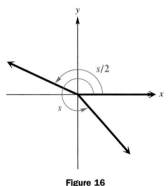

$$\sin \frac{s}{2} = \sqrt{\frac{1 - \frac{2}{3}}{2}} = \sqrt{\frac{1}{6}} = \frac{\sqrt{6}}{6}$$

$$\cos \frac{s}{2} = -\sqrt{\frac{1 + \frac{2}{3}}{2}} = -\sqrt{\frac{5}{6}} = -\frac{\sqrt{30}}{6}$$

$$\tan \frac{s}{2} = \frac{\sin \frac{s}{2}}{\cos \frac{s}{2}} = \frac{\frac{\sqrt{6}}{6}}{-\frac{\sqrt{30}}{6}} = -\frac{\sqrt{5}}{5}$$

Figure 16

Notice it is not necessary to use a half-angle identity for $\tan s/2$ once we find $\sin s/2$ and $\cos s/2$. However, using this identity would provide an excellent check.   ● ● ●

● ● ●   **Example 4**   Simplifying Expressions Using the Half-Angle Identities

Simplify each expression.

(a) $\pm \sqrt{\frac{1 + \cos 12x}{2}}$

This matches part of the identity for $\cos A/2$.

$$\cos \frac{A}{2} = \pm \sqrt{\frac{1 + \cos A}{2}}$$

Replace A with $12x$ to get

$$\pm \sqrt{\frac{1 + \cos 12x}{2}} = \cos \frac{12x}{2} = \cos 6x.$$

(b) $\dfrac{1 - \cos 5\alpha}{\sin 5\alpha}$

Use the third identity for $\tan A/2$ given earlier to get

$$\frac{1 - \cos 5\alpha}{\sin 5\alpha} = \tan \frac{5\alpha}{2}.$$

● ● ●

● ● ● **Example 5** Verifying an Identity

Verify that the following equation is an identity.

$$\left(\sin \frac{x}{2} + \cos \frac{x}{2}\right)^2 = 1 + \sin x$$

Work on the left.

$$\left(\sin \frac{x}{2} + \cos \frac{x}{2}\right)^2$$

$$= \sin^2 \frac{x}{2} + 2 \sin \frac{x}{2} \cos \frac{x}{2} + \cos^2 \frac{x}{2} \qquad (a + b)^2 = a^2 + 2ab + b^2$$

$$= 1 + 2 \sin \frac{x}{2} \cos \frac{x}{2} \qquad \sin^2 \frac{x}{2} + \cos^2 \frac{x}{2} = 1$$

$$= 1 + \sin 2\left(\frac{x}{2}\right) \qquad 2 \sin \frac{x}{2} \cos \frac{x}{2} = \sin 2\left(\frac{x}{2}\right)$$

$$= 1 + \sin x$$

● ● ●

5.6 Exercises

Concept Check   Determine whether the positive or negative square root should be selected.

1. $\sin 195° = \pm\sqrt{\dfrac{1 - \cos 390°}{2}}$

2. $\cos 58° = \pm\sqrt{\dfrac{1 + \cos 116°}{2}}$

3. $\tan 225° = \pm\sqrt{\dfrac{1 - \cos 450°}{1 + \cos 450°}}$

4. $\sin(-10°) = \pm\sqrt{\dfrac{1 - \cos(-20°)}{2}}$

Match each expression in Column I with its value in Column II. See Examples 1 and 2.

I		II	
5. $\sin 15°$	**8.** $\tan\left(-\dfrac{\pi}{8}\right)$	**A.** $2 - \sqrt{3}$	**D.** $\dfrac{\sqrt{2 + \sqrt{2}}}{2}$
6. $\tan 15°$	**9.** $\tan 67.5°$	**B.** $\dfrac{\sqrt{2 - \sqrt{2}}}{2}$	**E.** $1 - \sqrt{2}$
7. $\cos\dfrac{\pi}{8}$	**10.** $\cos 67.5°$	**C.** $\dfrac{\sqrt{2 - \sqrt{3}}}{2}$	**F.** $1 + \sqrt{2}$

Use the half-angle identities to find each exact value. See Examples 1 and 2.

11. $\sin 67.5°$ **12.** $\sin 195°$ **13.** $\cos 195°$ **14.** $\tan 195°$ **15.** $\cos 165°$ **16.** $\sin 165°$

Concept Check   In Exercises 17 and 18, the given graphing calculator screen was obtained for a particular stored value of X. What will the screen display for the value of the expression in the final line of the display?

17.
```
cos(X)
          .9682458366
sin(X)
                  .25
tan(X/2)
```

18.
```
cos(X)
               -.75
sin(X)
          .6614378278
tan(X/2)
```

19. Explain how you could use an identity of this section to find the exact value of sin 7.5°. (*Hint:* 7.5 = (1/2)(1/2)(30).)

20. The identity

$$\tan \frac{A}{2} = \pm \sqrt{\frac{1 - \cos A}{1 + \cos A}}$$

can be used to find $\tan 22.5° = \sqrt{3 - 2\sqrt{2}}$, and the identity

$$\tan \frac{A}{2} = \frac{\sin A}{1 + \cos A}$$

can be used to get $\tan 22.5° = \sqrt{2} - 1$. Show that these answers are the same, without using a calculator. (*Hint:* If $a > 0$ and $b > 0$ and $a^2 = b^2$, then $a = b$.)

Find each of the following. See Example 3.

21. $\cos \theta/2$, given $\cos \theta = 1/4$, with $0 < \theta < \pi/2$

22. $\sin \theta/2$, given $\cos \theta = -5/8$, with $\pi/2 < \theta < \pi$

23. $\tan \theta/2$, given $\sin \theta = 3/5$, with $90° < \theta < 180°$

24. $\cos \theta/2$, given $\sin \theta = -1/5$, with $180° < \theta < 270°$

25. $\sin \alpha/2$, given $\tan \alpha = 2$, with $0 < \alpha < \pi/2$

26. $\cos \alpha/2$, given $\cot \alpha = -3$, with $\pi/2 < \alpha < \pi$

27. $\tan \beta/2$, given $\tan \beta = \sqrt{7}/3$, with $180° < \beta < 270°$

28. $\cot \beta/2$, given $\tan \beta = -\sqrt{5}/2$, with $90° < \beta < 180°$

29. $\sin \theta$, given $\cos 2\theta = 3/5$ and θ terminates in quadrant I

30. $\cos \theta$, given $\cos 2\theta = 1/2$ and θ terminates in quadrant II

31. $\cos x$, given $\cos 2x = -5/12$ and $\pi/2 < x < \pi$

32. $\sin x$, given $\cos 2x = 2/3$ and $\pi < x < 3\pi/2$

Use an identity to write each expression as a single trigonometric function. See Example 4.

33. $\sqrt{\dfrac{1 - \cos 40°}{2}}$

34. $\sqrt{\dfrac{1 + \cos 76°}{2}}$

35. $\sqrt{\dfrac{1 - \cos 147°}{1 + \cos 147°}}$

36. $\sqrt{\dfrac{1 + \cos 165°}{1 - \cos 165°}}$

37. $\dfrac{1 - \cos 59.74°}{\sin 59.74°}$

38. $\dfrac{\sin 158.2°}{1 + \cos 158.2°}$

39. $\pm \sqrt{\dfrac{1 + \cos 18x}{2}}$

40. $\pm \sqrt{\dfrac{1 + \cos 20\alpha}{2}}$

41. $\pm \sqrt{\dfrac{1 - \cos 8\theta}{1 + \cos 8\theta}}$

42. $\pm \sqrt{\dfrac{1 - \cos 5A}{1 + \cos 5A}}$

43. $\pm \sqrt{\dfrac{1 + \cos x/4}{2}}$

44. $\pm \sqrt{\dfrac{1 - \cos 3\theta/5}{2}}$

Graph each expression and use the graph to conjecture an identity. Then verify your conjecture as in Example 5.

45. $\dfrac{\sin x}{1 + \cos x}$

46. $\dfrac{1 - \cos x}{\sin x}$

47. $\dfrac{\tan \dfrac{x}{2} + \cot \dfrac{x}{2}}{\cot \dfrac{x}{2} - \tan \dfrac{x}{2}}$

48. $1 - 8 \sin^2 \dfrac{x}{2} \cos^2 \dfrac{x}{2}$

Verify that each equation is an identity. See Example 5.

49. $\sec^2 \dfrac{x}{2} = \dfrac{2}{1 + \cos x}$

50. $\cot^2 \dfrac{x}{2} = \dfrac{(1 + \cos x)^2}{\sin^2 x}$

51. $\sin^2 \dfrac{x}{2} = \dfrac{\tan x - \sin x}{2 \tan x}$

52. $\dfrac{\sin 2x}{2 \sin x} = \cos^2 \dfrac{x}{2} - \sin^2 \dfrac{x}{2}$

53. $\dfrac{2}{1 + \cos x} - \tan^2 \dfrac{x}{2} = 1$

54. $\tan \dfrac{\gamma}{2} = \csc \gamma - \cot \gamma$

55. $1 - \tan^2 \dfrac{\theta}{2} = \dfrac{2 \cos \theta}{1 + \cos \theta}$

56. $\cos x = \dfrac{1 - \tan^2 \dfrac{x}{2}}{1 + \tan^2 \dfrac{x}{2}}$

57. In the text we derived the identity

$$\tan \frac{A}{2} = \frac{\sin A}{1 + \cos A}.$$

Multiply both numerator and denominator of the right side by $1 - \cos A$ to obtain the equivalent form

$$\tan \frac{A}{2} = \frac{1 - \cos A}{\sin A}.$$

Mach Number An airplane flying faster than sound sends out sound waves that form a cone, as shown in the figure. The cone intersects the ground to form a hyperbola. As this hyperbola passes over a particular point on the ground, a sonic boom is heard at that point. If α is the angle at the vertex of the cone, then

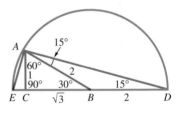

$$\sin \frac{\alpha}{2} = \frac{1}{m},$$

where m is the Mach number for the speed of the plane. (We assume $m > 1$.) The Mach number is the ratio of the speed of the plane and the speed of sound. Thus, a speed of Mach 1.4 means that the plane is flying at 1.4 times the speed of sound. Find α or m, as indicated.

58. $m = \frac{3}{2}$ **59.** $m = \frac{5}{4}$ **60.** $m = 2$ **61.** $m = \frac{5}{2}$ **62.** $\alpha = 30°$ **63.** $\alpha = 60°$

· **Relating Concepts** · · · · · · · · · · · · · · · ·

For individual or collaborative investigation
(Exercises 64–71)

These exercises use results from plane geometry, instead of the half-angle formulas, to obtain exact values of the trigonometric functions of 15°. **Work them in order.**

   Start with a right triangle ACB having a 60° *angle at A and a* 30° *angle at B. Let the hypotenuse of this triangle have length* 2. *Extend side BC and draw a semicircle with diameter along BC extended, center at B, and radius AB. Draw segment AE. (See the figure.) Since any angle inscribed in a semicircle is a right angle, triangle EAD is a right triangle.*

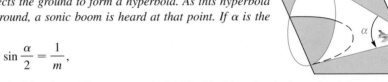

64. Why does $AB = BD$? Conclude that triangle *ABD* is isosceles.

65. Why does angle *ABD* have measure 150°?

66. Why do angles *DAB* and *ADB* both have measures of 15°?

67. What is the length *DC*?

68. Use the Pythagorean theorem to show that the length *AD* is $\sqrt{6} + \sqrt{2}$. (*Note:* $(\sqrt{6} + \sqrt{2})^2 = 8 + 4\sqrt{3}$.)

69. Use angle *ADB* of triangle *EAD* to find cos 15°.

70. Show that *AE* has length $\sqrt{6} - \sqrt{2}$ and find sin 15°.

71. Use triangle *ACD* to find tan 15°.

· ·

72. *Railroad Curves* In the United States, circular railroad curves are designated by the *degree of curvature*, the central angle subtended by a chord of 100 feet. See the figure. (*Source:* Hay, W. W., *Railroad Engineering*, John Wiley & Sons, 1982.)

(a) Use the figure to write an expression for $\cos \frac{\theta}{2}$.

(b) Use the result of part (a) and the third half-angle identity for tangent to write an expression for $\tan \frac{\theta}{4}$.

(c) If $b = 12$, what is the measure of angle θ to the nearest degree?

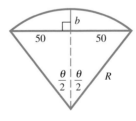

Chapter 5   Summary

Key Terms & Symbols	Key Ideas
5.1 Fundamental Identities	

Reciprocal Identities

$$\cot\theta = \frac{1}{\tan\theta} \qquad \sec\theta = \frac{1}{\cos\theta} \qquad \csc\theta = \frac{1}{\sin\theta}$$

Quotient Identities

$$\tan\theta = \frac{\sin\theta}{\cos\theta} \qquad \cot\theta = \frac{\cos\theta}{\sin\theta}$$

Pythagorean Identities

$$\sin^2\theta + \cos^2\theta = 1 \qquad \tan^2\theta + 1 = \sec^2\theta \qquad 1 + \cot^2\theta = \csc^2\theta$$

Negative-Angle Identities

$$\sin(-\theta) = -\sin\theta \qquad \cos(-\theta) = \cos\theta \qquad \tan(-\theta) = -\tan\theta$$

Key Terms & Symbols	Key Ideas
5.3 Sum and Difference Identities for Cosine **5.4 Sum and Difference Identities for Sine and Tangent**	

Cofunction Identities

$$\cos(90° - \theta) = \sin\theta \qquad \cot(90° - \theta) = \tan\theta$$
$$\sin(90° - \theta) = \cos\theta \qquad \sec(90° - \theta) = \csc\theta$$
$$\tan(90° - \theta) = \cot\theta \qquad \csc(90° - \theta) = \sec\theta$$

Sum and Difference Identities

$$\cos(A - B) = \cos A \cos B + \sin A \sin B$$
$$\cos(A + B) = \cos A \cos B - \sin A \sin B$$
$$\sin(A + B) = \sin A \cos B + \cos A \sin B$$
$$\sin(A - B) = \sin A \cos B - \cos A \sin B$$
$$\tan(A + B) = \frac{\tan A + \tan B}{1 - \tan A \tan B}$$
$$\tan(A - B) = \frac{\tan A - \tan B}{1 + \tan A \tan B}$$

Key Terms & Symbols	Key Ideas
5.5 Double-Angle Identities	

Double-Angle Identities

$$\cos 2A = \cos^2 A - \sin^2 A \qquad \cos 2A = 1 - 2\sin^2 A$$
$$\cos 2A = 2\cos^2 A - 1 \qquad \sin 2A = 2\sin A \cos A$$
$$\tan 2A = \frac{2\tan A}{1 - \tan^2 A}$$

Key Terms & Symbols	Key Ideas
5.6 Half-Angle Identities	

Half-Angle Identities

$$\sin \frac{A}{2} = \pm \sqrt{\frac{1 - \cos A}{2}} \qquad \tan \frac{A}{2} = \frac{1 - \cos A}{\sin A}$$

$$\cos \frac{A}{2} = \pm \sqrt{\frac{1 + \cos A}{2}} \qquad \tan \frac{A}{2} = \frac{\sin A}{1 + \cos A}$$

$$\tan \frac{A}{2} = \pm \sqrt{\frac{1 - \cos A}{1 + \cos A}}$$

(The sign is chosen based on the quadrant of $A/2$.)

Chapter 5 Review Exercises

Concept Check *For each expression in Column I, choose the expression from Column II that completes an identity.*

I

1. $\sec x$

2. $\csc x$

3. $\tan x$

4. $\cot x$

5. $\sin^2 x$

6. $\tan^2 x + 1$

7. $\tan^2 x$

II

A. $\dfrac{1}{\sin x}$

B. $\dfrac{1}{\cos x}$

C. $\dfrac{\sin x}{\cos x}$

D. $\dfrac{1}{\cot^2 x}$

E. $\dfrac{1}{\cos^2 x}$

F. $\dfrac{\cos x}{\sin x}$

G. $\dfrac{1}{\sin^2 x}$

H. $1 - \cos^2 x$

Use identities to write each expression in terms of $\sin \theta$ *and* $\cos \theta$, *and simplify.*

8. $\sec^2 \theta - \tan^2 \theta$

9. $\dfrac{\cot \theta}{\sec \theta}$

10. $\tan^2 \theta (1 + \cot^2 \theta)$

11. $\csc \theta + \cot \theta$

12. $\csc^2 \theta + \sec^2 \theta$

13. $\tan \theta - \sec \theta \csc \theta$

14. Give all the trigonometric functions that satisfy the condition $f(-x) = -f(x)$.

15. Give all the trigonometric functions that satisfy the condition $f(-x) = f(x)$.

16. Use the trigonometric identities to find the remaining five trigonometric functions of x, given $\cos x = 3/5$ and x is in quadrant IV.

17. Given $\tan x = -5/4$, where $\pi/2 < x < \pi$, use the trigonometric identities to find the other trigonometric functions of x.

18. Find the exact values of $\sin x$, $\cos x$, and $\tan x$, for $x = \pi/12$, using
(a) difference identities; (b) half-angle identities.

19. Find the exact values of the six trigonometric functions of $165°$.

Concept Check *For each expression in Column I, choose the expression from Column II that completes an identity.*

I   **II**

20. $\cos 210°$ **A.** $\sin(-35°)$ **F.** $\cot(-35°)$

21. $\sin 35°$ **B.** $\cos 55°$ **G.** $\cos^2 150° - \sin^2 150°$

22. $\tan(-35°)$ **C.** $\sqrt{\dfrac{1 + \cos 150°}{2}}$ **H.** $\sin 15° \cos 60° + \cos 15° \sin 60°$

23. $-\sin 35°$

24. $\cos 35°$ **D.** $2 \sin 150° \cos 150°$ **I.** $\cos(-35°)$

25. $\cos 75°$ **E.** $\cos 150° \cos 60° - \sin 150° \sin 60°$ **J.** $\cot 125°$

26. $\sin 75°$

27. $\sin 300°$

28. $\cos 300°$

29. Find $\tan \dfrac{x}{2}$, given $\sin x = .8$, $0 < x < \dfrac{\pi}{2}$.   **30.** Find $\sin 2x$, if $\sin x = .6$, $\dfrac{\pi}{2} < x < \pi$.

For each of the following, find $\sin(x + y)$, $\cos(x - y)$, $\tan(x + y)$, and the quadrant of $x + y$.

31. $\sin x = -1/4$, $\cos y = -4/5$, x and y in quadrant III

32. $\sin y = -2/3$, $\cos x = -1/5$, x in quadrant II, y in quadrant III

33. $\sin x = 1/10$, $\cos y = 4/5$, x in quadrant I, y in quadrant IV

34. $\cos x = 2/9$, $\sin y = -1/2$, x in quadrant IV, y in quadrant III

Find sine and cosine of each of the following.

35. θ, given $\cos 2\theta = -3/4$, $90° < 2\theta < 180°$ **36.** B, given $\cos 2B = 1/8$, B in quadrant IV

37. $2x$, given $\tan x = 3$, $\sin x < 0$ **38.** $2y$, given $\sec y = -5/3$, $\sin y > 0$

Find each of the following.

39. $\cos \theta/2$, given $\cos \theta = -1/2$, with $90° < \theta < 180°$ **40.** $\sin A/2$, given $\cos A = -3/4$, with $90° < A < 180°$

41. $\tan x$, given $\tan 2x = 2$, with $\pi < x < 3\pi/2$ **42.** $\sin y$, given $\cos 2y = -1/3$, with $\pi/2 < y < \pi$

Graph each expression and use the graph to conjecture an identity. Then verify your conjecture.

43. $-\dfrac{\sin 2x + \sin x}{\cos 2x - \cos x}$ **44.** $\dfrac{1 - \cos 2x}{\sin 2x}$ **45.** $\dfrac{\sin x}{1 - \cos x}$

46. $\dfrac{\cos x \sin 2x}{1 + \cos 2x}$ **47.** $\dfrac{2(\sin x - \sin^3 x)}{\cos x}$ **48.** $\csc x - \cot x$

Verify that each equation is an identity. (Hint: In Exercise 56, $\tan C = \tan(180° - (A + B)) = -\tan(A + B)$.)

49. $\sin^2 x - \sin^2 y = \cos^2 y - \cos^2 x$ **50.** $2 \cos^3 x - \cos x = \dfrac{\cos^2 x - \sin^2 x}{\sec x}$

51. $\dfrac{\sin^2 x}{2 - 2 \cos x} = \cos^2 \dfrac{x}{2}$ **52.** $\dfrac{\sin 2x}{\sin x} = \dfrac{2}{\sec x}$

53. $2 \cos A - \sec A = \cos A - \dfrac{\tan A}{\csc A}$ **54.** $\dfrac{2 \tan B}{\sin 2B} = \sec^2 B$

55. $1 + \tan^2 \alpha = 2 \tan \alpha \csc 2\alpha$

56. If $A + B + C = 180°$, $\tan A + \tan B + \tan C = (\tan A)(\tan B)(\tan C)$. (Assume triangle ABC is not a right triangle.)

57. $[\tan(45° + A)][\tan(45° - A)] = 1$ **58.** $\dfrac{2 \cot x}{\tan 2x} = \csc^2 x - 2$

59. $\tan \theta \sin 2\theta = 2 - 2 \cos^2 \theta$ **60.** $\csc A \sin 2A - \sec A = \cos 2A \sec A$

61. $2 \tan x \csc 2x - \tan^2 x = 1$

62. $2 \cos^2 \theta - 1 = \dfrac{1 - \tan^2 \theta}{1 + \tan^2 \theta}$

63. $\tan \theta \cos^2 \theta = \dfrac{2 \tan \theta \cos^2 \theta - \tan \theta}{1 - \tan^2 \theta}$

64. $-\cot \dfrac{x}{2} = \dfrac{\sin 2x + \sin x}{\cos 2x - \cos x}$

65. $2 \cos^3 x - \cos x = \dfrac{\cos^2 x - \sin^2 x}{\sec x}$

66. $\sin^3 \theta = \sin \theta - \cos^2 \theta \sin \theta$

67. $\cos^2 x + \sin^2 x(1 - 2 \sin^2 y) = \cos^2 y + \sin^2 y(1 - 2 \sin^2 x)$

68. $\tan \dfrac{7}{2} x = \dfrac{2 \tan \dfrac{7}{4} x}{1 - \tan^2 \dfrac{7}{4} x}$

69. $\sec^2 \alpha - 1 = \dfrac{\sec 2\alpha - 1}{\sec 2\alpha + 1}$

70. $\dfrac{\sin 3t + \sin 2t}{\sin 3t - \sin 2t} = \dfrac{\tan \dfrac{5t}{2}}{\tan \dfrac{t}{2}}$

71. $\tan 4\theta = \dfrac{2 \tan 2\theta}{2 - \sec^2 2\theta}$

72. $2 \cos^2 \dfrac{x}{2} \tan x = \tan x + \sin x$

73. $\tan \left(\dfrac{x}{2} + \dfrac{\pi}{4} \right) = \sec x + \tan x$

74. $\dfrac{1}{2} \cot \dfrac{x}{2} - \dfrac{1}{2} \tan \dfrac{x}{2} = \cot x$

75. Show that if A, B, and C are the three angles of an obtuse triangle, then $\tan A + \tan B + \tan C < 0$. (*Hint:* Use the identity in Exercise 56.)

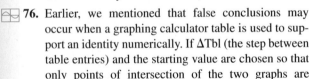 **76.** Earlier, we mentioned that false conclusions may occur when a graphing calculator table is used to support an identity numerically. If ΔTbl (the step between table entries) and the starting value are chosen so that only points of intersection of the two graphs are evaluated in the table, it will appear that the two expressions are equivalent. Decide whether each equation is an identity by using the table feature as indicated and then graphing the two expressions in the same window.

(a) $\cos x = \cos 4x$; TblStart $= 0$, ΔTbl $= 2\pi/3$

(b) $\sin(x + \pi) = .5 \sin x$; TblStart $= 0$, ΔTbl $= \pi$

(c) $4 \sin x \cos x = 2 \sin 2x$; TblStart $= 0$, ΔTbl $= \pi/2$

Chapter 5   Test

1. Given $\tan x = -5/6$, $3\pi/2 < x < 2\pi$, use trigonometric identities to find $\sin x$ and $\cos x$.

2. Express $\tan^2 x - \sec^2 x$ in terms of $\sin x$ and $\cos x$, and simplify.

3. Find $\sin(x + y)$, $\cos(x - y)$, and $\tan(x + y)$, if $\sin x = -1/3$, $\cos y = -2/5$, x is in quadrant III, and y is in quadrant II.

4. Use a half-angle identity to find $\sin(-22.5°)$.

Graph each expression and use the graph to conjecture an identity. Then verify your conjecture.

5. $\sec x - \sin x \tan x$

6. $\cot \dfrac{x}{2} - \cot x$

Verify that each equation is an identity.

7. $\sec^2 B = \dfrac{1}{1 - \sin^2 B}$

8. $\cos 2A = \dfrac{\cot A - \tan A}{\csc A \sec A}$

9. How do the cofunction identities given in this chapter differ from those given in Chapter 2?

10. *Voltage*   The voltage in common household current is expressed as $V = 163 \sin \omega t$, where ω is the angular velocity (in radians per second) of the generator at the electrical plant and t is time (in seconds).

(a) Use an identity to express V in terms of cosine.

(b) If $\omega = 120\pi$, what is the maximum voltage? Give the smallest positive value of t when the maximum voltage occurs.

Inverse Trigonometric Functions and Trigonometric Equations

6

6.1   Inverse Trigonometric
        Functions

6.2   Trigonometric Equations I

6.3   Trigonometric Equations II

6.4   Equations Involving Inverse
        Trigonometric Functions

Music is both art and science. In 500 B.C. Pythagoras, who is usually associated with the Pythagorean theorem, discovered mathematical relationships between lengths of strings and musical intervals. This discovery was the beginning of the science of musical sound. In the Middle Ages, music was studied together with arithmetic, geometry, and astronomy as part of the liberal arts curriculum. Later, in 1862, the psychologist and scientist Hermann von Helmholtz published a classic work that opened a new direction for music using mathematics and technology. This direction has had its greatest effect on music recording and reproduction, and culminated in 1957 when Max Mathews used a computer to create complex musical sounds.

When musicians tune instruments, they are able to compare like tones and accurately determine whether their pitches are the same frequency simply by listening, even though these tones vibrate hundreds or thousands of times per second. Some radios and telephones have small speakers that cannot vibrate slower than 200 times per second—yet 35 keys on a piano have frequencies below 200 and all of them can be clearly heard on these speakers.

Explanations of musical phenomena like these require a mathematical understanding of sound. Music is made up of sound waves that cause rapid increases and decreases in air pressure on a person's eardrum. Sound often involves periodic motion through the air, and this periodic motion can be modeled using trigonometric functions. Important aspects of music, the theme of this chapter, can be analyzed using trigonometric equations and graphs.*

6.1 Inverse Trigonometric Functions

• Review of Inverse Function Concepts • The Inverse Sine Function • The Inverse Cosine Function • The Inverse Tangent Function • Other Inverse Functions • Inverse Function Values • Expressions Involving Inverse Function Values

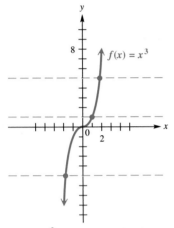

$f(x) = x^3$ is a one-to-one function. It satisfies the conditions of the horizontal line test.

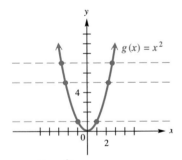

$g(x) = x^2$ is not one-to-one. It does not satisfy the conditions of the horizontal line test.

Figure 1

Review of Inverse Function Concepts Recall from Section 1.1 that for a function f, every element x in the domain corresponds to one and only one element y, or $f(x)$, in the range. This means that if point (a, b) lies on the graph of f, then there is no other point on the graph that has a as first coordinate. However, there may be other points having b as second coordinate, since the definition of function allows range elements to be used more than once.

If a function is defined so that each range element is used only once, then it is called a **one-to-one function.** For example, the function $f(x) = x^3$ is a one-to-one function because every real number has exactly one real cube root. On the other hand, $f(x) = x^2$ is not one-to-one because, for example, $f(2) = 4$ and $f(-2) = 4$. There are two domain elements, 2 and -2, that correspond to the range element 4.

The *horizontal line test* helps determine graphically whether a function is one-to-one.

Horizontal Line Test

Any horizontal line will intersect the graph of a one-to-one function in at most one point.

This test is applied to the graphs of $f(x) = x^3$ and $g(x) = x^2$ in Figure 1.

By interchanging the components of the ordered pairs of a one-to-one function f, we obtain a new set of ordered pairs that satisfies the definition of function. This new function is called the *inverse function,* or *inverse,* of f and is symbolized f^{-1}.

Inverse Function

The **inverse function** of the one-to-one function f is defined as

$$f^{-1} = \{(y, x) \mid (x, y) \text{ belongs to } f\}.$$

---

*Sources: Benade, Arthur, *Fundamentals of Musical Acoustics,* Oxford University Press, 1976.
Pierce, John, *The Science of Musical Sound,* Scientific American Books, 1992.

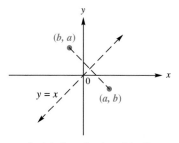

(b, a) is the reflection of (a, b)
across the line $y = x$.

(a)

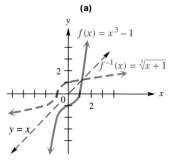

The graph of f^{-1} is the reflection of
the graph of f across the line $y = x$.

(b)

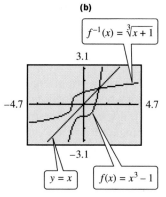

Most graphing calculators can draw
the inverse of a specified function
without the equation for the inverse
function being entered into the
calculator.

(c)

Figure 2

CAUTION The -1 in $f^{-1}(x)$ is not an exponent. That is,

$$f^{-1}(x) \neq \frac{1}{f(x)}.$$

Based on the definition of an inverse function, we can make the following statements.

General Statements about Inverses

1. If the point (a, b) lies on the graph of the one-to-one function f, then the point (b, a) lies on the graph of f^{-1}.
2. The domain of f is equal to the range of f^{-1}, and the range of f is equal to the domain of f^{-1}.
3. For all x in the domain of f, $f^{-1}[f(x)] = x$, and for all x in the domain of f^{-1}, $f[f^{-1}(x)] = x$.
4. Because the point (b, a) is the reflection of the point (a, b) across the line $y = x$, the graph of f^{-1} is the reflection of the graph of f across this line. See Figure 2.
5. To find the equation that defines the inverse of a one-to-one function f, follow these steps.

   Step 1 Let $y = f(x)$.
   Step 2 Interchange x and y in the equation.
   Step 3 Solve for y, and write $y = f^{-1}(x)$.

As we shall see later in this section, there may be a reason to restrict the domain of a function that is not one-to-one so that it becomes one-to-one. In this case, the range must be unchanged. For example, we saw that $f(x) = x^2$, with its natural domain $(-\infty, \infty)$, is not one-to-one. However, if we restrict its domain to be the set of nonnegative numbers $[0, \infty)$, we obtain a new function that is one-to-one and has the same range as before the restriction, $[0, \infty)$. See Figure 3.

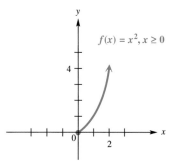

This is a one-to-one function.

Figure 3

N O T E We could also have chosen to restrict the domain of $f(x) = x^2$ to $(-\infty, 0]$ to obtain a one-to-one function. For important functions, such as the trigonometric functions, such choices are usually made based on general agreement by mathematicians.

C O N N E C T I O N S

Inverse functions are used by government agencies and other businesses to encode and decode information. These functions are usually very complicated. A simplified example involves the function defined by $f(x) = 3x - 2$. If each letter of the alphabet is assigned a numerical value according to its position ($A = 1, B = 2, \ldots, Z = 26$), the word ALGEBRA would be encoded using $f(x)$ as

$$1 \quad 34 \quad 19 \quad 13 \quad 4 \quad 52 \quad 1.$$

It could then be decoded using the inverse function with $f^{-1}(x) = \dfrac{x + 2}{3}$.

For Discussion or Writing

Use the alphabet assignment given above. The function with $f(x) = 2x + 5$ was used to encode the following message.

$$
\begin{array}{ccccccccc}
21 & 7 & 37 & 37 & 23 & 33 & 15 & 43 & 43 \\
23 & 43 & 43 & 45 & 7 & 55 & 23 & 33 & 19 \\
35 & 47 & 45 & 35 & 17 & 21 & 35 & 43 & 37 \\
23 & 45 & 7 & 29 & 43 & 7 & 33 & 13 & 11 \\
35 & 47 & 41 & 45 & 41 & 35 & 35 & 31 & 43.
\end{array}
$$

Find the inverse function using the steps given in Item 5 in the box above. Then, decode the message.

Looking Ahead to Calculus

The inverse trigonometric functions are used in calculus to express the solutions of trigonometric equations and integrate certain rational functions.

The Inverse Sine Function   Consider the function $y = \sin x$. From Figure 4 and the horizontal line test, it is clear that $y = \sin x$ is not a one-to-one function. By suitably restricting the domain of the sine function, however, a one-to-one function can be defined. It is common to restrict the domain of $y = \sin x$ to the interval $[-\pi/2, \pi/2]$, which gives the part of the graph shown in color in Figure 4. As Figure 4 shows, the range of $y = \sin x$ is $[-1, 1]$.

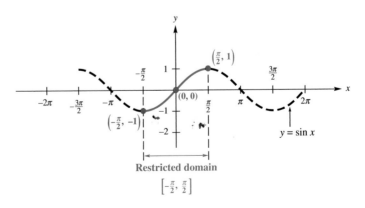

Figure 4

Reflecting the graph of $y = \sin x$ on the restricted domain across the line $y = x$ gives the graph of the inverse function, shown in Figure 5. Some key points are labeled on the graph.

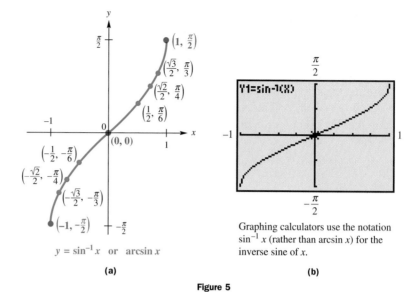

$y = \sin^{-1} x$ or arcsin x

(a)

Graphing calculators use the notation
$\sin^{-1} x$ (rather than arcsin x) for the
inverse sine of x.

(b)

Figure 5

The roles of x and y are interchanged in a pair of inverse functions. Therefore, the equation of the inverse of $y = \sin x$ is found by interchanging x and y to get $x = \sin y$. This equation is solved for y by writing **$y = \sin^{-1}x$**, read "inverse sine of x." (Note that $\sin^{-1} x$ does not mean $1/\sin x$.) As Figure 5 shows, the domain of $y = \sin^{-1} x$ is $[-1, 1]$, while the restricted domain of $y = \sin x$, $[-\pi/2, \pi/2]$, is the range of $y = \sin^{-1} x$. An alternative notation for $\sin^{-1} x$ is **arcsin x.**

$\sin^{-1}x$ or arcsin x

$y = \sin^{-1}x$ or $y = \arcsin x$ means $x = \sin y$, for y in $\left[-\pi/2, \pi/2\right]$.

Thus, we may think of $y = \sin^{-1}x$ or $y = \arcsin x$ as "y is the number in $[-\pi/2, \pi/2]$ whose sine is x." These two types of notation will be used in the rest of this book.

● ● ●   **Example 1**   Finding Inverse Sine Values

Find y in each equation.

Algebraic Solution

$$y = \sin^{-1}\frac{\sqrt{3}}{2}$$

(a) $y = \arcsin\dfrac{1}{2}$

The graph of the function $y = \arcsin x$ (Figure 5(a)) shows that the point $(1/2, \pi/6)$ lies on the graph. Therefore, $\arcsin(1/2) = \pi/6$. Alternatively, we may think of $y = \arcsin(1/2)$ as

y is the number in $\left[-\dfrac{\pi}{2}, \dfrac{\pi}{2}\right]$ whose sine is $\dfrac{1}{2}$.

Graphing Calculator Solution

To find these values with a graphing calculator, we graph $y = \sin^{-1}x$ and locate the points with x-values $1/2$ and -1. Figure 6(a) on the next page shows that when $x = 1/2$, $y = \pi/6 \approx .52359878$.

(continued)

Then we can rewrite the equation as $\sin y = 1/2$. Since $\sin \pi/6 = 1/2$ and $\pi/6$ is in the range of the arcsin function, $y = \pi/6$.

(b) $y = \sin^{-1}(-1)$

Writing the alternative equation, $\sin y = -1$, shows that $y = -\pi/2$. This can be verified by noticing that the point $(-1, -\pi/2)$ is on the graph of $y = \sin^{-1} x$.

Similarly, Figure 6(b) shows that when $x = -1$, $y = -\pi/2 \approx -1.570796$.

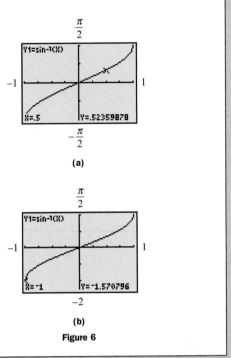

(a)

(b)

Figure 6

● ● ●

CAUTION In Example 1(b), it is tempting to give the value of $\sin^{-1}(-1)$ as $3\pi/2$, since $\sin(3\pi/2) = -1$. Notice, however, that $3\pi/2$ is not in the range of the inverse sine function. Be certain (in dealing with *all* inverse trigonometric functions) that the number given for an inverse function value is in the range of the particular inverse function being considered.

The Inverse Cosine Function The function $y = \cos^{-1} x$ (or $y = \arccos x$) is defined by restricting the domain of $y = \cos x$ to $[0, \pi]$. This domain becomes the range of $y = \cos^{-1} x$. The range of $y = \cos x$, the interval $[-1, 1]$, becomes the domain of $y = \cos^{-1} x$. The graph of $y = \cos x$ with domain $[0, \pi]$ is shown in Figure 7.

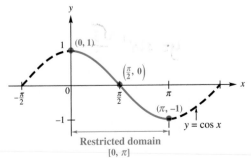

Figure 7

Using the same procedure we discussed for the inverse sine, we obtain the graph of the inverse cosine. See Figure 8.

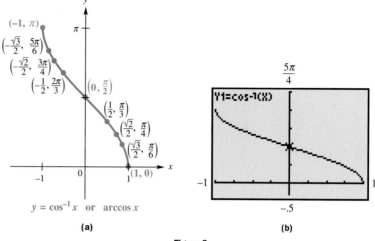

$$y = \cos^{-1} x \quad \text{or} \quad \arccos x$$

(a)

(b)

Figure 8

$\cos^{-1} x$ or arccos x

$y = \cos^{-1} x$ or $y = \arccos x$ means $x = \cos y$, for y in $[0, \pi]$.

● ● ● **Example 2** Finding Inverse Cosine Values

Find y in each equation.

Algebraic Solution

(a) $y = \arccos 1$

Since the point $(1, 0)$ lies on the graph of $y = \arccos x$ in Figure 8(a), the value of y is 0. Alternatively, we may think of $y = \arccos 1$ as "y is the number in $[0, \pi]$ whose cosine is 1," or $\cos y = 1$. Then $y = 0$, since $\cos 0 = 1$ and 0 is in the range of the arccos function.

(b) $y = \cos^{-1}\left(-\dfrac{\sqrt{2}}{2}\right)$

We must find the value of y that satisfies $\cos y = -\sqrt{2}/2$, where y is in the interval $[0, \pi]$, the range of the function $y = \cos^{-1} x$. The only value for y that satisfies these conditions is $3\pi/4$. Again, this can be verified from the graph in Figure 8(a).

Graphing Calculator Solution

Figure 9(a) shows that when $x = 1$, $y = 0$ on the graph of $y = \cos^{-1} x$. Similarly, Figure 9(b) shows that when $x = -\sqrt{2}/2 \approx -.7071068$, $y = 3\pi/4 \approx 2.3561945$.

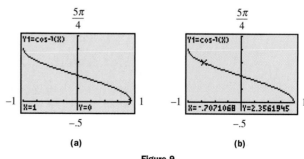

(a)

(b)

Figure 9

● ● ●

The Inverse Tangent Function The inverse tangent function is obtained by restricting the domain of the tangent function to $(-\pi/2, \pi/2)$ and then interchanging the roles of x and y. Figure 10 shows the restricted tangent function.

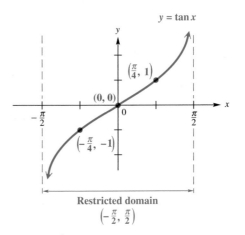

Figure 10

Figure 11 gives the graph of $y = \tan^{-1} x$.

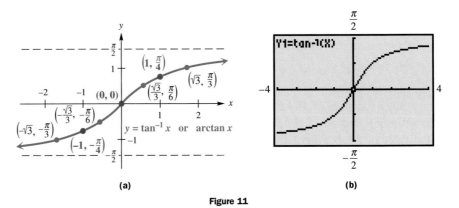

(a)   (b)

Figure 11

$\tan^{-1} x$ or $\arctan x$

$y = \tan^{-1} x$ or $y = \arctan x$ means $x = \tan y$, for y in $(-\pi/2, \pi/2)$.

From the graph we see that $\arctan 0 = 0$, $\arctan 1 = \pi/4$, $\arctan\left(-\sqrt{3}\right) = -\pi/3$, and so on.

Other Inverse Functions We now define the three remaining inverse trigonometric functions.

$y = \sec^{-1} x$ or $y = \operatorname{arcsec} x$ means $y = \cos^{-1}\left(\dfrac{1}{x}\right)$, where x is in $(-\infty, -1] \cup [1, \infty)$.

See Figure 12.

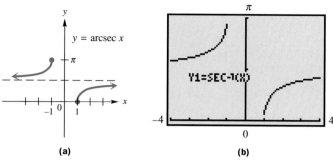

Figure 12

$y = \csc^{-1} x$ or $y = \text{arccsc } x$ means $y = \sin^{-1}\left(\dfrac{1}{x}\right)$, where x is in $(-\infty, -1] \cup [1, \infty)$.

See Figure 13.

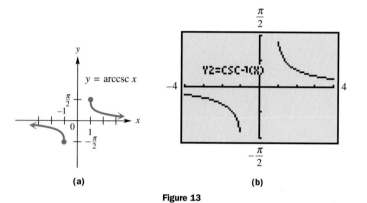

Figure 13

$y = \cot^{-1} x$ or $y = \text{arccot } x$ means $y = \tan^{-1}\left(\dfrac{1}{x}\right)$, if x is in $[0, \infty)$

or $y = \tan^{-1}\left(\dfrac{1}{x}\right) + \pi$, if x is in $(-\infty, 0)$.

See Figure 14.

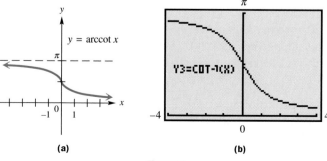

Figure 14

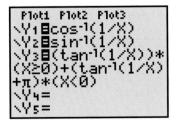

Figure 15

Figure 15 shows how the calculator graphs in Figures 12–14 were defined.

NOTE Sometimes the inverse secant and inverse cosecant functions are defined with different ranges than we have given here. We have elected to use intervals that match their reciprocal functions, except for one missing point. In calculus, different ranges are considered more convenient: for $\sec^{-1}x$, $[0, \pi/2) \cup [\pi, 3\pi/2)$; for $\csc^{-1}x$, $[\pi/2, \pi) \cup [3\pi/2, 2\pi)$; for $\cot^{-1}x$, $(-\pi/2, 0) \cup (0, \pi/2]$.

In summary, the six inverse trigonometric functions with their domains and ranges are given in the table.

Summary of the Inverse Trigonometric Functions

Function	Domain	Range	Quadrants of the Unit Circle from Which Range Values Come
$y = \sin^{-1}x$	$[-1, 1]$	$[-\pi/2, \pi/2]$	I and IV
$y = \cos^{-1}x$	$[-1, 1]$	$[0, \pi]$	I and II
$y = \tan^{-1}x$	$(-\infty, \infty)$	$(-\pi/2, \pi/2)$	I and IV
$y = \cot^{-1}x$	$(-\infty, \infty)$	$(0, \pi)$	I and II
$y = \sec^{-1}x$	$(-\infty, -1] \cup [1, \infty)$	$[0, \pi/2) \cup (\pi/2, \pi]$	I and II
$y = \csc^{-1}x$	$(-\infty, -1] \cup [1, \infty)$	$[-\pi/2, 0) \cup (0, \pi/2]$	I and IV

Inverse Function Values The inverse trigonometric functions are formally defined with real number ranges. However, there are times when it may be convenient to find the degree-measured angles equivalent to these real number values. It is also often convenient to think in terms of the unit circle and choose the inverse function values based on the quadrants given in the above table.

● ● ●   **Example 3**   Finding Inverse Values (Degree-Measured Angles)

Find the *degree measure* of θ in each of the following.

Algebraic Solution

(a) θ, if $\theta = \arctan 1$

Here θ must be in $(-90°, 90°)$, but since $1 > 0$, θ must be in quadrant I. The alternative statement, $\tan \theta = 1$, leads to $\theta = 45°$.

(b) θ, if $\theta = \sec^{-1} 2$

Write the equation as $\sec \theta = 2$. For $\sec^{-1}x$, θ is in quadrant I or II. Because 2 is positive, θ is in quadrant I and $\theta = 60°$, since $\sec 60° = 2$. Note that 60° (the degree equivalent of $\pi/3$) is in the range of the inverse secant function.

Graphing Calculator Solution

Figure 16 shows how a graphing calculator in degree mode displays the results found in the algebraic solution.

(continued)

(c) θ, if $\theta = \text{arccot}(-.3541)$

A calculator gives $\tan^{-1}(1/-.3541) \approx -70.500946°$. The restriction on the range of arccot means that θ must be in quadrant II.

$$\cot(-70.500946°) = -.3541$$

$$\cot(-70.500946° + 180°) = -.3541 \quad \text{\small Cotangent has period 180°.}$$

$$\cot(109.4990544°) = -.3541$$

$$\theta = 109.4990544°$$

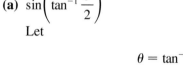

```
tan-1(1)
                     45
cos-1(1/2)
                     60
tan-1(1/-.3541)+1
80
          109.4990544
```

Degree mode

Figure 16

Expressions Involving Inverse Function Values

● ● ● **Example 4** Finding Function Values

Evaluate the following without using a calculator.

(a) $\sin\left(\tan^{-1}\dfrac{3}{2}\right)$

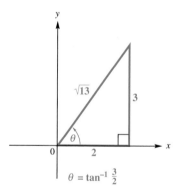

$\theta = \tan^{-1}\frac{3}{2}$

Figure 17

Let

$$\theta = \tan^{-1}\frac{3}{2}, \quad \text{so } \tan \theta = \frac{3}{2}.$$

Since $\tan^{-1} x$ is defined only in quadrants I and IV and since $3/2$ is positive, θ is in quadrant I. Sketch θ in quadrant I, and label a triangle as shown in Figure 17. The hypotenuse is $\sqrt{13}$, and the value of sine is the ratio of the side opposite and the hypotenuse, so

$$\sin\left(\tan^{-1}\frac{3}{2}\right) = \sin \theta = \frac{3}{\sqrt{13}} = \frac{3\sqrt{13}}{13}.$$

(b) $\tan\left(\cos^{-1}\left(-\dfrac{5}{13}\right)\right)$

Let $A = \cos^{-1}(-5/13)$. Then $\cos A = -5/13$. Since $\cos^{-1} x$ for a negative value of x is in quadrant II, sketch A in quadrant II, as shown in Figure 18.

From the triangle in Figure 18,

$$\tan\left(\cos^{-1}\left(-\frac{5}{13}\right)\right) = \tan A = -\frac{12}{5}.$$

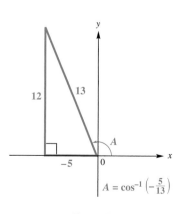

$A = \cos^{-1}\left(-\frac{5}{13}\right)$

Figure 18

(c) $\cos(\cos^{-1}(-.5))$

$\cos^{-1}(-.5)$ is the number, call it y, in $[0, \pi]$ whose cosine is $-.5$. Therefore, $\cos(\cos^{-1}(-.5)) = \cos y = -.5$.

(d) $\cos^{-1}\left(\cos\dfrac{5\pi}{4}\right) = \cos^{-1}\left(-\dfrac{\sqrt{2}}{2}\right) = \dfrac{3\pi}{4}$

Notice that in this case, $\cos^{-1}(\cos x) \ne x$.

● ● ●

● ● ● **Example 5** Finding Function Values Using Sum and Double-Angle Identities

Evaluate the following without using a calculator.

(a) $\cos\left(\arctan\sqrt{3} + \arcsin\dfrac{1}{3}\right)$

Let $A = \arctan\sqrt{3}$ and $B = \arcsin 1/3$, so $\tan A = \sqrt{3}$ and $\sin B = 1/3$. Sketch both A and B in quadrant I, as shown in Figure 19.

Now use the identity for $\cos(A + B)$.

$$\cos(A + B) = \cos A \cos B - \sin A \sin B$$

$$\cos\left(\arctan\sqrt{3} + \arcsin\dfrac{1}{3}\right) = \cos\left(\arctan\sqrt{3}\right)\cos\left(\arcsin\dfrac{1}{3}\right)$$

$$- \sin\left(\arctan\sqrt{3}\right)\sin\left(\arcsin\dfrac{1}{3}\right) \quad \textbf{(1)}$$

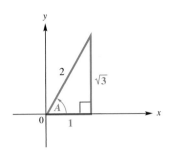

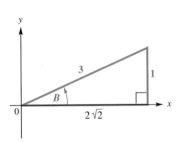

Figure 19

From the sketches in Figure 19,

$$\cos\left(\arctan\sqrt{3}\right) = \cos A = \dfrac{1}{2}, \qquad \cos\left(\arcsin\dfrac{1}{3}\right) = \cos B = \dfrac{2\sqrt{2}}{3},$$

$$\sin\left(\arctan\sqrt{3}\right) = \sin A = \dfrac{\sqrt{3}}{2}, \qquad \sin\left(\arcsin\dfrac{1}{3}\right) = \sin B = \dfrac{1}{3}.$$

Substitute these values into equation (1) to get

$$\cos\left(\arctan\sqrt{3} + \arcsin\dfrac{1}{3}\right) = \dfrac{1}{2}\cdot\dfrac{2\sqrt{2}}{3} - \dfrac{\sqrt{3}}{2}\cdot\dfrac{1}{3} = \dfrac{2\sqrt{2} - \sqrt{3}}{6}.$$

(b) $\tan\left(2\arcsin\dfrac{2}{5}\right)$

Let $\arcsin(2/5) = B$. Then, from the identity for the tangent of the double angle,

$$\tan\left(2\arcsin\dfrac{2}{5}\right) = \tan(2B) = \dfrac{2\tan B}{1 - \tan^2 B}.$$

Since $\arcsin(2/5) = B$, $\sin B = 2/5$. Sketch a triangle in quadrant I, find the length of the third side, and then find $\tan B$. From the triangle in Figure 20, $\tan B = 2/\sqrt{21}$, and

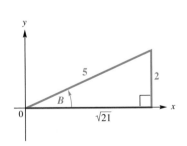

Figure 20

$$\tan\left(2\arcsin\dfrac{2}{5}\right) = \dfrac{2\left(\dfrac{2}{\sqrt{21}}\right)}{1 - \left(\dfrac{2}{\sqrt{21}}\right)^2} = \dfrac{\dfrac{4}{\sqrt{21}}}{1 - \dfrac{4}{21}} = \dfrac{\dfrac{4}{\sqrt{21}}}{\dfrac{17}{21}} = \dfrac{4\sqrt{21}}{17}.$$

● ● ●

N O T E In Exercises 55 and 56, we show how a graphing calculator can support the results in Examples 4 and 5.

Example 6 Writing Function Values in Terms of u

Write each function value as a non-trigonometric expression in u.

(a) $\sin(\tan^{-1} u)$

Let $\theta = \tan^{-1} u$, so $\tan \theta = u$. Here u may be positive or negative. Since $-\pi/2 < \tan^{-1} u < \pi/2$, sketch θ in quadrants I and IV and label two triangles as shown in Figure 21. Since sine is given by the ratio of the opposite side and the hypotenuse,

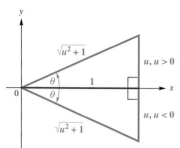

$$\sin(\tan^{-1} u) = \sin \theta = \frac{u}{\sqrt{u^2 + 1}} = \frac{u\sqrt{u^2 + 1}}{u^2 + 1}.$$

The result is positive when u is positive and negative when u is negative.

Figure 21

(b) $\cos(2 \sin^{-1} u)$

Let $\theta = \sin^{-1} u$, so $\sin \theta = u$. To find $\cos 2\theta$, use the identity $\cos 2\theta = 1 - 2 \sin^2 \theta$.

$$\cos 2\theta = 1 - 2 \sin^2 \theta = 1 - 2u^2$$

Example 7 Determining the Angle of Elevation of a Shot Put

Consider the model for shot-putter performance from Exercise 67 of Section 2.3. For given values of h and v, the greatest distance is achieved when the angle of elevation θ is

$$\theta = \arcsin\left(\sqrt{\frac{v^2}{2v^2 + 64h}}\right).$$

(*Source:* Townend, M. Stewart, *Mathematics in Sport*, Chichester, Ellis Horwood Limited, 1984.) Suppose a shot putter can consistently throw the steel ball with $h = 7.6$ feet and $v = 42$ feet per second. At what angle should he throw the ball to maximize distance?

To find this angle, substitute into the model and use a calculator in degree mode.

$$\theta = \arcsin\left(\sqrt{\frac{42^2}{2(42^2) + 64(7.6)}}\right) \qquad h = 7.6, v = 42$$

$$\theta \approx 41.5° \qquad\qquad \text{Use a calculator.}$$

6.1 Exercises

Concept Check   In Exercises 1–4, write short answers and fill in the blanks.

1. Consider the inverse sine function, defined by $y = \sin^{-1} x$ or $y = \arcsin x$.
  (a) What is its domain?
  (b) What is its range?
  (c) For this function, as x increases, y increases. Therefore, it is a(n) _____ function.
        (decreasing/increasing)
  (d) Why is $\arcsin(-2)$ not defined?

2. Consider the inverse cosine function, defined by $y = \cos^{-1} x$ or $y = \arccos x$.
  (a) What is its domain?
  (b) What is its range?
  (c) For this function, as x increases, y decreases. Therefore, it is a(n) _____ function.
        (decreasing/increasing)
  (d) $\text{Arccos}(-1/2) = 2\pi/3$. Why is $\arccos(-1/2)$ not equal to $-4\pi/3$?

3. Consider the inverse tangent function, defined by $y = \tan^{-1} x$ or $y = \arctan x$.

   (a) What is its domain?

   (b) What is its range?

   (c) For this function, as x increases, y increases. Therefore, it is a(n) _____ function.
   (decreasing/increasing)

   (d) Is there any real number x for which $\arctan x$ is not defined? If so, what is it (or what are they)?

4. Consider the three other inverse trigonometric functions, as defined in this section.

   (a) Give the domain and the range of the inverse cosecant function.

   (b) Give the domain and the range of the inverse secant function.

   (c) Give the domain and the range of the inverse cotangent function.

Find the exact value of the real number y in the following. Do not use a calculator. See Examples 1 and 2.

5. $y = \arcsin\left(-\dfrac{1}{2}\right)$

6. $y = \arccos\left(\dfrac{\sqrt{3}}{2}\right)$

7. $y = \tan^{-1} 1$

8. $y = \sin^{-1} 0$

9. $y = \cos^{-1}(-1)$

10. $y = \arctan(-1)$

11. $y = \sin^{-1}(-1)$

12. $y = \cos^{-1}\left(\dfrac{1}{2}\right)$

13. $y = \arctan 0$

14. $y = \arcsin\left(-\dfrac{\sqrt{3}}{2}\right)$

15. $y = \arccos 0$

16. $y = \tan^{-1}(-1)$

17. $y = \sin^{-1}\left(\dfrac{\sqrt{2}}{2}\right)$

18. $y = \cos^{-1}\left(-\dfrac{1}{2}\right)$

19. $y = \arccos\left(-\dfrac{\sqrt{3}}{2}\right)$

20. $y = \arcsin\left(-\dfrac{\sqrt{2}}{2}\right)$

21. $y = \cot^{-1}(-1)$

22. $y = \sec^{-1}(-\sqrt{2})$

23. $y = \csc^{-1}(-2)$

24. $y = \operatorname{arccot}(-\sqrt{3})$

25. $y = \operatorname{arcsec}\left(\dfrac{2\sqrt{3}}{3}\right)$

26. $y = \csc^{-1}\sqrt{2}$

27. $y = \operatorname{arccot}\left(\dfrac{\sqrt{3}}{3}\right)$

28. $y = \operatorname{arcsec} 2$

Give the degree measure of θ. Do not use a calculator.

29. $\theta = \arctan(-1)$

30. $\theta = \arccos\left(-\dfrac{1}{2}\right)$

31. $\theta = \arcsin\left(-\dfrac{\sqrt{3}}{2}\right)$

32. $\theta = \arcsin\left(-\dfrac{\sqrt{2}}{2}\right)$

33. $\theta = \cot^{-1}\left(-\dfrac{\sqrt{3}}{3}\right)$

34. $\theta = \sec^{-1}(-2)$

35. $\theta = \csc^{-1}(-2)$

36. $\theta = \csc^{-1}(-1)$

Use a calculator to give each value in decimal degrees. See Example 3.

37. $\theta = \sin^{-1}(-.13349122)$

38. $\theta = \cos^{-1}(-.13348816)$

39. $\theta = \arccos(-.39876459)$

40. $\theta = \arcsin .77900016$

41. $\theta = \csc^{-1} 1.9422833$

42. $\theta = \cot^{-1} 1.7670492$

Use a calculator to give each real number value. See Example 3, but be sure the calculator is in radian mode.

43. $y = \arctan 1.1111111$

44. $y = \arcsin .81926439$

45. $y = \cot^{-1}(-.92170128)$

46. $y = \sec^{-1}(-1.2871684)$

47. $y = \arcsin .92837781$

48. $y = \arccos .44624593$

Use a graphing calculator to graph each inverse function, and give the domain and the range.

49. $y = 2 \cot^{-1} x$

50. $y = \operatorname{arccsc} 2x$

51. $y = \operatorname{arcsec} \dfrac{1}{2} x$

· · · · · · · · · · · · **Relating Concepts** · · · · · · · · · · · ·
For individual or collaborative investigation
(Exercises 52–54*)

52. Consider the function defined by $f(x) = 3x - 2$ and its inverse $f^{-1}(x) = \dfrac{x + 2}{3}$. Simplify $f[f^{-1}(x)]$ and $f^{-1}[f(x)]$. What do you notice in each case? What would the graph look like in each case?

53. Use a graphing calculator to graph $y = \tan(\tan^{-1} x)$ in the standard viewing window, using radian mode. How

does this compare to the graph you described in Exercise 52?

54. Use a graphing calculator to graph $y = \tan^{-1}(\tan x)$ in the standard viewing window, using radian and dot modes. Why does this graph not agree with the graph you found in Exercise 53?

· ·

55. In Examples 4 and 5, we found function values without the use of a calculator. A calculator can be used, however, to support the results found there. For example, in Example 4(a) we showed that $\sin\left(\tan^{-1}\dfrac{3}{2}\right) = \dfrac{3\sqrt{13}}{13}$. The screen here supports this result, indicating the same approximations for the two expressions in the equation.

```
sin(tan⁻¹(3/2))
       .8320502943
3√(13)/13
       .8320502943
```

Use a calculator to support parts (b), (c), and (d) of Example 4.

56. The screen here shows how the result of Example 5(a) is supported with a graphing calculator. Use a graphing calculator to support the result of Example 5(b) similarly.

```
cos(tan⁻¹(√(3))+s
in⁻¹(1/3))
        .1827293862
(2√(2)-√(3))/6
        .1827293862
```

Find each value without using a calculator. See Examples 4 and 5.

57. $\tan\left(\arccos\dfrac{3}{4}\right)$

58. $\sin\left(\arccos\dfrac{1}{4}\right)$

59. $\cos(\tan^{-1}(-2))$

60. $\sec\left(\sin^{-1}\left(-\dfrac{1}{5}\right)\right)$

61. $\cot\left(\arcsin\left(-\dfrac{2}{3}\right)\right)$

62. $\cos\left(\arctan\dfrac{8}{3}\right)$

63. $\sec(\sec^{-1} 2)$

64. $\csc\left(\csc^{-1}\sqrt{2}\right)$

65. $\arccos\left(\cos\dfrac{\pi}{4}\right)$

66. $\arctan\left(\tan\left(-\dfrac{\pi}{4}\right)\right)$

67. $\arcsin\left(\sin\dfrac{\pi}{3}\right)$

68. $\arccos(\cos 0)$

69. $\sin\left(2\tan^{-1}\dfrac{12}{5}\right)$

70. $\cos\left(2\sin^{-1}\dfrac{1}{4}\right)$

71. $\cos\left(2\arctan\dfrac{4}{3}\right)$

72. $\tan\left(2\cos^{-1}\dfrac{1}{4}\right)$

73. $\sin\left(2\cos^{-1}\dfrac{1}{5}\right)$

74. $\cos(2\arctan(-2))$

75. $\tan\left(2\arcsin\left(-\dfrac{3}{5}\right)\right)$

76. $\sin\left(2\arccos\dfrac{2}{9}\right)$

77. $\sin\left(\sin^{-1}\dfrac{1}{2} + \tan^{-1}(-3)\right)$

78. $\cos\left(\tan^{-1}\dfrac{5}{12} - \cot^{-1}\dfrac{4}{3}\right)$

79. $\cos\left(\arcsin\dfrac{3}{5} + \arccos\dfrac{5}{13}\right)$

80. $\tan\left(\arccos\dfrac{\sqrt{3}}{2} - \arcsin\left(-\dfrac{3}{5}\right)\right)$

*The authors wish to thank Carol Walker of Hinds Community College for making a suggestion on which these exercises are based.

Use a calculator to find each value. Give answers as real numbers.

81. $\cos(\tan^{-1}.5)$
82. $\sin(\cos^{-1}.25)$
83. $\tan(\arcsin .12251014)$
84. $\cot(\arccos .58236841)$

Write each of the following as a non-trigonometric expression in u. See Example 6.

85. $\sin(\arccos u)$
86. $\tan(\arccos u)$
87. $\cot(\arcsin u)$
88. $\cos(\arcsin u)$

89. $\sin\left(\sec^{-1}\dfrac{u}{2}\right)$
90. $\cos\left(\tan^{-1}\dfrac{3}{u}\right)$
91. $\tan\left(\arcsin \dfrac{u}{\sqrt{u^2 + 2}}\right)$
92. $\cos\left(\arccos \dfrac{u}{\sqrt{u^2 + 5}}\right)$

Recall that a graphing calculator will return a 1 for a true statement and a 0 for a false statement. In Exercises 93 and 94, determine the possible values stored in X for which the given screen will come up on a graphing calculator in radian mode.

93.

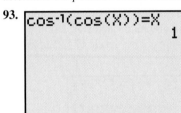

94.

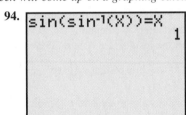

Solve each problem. See Example 7.

95. *(Modeling) Angle of Elevation of a Shot Put* Refer to Example 7.
  (a) What is the optimal angle when $h = 0$?
  (b) Fix h at 6 feet and regard θ as a function of v. As v gets larger and larger, the graph approaches a horizontal asymptote. Find the equation of that asymptote.

96. *(Modeling) Observation of a Painting* A painting 1 meter high and 3 meters from the floor will cut off an angle θ to an observer, where

$$\theta = \tan^{-1}\left(\frac{x}{x^2 + 2}\right).$$

Assume that the observer is x meters from the wall where the painting is displayed and that the eyes of the observer are 2 meters above the ground. (See the figure.)

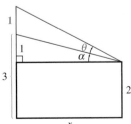

Find the value of θ for the following values of x. Round to the nearest degree.
  (a) 1   **(b)** 2   **(c)** 3
  (d) Derive the formula given above. (*Hint:* Use the identity for $\tan(\theta + \alpha)$. Use right triangles.)

(e) Graph the function for θ with a graphing calculator, and determine the distance that maximizes the angle.
  (f) The idea in part (e) was first investigated in 1471 by the astronomer Regiomontanus. (*Source:* Maor, E., *Trigonometric Delights,* Princeton University Press, 1998.) If the bottom of the picture is a meters above eye level and the top of the picture is b meters above eye level, then the optimum value of x is $\sqrt{ab}$ meters. Use this result to find the exact answer to part (e).

97. *(Modeling) Mach Number* Suppose an airplane flying faster than sound goes directly over you. Assume that the plane is flying level. At the instant you feel the sonic boom from the plane, the angle of elevation to the plane is modeled by

$$\alpha = 2\arcsin \frac{1}{m},$$

where m is the Mach number of the plane's speed. (See Exercises 58–63 in Section 5.6.) Find α to the nearest degree for each of the following values of m.
  (a) $m = 1.2$ **(b)** $m = 1.5$
  (c) $m = 2$ **(d)** $m = 2.5$

98. *Landscaping Formula* A shrub is planted in a 100-foot wide space between buildings measuring 75 feet and 150 feet tall. The location of the shrub determines how much sun it receives each day. Show that if θ is the angle in the figure and x is the distance of the shrub from the taller building, then the value of θ (in radians) is given by

$$f(x) = \pi - \arctan\left(\frac{75}{100 - x}\right) - \arctan\left(\frac{150}{x}\right).$$

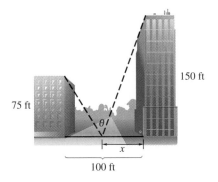

150 ft

75 ft

θ

x

100 ft

99. *Communications Satellite Coverage*   The figure shows a stationary communications satellite positioned 20,000 miles above the equator. What percent of the equator can be seen from the satellite? The diameter of Earth is 7927 miles at the equator.

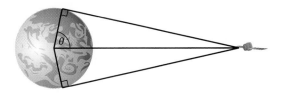

Quantitative Reasoning

100. *How can we determine the amount of oil in a submerged storage tank?* The level of oil in a storage tank buried in the ground can be found in much the same way a dipstick is used to determine the oil level in an automobile crankcase. The person in the figure on the left has lowered a calibrated rod into an oil storage tank. When the rod is removed, the reading on the rod can be used with the dimensions of the storage tank to calculate the amount of oil in the tank. Suppose the ends of the cylindrical storage tank in the figure are circles of radius 3 feet and the cylinder is 20 feet long. Determine the volume of oil in the tank if the rod shows a depth of 2 feet. (*Hint:* The volume will be 20 times the area of the shaded segment of the circle shown in the figure on the right.)

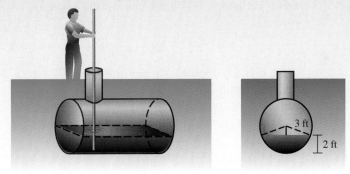

3 ft

2 ft

6.2   Trigonometric Equations I

- **Equations Solvable by Linear Methods** • **Equations Solvable by Factoring** • **Equations Solvable by the Quadratic Formula** • **An Application to Music**

In Chapter 5, we studied trigonometric equations that were identities. We now consider trigonometric equations that are *conditional;* that is, equations that are satisfied by some values but not others.

Equations Solvable by Linear Methods Conditional equations with trigonometric (or circular) functions can usually be solved using algebraic methods and trigonometric identities. For example, suppose we wish to find the solutions of the equation

$$2 \sin \theta + 1 = 0$$

for θ in the interval $[0°, 360°)$. Because $\sin \theta$ is the first power of a trigonometric function, we use the same method as we would in solving the linear equation $2x + 1 = 0$.

$$2 \sin \theta + 1 = 0$$
$$2 \sin \theta = -1 \quad \text{Subtract 1.}$$
$$\sin \theta = -\frac{1}{2} \quad \text{Divide by 2.}$$

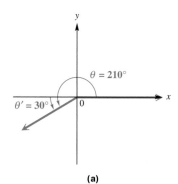

(a)

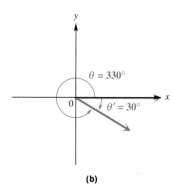

(b)

Figure 22

To find values of θ that satisfy $\sin \theta = -1/2$, we observe that θ must be in either quadrant III or IV since the sine function is negative in these two quadrants. Furthermore, the reference angle must be $30°$ since $\sin 30° = 1/2$. The sketches in Figure 22 show the two possible values of θ, $210°$ and $330°$.

CAUTION One value that satisfies $\sin \theta = -1/2$ is $\sin^{-1}(-1/2)$, which in degrees is $-30°$. However, when solving an equation such as this, we must pay close attention to the *domain*, which in this case is $[0°, 360°)$.

In some cases we are required to find *all* solutions of conditional trigonometric equations. All solutions of the equation $2 \sin \theta + 1 = 0$ would be written as

$$\theta = 210° + 360° \cdot n \qquad \text{or} \qquad \theta = 330° + 360° \cdot n,$$

where n is any integer. We add integer multiples of $360°$ to obtain all angles coterminal with $210°$ or $330°$. If we had been required to solve this equation for real numbers (or angles in radians) in the interval $[0, 2\pi)$, the two solutions would be $7\pi/6$ and $11\pi/6$, while all solutions would be written as

$$\theta = \frac{7\pi}{6} + 2n\pi \qquad \text{or} \qquad \theta = \frac{11\pi}{6} + 2n\pi,$$

where n is any integer.

In the examples in this section, we will find solutions in the intervals $[0°, 360°)$ or $[0, 2\pi)$. Remember that *all* solutions can be found using the methods described above.

In this book we will use two methods to support solutions of equations with a graphing calculator. The first of these is the **intersection-of-graphs method.**

Intersection-of-Graphs Method

The real solutions of the equation $f(x) = g(x)$ correspond to the x-coordinates of the points of intersection of the graphs of $y = f(x)$ and $y = g(x)$.

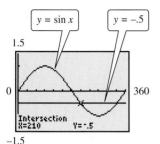

Degree mode

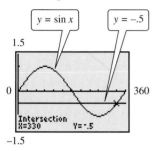

Degree mode

Figure 23

The screens in Figure 23 support the two solutions of $\sin x = -1/2$ in the interval $[0°, 360°)$. Notice that the x-coordinates of the points of intersection of the graphs of $y = \sin x$ and $y = -1/2 = -.5$ are 210 (degrees) and 330 (degrees), supporting the results found in the discussion above.

The other method we will use to support the solution of an equation with a graphing calculator is the **x-intercept method.**

---

x-Intercept Method

The real solutions of the equation $f(x) = 0$ are represented by the x-intercepts of the graph of $y = f(x)$.

The graphing calculator solution in Example 1 that follows shows how this method is applied. ■

Equations Solvable by Factoring   If an equation is of the form $a \cdot b = 0$, then either $a = 0$ or $b = 0$. This is called the *zero-factor property.*

● ● ● **Example 1**   Solving a Trigonometric Equation by Factoring

Solve $2 \cos^2 \theta - \cos \theta - 1 = 0$ in the interval $[0°, 360°)$.

Algebraic Solution

This equation is quadratic in the term $\cos \theta$.

$$2 \cos^2 \theta - \cos \theta - 1 = 0$$

$$(2 \cos \theta + 1)(\cos \theta - 1) = 0 \quad \text{Factor.}$$

$$2 \cos \theta + 1 = 0 \qquad \text{or} \qquad \cos \theta - 1 = 0$$

$$\cos \theta = -\frac{1}{2} \qquad \text{or} \qquad \cos \theta = 1$$

In the first case, we have $\cos \theta = -1/2$, indicating that θ must be in either quadrant II or III, with reference angle 60°. Using a sketch similar to those in Figure 22 would indicate that two solutions are 120° and 240°. The second case, $\cos \theta = 1$, has the quadrantal angle 0° as its only solution in the interval. (We do not include 360° since it is not in the stated interval.) Therefore, the solutions of this equation are 0°, 120°, and 240°. Check these solutions by substituting them in the given equation.

Graphing Calculator Solution

By graphing $y = 2 \cos^2 x - \cos x - 1$ (where $x = \theta$), we can support the solutions 0°, 120°, and 240° found in the algebraic solution. In this case, we will use the x-intercept method. The calculator must be in degree mode. Figure 24 indicates that 120° is a solution; the others can be supported similarly.

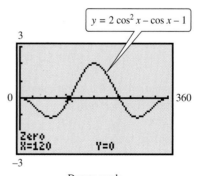

Degree mode

Figure 24

● ● ●

● ● ● **Example 2** Solving a Trigonometric Equation by Factoring

Solve $\sin x \tan x = \sin x$ in the interval $[0°, 360°)$.

$$\sin x \tan x = \sin x$$

$$\sin x \tan x - \sin x = 0 \qquad \text{Subtract } \sin x.$$

$$\sin x(\tan x - 1) = 0 \qquad \text{Factor.}$$

$$\sin x = 0 \quad \text{or} \quad \tan x - 1 = 0$$

$$\tan x = 1$$

$$x = 0° \quad \text{or} \quad x = 180° \qquad x = 45° \quad \text{or} \quad x = 225°$$

Verify these solutions by substitution. ● ● ●

C A U T I O N   There are four solutions in Example 2. Trying to solve the equation by dividing both sides by $\sin x$ would give just $\tan x = 1$, which would give $x = 45°$ or $x = 225°$. The other two solutions would not appear. The missing solutions are the ones that make the divisor, $\sin x$, equal 0. For this reason, it is best to avoid dividing by a variable expression.

Looking Ahead to Calculus

There are many instances in calculus where it is necessary to solve trigonometric equations. Examples include finding values for which a derivative is 0, solving related rate problems, and solving optimization problems.

Recall from algebra that squaring both sides of an equation, such as $\sqrt{x + 4} = x + 2$, will yield all solutions but may also give extraneous values. (In this equation, 0 is a solution, while -3 is extraneous. Verify this.) The same situation may occur when trigonometric equations are solved in this manner.

● ● ● **Example 3** Solving a Trigonometric Equation by Squaring

Solve $\tan x + \sqrt{3} = \sec x$ in the interval $[0, 2\pi)$.

Algebraic Solution

Since the tangent and secant functions are related by the identity $1 + \tan^2 x = \sec^2 x$, square both sides and express $\sec^2 x$ in terms of $\tan^2 x$.

$$\tan x + \sqrt{3} = \sec x$$

$$\tan^2 x + 2\sqrt{3}\tan x + 3 = \sec^2 x$$

$$\qquad\qquad (a + b)^2 = a^2 + 2ab + b^2$$

$$\tan^2 x + 2\sqrt{3}\tan x + 3 = 1 + \tan^2 x$$

$$\qquad\qquad \text{Substitute.}$$

$$2\sqrt{3}\tan x = -2 \quad \text{Subtract } 3 + \tan^2 x.$$

$$\tan x = -\frac{1}{\sqrt{3}} = -\frac{\sqrt{3}}{3}$$

The possible solutions in the given interval are $5\pi/6$ and $11\pi/6$. Now check them. Try $5\pi/6$ first.

Left side: $\quad \tan x + \sqrt{3} = \tan\dfrac{5\pi}{6} + \sqrt{3}$

$$= -\frac{\sqrt{3}}{3} + \sqrt{3} = \frac{2\sqrt{3}}{3}$$

Right side: $\quad \sec x = \sec\dfrac{5\pi}{6} = -\dfrac{2\sqrt{3}}{3}$ ←— Different

(continued)

Graphing Calculator Solution

Figure 25 shows that on the interval $[0, 2\pi)$, the only x-intercept of the graph of $y = \tan x + \sqrt{3} - \sec x$ is 5.7595865, which is an approximation for $\dfrac{11\pi}{6}$, the solution found algebraically.

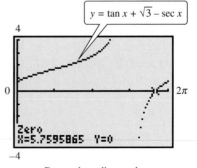

Dot mode; radian mode

Figure 25

The check shows that $5\pi/6$ is not a solution. Now check $11\pi/6$.

Left side: $\tan \dfrac{11\pi}{6} + \sqrt{3} = -\dfrac{\sqrt{3}}{3} + \sqrt{3} = \dfrac{2\sqrt{3}}{3}$

Right side: $\sec \dfrac{11\pi}{6} = \dfrac{2\sqrt{3}}{3}$ ⟵————————Same

This solution satisfies the equation, so $11\pi/6$ is the only solution of the given equation.

• • •

Example 4   Solving a Trigonometric Equation by Factoring

Solve $\tan^2 x + \tan x - 2 = 0$ in the interval $[0, 2\pi)$.

Algebraic Solution

Like Example 1, this equation is quadratic in form and may be solved for $\tan x$ by factoring.

$$\tan^2 x + \tan x - 2 = 0$$
$$(\tan x - 1)(\tan x + 2) = 0$$
$$\tan x - 1 = 0 \quad \text{or} \quad \tan x + 2 = 0$$
$$\tan x = 1 \quad \text{or} \quad \tan x = -2$$

The solutions for $\tan x = 1$ in the interval $[0, 2\pi)$ are $x = \pi/4$ and $5\pi/4$. To solve $\tan x = -2$ in that interval, we may use a scientific calculator set in *radian* mode. We find that $\tan^{-1}(-2) \approx -1.1071487$. This is a quadrant IV number, based on the range of the inverse tangent function. However, since we want solutions in the interval $[0, 2\pi)$, we must first add π to -1.1071487, and then add 2π.

$$x \approx -1.1071487 + \pi \approx 2.03444394$$
$$x \approx -1.1071487 + 2\pi \approx 5.1760366$$

The solutions in the required interval are

$$\underbrace{\dfrac{\pi}{4}, \quad \dfrac{5\pi}{4},}_{\substack{\text{Exact} \\ \text{values}}} \quad \underbrace{2.0, \quad 5.2.}_{\substack{\text{Approximate values} \\ \text{to the nearest tenth}}}$$

Graphing Calculator Solution

Figure 26 shows support for the solution $\dfrac{\pi}{4}$ found algebraically. The other three solutions in the interval $[0, 2\pi)$ can be found similarly. Note that in this situation, the calculator must be in radian mode.

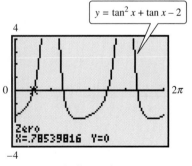

$y = \tan^2 x + \tan x - 2$

Radian mode

Figure 26

• • •

Equations Solvable by the Quadratic Formula

● ● ● **Example 5** Solving a Trigonometric Equation Using the Quadratic Formula

Solve $\cot^2 x + 3 \cot x = 1$ in $[0°, 360°)$.

Algebraic Solution

Write the equation in quadratic form, with 0 on one side.

$$\cot^2 x + 3 \cot x - 1 = 0$$

Since this quadratic equation cannot be solved by factoring, use the quadratic formula, with $a = 1$, $b = 3$, $c = -1$, and $\cot x$ as the variable.

$$\cot x = \frac{-3 \pm \sqrt{9 + 4}}{2} = \frac{-3 \pm \sqrt{13}}{2} \approx \frac{-3 \pm 3.6055513}{2}$$

$$\cot x \approx .30277564 \quad \text{or} \quad \cot x \approx -3.3027756$$

$$x \approx 73.2°, 253.2°, 163.2°, 343.2°$$

The final answers were obtained using a scientific calculator set in degree mode. A check shows that these answers are solutions of the given equation.

Graphing Calculator Solution

Figure 27 shows the graph of $y = \cot^2 x + 3 \cot x - 1$, with the calculator in degree mode, over the interval desired. The approximate degree solution $73.2°$ is indicated; the other three can be found similarly.

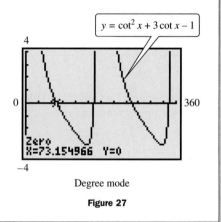

Degree mode

Figure 27

● ● ●

The methods for solving trigonometric equations illustrated in the examples can be summarized as follows.

Solving Trigonometric Equations Algebraically

Step 1 Decide whether the equation is linear or quadratic, so you can determine the solution method.

Step 2 If only one trigonometric function is present, first solve the equation for that function.

Step 3 If more than one trigonometric function is present, rearrange the equation so that one side equals 0. Then try to factor and set each factor equal to 0 to solve.

Step 4 If Step 3 does not work, try using identities to change the form of the equation. It may be helpful to square both sides of the equation first. If this is done, check for extraneous solutions.

Step 5 If the equation is quadratic in form, but not factorable, use the quadratic formula. Check for extraneous solutions.

> ### Solving Trigonometric Equations Graphically
>
> For an equation in the form $f(x) = g(x)$:
>
> **Step 1** Graph $y = f(x)$ and $y = g(x)$ over the required domain.
> **Step 2** Find the x-coordinates of the points of intersection of the graphs. (This is the intersection-of-graphs method.)
>
> For an equation in the form $f(x) = 0$:
>
> **Step 1** Graph $y = f(x)$ over the required domain.
> **Step 2** Find the x-intercepts of the graph. (This is the x-intercept method.)

An Application to Music

● ● ● **Example 6** Describing a Musical Tone by Interpreting a Graph

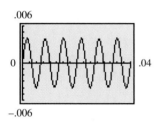

 A basic component of music is a pure tone. The calculator-generated graph in Figure 28 models the sinusoidal pressure P in pounds per square foot from a pure tone at time t seconds.

(a) The frequency of a pure tone is often measured in a unit called *hertz*. One hertz is equal to one cycle per second and is abbreviated *Hz*. What is the frequency f in hertz of the pure tone shown in the graph?

From the graph we can see that there are 6 cycles in .04 second. This is equivalent to $6/.04 = 150$ cycles per second. The pure tone has a frequency of $f = 150$ Hz.

Figure 28

(b) The time for the tone to produce one complete cycle is called the *period*. Approximate the period T of the pure tone in seconds.

Six periods cover a time of .04 second. One period would be equal to $T = .04/6 = 1/150$ or $.00\overline{6}$ second.

(c) An equation for the graph is $Y_1 = .004 \sin(300\pi x)$. Use a calculator to estimate all solutions to the equation $Y_1 = .004$ on the interval $[0, .02]$.

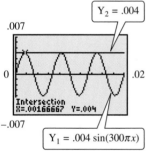

Figure 29

If we reproduce the graph in Figure 28 on a calculator and graph the second function $Y_2 = .004$, we determine that the approximate values of x at the points of intersection of the graphs are .0017, .0083, and .015. See Figure 29, which indicates the (approximate) solution .0017; the others can be found similarly.

● ● ●

6.2 Exercises

Concept Check *For the equations in Column I, a description of an approach for solving for* $\cos x$ *is found in Column II. Match each equation with the appropriate description.*

I	II
1. $2 \cos x - 1 = 0$	**A.** Factoring does not apply, so use the quadratic formula.
2. $2 \cos^2 x - \cos x - 1 = 0$	**B.** Add 1 to both sides, divide by 2, and apply the square root property.
3. $2 \cos^2 x - 3 \cos x - 1 = 0$	**C.** Factor the left side of the equation, set each factor equal to 0, and solve the two resulting linear equations.
4. $2 \cos^2 x - 1 = 0$	**D.** Add 1 to both sides of the equation, and divide both sides by 2.

5. Refer to Example 1. The solutions in the interval $[0°, 360°)$ are $0°$, $120°$, and $240°$. See the discussion at the beginning of this section, and express *all* solutions of this equation (in degrees).

6. Refer to Example 4. The solutions in the interval $[0, 2\pi)$ are $\pi/4$, $5\pi/4$, 2.0, and 5.2. See the discussion at the beginning of this section, and express *all* solutions of this equation (in real numbers).

Solve each equation for solutions in the interval $[0, 2\pi)$. Use algebraic methods and give exact values. You may wish to support your answers by finding approximations for these values using either of the two graphical methods discussed. See Examples 1–4.

7. $2 \cot x + 1 = -1$

8. $\sin x + 2 = 3$

9. $2 \sin x + 3 = 4$

10. $2 \sec x + 1 = \sec x + 3$

11. $\tan^2 x + 3 = 0$

12. $\sec^2 x + 2 = -1$

13. $(\cot x - 1)(\sqrt{3} \cot x + 1) = 0$

14. $(\csc x + 2)(\csc x - \sqrt{2}) = 0$

15. $\cos^2 x + 2 \cos x + 1 = 0$

16. $2 \cos^2 x - \sqrt{3} \cos x = 0$

17. $-2 \sin^2 x = 3 \sin x + 1$

18. $\tan^3 x = 3 \tan x$

19. $2 \cos^4 x = \cos^2 x$

20. $4(1 + \sin x)(1 - \sin x) = 3$

Solve each equation for exact solutions in the interval $[0°, 360°)$. Use either an algebraic method or a graphical method, according to the directions of your instructor. See Examples 1–4.

21. $(\cot \theta - \sqrt{3})(2 \sin \theta + \sqrt{3}) = 0$

22. $(\tan \theta - 1)(\cos \theta - 1) = 0$

23. $2 \sin \theta - 1 = \csc \theta$

24. $\tan \theta + 1 = \sqrt{3} + \sqrt{3} \cot \theta$

25. $\tan \theta - \cot \theta = 0$

26. $\cos^2 \theta = \sin^2 \theta + 1$

27. $\csc^2 \theta - 2 \cot \theta = 0$

28. $\sin^2 \theta \cos \theta = \cos \theta$

29. $2 \tan^2 \theta \sin \theta - \tan^2 \theta = 0$

30. $\sin^2 \theta \cos^2 \theta = 0$

31. $\sec^2 \theta \tan \theta = 2 \tan \theta$

32. $\cos^2 \theta - \sin^2 \theta = 0$

33. $\sin \theta + \cos \theta = 1$

34. $\sec \theta - \tan \theta = 1$

Solve each equation for solutions in the interval $[0°, 360°)$. Use a calculator and express approximate solutions to the nearest tenth of a degree. In Exercises 41–48, you will need to use the quadratic formula. See Examples 4 and 5.

35. $3 \sin^2 \theta - \sin \theta = 2$

36. $\dfrac{2 \tan \theta}{3 - \tan^2 \theta} = 1$

37. $\sec^2 \theta = 2 \tan \theta + 4$

38. $5 \sec^2 \theta = 6 \sec \theta$

39. $3 \cot^2 \theta = \cot \theta$

40. $8 \cos \theta = \cot \theta$

41. $9 \sin^2 \theta - 6 \sin \theta = 1$

42. $4 \cos^2 \theta + 4 \cos \theta = 1$

43. $\tan^2 \theta + 4 \tan \theta + 2 = 0$

44. $3 \cot^2 \theta - 3 \cot \theta - 1 = 0$

45. $\sin^2 \theta - 2 \sin \theta + 3 = 0$

46. $2 \cos^2 \theta + 2 \cos \theta - 1 = 0$

47. $\cot \theta + 2 \csc \theta = 3$

48. $2 \sin \theta = 1 - 2 \cos \theta$

Determine all solutions of each equation in radians.

49. $2 \sin^2 x - \sin x - 1 = 0$

50. $2 \cos^2 x + \cos x = 1$

51. $4 \cos^2 x - 1 = 0$

52. $2 \cos^2 x + 5 \cos x + 2 = 0$

53. $\cos^2 x + \cos x - 6 = 0$

54. $\sin^2 x - \sin x = 0$

The following equations cannot be solved by traditional algebraic methods. Use a graphing calculator to find all solutions in the interval $[0, 2\pi)$. Express solutions to as many decimal places as your calculator displays.

55. $x^2 + \sin x - x^3 - \cos x = 0$

56. $x^3 - \cos^2 x = \dfrac{1}{2}x - 1$

Recall that a graphing calculator will return a 1 for a true statement and a 0 for a false statement. In Exercises 57 and 58, determine the possible values for X in the interval $[0, 2\pi)$ that will produce the accompanying graphing calculator screen when in radian mode.

57.

```
2cos(X)-1=0
                1
```

58.

```
2sin(X)-1=0
                1
```

Solve each problem.

59. *(Modeling) Pressure on the Eardrum* No musical instrument can generate a true pure tone. A pure tone has a unique, constant frequency and amplitude that sounds rather dull and uninteresting. The pressures caused by pure tones on the eardrum are sinusoidal. The change in pressure P in pounds per square foot on a person's eardrum from a pure tone at time t in seconds can be modeled using the equation

$$P = A \sin(2\pi f t + \phi),$$

where f is the frequency in cycles per second, and ϕ is the phase angle. When P is positive, there is an increase in pressure and the eardrum is pushed inward; when P is negative, there is a decrease in pressure and the eardrum is pushed outward. (*Source:* Roederer, J., *Introduction to the Physics and Psychophysics of Music,* Second Edition, Springer-Verlag, 1975.)

(a) Middle C has a frequency of 261.63 cycles per second. Graph this tone with $A = .004$ and $\phi = \pi/7$ in the window $[0, .005]$ by $[-.005, .005]$.

(b) Determine algebraically the values of t for which $P = 0$ in $[0, .005]$, and support your answers graphically.

(c) Determine graphically the interval for which $P \le 0$ on $[0, .005]$.

(d) Would an eardrum hearing this tone be vibrating outward or inward when $P \le 0$?

60. *(Modeling) Frequency and Period of a Tone* Refer to Example 6. Determine a simple relationship between f and T.

61. *(Modeling) Accident Reconstruction* The model

$$.342D \cos \theta + h \cos^2 \theta = \frac{16D^2}{V_0^2}$$

is used to reconstruct accidents in which a vehicle vaults into the air after hitting an obstruction. V_0 is velocity in feet per second of the vehicle when it hits, D is distance (in feet) from the obstruction to the landing point, and h is the difference in height (in feet) between landing point and takeoff point. Angle θ is the takeoff angle, the angle between the horizontal and the

path of the vehicle. Find θ to the nearest degree if $V_0 = 60$, $D = 80$, and $h = 2$.

62. *(Modeling) Electromotive Force* In an electric circuit, let $V = \cos 2\pi t$ model the electromotive force in volts at t seconds. Find the smallest positive value of t where $0 \le t \le 1/2$ for the following values of V.
(a) $V = 0$ **(b)** $V = .5$ **(c)** $V = .25$

63. *(Modeling) Voltage Induced by a Coil of Wire* A coil of wire rotating in a magnetic field induces a voltage modeled by

$$e = 20 \sin\left(\frac{\pi t}{4} - \frac{\pi}{2}\right),$$

where t is time in seconds. Find the smallest positive time to produce the following voltages.
(a) 0   **(b)** $10\sqrt{3}$

64. *(Modeling) Movement of a Particle* A particle moves along a straight line. The distance of the particle from the origin at time t is modeled by

$$s(t) = \sin t + 2 \cos t.$$

Find a value of t that satisfies each equation.
(a) $s(t) = \dfrac{2 + \sqrt{3}}{2}$ **(b)** $s(t) = \dfrac{3\sqrt{2}}{2}$

65. Explain what is wrong with the following solution for all x in the interval $[0, 2\pi)$ of the equation $\sin^2 x - \sin x = 0$.

$$\sin^2 x - \sin x = 0$$
$$\sin x - 1 = 0 \qquad \text{Divide by } \sin x.$$
$$\sin x = 1 \qquad \text{Add 1.}$$

$$x = \frac{\pi}{2}$$

66. Explain what is wrong with the following solutions for all θ in the interval $[0°, 360°)$ of the equation $\tan^2 \theta - 1 = 0$.

$$\tan^2 \theta - 1 = 0$$
$$\tan^2 \theta = 1 \qquad \text{Add 1.}$$
$$\tan \theta = 1 \qquad \text{Take square roots on each side.}$$
$$\theta = 45°, 225°$$

6.3 Trigonometric Equations II

• **Equations with Half-Angles** • **Equations with Multiple Angles** • **An Application to Music**

In this section, we discuss trigonometric equations that involve functions of half-angles and multiple angles. Solving equations of these types often requires adjusting solution intervals to fit domain requirements.

Equations with Half-Angles

● ● ● **Example 1** Solving an Equation with a Half-Angle

Solve $2 \sin \dfrac{\theta}{2} = 1$ in the interval $[0°, 360°)$.

Algebraic Solution

As a compound inequality, the interval $[0°, 360°)$ is written

$$0° \le \theta < 360°.$$

Find the corresponding interval for $\theta/2$.

$$0° \le \dfrac{\theta}{2} < 180° \qquad \text{Divide both sides by 2.}$$

To find all values of $\theta/2$ in the interval $0°$ to $180°$ which satisfy the given equation, solve for the trigonometric function.

$$2 \sin \dfrac{\theta}{2} = 1$$

$$\sin \dfrac{\theta}{2} = \dfrac{1}{2} \qquad \text{Divide by 2.}$$

Both $\sin 30° = 1/2$ and $\sin 150° = 1/2$, and $30°$ and $150°$ are in the given interval for $\theta/2$, so

$$\dfrac{\theta}{2} = 30° \qquad \text{or} \qquad \dfrac{\theta}{2} = 150°$$

$$\theta = 60° \qquad \text{or} \qquad \theta = 300°. \qquad \text{Multiply by 2.}$$

The solutions in the given interval are $60°$ and $300°$.

Graphing Calculator Solution

Graph $y = 2 \sin \dfrac{x}{2} - 1$ in a window with x in $[0°, 360°]$ and the calculator in degree mode. The x-intercepts are the solutions found algebraically. The display in Figure 30 shows that $60°$ is one solution; the other x-intercept represents the other solution, $300°$.

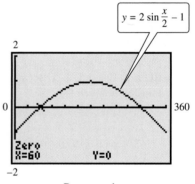

Degree mode

Figure 30

Using Xscl $= 30$ in Figure 30 makes it simple to support the algebraic solutions. We count the number of tick marks from 0 and find that there are two for $x = 60 = (2 \times 30)$ and ten for $x = 300 = (10 \times 30)$.

● ● ●

Equations with Multiple Angles

● ● ●   **Example 2**   Solving an Equation with a Double Angle

Solve $\cos 2x = \cos x$ in the interval $[0, 2\pi)$.

Algebraic Solution

First change $\cos 2x$ to a trigonometric function of x. Use the identity $\cos 2x = 2\cos^2 x - 1$ so that the equation involves only the cosine of x. Then factor as in the previous section.

$$\cos 2x = \cos x$$
$$2\cos^2 x - 1 = \cos x \quad \text{Substitute.}$$
$$2\cos^2 x - \cos x - 1 = 0$$
$$(2\cos x + 1)(\cos x - 1) = 0$$
$$2\cos x + 1 = 0 \qquad \text{or} \qquad \cos x - 1 = 0$$
$$\cos x = -\frac{1}{2} \qquad \text{or} \qquad \cos x = 1$$

In the required interval,

$$x = \frac{2\pi}{3} \qquad \text{or} \qquad x = \frac{4\pi}{3} \qquad \text{or} \qquad x = 0.$$

The solutions are 0, $2\pi/3$, and $4\pi/3$.

Graphing Calculator Solution

With the calculator in radian mode, graph $y = \cos 2x - \cos x$ in an appropriate window, and find the x-intercepts. The display in Figure 31 shows that one such x-intercept is 2.0943951, which is an approximation for $2\pi/3$. The other two x-intercepts correspond to 0 and $4\pi/3$.

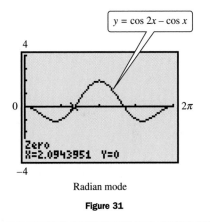

Radian mode

Figure 31

● ● ●

CAUTION   In the algebraic solution of Example 2, $\cos 2x$ cannot be changed to $\cos x$ by dividing by 2 since 2 is not a factor of $\cos 2x$.

$$\frac{\cos 2x}{2} \neq \cos x$$

The only way to change $\cos 2x$ to a trigonometric function of x is by using one of the identities for $\cos 2x$.

● ● ● **Example 3** Solving an Equation with a Multiple Angle

Solve $4 \sin \theta \cos \theta = \sqrt{3}$ in the interval $[0°, 360°)$.

Algebraic Solution

The identity $2 \sin \theta \cos \theta = \sin 2\theta$ is useful here.

$$4 \sin \theta \cos \theta = \sqrt{3}$$

$$2(2 \sin \theta \cos \theta) = \sqrt{3} \qquad 4 = 2 \cdot 2$$

$$2 \sin 2\theta = \sqrt{3} \qquad 2 \sin \theta \cos \theta = \sin 2\theta$$

$$\sin 2\theta = \frac{\sqrt{3}}{2} \qquad \text{Divide by 2.}$$

From the given domain $0° \le \theta < 360°$, the domain for 2θ is $0° \le 2\theta < 720°$. Now list all solutions in this interval.

$$2\theta = 60°, 120°, 420°, 480°$$

or $\qquad \theta = 30°, 60°, 210°, 240° \qquad$ Divide by 2.

The final two solutions for 2θ were found by adding $360°$ to $60°$ and $120°$, respectively.

Graphing Calculator Solution

We can use the intersection-of-graphs method. Figure 32 shows that the x-coordinate of one point of intersection is 30, supporting the algebraic solution 30°. The other three points of intersection give the other three solutions: 60, 210, and 240 (degrees).

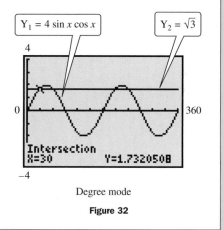

Degree mode

Figure 32

● ● ●

● ● ● **Example 4** Solving an Equation with a Multiple Angle

Solve $\tan 3x + \sec 3x = 2$ in the interval $[0, 2\pi)$.

Algebraic Solution

Since the tangent and secant functions are related by the identity $1 + \tan^2 \theta = \sec^2 \theta$, one way to begin is to express everything in terms of secant.

$$\tan 3x + \sec 3x = 2$$

$$\tan 3x = 2 - \sec 3x \qquad \text{Subtract sec } 3x.$$

$$\tan^2 3x = 4 - 4 \sec 3x + \sec^2 3x$$
$$\qquad \text{Square both sides;}$$
$$\qquad (a - b)^2 = a^2 - 2ab + b^2.$$

$$\sec^2 3x - 1 = 4 - 4 \sec 3x + \sec^2 3x$$
$$\qquad \text{Replace } \tan^2 3x \text{ with}$$
$$\qquad \sec^2 3x - 1.$$

Graphing Calculator Solution

The graphs of $Y_1 = \tan 3x + \sec 3x$ and $Y_2 = 2$ are shown in Figure 33, and the display indicates that 2.3088955 is one of the solutions. The remaining two can be found similarly.

(continued)

$$4 \sec 3x = 5$$

$$\sec 3x = \frac{5}{4}$$

$$\frac{1}{\cos 3x} = \frac{5}{4} \qquad \sec \theta = \frac{1}{\cos \theta}$$

$$\cos 3x = \frac{4}{5} \qquad \text{Use reciprocals.}$$

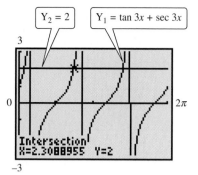

Connected mode; radian mode

Figure 33

Multiply the inequality $0 \le x < 2\pi$ by 3 to find the interval for $3x$: $[0, 6\pi)$. Using a calculator and the fact that cosine is positive in quadrants I and IV,

$$3x \approx .64350111, 5.6396842, 6.9266864, 11.922870,$$
$$13.209872, 18.206055.$$

Dividing by 3 gives

$$x \approx .21450037, 1.8798947, 2.3088955, 3.9742898,$$
$$4.4032906, 6.0686849.$$

Since both sides of the equation were squared, each of these proposed solutions must be checked. Verify by substitution in the given equation that the solutions are .21450037, 2.3088955, and 4.4032906.

● ● ●

An advantage of a graphing calculator solution, as seen in Example 4, is that extraneous values do not appear. On the other hand, a disadvantage is that an *exact* solution such as $2\pi/3$ in Example 2 is not displayed; the calculator gives only an approximation. ▪

An Application to Music   A piano string can vibrate at more than one frequency when it is struck. It produces a complex wave that can mathematically be modeled by a sum of several pure tones. If a piano key with a frequency of f_1 is played, then the corresponding string will not only vibrate at f_1 but it will also vibrate at the higher frequencies of $2f_1, 3f_1, 4f_1, \ldots, nf_1$. f_1 is called the **fundamental frequency** of the string, and higher frequencies are called the **upper harmonics.** The human ear will hear the sum of these frequencies as one complex tone. (*Source:* Roederer, J., *Introduction to the Physics and Psychophysics of Music,* Second Edition, Springer-Verlag, 1975.)

● ● ●   **Example 5**   Analyzing Pressures of Upper Harmonics

Suppose that the A key above middle C is played. Its fundamental frequency is $f_1 = 440$ Hz, and its associated pressure is expressed as

$$P_1 = .002 \sin 880\pi t.$$

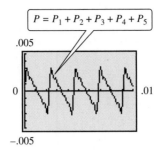

Figure 34

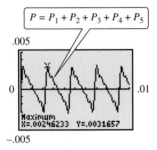

Figure 35

The string will also vibrate at 880, 1320, 1760, . . . Hz. The corresponding pressures of these upper harmonics are

$$P_2 = \frac{.002}{2} \sin 1760\pi t, \qquad P_3 = \frac{.002}{3} \sin 2640\pi t,$$

$$P_4 = \frac{.002}{4} \sin 3520\pi t, \qquad \text{and} \qquad P_5 = \frac{.002}{5} \sin 4400\pi t.$$

The graph of

$$P = P_1 + P_2 + P_3 + P_4 + P_5$$

is shown in Figure 34 and is "sawtoothed."

(a) What is the maximum value of P?

A graphing calculator shows that the maximum value of P is approximately .00317. See Figure 35.

(b) At what values of x does this maximum occur in the interval $[0, .01]$?

The maximum occurs at $x \approx .000188, .00246, .00474, .00701,$ and $.00928$. Figure 35 shows how the second value is found; the others are found similarly.

● ● ●

6.3 Exercises

Concept Check   Use the concepts of this section to answer each question.

1. Suppose you are solving a trigonometric equation for solutions in $[0, 2\pi)$, and your work leads to

$$2x = \frac{2\pi}{3}, 2\pi, \frac{8\pi}{3}.$$

What are the corresponding values of x?

2. Suppose you are solving a trigonometric equation for solutions in $[0, 2\pi)$, and your work leads to

$$\frac{1}{2}x = \frac{\pi}{16}, \frac{5\pi}{12}, \frac{5\pi}{8}.$$

What are the corresponding values of x?

3. Suppose you are solving a trigonometric equation for solutions in $[0°, 360°)$, and your work leads to

$$3\theta = 180°, 630°, 720°, 930°.$$

What are the corresponding values of θ?

4. Suppose you are solving a trigonometric equation for solutions in $[0°, 360°)$, and your work leads to

$$\frac{1}{3}\theta = 45°, 60°, 75°, 90°.$$

What are the corresponding values of θ?

Solve each equation for solutions in the interval $[0, 2\pi)$. Use algebraic methods and give exact values. You may wish to support your answers by finding approximations for these values using either of the two graphical methods discussed. See Examples 1–4.

5. $\cos 2x = \dfrac{\sqrt{3}}{2}$

6. $\cos 2x = -\dfrac{1}{2}$

7. $\sin 3x = -1$

8. $\sin 3x = 0$

9. $3 \tan 3x = \sqrt{3}$

10. $\cot 3x = \sqrt{3}$

11. $\sqrt{2} \cos 2x = -1$

12. $2\sqrt{3} \sin 2x = \sqrt{3}$

13. $\sin \dfrac{x}{2} = \sqrt{2} - \sin \dfrac{x}{2}$

14. $\sin x = \sin 2x$

15. $\tan 4x = 0$

16. $\cos 2x - \cos x = 0$

17. $8 \sec^2 \dfrac{x}{2} = 4$

18. $\sin^2 \dfrac{x}{2} - 2 = 0$

19. $\sin \dfrac{x}{2} = \cos \dfrac{x}{2}$

20. $\sec \dfrac{x}{2} = \cos \dfrac{x}{2}$

21. $\cos 2x + \cos x = 0$

22. $\sin x \cos x = \dfrac{1}{4}$

Solve each equation for exact solutions in the interval $[0°, 360°)$. *Use either an algebraic or a graphical method, according to the directions of your instructor. See Examples 1–4.*

23. $\sqrt{2} \sin 3\theta - 1 = 0$

24. $-2 \cos 2\theta = \sqrt{3}$

25. $\cos \dfrac{\theta}{2} = 1$

26. $\sin \dfrac{\theta}{2} = 1$

27. $2\sqrt{3} \sin \dfrac{\theta}{2} = 3$

28. $2\sqrt{3} \cos \dfrac{\theta}{2} = -3$

29. $2 \sin \theta = 2 \cos 2\theta$

30. $\cos \theta - 1 = \cos 2\theta$

31. $1 - \sin \theta = \cos 2\theta$

32. $\sin 2\theta = 2 \cos^2 \theta$

33. $\csc^2 \dfrac{\theta}{2} = 2 \sec \theta$

34. $\cos \theta = \sin^2 \dfrac{\theta}{2}$

35. $2 - \sin 2\theta = 4 \sin 2\theta$

36. $4 \cos 2\theta = 8 \sin \theta \cos \theta$

37. $2 \cos^2 2\theta = 1 - \cos 2\theta$

38. $\sin \theta - \sin 2\theta = 0$

 The following equations cannot be solved by traditional algebraic methods. Use a graphing calculator to find all solutions in the interval $[0, 2\pi)$. *Express solutions to as many decimal places as your calculator displays.*

39. $2 \sin 2x - x^3 + 1 = 0$

40. $3 \cos \dfrac{x}{2} + \sqrt{x} - 2 = -\dfrac{1}{2}x + 2$

· · · · · · · · · · · · · · **Relating Concepts** · · · · · · · · · · · · · ·

For individual or collaborative investigation
(Exercises 41 and 42)

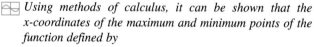 Using methods of calculus, it can be shown that the *x*-coordinates of the maximum and minimum points of the function defined by

$$f(x) = \sin^2 2x + \cos \frac{1}{2}x$$

are the same as the x-intercepts of the function defined by

$$f'(x) = 4 \sin 2x \cos 2x - \frac{1}{2} \sin \frac{1}{2}x.$$

Work Exercises 41 and 42 in order.

41. Graph $Y_1 = f(x)$ and $Y_2 = f'(x)$ with a calculator over the interval $[0, 2\pi]$. Use the same screen for both.

42. Verify that the least positive *x*-intercept of Y_2 in $[0, 2\pi]$ corresponds to the *x*-coordinate of the first maximum or minimum point of Y_1 in this interval.

Solve each problem. See Example 5.

43. *(Modeling) Pressure of a Plucked String* If a string with a fundamental frequency of 110 Hz is plucked in the middle, it will vibrate at the odd harmonics of 110, 330, 550, ... Hz but not at the even harmonics of 220, 440, 660, ... Hz. The resulting pressure *P* caused by the string can be modeled by the equation

$$P = .003 \sin 220\pi t + \frac{.003}{3} \sin 660\pi t$$
$$+ \frac{.003}{5} \sin 1100\pi t + \frac{.003}{7} \sin 1540\pi t.$$

(*Sources:* Benade, A., *Fundamentals of Musical Acoustics,* Dover Publications, 1990. Roederer, J., *Introduction to the Physics and Psychophysics of Music,* Second Edition, Springer-Verlag, 1975.)

(a) Graph *P* in the window $[0, .03]$ by $[-.005, .005]$.

(b) Use the graph to describe the shape of the sound wave that is produced.

(c) See Exercise 59 in Section 6.2. At lower frequencies, the inner ear will hear a tone only when the eardrum is moving outward. Determine the times in the interval $[0, .03]$ when this will occur.

44. *(Modeling) Hearing Beats in Music* Musicians sometimes tune instruments by playing the same tone on two different instruments and listening for a phenomenon known as *beats.* Beats occur when two tones vary in frequency by only a few hertz. When the two instruments are in tune, the beats disappear. The ear hears beats because the pressure slowly rises and falls as a result of this slight variation in the frequency. This phenomenon can be seen using a graphing calculator. (*Source:* Pierce, J., *The Science of Musical Sound,* Scientific American Books, 1992.)

(a) Consider two tones with frequencies of 220 and 223 Hz and pressures $P_1 = .005 \sin 440\pi t$ and $P_2 = .005 \sin 446\pi t$, respectively. Graph the pressure $P = P_1 + P_2$ felt by an eardrum over the one-second interval $[.15, 1.15]$. How many beats are there in one second?

(b) Repeat part (a) with frequencies of 220 and 216.

(c) Determine a simple way to find the number of beats per second if the frequency of each tone is given.

 45. 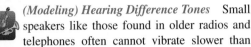 *(Modeling) Hearing Difference Tones* Small speakers like those found in older radios and telephones often cannot vibrate slower than 200 Hz—yet 35 keys on a piano have frequencies below 200 Hz. When a musical instrument creates a tone of 110 Hz, it also creates tones at 220, 330, 440, 550, 660, ... Hz. A small speaker cannot reproduce the 110-Hz vibration but it can reproduce the higher frequencies, which are called the upper harmonics. The low tones can still be heard because the speaker produces *difference tones* of the upper harmonics. The difference between consecutive frequencies is 110 Hz, and this difference tone will be heard by a listener. We can model this phenomenon using a graphing calculator. (*Source:* Benade, A., *Fundamentals of Musical Acoustics,* Dover Publications, 1990.)

(a) In the window $[0, .03]$ by $[-2, 2]$, graph the upper harmonics represented by the pressure

$$P = \frac{1}{2}\sin[2\pi(220)t] + \frac{1}{3}\sin[2\pi(330)t]$$

$$+ \frac{1}{4}\sin[2\pi(440)t].$$

(b) Estimate all *t*-coordinates where *P* is maximum.

(c) What does a person hear in addition to the frequencies of 220, 330, and 440 Hz?

(d) Graph the pressure produced by a speaker that can vibrate at 110 Hz and above.

46. *(Modeling) Daylight Hours in New Orleans* The seasonal variation in length of daylight can be modeled by a sine function. For example, the daily number of hours of daylight in New Orleans is given by

$$h = \frac{35}{3} + \frac{7}{3}\sin\frac{2\pi x}{365},$$

where *x* is the number of days after March 21 (disregarding leap year). (*Source:* Bushaw, Donald et al., *A Sourcebook of Applications of School Mathematics.* Copyright © 1980 by The Mathematical Association of America.)

(a) On what date will there be about 14 hours of daylight?

(b) What date has the least number of hours of daylight?

(c) When will there be about 10 hours of daylight?

(Modeling) Alternating Electric Current The study of alternating electric current requires the solutions of equations of the form

$$i = I_{\max}\sin 2\pi ft,$$

for time t in seconds, where i is instantaneous current in amperes, $I_{\max}$ is maximum current in amperes, and f is the number of cycles per second. (*Source:* Hannon, R. H., *Basic Technical Mathematics with Calculus,* W. B. Saunders Company, 1978.) *Find the smallest positive value of t, given the following data.*

47. $i = 40, I_{\max} = 100,$
    $f = 60$

48. $i = 50, I_{\max} = 100,$
    $f = 120$

49. $i = I_{\max}, f = 60$

50. $i = \frac{1}{2}I_{\max}, f = 60$

51. *Concept Check* What is wrong with the following solution? Solve $2\sin(1/2)x = -1$ in the interval $[0°, 360°)$.

$$2\sin\frac{1}{2}x = -1$$

$$\sin\frac{1}{2}x = -\frac{1}{2}$$

$$\frac{1}{2}x = 210° \quad \text{or} \quad \frac{1}{2}x = 330°$$

$$x = 420° \quad \text{or} \quad x = 660°$$

The solutions are 420° and 660°.

52. *Concept Check* What is wrong with the following solution? Solve $\tan 2\theta = 2$ in the interval $[0, 2\pi)$.

$$\tan 2\theta = 2$$

$$\frac{\tan 2\theta}{2} = \frac{2}{2}$$

$$\tan\theta = 1$$

$$\theta = \frac{\pi}{4} \quad \text{or} \quad \theta = \frac{5\pi}{4}$$

The solutions are $\pi/4$ and $5\pi/4$.

6.4 Equations Involving Inverse Trigonometric Functions

• Solving for *x* in Terms of *y* Using Inverse Functions   • Solving Inverse Trigonometric Equations

Until now, the equations in this chapter have involved trigonometric functions of angles or multiples of angles. Now we examine equations involving *inverse* trigonometric functions.

Solving for x in Terms of y Using Inverse Functions

● ● ● **Example 1** Solving an Equation for a Variable Using Inverse Notation

Solve $y = 3 \cos 2x$ for x.

We want $\cos 2x$ alone on one side of the equation so we can solve for $2x$ and then for x. First, divide both sides of the equation by 3.

$$y = 3 \cos 2x$$

$$\frac{y}{3} = \cos 2x$$

Now write the statement in alternative form.

$$2x = \arccos \frac{y}{3}$$

$$x = \frac{1}{2} \arccos \frac{y}{3} \qquad \text{Multiply both sides by } 1/2.$$

● ● ●

Solving Inverse Trigonometric Equations

● ● ● **Example 2** Solving an Equation Involving an Inverse Trigonometric Function

Solve $2 \arcsin x = \pi$.

Algebraic Solution

First solve for $\arcsin x$.

$$2 \arcsin x = \pi$$

$$\arcsin x = \frac{\pi}{2} \qquad \text{Divide by 2.}$$

Use the definition of $\arcsin x$ to get

$$x = \sin \frac{\pi}{2}$$

$$x = 1.$$

Verify that the solution satisfies the given equation.

Graphing Calculator Solution

The graph of $y = 2 \arcsin x - \pi$ has x-intercept 1; this is the same solution found algebraically. See Figure 36.

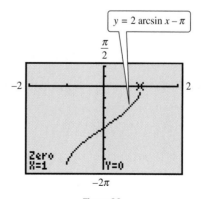

Figure 36

● ● ●

● ● ● **Example 3** Solving an Equation Involving Inverse Trigonometric Functions

Solve $\cos^{-1} x = \sin^{-1}(1/2)$.

Algebraic Solution

Let $\sin^{-1}(1/2) = u$. Then $\sin u = 1/2$ and the equation becomes

$$\cos^{-1} x = u,$$

for u in quadrant I. This can be written

$$\cos u = x.$$

Sketch a triangle and label it using the facts that u is in quadrant I and $\sin u = 1/2$. See Figure 37.

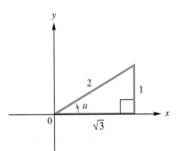

Figure 37

Since $x = \cos u$,

$$x = \frac{\sqrt{3}}{2}.$$

Check the solution by substitution.

Graphing Calculator Solution

In Figure 38, we see that .8660254, an approximation for $\frac{\sqrt{3}}{2}$, is the x-coordinate of the intersection of the graphs of $Y_1 = \cos^{-1} x$ and $Y_2 = \sin^{-1}(1/2)$.

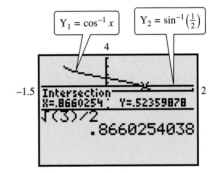

Figure 38

● ● ●

● ● ● **Example 4** Solving an Inverse Trigonometric Equation Using an Identity

Solve $\arcsin x - \arccos x = \pi/6$.

Begin by adding $\arccos x$ to both sides of the equation so that one inverse function is alone on one side of the equation.

$$\arcsin x - \arccos x = \frac{\pi}{6}$$

$$\arcsin x = \arccos x + \frac{\pi}{6} \qquad \textbf{(1)}$$

Use the definition of arcsin to write this statement as

$$\sin\left(\arccos x + \frac{\pi}{6}\right) = x.$$

Let $u = \arccos x$, so $0 \le u \le \pi$ by definition. Then

$$\sin\left(u + \frac{\pi}{6}\right) = x. \tag{2}$$

Using the identity for $\sin(A + B)$,

$$\sin\left(u + \frac{\pi}{6}\right) = \sin u \cos \frac{\pi}{6} + \cos u \sin \frac{\pi}{6}.$$

Substitute this result into equation (2) to get

$$\sin u \cos \frac{\pi}{6} + \cos u \sin \frac{\pi}{6} = x. \tag{3}$$

From equation (1) and by the definition of the arcsin function,

$$-\frac{\pi}{2} \le \arccos x + \frac{\pi}{6} \le \frac{\pi}{2}$$

$$-\frac{2\pi}{3} \le \arccos x \le \frac{\pi}{3}. \qquad \text{Subtract } \pi/6 \text{ from each expression.}$$

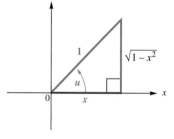

Figure 39

Since $0 \le \arccos x \le \pi$, it follows that here we must have $0 \le \arccos x \le \pi/3$. Thus $x > 0$, and we can sketch the triangle in Figure 39. From this triangle we find that $\sin u = \sqrt{1 - x^2}$. Now substitute into equation (3) using $\sin u = \sqrt{1 - x^2}$, $\sin \pi/6 = 1/2$, $\cos \pi/6 = \sqrt{3}/2$, and $\cos u = x$.

$$\left(\sqrt{1 - x^2}\right)\frac{\sqrt{3}}{2} + x \cdot \frac{1}{2} = x$$

$$\left(\sqrt{1 - x^2}\right)\sqrt{3} + x = 2x$$

$$\left(\sqrt{3}\right)\sqrt{1 - x^2} = x$$

$$3(1 - x^2) = x^2 \qquad \text{Square both sides.}$$

$$3 - 3x^2 = x^2$$

$$3 = 4x^2$$

$$x = \sqrt{\frac{3}{4}} \qquad \text{Choose the positive square root because } x > 0.$$

$$x = \frac{\sqrt{3}}{2}$$

To check, replace x with $\sqrt{3}/2$ in the original equation:

$$\arcsin \frac{\sqrt{3}}{2} - \arccos \frac{\sqrt{3}}{2} = \frac{\pi}{3} - \frac{\pi}{6} = \frac{\pi}{6},$$

as required. The solution is $\sqrt{3}/2$. ● ● ●

Exercises 39 and 40 suggest methods for graphing calculator solutions for the equation in Example 4.

6.4 Exercises

Concept Check *Use the concepts of this section to answer each question.*

1. Which one of the following equations has solution 0?
    A. $\arctan 1 = x$ **B.** $\arccos 0 = x$ **C.** $\arcsin 0 = x$

2. Which one of the following equations has solution $\dfrac{\pi}{4}$?

    A. $\arcsin\left(\dfrac{\sqrt{2}}{2}\right) = x$ **B.** $\arccos\left(-\dfrac{\sqrt{2}}{2}\right) = x$ **C.** $\arctan\left(\dfrac{\sqrt{3}}{3}\right) = x$

3. Which one of the following equations has solution $\dfrac{3\pi}{4}$?

    A. $\arctan 1 = x$ **B.** $\arcsin\left(\dfrac{\sqrt{2}}{2}\right) = x$ **C.** $\arccos\left(-\dfrac{\sqrt{2}}{2}\right) = x$

4. Which one of the following equations has solution $-\dfrac{\pi}{6}$?

    A. $\arctan\left(\dfrac{\sqrt{3}}{3}\right) = x$ **B.** $\arccos\left(-\dfrac{1}{2}\right) = x$ **C.** $\arcsin\left(-\dfrac{1}{2}\right) = x$

Solve each equation for x. See Example 1.

5. $y = 5 \cos x$ 6. $4y = \sin x$ 7. $2y = \cot 3x$ 8. $6y = \dfrac{1}{2}\sec x$

9. $y = 3 \tan 2x$ 10. $y = 3 \sin \dfrac{x}{2}$ 11. $y = 6 \cos \dfrac{x}{4}$ 12. $y = -\sin \dfrac{x}{3}$

13. $y = -2 \cos 5x$ 14. $y = 3 \cot 5x$ 15. $y = \cos(x + 3)$ 16. $y = \tan(2x - 1)$

17. $y = \sin x - 2$ 18. $y = \cot x + 1$ 19. $y = 2 \sin x - 4$ 20. $y = 4 + 3 \cos x$

21. Refer to Exercise 17. A student attempting to solve this equation wrote as the first step

    $$y = \sin(x - 2),$$

    inserting parentheses as shown. Explain why this is incorrect.

22. Explain why the equation

    $$\sin^{-1}x = \cos^{-1}2$$

    cannot have a solution. (No work needs to be shown here.)

Solve each equation for exact solutions. If the solution is irrational, you may wish to find an approximation and use a graphing calculator to support it. If the solution is rational, you may wish to use a graphing calculator to find the exact decimal value to support it. See Examples 2 and 3.

23. $\dfrac{4}{3}\cos^{-1}\dfrac{y}{4} = \pi$ 24. $4\pi + 4\tan^{-1}y = \pi$ 25. $2\arccos\left(\dfrac{y - \pi}{3}\right) = 2\pi$

26. $\arccos\left(y - \dfrac{\pi}{3}\right) = \dfrac{\pi}{6}$ 27. $\arcsin x = \arctan \dfrac{3}{4}$ 28. $\arctan x = \arccos \dfrac{5}{13}$

29. $\cos^{-1}x = \sin^{-1}\dfrac{3}{5}$ 30. $\cot^{-1}x = \tan^{-1}\dfrac{4}{3}$

Solve each equation for exact solutions. Refer to the directions for Exercises 23–30 regarding graphical support. See Example 4.

31. $\sin^{-1}x - \tan^{-1}1 = -\dfrac{\pi}{4}$ 32. $\sin^{-1}x + \tan^{-1}\sqrt{3} = \dfrac{2\pi}{3}$

33. $\arccos x + 2\arcsin \dfrac{\sqrt{3}}{2} = \pi$ 34. $\arccos x + 2\arcsin \dfrac{\sqrt{3}}{2} = \dfrac{\pi}{3}$

35. $\arcsin 2x + \arccos x = \dfrac{\pi}{6}$

36. $\arcsin 2x + \arcsin x = \dfrac{\pi}{2}$

37. $\cos^{-1}x + \tan^{-1}x = \dfrac{\pi}{2}$

38. $\sin^{-1}x + \tan^{-1}x = 0$

39. Provide graphical support for the solution in Example 4 by showing that the graph of $y = \arcsin x - \arccos x - \pi/6$ has x-intercept $\sqrt{3}/2 \approx .8660254$.

40. Provide graphical support for the solution in Example 4 by showing that the x-coordinate of the point of intersection of the graphs of $Y_1 = \arcsin x - \arccos x$ and $Y_2 = \pi/6$ is $\sqrt{3}/2 \approx .8660254$.

The following equations cannot be solved by traditional algebraic methods. Use a graphing calculator to find all solutions in the interval $[0, 6]$. *Express solutions to as many decimal places as your calculator displays.*

41. $(\arctan x)^3 - x + 2 = 0$

42. $\pi \sin^{-1}(.2x) - 3 = -\sqrt{x}$

Solve each problem.

43. 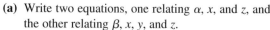 *(Modeling) Tone Heard by a Listener* When two sources located at different positions produce the same pure tone, the human ear will often hear one sound that is equal to the sum of the individual tones. Since the sources are at different locations, they will have different phase angles ϕ. If two speakers located at different positions produce pure tones $P_1 = A_1 \sin(2\pi ft + \phi_1)$ and $P_2 = A_2 \sin(2\pi ft + \phi_2)$, where $-\pi/4 \le \phi_1, \phi_2 \le \pi/4$, then the resulting tone heard by a listener can be written as $P = A \sin(2\pi ft + \phi)$, where

$$A = \sqrt{(A_1 \cos \phi_1 + A_2 \cos \phi_2)^2 + (A_1 \sin \phi_1 + A_2 \sin \phi_2)^2}$$

and

$$\phi = \arctan\left(\frac{A_1 \sin \phi_1 + A_2 \sin \phi_2}{A_1 \cos \phi_1 + A_2 \cos \phi_2}\right).$$

(*Source:* Fletcher, N. and T. Rossing, *The Physics of Musical Instruments,* Second Edition, Springer-Verlag, 1998.)

(a) Calculate A and ϕ if $A_1 = .0012$, $\phi_1 = .052$, $A_2 = .004$, and $\phi_2 = .61$. Also find an expression for $P = A \sin(2\pi ft + \phi)$ if $f = 220$.

(b) Graph $Y_1 = P$ and $Y_2 = P_1 + P_2$ on the same coordinate axes on the interval $[0, .01]$. Are the two graphs the same?

44.  *(Modeling) Tone Heard by a Listener* Repeat Exercise 43, with $A_1 = .0025$, $\phi_1 = \pi/7$, $A_2 = .001$, $\phi_2 = \pi/6$, and $f = 300$.

45. *(Modeling) Depth of Field* When a large-view camera is used to take a picture of an object that is not parallel to the film, the lens board should be tilted so that the planes containing the subject, the lens board, and the film intersect in a line. (See the figure.) This gives the best "depth of field." (*Source:* Bushaw, Donald et al., *A Sourcebook of Applications of School Mathematics.* Copyright © 1980 by The Mathematical Association of America.)

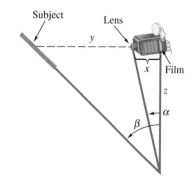

(a) Write two equations, one relating α, x, and z, and the other relating β, x, y, and z.

(b) Eliminate z from the equations in part (a) to get one equation relating α, β, x, and y.

(c) Solve the equation from part (b) for α.

(d) Solve the equation from part (b) for β.

46. *(Modeling) Programming Language for Inverse Functions* In Visual Basic, the most widely used programming language for PCs, only the arctangent function is available. To use the other inverse trigonometric functions, it is necessary to express them in terms of arctangent. This can be done as follows.

(a) Let $u = \arcsin x$. Solve the equation for x in terms of u.

(b) Use the result of part (a) to label the three sides of the triangle in the figure in terms of x.

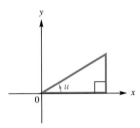

(c) Use the triangle from part (b) to write an equation for $\tan u$ in terms of x.

(d) Solve the equation from part (c) for u.

47. *(Modeling) Alternating Electric Current* In the study of alternating electric current, instantaneous voltage is modeled by

$$e = E_{max} \sin 2\pi f t,$$

where f is the number of cycles per second, E_{max} is the maximum voltage, and t is time in seconds.

(a) Solve the equation for t.

(b) Find the smallest positive value of t if $E_{max} = 12$, $e = 5$, and $f = 100$. Use a calculator.

48. *(Modeling) Viewing Angle of an Observer* While visiting a museum, Marsha Langlois views a painting that is 3 feet high and hanging 6 feet above the ground. See the figure. Assume her eyes are 5 feet above the ground, and let x be the distance from the spot where she is standing to the wall displaying the painting.

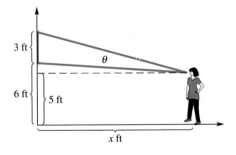

(a) Show that θ, the viewing angle subtended by the painting, is given by

$$\theta = \tan^{-1}\left(\frac{4}{x}\right) - \tan^{-1}\left(\frac{1}{x}\right).$$

(b) Find the value of x for each value of θ.

(i) $\theta = \dfrac{\pi}{6}$ (ii) $\theta = \dfrac{\pi}{8}$

(c) Find the value of θ for each value of x.

(i) $x = 4$ (ii) $x = 3$

49. *(Modeling) Movement of an Arm* In the exercises for Section 4.1 we found the equation

$$y = \frac{1}{3} \sin \frac{4\pi t}{3},$$

where t is time (in seconds) and y is the angle formed by a rhythmically moving arm.

(a) Solve the equation for t.

(b) At what time(s) does the arm form an angle of .3 radian?

50. The function $y = \sec^{-1} x$ is not found on graphing calculators. However, with some models it can be graphed as

$$y = \frac{\pi}{2} - ((x > 0) - (x < 0))$$
$$\times \left(\frac{\pi}{2} - \tan^{-1}\left(\sqrt{(x^2 - 1)}\right)\right).$$

(This formula appears as Y_1 in the screen here.) Use the formula to obtain the graph of $y = \sec^{-1} x$ in the window $[-4, 4]$ by $[0, \pi]$.

(In Exercise 77 of the Chapter Review, an alternative way of graphing $y = \csc^{-1} x$ is given.)

Chapter 6 Summary

Key Terms & Symbols	Key Ideas

6.1 Inverse Trigonometric Functions

one-to-one function
horizontal line test
inverse function (inverse) f^{-1}
$\sin^{-1} x$ or arcsin x
$\cos^{-1} x$ or arccos x
$\tan^{-1} x$ or arctan x
$\sec^{-1} x$ or arcsec x
$\csc^{-1} x$ or arccsc x
$\cot^{-1} x$ or arccot x

Inverse Trigonometric Functions

Function	Domain	Range	Quadrants of the Unit Circle from Which Range Values Come
$y = \sin^{-1} x$	$[-1, 1]$	$\left[-\dfrac{\pi}{2}, \dfrac{\pi}{2}\right]$	I and IV
$y = \cos^{-1} x$	$[-1, 1]$	$[0, \pi]$	I and II
$y = \tan^{-1} x$	$(-\infty, \infty)$	$\left(-\dfrac{\pi}{2}, \dfrac{\pi}{2}\right)$	I and IV
$y = \cot^{-1} x$	$(-\infty, \infty)$	$(0, \pi)$	I and II
$y = \sec^{-1} x$	$(-\infty, -1] \cup [1, \infty)$	$\left[0, \dfrac{\pi}{2}\right) \cup \left(\dfrac{\pi}{2}, \pi\right]$	I and II
$y = \csc^{-1} x$	$(-\infty, -1] \cup [1, \infty)$	$\left[-\dfrac{\pi}{2}, 0\right) \cup \left(0, \dfrac{\pi}{2}\right]$	I and IV

Graphs

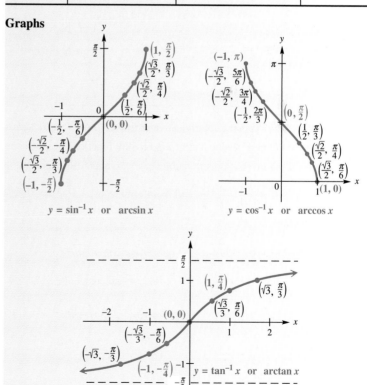

$y = \sin^{-1} x$ or arcsin x

$y = \cos^{-1} x$ or arccos x

$y = \tan^{-1} x$ or arctan x

See page 245 for graphs of the other inverse trigonometric functions.

Key Terms & Symbols	Key Ideas
6.2 Trigonometric Equations I 6.3 Trigonometric Equations II	**Solving Trigonometric Equations Algebraically** **Step 1** Decide whether the equation is linear or quadratic, so you can determine the solution method. **Step 2** If only one trigonometric function is present, first solve the equation for that function. **Step 3** If more than one trigonometric function is present, rearrange the equation so that one side equals 0. Then try to factor and set each factor equal to 0 to solve. **Step 4** If Step 3 does not work, try using identities to change the form of the equation. It may be helpful to square both sides of the equation first. If this is done, check for extraneous solutions. **Step 5** If the equation is quadratic in form, but not factorable, use the quadratic formula. Check for extraneous solutions. **Solving Trigonometric Equations Graphically** For an equation of the form $f(x) = g(x)$: **Step 1** Graph $y = f(x)$ and $y = g(x)$ over the required domain. **Step 2** Find the x-coordinates of the points of intersection of the graphs. (This is the intersection-of-graphs method.) For an equation of the form $f(x) = 0$: **Step 1** Graph $y = f(x)$ over the required domain. **Step 2** Find the x-intercepts of the graph. (This is the x-intercept method.)

Chapter 6 Review Exercises

Concept Check   *Tell whether each statement is true or false. If false, tell why.*

1. The ranges of the inverse sine and inverse cosine functions are the same.

2. The ranges of the inverse tangent and inverse cotangent functions are the same.

3. It is true that $\sin\left(\dfrac{11\pi}{6}\right) = -\dfrac{1}{2}$, and therefore $\arcsin\left(-\dfrac{1}{2}\right) = \dfrac{11\pi}{6}$.

4. For all x, $\tan(\tan^{-1} x) = x$.

Give the exact real number value of y. Do not use a calculator.

5. $y = \sin^{-1}\left(\dfrac{\sqrt{2}}{2}\right)$ **6.** $y = \arccos\left(-\dfrac{1}{2}\right)$ **7.** $y = \tan^{-1}\left(-\sqrt{3}\right)$

8. $y = \arcsin(-1)$ **9.** $y = \cos^{-1}\left(-\dfrac{\sqrt{2}}{2}\right)$ **10.** $y = \arctan\left(\dfrac{\sqrt{3}}{3}\right)$

11. $y = \sec^{-1}(-2)$ **12.** $y = \operatorname{arccsc}\left(\dfrac{2\sqrt{3}}{3}\right)$ **13.** $y = \operatorname{arccot}(-1)$

Give the degree measure of θ. Do not use a calculator.

14. $\theta = \arccos\left(\dfrac{1}{2}\right)$ **15.** $\theta = \arcsin\left(-\dfrac{\sqrt{3}}{2}\right)$ **16.** $\theta = \tan^{-1} 0$

Use a calculator to give the degree measure of θ.

17. $\theta = \arctan 1.7804675$

18. $\theta = \sin^{-1}(-.66045320)$

19. $\theta = \cos^{-1}.80396577$

20. $\theta = \cot^{-1} 4.5046388$

21. $\theta = \text{arcsec } 3.4723155$

22. $\theta = \csc^{-1} 7.4890096$

23. Explain why $\sin^{-1} 3$ cannot be defined.

24. $\text{Arcsin}(\sin 5\pi/6) \neq 5\pi/6$. Explain why.

25. What is the domain of the arccotangent function?

26. What is the range of the arcsecant function as defined in this text?

Evaluate the following without using a calculator.

27. $\sin\left(\sin^{-1}\dfrac{1}{2}\right)$

28. $\tan\left(\tan^{-1}\dfrac{2}{3}\right)$

29. $\cos(\arccos(-1))$

30. $\sin\left(\arcsin\left(-\dfrac{\sqrt{3}}{2}\right)\right)$

31. $\arccos\left(\cos\dfrac{3\pi}{4}\right)$

32. $\text{arcsec}(\sec \pi)$

33. $\tan^{-1}\left(\tan\dfrac{\pi}{4}\right)$

34. $\cos^{-1}(\cos 0)$

35. $\sin\left(\arccos\dfrac{3}{4}\right)$

36. $\cos(\arctan 3)$

37. $\cos(\csc^{-1}(-2))$

38. $\sec\left(2\sin^{-1}\left(-\dfrac{1}{3}\right)\right)$

39. $\tan\left(\arcsin\dfrac{3}{5} + \arccos\dfrac{5}{7}\right)$

Write each of the following as a non-trigonometric expression in u.

40. $\sin(\tan^{-1} u)$

41. $\cos\left(\arctan\dfrac{u}{\sqrt{1-u^2}}\right)$

42. $\tan\left(\text{arcsec}\dfrac{\sqrt{u^2+1}}{u}\right)$

Graph each of the following, and give the domain and range.

43. $y = \sin^{-1} x$

44. $y = \cos^{-1} x$

45. $y = \text{arccot } x$

In Exercises 46–70, you may wish to support your answers with a graphing calculator.

Solve each equation for solutions in the interval $[0, 2\pi)$. Use a calculator in Exercises 47 and 48.

46. $\sin^2 x = 1$

47. $2 \tan x - 1 = 0$

48. $3 \sin^2 x - 5 \sin x + 2 = 0$

49. $\tan x = \cot x$

50. $\sec^4 2x = 4$

51. $\tan^2 2x - 1 = 0$

52. $\sec\dfrac{x}{2} = \cos\dfrac{x}{2}$

53. $\cos 2x + \cos x = 0$

54. $4 \sin x \cos x = \sqrt{3}$

Solve each equation for solutions in the interval $[0°, 360°)$. When appropriate, use a calculator and express solutions to the nearest tenth of a degree.

55. $\sin^2 \theta + 3 \sin \theta + 2 = 0$

56. $2 \tan^2 \theta = \tan \theta + 1$

57. $\sin 2\theta = \cos 2\theta + 1$

58. $2 \sin 2\theta = 1$

59. $3 \cos^2 \theta + 2 \cos \theta - 1 = 0$

60. $5 \cot^2 \theta - \cot \theta - 2 = 0$

61. $\sin 2\theta + \sin 4\theta = 0$

62. $\cos \theta - \cos 2\theta = 2 \cos \theta$

Solve each equation for x.

63. $4y = 2 \sin x$

64. $y = 3 \cos\dfrac{x}{2}$

65. $2y = \tan(3x + 2)$

66. $5y = 4 \sin x - 3$

67. $\dfrac{4}{3} \arctan\dfrac{x}{2} = \pi$

68. $\arccos x = \arcsin\dfrac{2}{7}$

69. $\arccos x + \arctan 1 = \dfrac{11\pi}{12}$

70. $\text{arccot } x = \arcsin\left(\dfrac{-\sqrt{2}}{2}\right) + \dfrac{3\pi}{4}$

Solve each equation for t in terms of d.

71. $d = 550 + 450 \cos\left(\dfrac{\pi}{50}t\right)$

72. $d = 40 + 60 \cos\left[\dfrac{\pi}{6}(t - 2)\right]$

Solve each problem.

73. *(Modeling) Viewing Angle of an Observer* A 10-foot wide blackboard is situated 5 feet from the left wall of a classroom. See the figure. A student sitting next to the wall x feet from the front of the classroom has a viewing angle of θ radians.

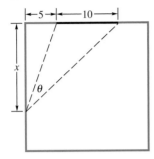

(a) Show that the value of θ is given by the function defined by

$$f(x) = \arctan\left(\frac{15}{x}\right) - \arctan\left(\frac{5}{x}\right).$$

(b) Graph $f(x)$ with a graphing calculator to estimate the value of x that maximizes the viewing angle.

(c) Refer to part (f) of Exercise 96 in Section 6.1. Use the result to find the exact value of x in part (b) here.

74. *Snell's Law* Recall Snell's law from Exercises 43–46 of Section 2.3:

$$\frac{c_1}{c_2} = \frac{\sin\theta_1}{\sin\theta_2},$$

where c_1 is the speed of light in one medium, c_2 is the speed of light in a second medium, and θ_1 and θ_2 are the angles shown in the figure. Suppose a light is shining up through water into the air as in the figure.

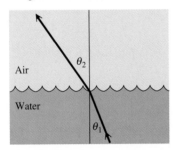

As θ_1 increases, θ_2 approaches 90°, at which point no light will emerge from the water. Assume the ratio c_1/c_2 in this case is .752. For what value of θ_1 does $\theta_2 = 90°$? This value of θ_1 is called the *critical angle* for water.

75. *Snell's Law* Refer to Exercise 74. What happens when θ_1 is greater than the critical angle?

76. *British Nautical Mile* The British nautical mile is defined as the length of a minute of arc of a meridian. Since Earth is flat at its poles, the nautical mile, in feet, is given by

$$L = 6077 - 31\cos 2\theta,$$

where θ is the latitude in degrees. See the figure. (*Source:* Bushaw, Donald et al., *A Sourcebook of Applications of School Mathematics.* Copyright © 1980 by The Mathematical Association of America.)

A nautical mile is the length on any of the meridians cut by a central angle of measure 1 minute.

(a) Find the latitude between 0° and 90° at which the nautical mile is 6074 feet.

(b) At what latitude between 0° and 180° is the nautical mile 6108 feet?

(c) In the United States, the nautical mile is defined everywhere as 6080.2 feet. At what latitude between 0° and 90° does this agree with the British nautical mile?

77. The function $y = \csc^{-1} x$ is not found on graphing calculators. However, with some models it can be graphed as

$$y = ((x > 0) - (x < 0))\left(\frac{\pi}{2} - \tan^{-1}\left(\sqrt{(x^2 - 1)}\right)\right).$$

(This formula appears as Y_1 in the screen here.) Use the formula to obtain the graph of $y = \csc^{-1} x$ in the window $[-4, 4]$ by $[-\pi/2, \pi/2]$.

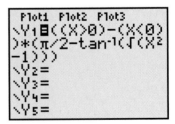

78. **(a)** Use the graph of $y = \sin^{-1} x$ to approximate $\sin^{-1}(.4)$.

(b) Use the inverse sine key of a graphing calculator to approximate $\sin^{-1}(.4)$.

Chapter 6   Test

You may wish to use a graphing calculator to support your answers, as explained in this chapter.

1. Graph $y = \sin^{-1} x$, and indicate the coordinates of three points on the graph. Give the domain and the range.

2. Find the exact value of y for each equation.

   (a) $y = \arccos\left(-\dfrac{1}{2}\right)$ **(b)** $y = \sin^{-1}\left(-\dfrac{\sqrt{3}}{2}\right)$ **(c)** $y = \tan^{-1} 0$ **(d)** $y = \operatorname{arcsec}(-2)$

   (e) $y = \csc^{-1}\left(\dfrac{2\sqrt{3}}{3}\right)$ **(f)** $y = \operatorname{arccot}\left(\sqrt{3}\right)$

3. Find each exact value.

   (a) $\cos\left(\arcsin\dfrac{2}{3}\right)$ **(b)** $\sin\left(2\cos^{-1}\dfrac{1}{3}\right)$

4. Write $\tan(\arcsin u)$ as a non-trigonometric expression in u.

Solve each equation in Exercises 5–8 algebraically.

5. Solve $\sin^2 \theta = \cos^2 \theta + 1$ for solutions in the interval $[0°, 360°)$.

6. Solve $\csc^2 \theta - 2 \cot \theta = 4$ for solutions in the interval $[0°, 360°)$. Using a calculator, express approximate solutions to the nearest tenth of a degree.

7. Solve $\cos x = \cos 2x$ for solutions in the interval $[0, 2\pi)$.

8. Solve $2\sqrt{3}\sin\left(\dfrac{\theta}{2}\right) = 3$ for solutions in the interval $[0°, 360°)$.

9. Solve each equation for x.

   (a) $y = \cos(3x)$ **(b)** $\arcsin x = \arctan\left(\dfrac{4}{3}\right)$

10. *(Modeling) Movement of a Runner's Arm*   A runner's arm swings rhythmically according to the model

$$y = \left(\frac{\pi}{8}\right)\cos\left[\pi\left(t - \frac{1}{3}\right)\right],$$

where y represents the angle between the actual position of the upper arm and the downward vertical position and t represents time in seconds. At what times in the interval $[0, \pi)$ is the angle y equal to 0?

Chapter 6    Internet Project

Modeling a Damped Pendulum

The Chapter 4 Internet Project investigated trigonometric models for the periodic phenomenon of sunset time. We now extend the idea of using trigonometric functions to model data using a damped pendulum. A damped pendulum oscillates back and forth. The period (time of each swing) remains constant, but the amplitude diminishes at a constant rate. This results in a model that consists of an exponential function (that is, a function of the form $f(x) = r^x$, where $r > 0$, $r \neq 1$) multiplied by a sine function. While several forms are possible, a popular one is

$$y = r^x a \sin[b(x - d)] + c,$$

where x represents time, a represents amplitude, b represents 2π divided by period, c represents vertical shift, d represents horizontal phase shift, and r represents rate at which amplitude diminishes.

On the Web site for this text (www.awl.com/lhs), we give data gathered from an object oscillating for 10 seconds. A scatter diagram of the data is shown in the figure here. In this case, x represents time elapsed in seconds and y represents distance in feet of the object from the probe of the data collector.

The Web project helps you to find an equation that models the data. As an extension, you will find information and links so that you can investigate the relationships between music and periodic functions presented in this chapter.

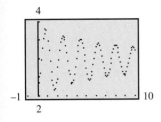

Applications of
7 Trigonometry and Vectors

7.1 **Oblique Triangles and the Law of Sines**

7.2 **The Ambiguous Case of the Law of Sines**

7.3 **The Law of Cosines**

7.4 **Vectors and the Dot Product**

7.5 **Applications of Vectors**

Aerial photography first began in 1858 when French photographer Gaspard Tournachon took pictures of Paris from a hot-air balloon that had a makeshift darkroom. Since then hot-air balloons have been replaced by airplanes, helicopters, and satellites, and aerial photography has been used in surveying, road design, weather forecasting, military surveillance, topographic maps, and even archaeology. The first archaeological aerial photographs were taken of Stonehenge in 1906. By searching these photographs for unusual soil and marks caused by structures lying below the ground, Stonehenge Avenue was discovered.

In aerial photography a series of photographs are usually taken with sufficient overlap to allow for the stereoscopic vision necessary to obtain accurate ground measurements. These photographs are often used to construct a map that gives both the coordinates and elevations of important features located on the ground. The perspective of these photographs can be affected if the airplane is not perfectly horizontal, the ground below is not level, or the camera is tilted. For

example, if the distance between two objects in a photograph is 6 inches, what other information do we need to know in order to determine the actual ground distance between these objects? See the figure on the next page. Being able to determine the measurements of the sides and angles in a triangle is essential to solving applications involving aerial photography, the theme of this chapter.*

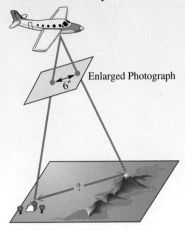

Enlarged Photograph

7.1   Oblique Triangles and the Law of Sines

- **Congruency and Oblique Triangles**   •   **Derivation of the Law of Sines**   •   **Solving SAA or ASA Triangles (Case 1)**
- **Area of a Triangle**

Until now, our applied work with trigonometry has been limited to right triangles. However, the concepts developed in earlier chapters can be extended to apply to *all* triangles. Every triangle has three sides and three angles. In this chapter we show that if any three of the six measures of a triangle (provided at least one measure is a side) are known, then the other three measures can be found. This process is called *solving a triangle.*

Congruency and Oblique Triangles   The following axioms from geometry allow us to prove that two triangles are congruent (that is, their corresponding sides and angles are equal).

Congruence Axioms

Side-Angle-Side (SAS)   If two sides and the included angle of one triangle are equal, respectively, to two sides and
the included angle of a second triangle, then the
triangles are congruent.

Angle-Side-Angle (ASA)   If two angles and the included side of one triangle are equal, respectively, to two angles and
the included side of a second triangle, then the
triangles are congruent.

Side-Side-Side (SSS)   If three sides of one triangle are equal, respectively, to three sides of a second triangle, then
the triangles are congruent.

---

*Sources: Brooks, R. and D. Johannes, *Phytoarchaeology,* Dioscorides Press, 1990.
Moffitt, F. and E. Mikhail, *Photogrammetry,* Third Edition, Harper & Row, 1980.

Throughout this chapter keep in mind that whenever any of the groups of data described above are given, the triangle is uniquely determined; that is, all other data in the triangle are given by one and only one set of measures. We will continue to label triangles as we did earlier with right triangles: side a opposite angle A, side b opposite angle B, and side c opposite angle C.

A triangle that is not a right triangle is called an **oblique triangle.** The measures of the three sides and the three angles of a triangle can be found if at least one side and any other two measures are known. There are four possible cases.

Data Required for Solving Oblique Triangles

Case 1 One side and two angles are known (SAA or ASA).

Case 2 Two sides and one angle not included between the two sides are known (SSA). This case may lead to more than one triangle.

Case 3 Two sides and the angle included between the two sides are known (SAS).

Case 4 Three sides are known (SSS).

N O T E   If we know three angles of a triangle, we cannot find unique side lengths since AAA assures us only of similarity, not congruence. For example, there are infinitely many triangles ABC with $A = 35°$, $B = 65°$, and $C = 80°$.

Case 1, discussed in this section, and Case 2, discussed in the next section, require the *law of sines.* Cases 3 and 4, discussed in Section 7.3, require the *law of cosines.*

Derivation of the Law of Sines   To derive the law of sines, we start with an oblique triangle, such as the acute triangle in Figure 1(a) or the obtuse triangle in Figure 1(b). (Recall that these terms were defined in Section 1.3.) The following discussion applies to both triangles. First, construct the perpendicular from B to side AC. Let h be the length of this perpendicular. Then c is the hypotenuse of right triangle ADB, and a is the hypotenuse of right triangle BDC. By results from Chapter 2,

$$\text{in } \triangle ADB, \qquad \sin A = \frac{h}{c} \quad \text{or} \quad h = c \sin A,$$

$$\text{in } \triangle BDC, \qquad \sin C = \frac{h}{a} \quad \text{or} \quad h = a \sin C.$$

Since $h = c \sin A$ and $h = a \sin C$,

$$a \sin C = c \sin A,$$

or, upon dividing both sides by $\sin A \sin C$,

$$\frac{a}{\sin A} = \frac{c}{\sin C}.$$

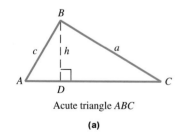

Acute triangle ABC

(a)

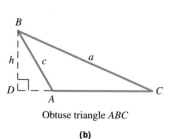

Obtuse triangle ABC

(b)

Figure 1

In a similar way, by constructing the perpendiculars from other vertices, it can be shown that

$$\frac{a}{\sin A} = \frac{b}{\sin B} \quad \text{and} \quad \frac{b}{\sin B} = \frac{c}{\sin C}.$$

This discussion proves the following theorem.

Law of Sines

In any triangle ABC, with sides a, b, and c,

$$\frac{a}{\sin A} = \frac{b}{\sin B}, \quad \frac{a}{\sin A} = \frac{c}{\sin C}, \quad \text{and} \quad \frac{b}{\sin B} = \frac{c}{\sin C}.$$

This can be written in compact form as

$$\frac{a}{\sin A} = \frac{b}{\sin B} = \frac{c}{\sin C}.$$

Sometimes it is more convenient to use an alternative form of the law of sines,

$$\frac{\sin A}{a} = \frac{\sin B}{b} = \frac{\sin C}{c}.$$

Solving SAA or ASA Triangles (Case 1)   If two angles and the side opposite one of the angles are known (Case 1 SAA), the law of sines can be used directly to solve for the side opposite the other known angle. The triangle can then be solved completely, as shown in the first example.

● ● ●   **Example 1   Using the Law of Sines to Solve a Triangle Involving SAA**

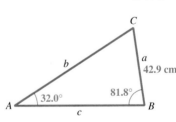

Figure 2

Solve $\triangle ABC$ if $A = 32.0°$, $B = 81.8°$, and $a = 42.9$ cm.

Start by drawing a triangle, roughly to scale, and labeling the given parts as in Figure 2. Since the values of A, B, and a are known, use the form of the law of sines that involves these variables.

$$\frac{a}{\sin A} = \frac{b}{\sin B}$$

Substituting the known values gives

$$\frac{42.9}{\sin 32.0°} = \frac{b}{\sin 81.8°}$$

$$b = \frac{42.9 \sin 81.8°}{\sin 32.0°}. \quad \text{Multiply by } \sin 81.8°.$$

When using a calculator to find b, keep intermediate answers in the calculator until the final result is found. Then round to the proper number of significant digits. In this case, find $\sin 81.8°$, and then multiply that number by 42.9. Keep the result in the calculator while you find $\sin 32.0°$, and then divide. Since the given information is accurate to three significant digits, round the value of b to get

$$b \approx 80.1 \text{ cm.}$$

Find C from the fact that the sum of the angles of any triangle is 180°.

$$A + B + C = 180°$$
$$C = 180° - A - B$$
$$C = 180° - 32.0° - 81.8°$$
$$C = 66.2°$$

Now use the law of sines again to find c. (Why does the Pythagorean theorem not apply?)

$$\frac{a}{\sin A} = \frac{c}{\sin C}$$

$$\frac{42.9}{\sin 32.0°} = \frac{c}{\sin 66.2°} \qquad \text{Substitute.}$$

$$c = \frac{42.9 \sin 66.2°}{\sin 32.0°} \qquad \text{Multiply by } \sin 66.2°.$$

$$c \approx 74.1 \text{ cm} \qquad \text{Use a calculator.} \qquad \bullet \bullet \bullet$$

CAUTION In applications of oblique triangles, such as the one in Example 1, be sure to carefully label a sketch to set up the correct equation.

$\bullet \bullet \bullet$ **Example 2**   Using the Law of Sines in an Application

Ben Sultenfuss wishes to measure the distance across the Big Muddy River. See Figure 3. He finds that $C = 112.90°$, $A = 31.10°$, and $b = 347.6$ feet. Find the required distance.

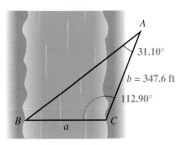

Figure 3

Algebraic Solution

To use the law of sines, one side and the angle opposite it must be known. Since b is the only side whose length is given, angle B must be found before the law of sines can be used.

$$B = 180° - A - C$$
$$B = 180° - 31.10° - 112.90° = 36.00°$$

Graphing Calculator Solution

Triangle ABC in Figure 3 can be solved using a graphing calculator program, as shown in Figure 4 on the next page. Programs such as this

(continued)

Now use the form of the law of sines involving A, B, and b to find a.

$$\frac{a}{\sin A} = \frac{b}{\sin B}$$

$$\frac{a}{\sin 31.10°} = \frac{347.6}{\sin 36.00°} \qquad \text{Substitute.}$$

$$a = \frac{347.6 \sin 31.10°}{\sin 36.00°} \qquad \text{Multiply by } \sin 31.10°.$$

$$a \approx 305.5 \text{ feet} \qquad \text{Use a calculator.}$$

one are available from users' groups or from the Web site for this text.

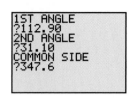

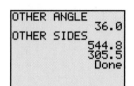

Figure 4

The next example involves the use of bearing, first discussed in Section 2.5.

● ● ● **Example 3** Using the Law of Sines in an Application

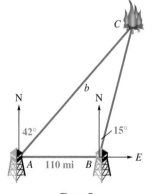

Figure 5

Two tracking stations are on an east-west line 110 miles apart. A large forest fire is located on a bearing of N 42° E from the western station and a bearing of N 15° E from the eastern station. How far is the fire from the western station?

Figure 5 shows the two stations at points A and B and the fire at point C. Angle $BAC = 90° - 42° = 48°$, the obtuse angle at B equals $90° + 15° = 105°$, and the third angle, C, equals $180° - 105° - 48° = 27°$. Use the law of sines to find side b.

$$\frac{b}{\sin 105°} = \frac{110}{\sin 27°}$$

$$b = \frac{110 \sin 105°}{\sin 27°}$$

$$b \approx 234$$

The fire is 230 miles (to two significant digits) from the western station.

● ● ●

Area of a Triangle The method used to derive the law of sines can also be used to derive a useful formula to find the area of a triangle. A familiar formula for the area of a triangle is $\mathcal{A} = (1/2)bh$, where $\mathcal{A}$ represents area, b base, and h height. This formula cannot always be used easily since in practice h is often unknown. To find a more useful formula, refer to acute triangle ABC in Figure 6(a) or obtuse triangle ABC in Figure 6(b).

A perpendicular has been drawn from B to the base of the triangle (or the extension of the base). This perpendicular forms two right triangles. Using $\triangle ABD$,

$$\sin A = \frac{h}{c}$$

$$h = c \sin A .$$

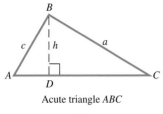

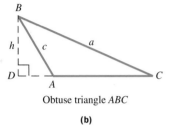

Acute triangle ABC

(a)

Obtuse triangle ABC

(b)

Figure 6

Substituting into the formula $\mathscr{A} = (1/2)bh$,

$$\mathscr{A} = \frac{1}{2}b(c \sin A)$$

$$\mathscr{A} = \frac{1}{2}bc \sin A.$$

Any other pair of sides and the angle between them could have been used, as stated in the next theorem.

Area of a Triangle

In any triangle ABC, the area $\mathscr{A}$ is given by any of the following formulas:

$$\mathscr{A} = \frac{1}{2}bc \sin A, \qquad \mathscr{A} = \frac{1}{2}ab \sin C, \qquad \mathscr{A} = \frac{1}{2}ac \sin B.$$

In words, the area is given by half the product of the lengths of two sides and the sine of the angle included between them.

NOTE If the included angle measures 90°, its sine is 1, and the formula becomes the familiar $\mathscr{A} = (1/2)bh$.

● ● ● **Example 4** Finding the Area of a Triangle Using $\mathscr{A} = (1/2)ab \sin C$

Find the area of $\triangle ABC$ if $A = 24° \ 40'$, $b = 27.3$ cm, and $C = 52° \ 40'$.

Algebraic Solution

Before we can use the formula given above, we must use the law of sines to find either a or c. Since the sum of the measures of the angles of any triangle is 180°,

$$B = 180° - 24° \ 40' - 52° \ 40' = 102° \ 40'.$$

We use the form of the law of sines that relates a, b, A, and B to find a.

$$\frac{a}{\sin A} = \frac{b}{\sin B}$$

$$\frac{a}{\sin 24° \ 40'} = \frac{27.3}{\sin 102° \ 40'}$$

$$a \approx 11.7 \text{ cm}$$

Now, find the area.

$$\mathscr{A} = \frac{1}{2}ab \sin C = \frac{1}{2}(11.7)(27.3) \sin 52° \ 40' \approx 127$$

The area of $\triangle ABC$ is 127 square cm, to three significant digits.

Graphing Calculator Solution

Figure 7 shows how a graphing calculator program supports the algebraic solution. (See the Web site for this text for a sample program.)

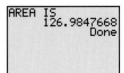

Figure 7

● ● ●

N O T E   Whenever possible, use given values in solving triangles or finding areas rather than values obtained in intermediate steps, to avoid possible rounding errors.

C O N N E C T I O N S   As mentioned in the chapter introduction, aerial photographs can be used to provide coordinates of ordered pairs to determine distances on the ground. Suppose we assign coordinates as shown in Figure 8.

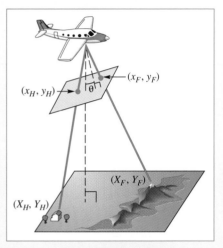

Figure 8

If an object's photographic coordinates are (x, y), then its ground coordinates (X, Y) in feet can be computed using the following formulas.

$$X = \frac{(a - h)x}{f \sec \theta - y \sin \theta}, \qquad Y = \frac{(a - h)y \cos \theta}{f \sec \theta - y \sin \theta}$$

Here, f is focal length of the camera in inches, a is altitude in feet of the airplane, and h is elevation in feet of the object. Suppose that a house has photographic coordinates $(x_H, y_H) = (.9, 3.5)$ with elevation 150 feet, while a nearby forest fire has photographic coordinates $(x_F, y_F) = (2.1, -2.4)$ and is at elevation 690 feet. If the photograph was taken at 7400 feet by a camera with focal length 6 inches and tilt angle $\theta = 4.1°$, we can use these formulas to find the distance on the ground in feet between the house and the forest fire. (*Source:* Moffitt, F. and E. Mikhail, *Photogrammetry,* Third Edition, Harper & Row, 1980.)

For Discussion or Writing

1. Use the formulas to find the ground coordinates of the house and the fire.
2. Use the distance formula given in Section 1.1 to find the required distance on the ground to the nearest tenth of a foot.

7.1 Exercises

1. *Concept Check* Consider this oblique triangle.

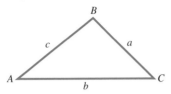

Which one of the following proportions is *not* valid?

A. $\dfrac{a}{b} = \dfrac{\sin A}{\sin B}$ **B.** $\dfrac{a}{\sin A} = \dfrac{b}{\sin B}$ **C.** $\dfrac{\sin A}{a} = \dfrac{b}{\sin B}$ **D.** $\dfrac{\sin A}{a} = \dfrac{\sin B}{b}$

2. *Concept Check* Which two of the following situations do not provide sufficient information for solving a triangle by the law of sines?
   A. You are given two angles and the side included between them.
   B. You are given two angles and a side opposite one of them.
   C. You are given two sides and the angle included between them.
   D. You are given three sides.

Find the length of each side a. Do not use a calculator.

3.
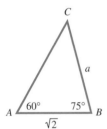

4.
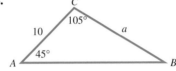

Determine the remaining sides and angles of each △ABC. See Example 1.

5.
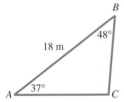

6.
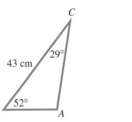

7.
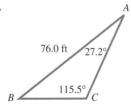

8.
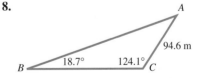

9. $A = 68.41°, B = 54.23°, a = 12.75$ ft
10. $C = 74.08°, B = 69.38°, c = 45.38$ m
11. $A = 87.2°, b = 75.9$ yd, $C = 74.3°$
12. $B = 38° \, 40', a = 19.7$ cm, $C = 91° \, 40'$
13. $B = 20° \, 50', C = 103° \, 10', AC = 132$ ft
14. $A = 35.3°, B = 52.8°, AC = 675$ ft
15. $A = 39.70°, C = 30.35°, b = 39.74$ m
16. $C = 71.83°, B = 42.57°, a = 2.614$ cm
17. $B = 42.88°, C = 102.40°, b = 3974$ ft
18. $A = 18.75°, B = 51.53°, c = 2798$ yd
19. $A = 39° \, 54', a = 268.7$ m, $B = 42° \, 32'$
20. $C = 79° \, 18', c = 39.81$ mm, $A = 32° \, 57'$

21. Explain why the law of sines cannot be used to solve a triangle if we are given the lengths of the three sides of the triangle.

22. *Concept Check* In Example 1, we asked the question, "Why does the Pythagorean theorem not apply?" Answer this question.

23. Kathleen Burk, a perceptive trigonometry student, makes the statement, "If we know *any* two angles and one side of a triangle, then the triangle is uniquely determined." Is this a valid statement? Explain, referring to the congruence axioms given in this section.

24. *Concept Check* If *a* is twice as long as *b*, is *A* necessarily twice as large as *B*?

Solve each problem. See Examples 2 and 3.

25. *Distance across a River* To find the distance *AB* across a river, a distance *BC* = 354 meters is laid off on one side of the river. It is found that *B* = 112° 10′ and *C* = 15° 20′. Find *AB*.

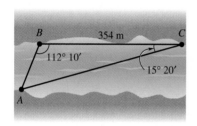

26. *Distance across a Canyon* To determine the distance *RS* across a deep canyon, Joanna lays off a distance *TR* = 582 yards. She then finds that *T* = 32° 50′ and *R* = 102° 20′. Find *RS*.

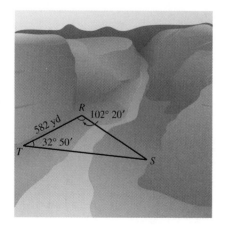

27. *Distance between Radio Direction Finders* Radio direction finders are placed at points *A* and *B*, which are 3.46 miles apart on an east-west line, with *A* west of *B*. From *A* the bearing of a certain radio transmitter is 47.7°, and from *B* the bearing is 302.5°. Find the distance of the transmitter from *A*.

28. *Distance a Ship Travels* A ship is sailing due north. At a certain point the bearing of a lighthouse 12.5 km distant is N 38.8° E. Later on, the captain notices that the bearing of the lighthouse has become S 44.2° E. How far did the ship travel between the two observations of the lighthouse?

29. *Measurement of a Folding Chair* A folding chair is to have a seat 12.0 inches deep with angles as shown in the figure. How far down from the seat should the crossing legs be joined? (Find *x* in the figure.)

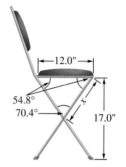

30. *Distance across a River* Mark notices that the bearing of a tree on the opposite bank of a river flowing north is 115.45°. Lisa is on the same bank as Mark, but 428.3 meters away. She notices that the bearing of the tree is 45.47°. The two banks are parallel. What is the distance across the river?

31. *Angle Formed by Radii of Gears* Three gears are arranged as shown in the figure. Find angle *θ*.

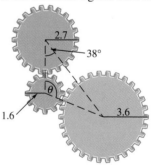

32. *Distance between Atoms* Three atoms with atomic radii of 2.0, 3.0, and 4.5 are arranged as in the figure. Find the distance between the centers of atoms *A* and *C*.

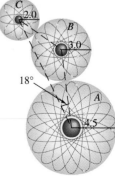

33. *Distance between a Ship and a Lighthouse* The bearing of a lighthouse from a ship was found to be N 37° E. After the ship sailed 2.5 miles due south, the new bearing was N 25° E. Find the distance between the ship and the lighthouse at each location.

34. *Height of a Balloon* A balloonist is directly above a straight road 1.5 miles long that joins two villages. She finds that the town closer to her is at an angle of depression of 35°, and the farther town is at an angle of depression of 31°. How high above the ground is the balloon?

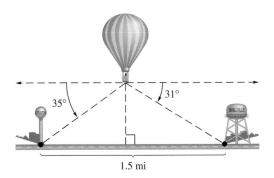

35. *Distance to the Moon* Since the moon is a relatively close celestial object, its distance can be measured directly using trigonometry. To find this distance, two different photographs of the moon are taken at precisely the same time in two different locations with a known distance between them. The moon will have a different angle of elevation at each location. On April 29, 1976, at 11:35 A.M., the lunar angles of elevation during a partial solar eclipse at Bochum in upper Germany and at Donaueschingen in lower Germany were measured as 52.6997° and 52.7430°, respectively. The two cities are 398 km apart. Calculate the distance to the moon from Bochum on this day, and compare it with the actual value of 406,000 km. Disregard the curvature of Earth in this calculation. (*Source:* Scholosser, W., T. Schmidt-Kaler, and E. Milone, *Challenges of Astronomy,* Springer-Verlag, 1991.)

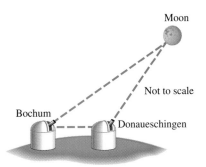

36. *Ground Distances Measured by Aerial Photography* The distance covered by an aerial photograph is determined by both the focal length of the camera and the tilt of the camera from the perpendicular to the ground. Although the tilt is usually small, both archaeological and Canadian photographs often use larger tilts. A camera lens with a 12-inch focal length will have an angular coverage of 60°. If an aerial photograph is taken with this camera tilted $\theta = 35°$ at an altitude of 5000 feet, calculate the ground distance d in miles that will be shown in this photograph. (*Sources:* Brooks, R. and D. Johannes, *Phytoarchaeology,* Dioscorides Press, 1990; Moffitt, F. and E. Mikhail, *Photogrammetry,* Third Edition, Harper & Row, 1980.)

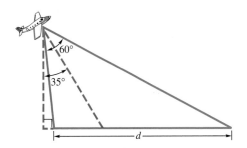

37. *Ground Distances Measured by Aerial Photography* Refer to the previous exercise. A camera lens with a 6-inch focal length has an angular coverage of 86°. Suppose an aerial photograph is taken vertically with no tilt at an altitude of 3500 feet over ground with an increasing slope of 5°, as shown in the figure. Calculate the ground distance CB that would appear in the resulting photograph. (*Source:* Moffitt, F. and E. Mikhail, *Photogrammetry,* Third Edition, Harper & Row, 1980.)

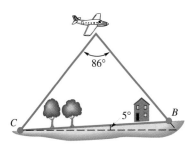

38. *Ground Distances Measured by Aerial Photography* Repeat the previous exercise if the camera lens has an 8.25-inch focal length with an angular coverage of 72°.

Find the area of each triangle using the formula $\mathcal{A} = \dfrac{1}{2}bh$ *and then verify that the formula* $\mathcal{A} = \dfrac{1}{2}ab \sin C$ *gives the same result.*

39.
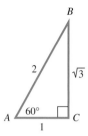

40.
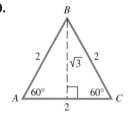

41.

42.
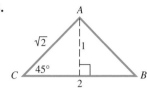

Find the area of each $\triangle ABC$. *See Example 4.*

43. $A = 42.5°, b = 13.6 \text{ m}, c = 10.1 \text{ m}$

44. $C = 72.2°, b = 43.8 \text{ ft}, a = 35.1 \text{ ft}$

45. $B = 124.5°, a = 30.4 \text{ cm}, c = 28.4 \text{ cm}$

46. $C = 142.7°, a = 21.9 \text{ km}, b = 24.6 \text{ km}$

47. $A = 56.80°, b = 32.67 \text{ in.}, c = 52.89 \text{ in.}$

48. $A = 34.97°, b = 35.29 \text{ m}, c = 28.67 \text{ m}$

Solve each problem.

49. *Area of a Metal Plate* A painter is going to apply a special coating to a triangular metal plate on a new building. Two sides measure 16.1 meters and 15.2 meters. She knows that the angle between these sides is 125°. What is the area of the surface she plans to cover with the coating?

50. *Area of a Triangular Lot* A real estate agent wants to find the area of a triangular lot. A surveyor takes measurements and finds that two sides are 52.1 meters and 21.3 meters, and the angle between them is 42.2°. What is the area of the triangular lot?

· · · · · · · · · · · · · **Relating Concepts** · · · · · · · · · · · · ·

For individual or collaborative investigation
(Exercises 51–55)

*In any triangle, the longest side is opposite the largest angle. This result from geometry can be proven using trigonometry. To prove it for acute triangles, **work Exercises 51–55 in order.** (The case for obtuse triangles will be considered in Section 7.3 Exercises.)*

51. Is the graph of the function $y = \sin x$ increasing or decreasing over the interval $[0, \pi/2]$?

52. Suppose angle A is the largest angle of an acute triangle, and let B be an angle smaller than A. Explain why $\dfrac{\sin B}{\sin A} < 1$.

53. Solve for b in the first form of the law of sines.

54. Use the result in Exercise 52 to show that $b < a$.

55. Use the result proved in Exercises 51–54 to explain why no $\triangle ABC$ satisfies $A = 83°, a = 14, b = 20$.

· ·

56. For a triangle inscribed in a circle of radius r, each of the law of sines ratios $\dfrac{a}{\sin A}$, $\dfrac{b}{\sin B}$, and $\dfrac{c}{\sin C}$ have value $2r$.

(a) The circle in the figure at the right has diameter 1. What are the values of a, b, and c? (*Note:* This result provides an alternate way to define the sine function for angles between 0° and 180°. It was used nearly 2000 years ago by the mathematician Ptolemy to construct one of the earliest trigonometry tables.)

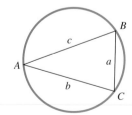

(b) The following theorem is also attributed to Ptolemy: *In a quadrilateral inscribed in a circle, the product of the diagonals is equal to the sum of the products of the opposite sides.* (*Source:* Eves, H., *An Introduction to the History of Mathematics,* Sixth Edition, Saunders College Publishing, 1990.) The circle in the figure below has diameter 1. Explain why the lengths of the line segments are as shown, and then apply Ptolemy's theorem to derive the formula for the sine of the sum of two angles.

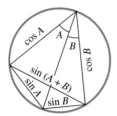

57. Several of the exercises on right triangle applications involved a figure similar to the one shown here, in which angles α and β and the length of line segment AB are known, and the length of side CD is to be determined. Use the law of sines to obtain x in terms of α, β, and d.

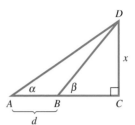

Quantitative Reasoning*

58. *Just how much does the U.S. flag "show its colors"?* The flag of the United States includes the colors red, white, and blue. Which color is predominant? Clearly the answer is either red or white. (It can be shown that only 18.73% of the total area is blue.)

(a) Let R denote the radius of the circumscribing circle of a five-pointed star appearing on the American flag. The star can be decomposed into ten congruent triangles. In the figure, r is the radius of the circumscribing circle of the pentagon in the interior of the star. Show that the area of a star is

$$A = \left[5\frac{\sin A \sin B}{\sin(A + B)} \right] R^2.$$

(*Hint:* $\sin C = \sin[180° - (A + B)] = \sin(A + B)$.)

(b) Angles A and B have values 18° and 36°, respectively. (See Exercise 80 of Section 1.2.) Express the area of a star in terms of its radius, R.

(c) To determine whether red or white is predominant, we must know the measurements of the flag. Consider a flag of width 10 inches, length 19 inches, length of each upper stripe 11.4 inches, and radius R of the circumscribing circle of each star .308 inch. The thirteen stripes consist of six matching pairs of red and white stripes and one additional red, upper stripe. Therefore, we must compare the area of a red, upper stripe with the total area of the 50 white stars.

(i) Compute the area of the red upper stripe.

(ii) Compute the total area of the 50 white stars.

(iii) Which color occupies the greatest area on the flag?

---

*The data for this exercise (along with further interesting mathematics involving the American flag) can be found in *Slicing Pizzas, Racing Turtles, and Further Adventures in Applied Mathematics* by Robert B. Banks, Princeton University Press, 1999.

7.2 The Ambiguous Case of the Law of Sines

• Description of the Ambiguous Case • Solving SSA Triangles (Case 2) • Analyzing Data for Possible Number of Triangles

Description of the Ambiguous Case The law of sines can be used when two angles and the side opposite one of these angles (SAA) are given. Also, if two angles and the included side are known (ASA), then the third angle can be found by using the fact that the sum of the angles of a triangle is 180°, and then the law of sines can be applied. However, if we are given the lengths of two sides and the angle opposite one of them (Case 2 SSA from Section 7.1), it is possible that 0, 1, or 2 such triangles exist. (Recall that there is no SSA congruence theorem.)

To illustrate these facts, suppose that the measure of acute angle A of $\triangle ABC$, the length of side a, and the length of side b are given. See the sketches below. Draw angle A having a terminal side of length b. Now draw a side of length a opposite angle A. The following chart shows that there might be more than one possible outcome. This situation is called the **ambiguous case of the law of sines.**

If angle A is acute, there are four possible outcomes.

Number of Possible Triangles	Sketch	Condition Necessary for Case to Hold
0	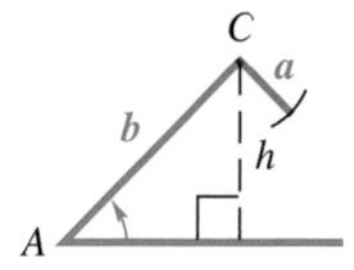	$a < h$ $(h = b \sin A)$
1	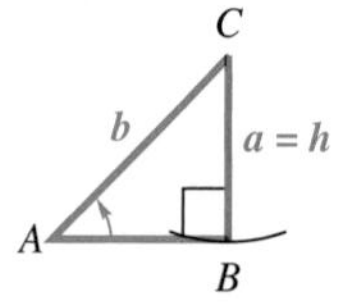	$a = h$
1	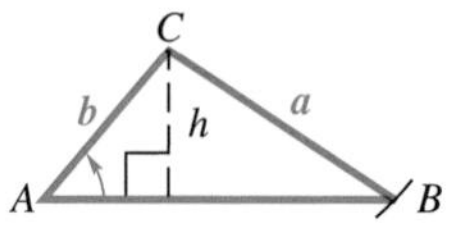	$a \geq b$
2	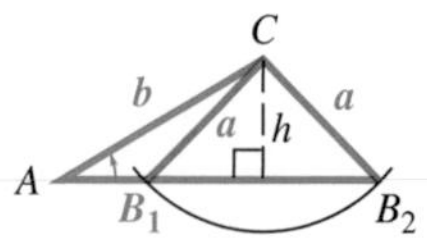	$b > a > h$

If angle A is obtuse, there are two possible outcomes.

Number of Possible Triangles	Sketch	Condition Necessary for Case to Hold
0		$a \le b$
1		$a > b$

We can apply the law of sines to the values of a, b, and A and use some basic properties of geometry and trigonometry to determine which situation applies. The following facts should be kept in mind.

1. For any angle θ of a triangle, $0 < \sin\theta \le 1$. If $\sin\theta = 1$, then $\theta = 90°$ and the triangle is a right triangle.
2. $\sin\theta = \sin(180° - \theta)$ (That is, supplementary angles have the same sine value.)
3. The smallest angle is opposite the shortest side, the largest angle is opposite the longest side, and the middle-valued angle is opposite the medium side (assuming the triangle is scalene).

Solving SSA Triangles (Case 2)

● ● ●   **Example 1   Solving a Triangle Involving SSA Using the Law of Sines (No Such Triangle)**

Solve $\triangle ABC$ if $B = 55° \, 40'$, $b = 8.94$ meters, and $a = 25.1$ meters.
   Since we are given B, b, and a, use the law of sines to find A.

$$\frac{\sin A}{a} = \frac{\sin B}{b}$$

$$\frac{\sin A}{25.1} = \frac{\sin 55° \, 40'}{8.94} \qquad \text{Substitute.}$$

$$\sin A = \frac{25.1 \sin 55° \, 40'}{8.94}$$

$$\sin A \approx 2.3184379$$

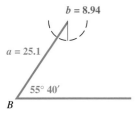

$b = 8.94$

$a = 25.1$

$55° \, 40'$

B

Figure 9

Since $\sin A$ cannot be greater than 1, there can be no such angle A and thus no triangle with the given information. An attempt to sketch such a triangle leads to the situation seen in Figure 9.   ● ● ●

• • • **Example 2** Solving a Triangle Involving SSA Using the Law of Sines (Two Triangles)

Solve $\triangle ABC$ if $A = 55.3°$, $a = 22.8$ feet, and $b = 24.9$ feet.

To begin, use the law of sines to find angle B.

$$\frac{a}{\sin A} = \frac{b}{\sin B}$$

$$\frac{22.8}{\sin 55.3°} = \frac{24.9}{\sin B}$$

$$\sin B = \frac{24.9 \sin 55.3°}{22.8}$$

$$\sin B \approx .8978678$$

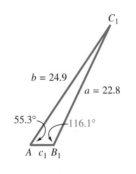

Since $\sin B \approx .8978678$, to the nearest tenth we have one value of B as

$$B = 63.9°.$$

Supplementary angles have the same sine value, so another *possible* value of B is

$$B = 180° - 63.9° = 116.1°.$$

To see if $B = 116.1°$ is a valid possibility, simply add $116.1°$ to the measure of the given value of A, $55.3°$. Since $116.1° + 55.3° = 171.4°$, and this sum is less than $180°$ (the sum of the angles of a triangle), we know that it is a valid angle measure for this triangle.

To keep track of these two different values of B, let

$$B_1 = 116.1° \quad \text{and} \quad B_2 = 63.9°.$$

Now separately solve triangles AB_1C_1 and AB_2C_2 shown in Figure 10. We begin with $\triangle AB_1C_1$. Find C_1 first.

$$C_1 = 180° - A - B_1 = 8.6°$$

Now, use the law of sines to find c_1.

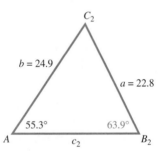

Figure 10

$$\frac{a}{\sin A} = \frac{c_1}{\sin C_1}$$

$$\frac{22.8}{\sin 55.3°} = \frac{c_1}{\sin 8.6°}$$

$$c_1 = \frac{22.8 \sin 8.6°}{\sin 55.3°}$$

$$c_1 \approx 4.15 \text{ feet}$$

To solve $\triangle AB_2C_2$, first find C_2.

$$C_2 = 180° - A - B_2 = 60.8°$$

By the law of sines,

$$\frac{22.8}{\sin 55.3°} = \frac{c_2}{\sin 60.8°}$$

$$c_2 = \frac{22.8 \sin 60.8°}{\sin 55.3°}$$

$$c_2 \approx 24.2 \text{ feet}.$$

• • •

CAUTION When solving a triangle using the type of data given in Example 2, remember to find the possible obtuse angle. The inverse sine function of a calculator will not give it directly. As we shall see in the next example, it is possible that the obtuse angle will not be a valid measure.

● ● ●   **Example 3** Solving a Triangle Involving SSA Using the Law of Sines (One Triangle)

Solve $\triangle ABC$ given $A = 43.5°$, $a = 10.7$ inches, and $b = 7.2$ inches.

To find angle B, use the law of sines.

$$\frac{\sin B}{7.2} = \frac{\sin 43.5°}{10.7}$$

$$\sin B = \frac{7.2 \sin 43.5°}{10.7}$$

$$\sin B \approx .46319186$$

$$B \approx 27.6° \qquad \text{Use the inverse sine function of a calculator.}$$

This is the acute angle. The other possible value of B is

$$B = 180° - 27.6° = 152.4°.$$

However, when we add this possible obtuse angle to the given angle $A = 43.5°$, we get

$$152.4° + 43.5° = 195.9°,$$

which is greater than $180°$. So there can be only one triangle. (Notice that this is the third situation listed in the chart at the beginning of this section.) Then

$$C = 180° - 27.6° - 43.5° = 108.9°,$$

and side c can be found with the law of sines.

$$\frac{c}{\sin 108.9°} = \frac{10.7}{\sin 43.5°}$$

$$c = \frac{10.7 \sin 108.9°}{\sin 43.5°}$$

$$c \approx 14.7 \text{ inches} \qquad\qquad\qquad ●\ ●\ ●$$

Analyzing Data for Possible Number of Triangles

● ● ●   **Example 4** Analyzing Data Involving an Obtuse Angle

Without using the law of sines, explain why the data

$$A = 104°, \ a = 26.8 \text{ meters}, \ b = 31.3 \text{ meters}$$

cannot be valid for a $\triangle ABC$.

Since A is an obtuse angle, the largest side of the triangle must be a, the side opposite A. However, we are given $b > a$, which is impossible if A is obtuse. Therefore, no such $\triangle ABC$ exists. ● ● ●

7.2 Exercises

1. *Concept Check* Which one of the following sets of data does not determine a unique triangle?
   A. $A = 40°, B = 60°, C = 80°$ **B.** $a = 5, b = 12, c = 13$
   C. $a = 3, b = 7, C = 50°$ **D.** $a = 2, b = 2, c = 2$

2. *Concept Check* Which one of the following sets of data determines a unique triangle?
   A. $A = 50°, B = 50°, C = 80°$ **B.** $a = 3, b = 5, c = 20$
   C. $A = 40°, B = 20°, C = 30°$ **D.** $a = 7, b = 24, c = 25$

3. *Concept Check* In the figure below, a line of length h is to be drawn from the point $(3, 4)$ to the positive x-axis in order to form a triangle. For what value(s) of h can you draw the following?
   (a) two triangles **(b)** exactly one triangle
   (c) no triangle

4. *Concept Check* In the figure below, a line of length h is to be drawn from the point $(-3, 4)$ to the positive x-axis in order to form a triangle. For what value(s) of h can you draw the following?
   (a) two triangles **(b)** exactly one triangle
   (c) no triangle

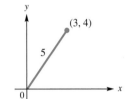

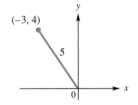

Determine the number of triangles ABC possible with the given parts. See Examples 1–4.

5. $a = 50, b = 26, A = 95°$

6. $b = 60, a = 82, B = 100°$

7. $a = 31, b = 26, B = 48°$

8. $a = 35, b = 30, A = 40°$

9. $a = 50, b = 61, A = 58°$

10. $B = 54°, c = 28, b = 23$

Find each angle B. Do not use a calculator.

11.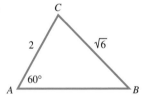

12.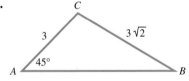

Find the unknown angles in △ABC for each triangle that exists. See Examples 1–3.

13. $A = 29.7°, b = 41.5$ ft, $a = 27.2$ ft

14. $B = 48.2°, a = 890$ cm, $b = 697$ cm

15. $C = 41° 20', b = 25.9$ m, $c = 38.4$ m

16. $B = 48° 50', a = 3850$ in., $b = 4730$ in.

17. $B = 74.3°, a = 859$ m, $b = 783$ m

18. $C = 82.2°, a = 10.9$ km, $c = 7.62$ km

19. $A = 142.13°, b = 5.432$ ft, $a = 7.297$ ft

20. $B = 113.72°, a = 189.6$ yd, $b = 243.8$ yd

Solve each △ABC that exists. See Examples 1–3.

21. $A = 42.5°, a = 15.6$ ft, $b = 8.14$ ft

22. $C = 52.3°, a = 32.5$ yd, $c = 59.8$ yd

23. $B = 72.2°, b = 78.3$ m, $c = 145$ m

24. $C = 68.5°, c = 258$ cm, $b = 386$ cm

25. $A = 38° 40', a = 9.72$ km, $b = 11.8$ km

26. $C = 29° 50', a = 8.61$ m, $c = 5.21$ m

27. $A = 96.80°, b = 3.589$ ft, $a = 5.818$ ft

28. $C = 88.70°, b = 56.87$ yd, $c = 112.4$ yd

29. $B = 39.68°, a = 29.81$ m, $b = 23.76$ m

30. $A = 51.20°, c = 7986$ cm, $a = 7208$ cm

31. Apply the law of sines to the following: $a = \sqrt{5}$, $c = 2\sqrt{5}$, $A = 30°$. What is the value of sin C? What is the measure of C? Based on its angle measures, what kind of triangle is $\triangle ABC$?

32. In your own words, explain the condition that must exist to determine that there is no triangle satisfying the given values of a, b, and B, once the value of sin B is found.

33. Without using the law of sines, explain why no $\triangle ABC$ exists satisfying $A = 103° 20'$, $a = 14.6$ ft, $b = 20.4$ ft.

34. Apply the law of sines to the data given in Example 4. Describe in your own words what happens when you try to find the measure of angle B using a calculator.

35. *Property Survey* A surveyor reported the following data about a piece of property: "The property is triangular in shape, with dimensions as shown in the figure." Use the law of sines to see whether such a piece of property could exist.

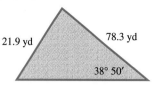

Can such a triangle exist?

36. *Property Survey* The surveyor tries again: "A second triangular piece of property has dimensions as shown." This time it turns out that the surveyor did not consider every possible case. Use the law of sines to show why.

Use the law of sines to prove that each statement is true for any $\triangle ABC$, with corresponding sides a, b, and c.

37. $\dfrac{a + b}{b} = \dfrac{\sin A + \sin B}{\sin B}$

38. $\dfrac{a - b}{a + b} = \dfrac{\sin A - \sin B}{\sin A + \sin B}$

7.3 The Law of Cosines

- Derivation of the Law of Cosines • Solving SAS Triangles (Case 3) • Solving SSS Triangles (Case 4)
- Heron's Formula for the Area of a Triangle

Derivation of the Law of Cosines As mentioned in Section 7.1, if we are given two sides and the included angle (Case 3) or three sides (Case 4) of a triangle, a unique triangle is formed. These are the SAS and SSS cases, respectively. In these cases, however, we cannot begin the solution of the triangle by using the law of sines because we are not given a side and the angle opposite it. Both cases require using the law of cosines, introduced in this section.

Remember the following property of triangles when applying the law of cosines.

Triangle Side Length Restriction
In any triangle, the sum of the lengths of any two sides must be greater than the length of the remaining side.

For example, it would be impossible to construct a triangle with sides of lengths 3, 4, and 10. See Figure 11.

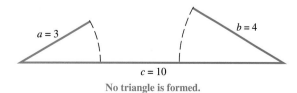

No triangle is formed.

Figure 11

To derive the law of cosines, let ABC be any oblique triangle. Choose a coordinate system so that vertex B is at the origin and side BC is along the positive x-axis. See Figure 12.

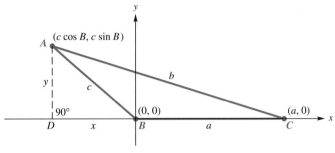

Figure 12

Let (x, y) be the coordinates of vertex A of the triangle. Verify that for angle B, whether obtuse or acute,

$$\sin B = \frac{y}{c} \qquad \text{and} \qquad \cos B = \frac{x}{c}.$$

(Here x is negative when B is obtuse.) From these results

$$y = c \sin B \qquad \text{and} \qquad x = c \cos B,$$

so the coordinates of point A become

$$(c \cos B, c \sin B).$$

Point C has coordinates $(a, 0)$, and AC has length b. By the distance formula,

$$b = \sqrt{(c \cos B - a)^2 + (c \sin B)^2}$$
$$b^2 = (c \cos B - a)^2 + (c \sin B)^2 \qquad \text{Square both sides.}$$
$$= c^2 \cos^2 B - 2ac \cos B + a^2 + c^2 \sin^2 B$$
$$= a^2 + c^2(\cos^2 B + \sin^2 B) - 2ac \cos B$$
$$= a^2 + c^2(1) - 2ac \cos B$$
$$b^2 = a^2 + c^2 - 2ac \cos B.$$

This result is one form of the law of cosines. In the work above, we could just as easily have placed A or C at the origin. This would have given the same result, but with the variables rearranged.

The various forms of the law of cosines are summarized in the following theorem.

Law of Cosines

In any $\triangle ABC$, with sides a, b, and c,

$$a^2 = b^2 + c^2 - 2bc \cos A$$
$$b^2 = a^2 + c^2 - 2ac \cos B$$
$$c^2 = a^2 + b^2 - 2ab \cos C.$$

In words, the law of cosines says that the square of a side of a triangle is equal to the sum of the squares of the other two sides, minus twice the product of those two sides and the cosine of the angle included between them.

NOTE If we let $C = 90°$ in the third form of the law of cosines given above, we have $\cos C = \cos 90° = 0$, and the formula becomes

$$c^2 = a^2 + b^2,$$

the familiar equation of the Pythagorean theorem. Thus, the Pythagorean theorem is a special case of the law of cosines.

Solving SAS Triangles (Case 3)

● ● ●   **Example 1**   Using the Law of Cosines in an Application

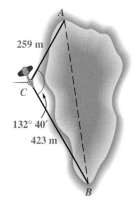

259 m

132° 40′

423 m

Figure 13

A surveyor wishes to find the distance between two inaccessible points A and B on opposite sides of a lake. While standing at point C, she finds that $AC = 259$ meters, $BC = 423$ meters, and angle ACB measures $132°\ 40'$. Find the distance AB. See Figure 13.

    The law of cosines can be used here since we know the lengths of two sides of the triangle and the measure of the included angle (SAS).

$$AB^2 = 259^2 + 423^2 - 2(259)(423)\cos 132°\ 40'$$

$AB^2 \approx 394{,}510.6$ Use a calculator.

$AB \approx 628$ Take the square root and round to three significant digits.

The distance between the points is approximately 628 meters.   ● ● ●

● ● ●   **Example 2**   Using the Law of Cosines to Solve a Triangle Involving SAS

Solve $\triangle ABC$ if $A = 42.3°$, $b = 12.9$ meters, and $c = 15.4$ meters. See Figure 14.

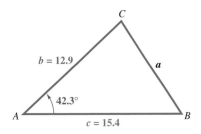

Figure 14

Algebraic Solution

Start by finding a with the law of cosines.

$$a^2 = b^2 + c^2 - 2bc \cos A$$

$$a^2 = 12.9^2 + 15.4^2 - 2(12.9)(15.4)\cos 42.3°$$

$$a^2 \approx 109.7$$

$$a \approx 10.5 \text{ meters}$$

Graphing Calculator Solution

This triangle can be solved using a program on a graphing calculator. The inputs are the two sides and the included angle. See Figure 15.

(continued)

We now must find the measures of angles B and C. There are several approaches we can use. Let us use the law of sines to find one of these angles. Of the two remaining angles, B must be the smaller since it is opposite the shorter of the two sides b and c. Therefore, it cannot be obtuse, and we will avoid any ambiguity when we find its sine.

$$\frac{\sin 42.3°}{10.5} = \frac{\sin B}{12.9}$$

$$\sin B = \frac{12.9 \sin 42.3°}{10.5}$$

$$B \approx 55.8° \qquad \text{Use the inverse sine function of a calculator.}$$

The easiest way to find C is to subtract the measures of A and B from 180°.

$$C = 180° - A - B = 81.9°$$

(There is a slight discrepancy due to rounding.)

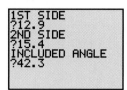

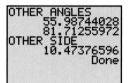

Figure 15

CAUTION Had we chosen to use the law of sines to find C rather than B in the algebraic solution in Example 2, we would not have known whether C equals 81.9° or its supplement, 98.1°.

Solving SSS Triangles (Case 4)

● ● ● **Example 3** Using the Law of Cosines to Solve a Triangle Involving SSS

Solve $\triangle ABC$ if $a = 9.47$ feet, $b = 15.9$ feet, and $c = 21.1$ feet.

Algebraic Solution

We are given the lengths of three sides of the triangle, so we may use the law of cosines to solve for any angle of the triangle. We solve for C, the largest angle, using the law of cosines. We will know that C is obtuse if $\cos C < 0$. Use the form of the law of cosines that involves C.

$$c^2 = a^2 + b^2 - 2ab \cos C$$

$$\cos C = \frac{a^2 + b^2 - c^2}{2ab}$$

$$\cos C = \frac{(9.47)^2 + (15.9)^2 - (21.1)^2}{2(9.47)(15.9)} \qquad \text{Substitute.}$$

$$\cos C \approx -.34109402 \qquad \text{Use a calculator.}$$

Graphing Calculator Solution

This triangle can also be solved using a program on a graphing calculator. In this case, the inputs are the three sides. See Figure 16.

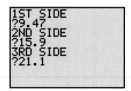

(a)

Figure 16

(continued)

Using the inverse cosine function of a calculator, we get obtuse angle C.

$$C \approx 109.9°$$

We can use either the law of sines or the law of cosines to find $B \approx 45.1°$. (Verify this.) Since $A = 180° - B - C$,

$$A \approx 25.0°.$$

Figure 16(b) shows the output.

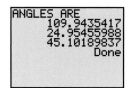

(b)

Figure 16

● ● ●

As shown in this chapter, four possible cases can occur when solving an oblique triangle. These cases are summarized in the following chart, along with a suggested procedure for solving in each case. There are other procedures that work, but we give the one that is most efficient. In all four cases, it is assumed that the given information actually produces a triangle.

Oblique Triangle	Suggested Procedure for Solving
Case 1: One side and two angles are known. (SAA or ASA)	**Step 1** Find the remaining angle using the angle sum formula ($A + B + C = 180°$). **Step 2** Find the remaining sides using the law of sines.
Case 2: Two sides and one angle (not included between the two sides) are known. (SSA)	*Be aware of the ambiguous case; there may be no triangle or two triangles.* **Step 1** Find an angle using the law of sines. **Step 2** Find the remaining angle using the angle sum formula. **Step 3** Find the remaining side using the law of sines. *If two triangles exist, repeat Steps 2 and 3.*
Case 3: Two sides and the included angle are known. (SAS)	**Step 1** Find the third side using the law of cosines. **Step 2** Find the smaller of the two remaining angles using the law of sines. **Step 3** Find the remaining angle using the angle sum formula.
Case 4: Three sides are known. (SSS)	**Step 1** Find the largest angle using the law of cosines. **Step 2** Find either remaining angle using the law of sines. **Step 3** Find the remaining angle using the angle sum formula.

C O N N E C T I O N S   The law of cosines was known to Greek mathematicians. Propositions 12 and 13 of Book II of Euclid's *Elements* (300 B.C.), stated together in modern language, say: *In an obtuse-angled (acute-angled) triangle, the square of the side opposite the obtuse*

(continued)

(acute) angle is equal to the sum of the squares of the other two sides increased (decreased) by twice the product of one of these sides and the projection of the other on it.

The law of sines was known in an equivalent form by Ptolemy (A.D. 150), but it was the Hindu mathematician al-Biruni (973–1048) who first set it forth clearly and proved that the sides of a triangle have the same ratio as the sines of the opposite angles. (*Sources:* Eves, H., *An Introduction to the History of Mathematics,* Sixth Edition, Saunders College Publishing, 1990; *Historical Topics for the Mathematics Classroom, Thirty-first Yearbook of the National Council of Teachers of Mathematics,* Washington, D.C., 1969.)

For Discussion or Writing

1. Refer to the law of tangents in Exercise 60 of this section, and state it in your own words.
2. Refer to Newton's Formula just above Exercise 25 of the Review Exercises for this chapter, and state it in your own words.

Heron's Formula for the Area of a Triangle   The law of cosines can be used to derive a formula for the area of a triangle when only the lengths of the three sides are known. This formula is known as *Heron's formula,* named after the Greek mathematician Heron of Alexandria, who lived around A.D. 75. It is found in his work *Metrica.*

Heron's Area Formula

If a triangle has sides of lengths a, b, and c, and if the **semiperimeter** is

$$s = \frac{1}{2}(a + b + c),$$

then the area of the triangle is

$$\mathcal{A} = \sqrt{s(s - a)(s - b)(s - c)}.$$

● ● ●   **Example 4**   Finding Area Using Heron's Formula

Find the area of the triangle having sides of lengths $a = 29.7$ feet, $b = 42.3$ feet, and $c = 38.4$ feet.

Algebraic Solution

To use Heron's area formula, first find s.

$$s = \frac{1}{2}(a + b + c)$$

$$s = \frac{1}{2}(29.7 + 42.3 + 38.4)$$

$$s = 55.2$$

Graphing Calculator Solution

The area of this triangle can be found with a program based on Heron's formula. See Figure 17.

(continued)

The area is

$$\mathcal{A} = \sqrt{s(s - a)(s - b)(s - c)}$$

$$\mathcal{A} = \sqrt{55.2(55.2 - 29.7)(55.2 - 42.3)(55.2 - 38.4)}$$

$$\mathcal{A} = \sqrt{55.2(25.5)(12.9)(16.8)}$$

$$\mathcal{A} \approx 552 \text{ square feet.}$$

```
ENTER SIDES
A
?29.7
B
?42.3
C
?38.4
```

```
AREA =
        552.3179085
            Done
```

Figure 17

● ● ●

C O N N E C T I O N S   We have introduced two new formulas for the area of a triangle in this chapter. You should now be able to find the area $\mathcal{A}$ of a triangle using one of three formulas:

(a) $\mathcal{A} = (1/2)bh$
(b) $\mathcal{A} = (1/2)ab \sin C$ (or $\mathcal{A} = (1/2)ac \sin B$ or $\mathcal{A} = (1/2)bc \sin A$)
(c) $\mathcal{A} = \sqrt{s(s - a)(s - b)(s - c)}$.

Another area formula can be used when the coordinates of the vertices of a triangle are given. If the vertices are the ordered pairs (x_1, y_1), (x_2, y_2), and (x_3, y_3), then

$$\mathcal{A} = \frac{1}{2} \left| (x_1 y_2 - y_1 x_2 + x_2 y_3 - y_2 x_3 + x_3 y_1 - y_3 x_1) \right|.$$

For Discussion or Writing

Consider $\triangle PQR$ with vertices $P(2, 5)$, $Q(-1, 3)$, and $R(4, 0)$. (*Hint:* Draw a sketch first.)

1. Find the area of the triangle using the new formula just introduced.
2. Find the area of the triangle using (c) above. Use the distance formula to find the lengths of the three sides.
3. Find the area of the triangle using (b) above. First use the law of cosines to find the measure of an angle.

7.3   Exercises

1. *Concept Check*   Decide whether the law of sines, law of cosines, or neither is applicable for solving $\triangle ABC$ for the required value. Then solve for the indicated value.
  (a) $a = 342$, $b = 116$, $c = 401$;   solve for C
  (b) $a = 12.2$, $b = 13.1$, $C = 20.2°$;   solve for c
  (c) $a = 21.13$, $B = 48° \ 13'$, $C = 81° \ 42'$;   solve for b
  (d) $A = 40°$, $B = 60°$, $C = 80°$;   solve for a

2. *Concept Check*   If you are given two angles and the side included between them and you want to solve the triangle, there is a relatively easy first step. What is it?

Find the length of the remaining side of each triangle. Do not use a calculator.

3.

4.

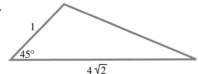

Find the value of θ in each triangle. Do not use a calculator.

5.

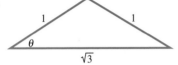

6.

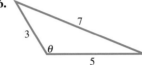

Solve each △ABC. See Example 2.

7. $C = 28.3°$, $b = 5.71$ in., $a = 4.21$ in.

8. $A = 41.4°$, $b = 2.78$ yd, $c = 3.92$ yd

9. $C = 45.6°$, $b = 8.94$ m, $a = 7.23$ m

10. $A = 67.3°$, $b = 37.9$ km, $c = 40.8$ km

11. $A = 80° \, 40'$, $b = 143$ cm, $c = 89.6$ cm

12. $C = 72° \, 40'$, $a = 327$ ft, $b = 251$ ft

13. $B = 74.80°$, $a = 8.919$ in., $c = 6.427$ in.

14. $C = 59.70°$, $a = 3.725$ mi, $b = 4.698$ mi

15. $A = 112.8°$, $b = 6.28$ m, $c = 12.2$ m

16. $B = 168.2°$, $a = 15.1$ cm, $c = 19.2$ cm

Find all the angles in each △ABC. See Example 3.

17. $a = 3.0$ ft, $b = 5.0$ ft, $c = 6.0$ ft

18. $a = 4.0$ ft, $b = 5.0$ ft, $c = 8.0$ ft

19. $a = 9.3$ cm, $b = 5.7$ cm, $c = 8.2$ cm

20. $a = 28$ ft, $b = 47$ ft, $c = 58$ ft

21. $a = 42.9$ m, $b = 37.6$ m, $c = 62.7$ m

22. $a = 187$ yd, $b = 214$ yd, $c = 325$ yd

23. $AB = 1240$ ft, $AC = 876$ ft, $BC = 918$ ft

24. $AB = 298$ m, $AC = 421$ m, $BC = 324$ m

Solve each problem, using the law of sines or the law of cosines. See Example 1.

25. *Distance across a Lake* Points A and B are on opposite sides of Lake Yankee. From a third point, C, the angle between the lines of sight to A and B is $46.3°$. If AC is 350 meters long and BC is 286 meters long, find AB.

26. *Diagonals of a Parallelogram* The sides of a parallelogram are 4.0 cm and 6.0 cm. One angle is $58°$ while another is $122°$. Find the lengths of the diagonals of the parallelogram.

27. *Playhouse Layout* The layout for a child's playhouse in her backyard shows the dimensions given in the figure. Find x.

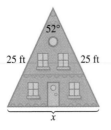

28. *Distance between Two Ships* Two ships leave a harbor together, traveling on courses that have an angle of $135° \, 40'$ between them. If they each travel 402 miles, how far apart are they?

29. *Flight Distance* Airports A and B are 450 km apart, on an east-west line. Tom flies in an approximately northeast direction from A to airport C. From C he flies 359 km on a bearing of $128° \, 40'$ to B. How far is C from A?

30. *Length of a Rope* A hill slopes at an angle of $12.47°$ with the horizontal. From the base of the hill, the angle of elevation of a 459.0-foot tower at the top of the hill is $35.98°$. How much rope would be required to reach from the top of the tower to the bottom of the hill?

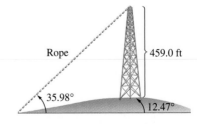

31. *Distance between Points on a Crane* A crane with a counterweight is shown in the figure. Find the horizontal distance between points *A* and *B*.

32. *Angles between a Beam and Cables* A weight is supported by cables attached to both ends of a balance beam, as shown in the figure. What angles are formed between the beam and the cables?

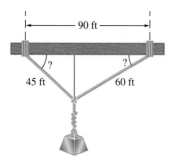

33. *Distance between a Satellite and a Tracking Station* A satellite traveling in a circular orbit 1600 km above Earth is due to pass directly over a tracking station at noon. See the figure. Assume that the satellite takes 2 hours to make an orbit and that the radius of Earth is 6400 km. Find the distance between the satellite and the tracking station at 12:03 P.M. (*Source:* Kastner, B., *Space Mathematics*, National Aeronautics and Space Administration (NASA), 1985.)

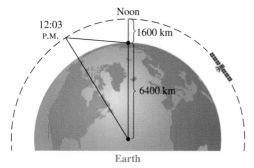

Not to scale

34. *Measurement Using Triangulation* Surveyors are often confronted with obstacles, such as trees, when measuring the boundary of a lot. One technique used to obtain an accurate measurement is the so-called *triangulation method*. In this technique, a triangle is constructed around the obstacle and one angle and two sides of the triangle are measured. Use this technique to find the length of the property line (the straight line between the two markers) in the figure. (*Source:* Kavanagh, B. and S. Bird, *Surveying Principles and Applications*, Fifth Edition, Prentice-Hall, 2000.)

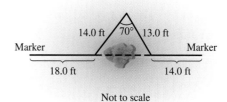

Not to scale

35. *Distance between Two Factories* Two factories blow their whistles at exactly 5:00 P.M. A man hears the two blasts at 3 seconds and 6 seconds after 5:00, respectively. The angle between his lines of sight to the two factories is 42.2°. If sound travels 344 meters per second, how far apart are the factories?

36. *Distance between an Airplane and a Mountain* A person in a plane flying a straight course observes a mountain at a bearing of 24.1° to the right of its course. At that time the plane is 7.92 km from the mountain. A short time later, the bearing to the mountain becomes 32.7°. How far is the airplane from the mountain when the second bearing is taken?

Find the measure of each angle θ to two decimal places.

37.

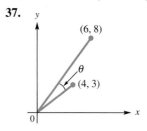

38.

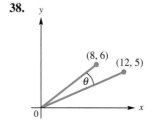

Find the exact area of each triangle using the formula $\mathcal{A} = \frac{1}{2}bh$, and then verify that Heron's formula gives the same result.

39.

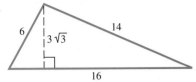

40.

Find the area of each $\triangle ABC$. See Example 4.

41. $a = 12$ m, $b = 16$ m, $c = 25$ m

42. $a = 22$ in., $b = 45$ in., $c = 31$ in.

43. $a = 154$ cm, $b = 179$ cm, $c = 183$ cm

44. $a = 25.4$ yd, $b = 38.2$ yd, $c = 19.8$ yd

45. $a = 76.3$ ft, $b = 109$ ft, $c = 98.8$ ft

46. $a = 15.89$ in., $b = 21.74$ in., $c = 10.92$ in.

Volcano Movement To help predict eruptions from the volcano Mauna Loa on the island of Hawaii, scientists keep track of the volcano's movement by using a "super triangle" with vertices on the three volcanoes shown on the map at the right. (For example, in a recent year, Mauna Loa moved 6 inches, a result of increasing internal pressure.) Refer to the map to work Exercises 47 and 48.

47. $AB = 22.47928$ miles, $AC = 28.14276$ miles, $A = 58.56989°$; find BC

48. $AB = 22.47928$ miles, $BC = 25.24983$ miles, $A = 58.56989°$; find B

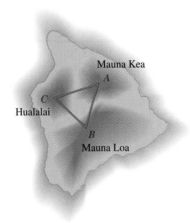

Solve each problem. See Example 4.

49. *Required Amount of Paint* A painter needs to cover a triangular region 75 meters by 68 meters by 85 meters. A can of paint covers 75 square meters of area. How many cans (to the next higher number of cans) will be needed?

50. *Area of the Bermuda Triangle* Find the area of the Bermuda Triangle if the sides of the triangle have approximate lengths 850 miles, 925 miles, and 1300 miles.

51. *Perfect Triangles* A *perfect triangle* is a triangle whose sides have whole number lengths and whose area is numerically equal to its perimeter. Show that the triangle with sides of lengths 9, 10, and 17 is perfect.

52. *Heron Triangles* A *Heron triangle* is a triangle having integer sides and area. Show that each of the following is a Heron triangle.
   (a) $a = 11, b = 13, c = 20$
   (b) $a = 13, b = 14, c = 15$
   (c) $a = 7, b = 15, c = 20$

53. *Concept Check* Refer to Figure 11. If you attempt to find any angle of a triangle using the values $a = 3$, $b = 4$, and $c = 10$ with the law of cosines, what happens?

54. A familiar saying is "The shortest distance between two points is a straight line." Explain how this relates to the geometric property that states that the sum of the lengths of any two sides of a triangle must be greater than the remaining side.

55. Consider $\triangle ABC$ shown below.
   (a) Use the law of sines to find candidates for the value of angle C.
   (b) Rework part (a) using the law of cosines.
   (c) Why is the law of cosines a better method in this case?

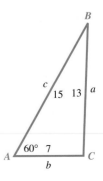

56. Show that the measure of angle A is twice the measure of angle B. (*Hint:* Use the law of cosines to find $\cos A$ and $\cos B$, and then show that $\cos A = 2\cos^2 B - 1$.)

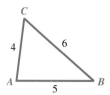

57. In Example 2 of Section 7.2, we used the law of sines to solve $\triangle ABC$ given that $A = 55.3°$, $a = 22.8$ feet, and $b = 24.9$ feet. Use the given three pieces of information and the law of cosines to find the two possible values of c. (*Hint:* Your solution will require that you solve a quadratic equation in c.)

58. Let a and b be the equal sides of an isosceles triangle. Prove that $c^2 = 2a^2(1 - \cos C)$.

59. Let point D on side AB of $\triangle ABC$ be such that CD bisects angle C. Show that $AD/DB = b/a$.

60. In addition to the law of sines and the law of cosines, there is a **law of tangents.** In any triangle ABC,

$$\frac{\tan \dfrac{1}{2}(A - B)}{\tan \dfrac{1}{2}(A + B)} = \frac{a - b}{a + b}.$$

Verify this law for the $\triangle ABC$ with $a = 2$, $b = 2\sqrt{3}$, $A = 30°$, and $B = 60°$.

· **Relating Concepts** · · · · · · · · · · · · · · ·
For individual or collaborative investigation
(Exercises 61–64)

In any triangle, the longest side is opposite the largest angle. This result from geometry was proven for acute triangles in the exercises for Section 7.1. To prove it for obtuse triangles, **work Exercises 61–64 in order.**

61. Suppose angle A is the largest angle of an obtuse triangle. Why is $\cos A$ negative?

62. Consider the law of cosines expression for a^2, and show that $a^2 > b^2 + c^2$.

63. Use the result of Exercise 62 to show that $a > b$ and $a > c$.

64. Use the result of Exercise 63 to explain why no $\triangle ABC$ satisfies $A = 103°$, $a = 25$, and $c = 30$.

7.4 Vectors and the Dot Product

* **Basic Terminology** • **Finding Components and Magnitudes** • **Algebraic Interpretation of Vectors**
* **Operations with Vectors** • **Dot Product and the Angle between Vectors**

Basic Terminology Many quantities in mathematics involve magnitudes, such as 45 pounds or 60 miles per hour. These quantities are called **scalars.** Other quantities, called **vector quantities,** involve both magnitude and direction. Typical vector quantities are velocity, acceleration, and force.

A vector quantity is often represented with a directed line segment (a segment that uses an arrowhead to indicate direction), called a **vector.** The length of the vector represents the **magnitude** of the vector quantity. The direction of the vector, as indicated by the arrowhead, represents the direction of the quantity. For example, the vector in Figure 18 represents a force of 10 pounds applied at an angle of 30° from the horizontal.

The symbol for a vector is often printed in boldface type. When writing vectors by hand, it is customary to use an arrow over the letter or letters. Thus **OP** and $\overrightarrow{OP}$ both represent the vector **OP**. Vectors may be named with either one lowercase or uppercase letter, or two uppercase letters. When two letters are used, the first indicates the **initial point** and the second indicates the **terminal point** of the vector. Knowing these points gives the direction of the vector. For

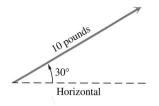

Figure 18

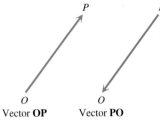

Vector **OP**   Vector **PO**

Figure 19

example, vectors **OP** and **PO** in Figure 19 are not the same vector. They have the same magnitude, but they have opposite directions. The magnitude of vector **OP** is written $|\mathbf{OP}|$.

Two vectors are *equal* if and only if they have the same direction and the same magnitude. In Figure 20 vectors **A** and **B** are equal, as are vectors **C** and **D**. As Figure 20 shows, equal vectors need not coincide, but they must be parallel. Vectors **A** and **E** are unequal because they do not have the same direction, while $\mathbf{A} \neq \mathbf{F}$ because they have different magnitudes.

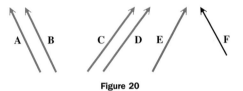

Figure 20

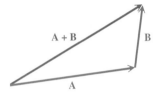

Figure 21

To find the *sum* of two vectors **A** and **B**, written $\mathbf{A} + \mathbf{B}$, we place the initial point of vector **B** at the terminal point of vector **A**, as shown in Figure 21. The vector with the same initial point as **A** and the same terminal point as **B** is the sum $\mathbf{A} + \mathbf{B}$. The sum of two vectors is also a vector.

Another way to find the sum of two vectors is to use the **parallelogram rule.** Place vectors **A** and **B** so that their initial points coincide. Then complete a parallelogram that has **A** and **B** as two adjacent sides. The diagonal of the parallelogram with the same initial point as **A** and **B** is the same vector sum $\mathbf{A} + \mathbf{B}$ found by the definition. See Figure 22.

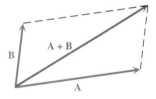

Figure 22

The vector sum $\mathbf{A} + \mathbf{B}$ is the **resultant** of vectors **A** and **B**. Each of the vectors **A** and **B** is a **component** of vector $\mathbf{A} + \mathbf{B}$. In many practical applications, such as surveying, it is necessary to break a vector into its **vertical** and **horizontal components.** These components are two vectors, one vertical and one horizontal, whose resultant is the original vector. As shown in Figure 23, vector **OR** is the vertical component and vector **OS** is the horizontal component of **OP**.

For every vector **v** there is a vector $-\mathbf{v}$ with the same magnitude as **v** but opposite direction. Vector $-\mathbf{v}$ is the **opposite** of **v**. See Figure 24. The sum of **v** and $-\mathbf{v}$, called the **zero vector 0,** has magnitude 0. The zero vector has arbitrary direction. As with real numbers, to *subtract* vector **B** from vector **A**, we find the vector sum $\mathbf{A} + (-\mathbf{B})$. See Figure 25.

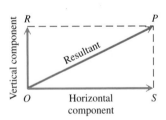

Figure 23

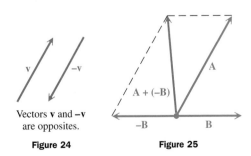

Vectors **v** and $-\mathbf{v}$
are opposites.

Figure 24   **Figure 25**

The **scalar product** of a real number (or scalar) k and a vector **u** is the vector $k\mathbf{u}$, which has magnitude $|k|$ times the magnitude of **u**. As shown in Figure 26, $k\mathbf{u}$ has the same direction as **u** if $k > 0$, and the opposite direction if $k < 0$.

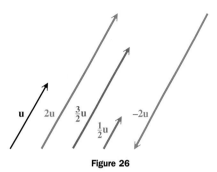

Figure 26

Finding Components and Magnitudes

● ● ●   **Example 1**   Finding Magnitudes of Vertical and Horizontal Components

Vector **w** has magnitude 25.0 and is inclined at an angle of 40° from the horizontal. Find the magnitudes of the horizontal and vertical components of the vector.

Algebraic Solution

In Figure 27, the vertical component is labeled **v** and the horizontal component is labeled **u**.

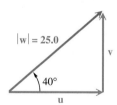

Figure 27

Vectors **u**, **v**, and **w** form a right triangle. In this right triangle,

$$\sin 40° = \frac{|\mathbf{v}|}{|\mathbf{w}|} = \frac{|\mathbf{v}|}{25.0},$$

and

$$|\mathbf{v}| = 25.0 \sin 40° \approx 16.1.$$

In the same way,

$$\cos 40° = \frac{|\mathbf{u}|}{25.0}$$

$$|\mathbf{u}| = 25.0 \cos 40° \approx 19.2.$$

Graphing Calculator Solution

The screen in Figure 28 shows how the *vertical (y) component* and the *horizontal (x) component* can be found using the *polar-to-rectangular* conversion capability of a graphing calculator. (*Polar coordinates* are covered in Chapter 8.)

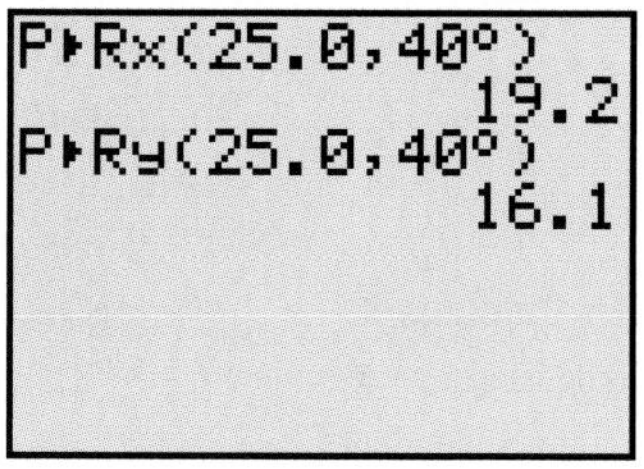

```
P▸Rx(25.0,40°)
            19.2
P▸Ry(25.0,40°)
            16.1
```

Figure 28

● ● ●

Properties of parallelograms are helpful when studying vectors.

1. A parallelogram is a quadrilateral whose opposite sides are parallel.
2. The opposite sides and opposite angles of a parallelogram are equal, and adjacent angles of a parallelogram are supplementary.
3. The diagonals of a parallelogram bisect each other, but do not necessarily bisect the angles of the parallelogram.

● ● ● **Example 2** Finding the Magnitude of the Resultant of Two Vectors in an Application

Two forces of 15 newtons and 22 newtons act at a point in the plane. (A newton is a unit of force used in physics.) If the angle between the forces is 100°, find the magnitude of the resultant force.

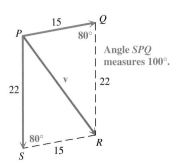

Figure 29

As shown in Figure 29, a parallelogram that has the forces as adjacent sides can be formed. The angles of the parallelogram adjacent to angle P each measure 80°, since adjacent angles of a parallelogram are supplementary. (Angle SPQ measures 100°.) Opposite sides of the parallelogram are equal in length. The resultant force divides the parallelogram into two triangles. Use the law of cosines to get

$$|\mathbf{v}|^2 = 15^2 + 22^2 - 2(15)(22)\cos 80°$$
$$|\mathbf{v}| \approx 24.$$

● ● ●

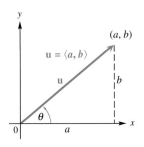

Figure 30

Algebraic Interpretation of Vectors We now consider vectors in conjunction with a rectangular coordinate system. A vector with its initial point at the origin is called a **position vector.** A position vector $\mathbf{u}$ with its endpoint at the point (a,b) is written $\langle a,b \rangle$, so $\mathbf{u} = \langle a,b \rangle$. This means that every vector in the real plane corresponds to an ordered pair of real numbers. Thus, geometrically a vector is a directed line segment; algebraically, it is an ordered pair. The numbers a and b are the **x-component** and **y-component** of vector $\mathbf{u}$. Figure 30 shows the vector $\mathbf{u} = \langle a,b \rangle$. The positive angle between the x-axis and a position vector is the **direction angle** for the vector. In Figure 30, θ is the direction angle for vector $\mathbf{u}$.

From Figure 30, we can see that the magnitude and direction of a vector are related to its x- and y-components.

Looking Ahead to Calculus

In addition to two-dimensional vectors in a plane, calculus courses introduce three-dimensional vectors in space. The magnitude of the two-dimensional vector $\langle a,b \rangle$ is given by $\sqrt{a^2 + b^2}$. If this is extended to the three-dimensional vector $\langle a,b,c \rangle$, the expression becomes $\sqrt{a^2 + b^2 + c^2}$. Similar extensions are made for other concepts.

Magnitude and Direction Angle of a Vector $\langle a,b \rangle$

The magnitude (length) of vector $\mathbf{u} = \langle a,b \rangle$ is given by

$$|\mathbf{u}| = \sqrt{a^2 + b^2}.$$

The direction angle θ satisfies $\tan \theta = b/a$, where $a \neq 0$.

● ● ● **Example 3** Finding the Magnitude and Direction Angle

Find the magnitude and direction angle for $\mathbf{u} = \langle 3, -2 \rangle$.

Algebraic Solution

The magnitude of $\mathbf{u} = \sqrt{3^2 + (-2)^2} = \sqrt{13}$. To find the direction angle θ, start with

$$\tan \theta = \frac{y}{x} = \frac{-2}{3} = -\frac{2}{3}.$$

Vector $\mathbf{u}$ has positive x-component and negative y-component, placing the vector in quadrant IV. A calculator gives

$$\tan^{-1}\left(-\frac{2}{3}\right) \approx -33.7°.$$

Adding $360°$ yields the direction angle $\theta = 326.3°$. This is shown in Figure 31.

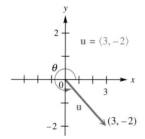

Figure 31

Graphing Calculator Solution

The calculator screen returns the magnitude and direction angle, given the x- and y-components. Notice that an approximation for $\sqrt{13}$ is given, and the direction angle is returned as a measure with smallest possible absolute value. See Figure 32.

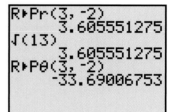

Figure 32

● ● ●

Horizontal and Vertical Components

The horizontal and vertical components, respectively, of a vector $\mathbf{u}$ having magnitude $|\mathbf{u}|$ and direction angle θ are given by

$$a = |\mathbf{u}| \cos \theta \qquad \text{and} \qquad b = |\mathbf{u}| \sin \theta.$$

● ● ● **Example 4** Finding Horizontal and Vertical Components

Vector $\mathbf{w}$ in Figure 33 has magnitude 25.0 and direction angle 41.7°. Find the horizontal and vertical components.

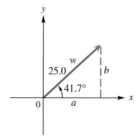

Figure 33

Algebraic Solution

Use the two formulas above with $|\mathbf{w}| = 25.0$ and $\theta = 41.7°$.

$$a = 25.0 \cos 41.7° \qquad b = 25.0 \sin 41.7°$$
$$a \approx 18.7 \qquad\qquad b \approx 16.6$$

Therefore, $\mathbf{w} = \langle 18.7, 16.6 \rangle$.

Graphing Calculator Solution

See Figure 34. The results support the algebraic solution.

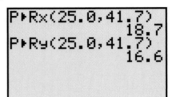

Figure 34

• • •

Operations with Vectors In Figure 35, $\mathbf{m} = \langle a, b \rangle$, $\mathbf{n} = \langle c, d \rangle$, and $\mathbf{p} = \langle a + c, b + d \rangle$. Using geometry, we can show that the endpoints of the three vectors and the origin form a parallelogram. Since a diagonal of this parallelogram gives the resultant of $\mathbf{m}$ and $\mathbf{n}$, we have $\mathbf{p} = \mathbf{m} + \mathbf{n}$ or

$$\langle a + c, b + d \rangle = \langle a, b \rangle + \langle c, d \rangle.$$

Similarly, we could verify the following vector operations.

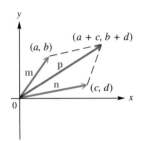

Figure 35

Vector Operations

For any real numbers a, b, c, d, and k,

$$\langle a, b \rangle + \langle c, d \rangle = \langle a + c, b + d \rangle$$
$$k \cdot \langle a, b \rangle = \langle ka, kb \rangle.$$
$$\text{If } \mathbf{a} = \langle a_1, a_2 \rangle, \text{ then } -\mathbf{a} = \langle -a_1, -a_2 \rangle.$$
$$\langle a, b \rangle - \langle c, d \rangle = \langle a, b \rangle + -\langle c, d \rangle$$

• • • **Example 5** Performing Vector Operations

Consider the vectors shown in Figure 36, and perform the operations.

(a) $\mathbf{u} + \mathbf{v} = \langle -2, 1 \rangle + \langle 4, 3 \rangle = \langle -2 + 4, 1 + 3 \rangle = \langle 2, 4 \rangle$

(b) $-2\mathbf{u} = -2 \cdot \langle -2, 1 \rangle = \langle -2(-2), -2(1) \rangle = \langle 4, -2 \rangle$

(c) $4\mathbf{u} - 3\mathbf{v} = 4 \cdot \langle -2, 1 \rangle - 3 \cdot \langle 4, 3 \rangle = \langle -8, 4 \rangle - \langle 12, 9 \rangle$
$$= \langle -8 - 12, 4 - 9 \rangle = \langle -20, -5 \rangle$$

• • •

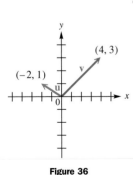

Figure 36

The equations in the box preceding Example 4 allow us to make the following statement: If $\mathbf{u} = \langle a, b \rangle$ has direction angle θ, then

$$\mathbf{u} = \langle |\mathbf{u}| \cos \theta, |\mathbf{u}| \sin \theta \rangle.$$

We use this in the next example.

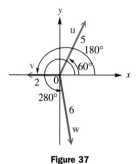

Figure 37

• • • **Example 6** Writing Vectors in the Form $\langle a, b \rangle$

Write each vector in Figure 37 in the form $\langle a, b \rangle$.

$$\mathbf{u} = \langle 5 \cos 60°, 5 \sin 60° \rangle = \left\langle 5 \cdot \frac{1}{2}, 5 \cdot \frac{\sqrt{3}}{2} \right\rangle = \left\langle \frac{5}{2}, \frac{5\sqrt{3}}{2} \right\rangle$$

$$\mathbf{v} = \langle 2 \cos 180°, 2 \sin 180° \rangle = \langle 2(-1), 2(0) \rangle = \langle -2, 0 \rangle$$

$$\mathbf{w} = \langle 6 \cos 280°, 6 \sin 280° \rangle \approx \langle 1.0419, -5.9088 \rangle$$

• • •

A **unit vector** is a vector that has magnitude 1. Two very important vectors are defined as follows.

Unit Vectors

$$\mathbf{i} = \langle 1, 0 \rangle \qquad \mathbf{j} = \langle 0, 1 \rangle$$

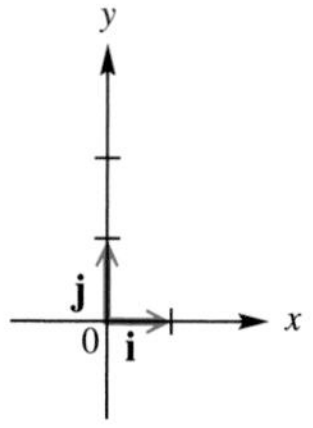

See Figure 38(a).

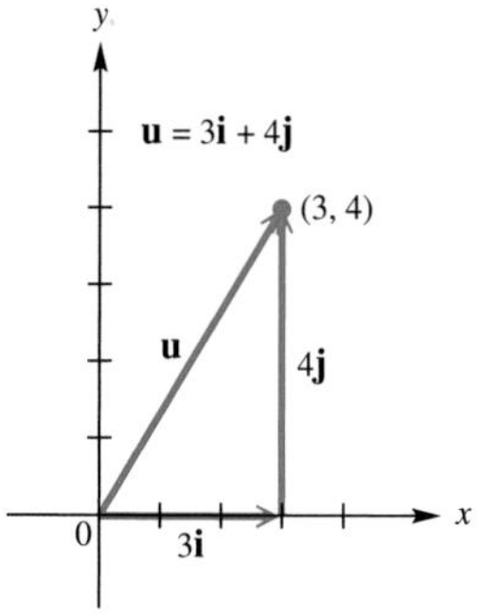

With the unit vectors $\mathbf{i}$ and $\mathbf{j}$, we can express any other vector $\langle a, b \rangle$ in the form $a\mathbf{i} + b\mathbf{j}$, as shown for $\langle 3, 4 \rangle$ in Figure 38(b). The vector operations given above can be restated, using the $a\mathbf{i} + b\mathbf{j}$ notation.

(a)

---

i, j Form for Vectors

If $\mathbf{v} = \langle a, b \rangle$, then $\mathbf{v} = a\mathbf{i} + b\mathbf{j}$.

(b)

Figure 38

Dot Product and the Angle Between Vectors   The *dot product* of two vectors is a real number, not a vector. It is also known as the *inner product* or *scalar product*. Dot products are used to determine the angle between two vectors, derive geometric theorems, and solve physics problems.

Dot Product

The **dot product** of the two vectors $\mathbf{u} = \langle a, b \rangle$ and $\mathbf{v} = \langle c, d \rangle$ is denoted by $\mathbf{u} \cdot \mathbf{v}$, read "$\mathbf{u}$ dot $\mathbf{v}$," and given by

$$\mathbf{u} \cdot \mathbf{v} = ac + bd.$$

• • • **Example 7** Finding the Dot Product

Find each dot product.

(a) $\langle 2, 3 \rangle \cdot \langle 4, -1 \rangle = 2(4) + 3(-1) = 5$

(b) $\langle 6, 4 \rangle \cdot \langle -2, 3 \rangle = 6(-2) + 4(3) = 0$

• • •

The following properties of dot products are easily verified.

Properties of the Dot Product

For all vectors **u**, **v**, and **w** and real numbers k,

(a) $\mathbf{u} \cdot \mathbf{v} = \mathbf{v} \cdot \mathbf{u}$

(b) $\mathbf{u} \cdot (\mathbf{v} + \mathbf{w}) = \mathbf{u} \cdot \mathbf{v} + \mathbf{u} \cdot \mathbf{w}$

(c) $(\mathbf{u} + \mathbf{v}) \cdot \mathbf{w} = \mathbf{u} \cdot \mathbf{w} + \mathbf{v} \cdot \mathbf{w}$

(d) $(k\mathbf{u}) \cdot \mathbf{v} = k(\mathbf{u} \cdot \mathbf{v}) = \mathbf{u} \cdot (k\mathbf{v})$

(e) $\mathbf{0} \cdot \mathbf{u} = 0$

(f) $\mathbf{u} \cdot \mathbf{u} = |\mathbf{u}|^2.$

To prove the first part of property (d), let $\mathbf{u} = \langle a, b \rangle$ and $\mathbf{v} = \langle c, d \rangle$. Then,

$$(k\mathbf{u}) \cdot \mathbf{v} = (k\langle a, b \rangle) \cdot \langle c, d \rangle = \langle ka, kb \rangle \cdot \langle c, d \rangle = kac + kbd$$
$$= k(ac + bd) = k(\langle a, b \rangle \cdot \langle c, d \rangle) = k(\mathbf{u} \cdot \mathbf{v}).$$

The proofs of the remaining properties are similar.

As shown in Example 7, the dot product of two vectors can be positive or 0. It may also be negative. There is a geometric interpretation of the dot product that explains when each of these cases occurs. This interpretation involves the angle between the two vectors. Consider the vectors $\mathbf{u} = \langle a, b \rangle$ and $\mathbf{v} = \langle c, d \rangle$ drawn from the origin of a rectangular coordinate system, as shown in Figure 39. The **angle θ between u and v** is defined to be the angle having the two vectors as its sides for which $0° \le \theta \le 180°$. The following theorem, which we state without proof, relates the dot product to the angle between the vectors.

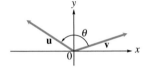

Figure 39

Geometric Interpretation of Dot Product

If θ is the angle between the two nonzero vectors **u** and **v**, where $0° \le \theta \le 180°$, then

$$\mathbf{u} \cdot \mathbf{v} = |\mathbf{u}| |\mathbf{v}| \cos \theta.$$

● ● ● **Example 8   Finding the Angle between Two Vectors**

Find the angle between the two vectors $\mathbf{u} = \langle 3, 4 \rangle$ and $\mathbf{v} = \langle 2, 1 \rangle$.

$$\cos \theta = \frac{\mathbf{u} \cdot \mathbf{v}}{|\mathbf{u}| |\mathbf{v}|} = \frac{\langle 3, 4 \rangle \cdot \langle 2, 1 \rangle}{|\langle 3, 4 \rangle| |\langle 2, 1 \rangle|}$$
$$= \frac{3(2) + 4(1)}{\sqrt{9 + 16} \, \sqrt{4 + 1}} = \frac{10}{5\sqrt{5}} \approx .894427191$$

Therefore, $\theta = \cos^{-1}(.894427191) \approx 26.57°.$ ● ● ●

For angles θ between $0°$ and $180°$, $\cos \theta$ is positive, 0, or negative when θ is less than, equal to, or greater than $90°$, respectively. Therefore, the dot product is positive, 0, or negative according to this table.

Dot Product	Angle between Vectors
Positive	Acute
0	Right
Negative	Obtuse

7.4 Exercises

1. In your own words, write a few sentences describing how a vector differs from a scalar.

2. Is a scalar product a vector or a scalar? Explain.

*Concept Check Exercises 3–6 refer to the vectors **m**–**t** at the right.*

3. Name all pairs of vectors that appear to be equal.

4. Name all pairs of vectors that are opposites.

5. Name all pairs of vectors where the first is a scalar multiple of the other, with the scalar positive.

6. Name all pairs of vectors where the first is a scalar multiple of the other, with the scalar negative.

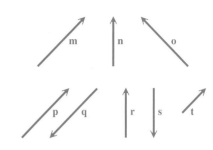

*Exercises 7–24 refer to the vectors **a**–**h** at the right. Draw a sketch to represent each vector. For example, find **a** + **e** by placing **a** and **e** so that their initial points coincide. Then use the parallelogram rule to find the resultant, as shown in the figure below.*

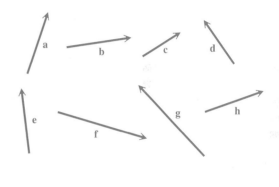

7. −**b**	**8.** −**g**	**9.** 3**a**	**10.** 2**h**	**11.** **a** + **c**
12. **a** + **b**	**13.** **h** + **g**	**14.** **e** + **f**	**15.** **a** + **h**	**16.** **b** + **d**
17. **h** + **d**	**18.** **a** + **f**	**19.** **a** − **c**	**20.** **d** − **e**	**21.** **a** + (**b** + **c**)
22. (**a** + **b**) + **c**	**23.** **c** + **d**	**24.** **d** + **c**		

25. From the results of Exercises 21 and 22, do you think vector addition is associative?

26. From the results of Exercises 23 and 24, do you think vector addition is commutative?

*For each pair of vectors **u** and **w** with angle θ between them, sketch the resultant.*

27. $|\mathbf{u}| = 12, |\mathbf{w}| = 20, \theta = 27°$ **28.** $|\mathbf{u}| = 8, |\mathbf{w}| = 12, \theta = 20°$ **29.** $|\mathbf{u}| = 20, |\mathbf{w}| = 30, \theta = 30°$

30. $|\mathbf{u}| = 27, |\mathbf{w}| = 50, \theta = 12°$ **31.** $|\mathbf{u}| = 50, |\mathbf{w}| = 70, \theta = 40°$

*For each of the following, vector **v** has the given magnitude and direction. Find the magnitudes of the horizontal and vertical components of **v**, if α is the angle of inclination of **v** from the horizontal. See Example 1.*

32. $\alpha = 20°, |\mathbf{v}| = 50$ **33.** $\alpha = 38°, |\mathbf{v}| = 12$ **34.** $\alpha = 70°, |\mathbf{v}| = 150$

35. $\alpha = 50°, |\mathbf{v}| = 26$ **36.** $\alpha = 35° \, 50', |\mathbf{v}| = 47.8$ **37.** $\alpha = 27° \, 30', |\mathbf{v}| = 15.4$

38. $\alpha = 128.5°, |\mathbf{v}| = 198$ **39.** $\alpha = 146.3°, |\mathbf{v}| = 238$

40. *Concept Check* Suppose that a calculator shows that a vector has direction angle −131°. What would be the smallest positive angle coterminal with this?

Two forces act at a point in the plane. The angle between the two forces is given. Find the magnitude of the resultant force. See Example 2.

41. forces of 250 and 450 newtons, forming an angle of 85°

42. forces of 19 and 32 newtons, forming an angle of 118°

43. forces of 17.9 and 25.8 pounds, forming an angle of 105.5°

44. forces of 75.6 and 98.2 pounds, forming an angle of 82° 50′

45. forces of 116 and 139 pounds, forming an angle of 140° 50′

46. forces of 37.8 and 53.7 pounds, forming an angle of 68.5°

Find the magnitude and direction angle (to the nearest tenth) for each vector. Give the measure of the direction angle as an angle in $[0, 360°)$. See Example 3.

47. $\langle 1, 1 \rangle$ **48.** $\langle -4, 4\sqrt{3} \rangle$ **49.** $\langle 8\sqrt{2}, -8\sqrt{2} \rangle$ **50.** $\langle \sqrt{3}, -1 \rangle$

51. $\langle 15, -8 \rangle$ **52.** $\langle -7, 24 \rangle$ **53.** $\langle -6, 0 \rangle$ **54.** $\langle 0, -12 \rangle$

*In each of the following exercises, **v** has the given direction angle and magnitude. Find the x- and y-components of **v** to three decimal places, if necessary, and write **v** as an ordered pair. See Examples 4 and 6.*

55. $\theta = 45°, |\mathbf{v}| = 20$ **56.** $\theta = 75°, |\mathbf{v}| = 100$ **57.** $\theta = 128° 30', |\mathbf{v}| = 198$

58. $\theta = 146° 10', |\mathbf{v}| = 238$ **59.** $\theta = 251° 20', |\mathbf{v}| = 69.1$ **60.** $\theta = 302° 40', |\mathbf{v}| = 7890$

Given $\mathbf{u} = \langle -2, 5 \rangle$ and $\mathbf{v} = \langle 4, 3 \rangle$, find the following. See Example 5.

61. $\mathbf{u} + \mathbf{v}$ **62.** $\mathbf{u} - \mathbf{v}$ **63.** $\mathbf{v} - \mathbf{u}$ **64.** $5\mathbf{v}$ **65.** $-5\mathbf{v}$ **66.** $3\mathbf{u} + 6\mathbf{v}$

Write each vector in the form $a\mathbf{i} + b\mathbf{j}$. Round a and b to three decimal places, if necessary. See Figure 38.

67. $\langle -5, 8 \rangle$ **68.** $\langle 6, -3 \rangle$ **69.** $\langle 2, 0 \rangle$ **70.** $\langle 0, -4 \rangle$

71. direction angle 45°, magnitude 8   **72.** direction angle 210°, magnitude 3

73. direction angle 115°, magnitude .6   **74.** direction angle 208°, magnitude .9

Find the dot product for each pair of vectors. See Example 7.

75. $\langle 6, -1 \rangle, \langle 2, 5 \rangle$ **76.** $\langle -3, 8 \rangle, \langle 7, -5 \rangle$ **77.** $\langle 2, -3 \rangle, \langle 6, 5 \rangle$

78. $\langle 1, 2 \rangle, \langle 3, -1 \rangle$ **79.** $\langle 4, 0 \rangle, \langle 5, -9 \rangle$ **80.** $\langle 2, 4 \rangle, \langle 0, -1 \rangle$

Find the angle between each pair of vectors. See Example 8.

81. $\langle 2, 1 \rangle, \langle -3, 1 \rangle$ **82.** $\langle 1, 7 \rangle, \langle 1, 1 \rangle$ **83.** $\langle 1, 2 \rangle, \langle -6, 3 \rangle$

84. $\langle 4, 0 \rangle, \langle 2, 2 \rangle$ **85.** $\langle 3, 4 \rangle, \langle 0, 1 \rangle$ **86.** $\langle -5, 12 \rangle, \langle 3, 2 \rangle$

Let $\mathbf{u} = \langle -2, 1 \rangle$, $\mathbf{v} = \langle 3, 4 \rangle$, and $\mathbf{w} = \langle -5, 12 \rangle$. Use properties of the dot product to evaluate each of the following.

87. $(3\mathbf{u}) \cdot \mathbf{v}$ **88.** $\mathbf{u} \cdot (\mathbf{v} - \mathbf{w})$ **89.** $\mathbf{u} \cdot \mathbf{v} - \mathbf{u} \cdot \mathbf{w}$ **90.** $\mathbf{u} \cdot (3\mathbf{v})$

Two vectors are said to be orthogonal *when the angle between them is a right angle. This occurs when the dot product of the vectors is 0.*

Orthogonal Vectors

Two nonzero vectors **u** and **v** are **orthogonal vectors** if and only if $\mathbf{u} \cdot \mathbf{v} = 0$.

Use this property to determine whether each pair of vectors is orthogonal.

91. $\langle 1, 2 \rangle, \langle -6, 3 \rangle$ **92.** $\langle 3, 4 \rangle, \langle 6, 8 \rangle$ **93.** $\langle 1, 0 \rangle, \langle \sqrt{2}, 0 \rangle$

94. $\langle 1, 1 \rangle, \langle 1, -1 \rangle$ **95.** $\langle \sqrt{5}, -2 \rangle, \langle -5, 2\sqrt{5} \rangle$ **96.** $\langle -4, 3 \rangle, \langle 8, -6 \rangle$

Relating Concepts

For individual or collaborative investigation
(Exercises 97–102)

Consider the two vectors **v** *and* **u** *shown. Assume all values are exact.* **Work Exercises 97–102 in order.**

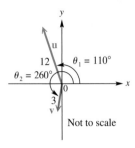

Not to scale

97. Use trigonometry alone (without using vector notation) to find the magnitude and direction angle of

u + **v**. You should use the law of cosines and the law of sines in your work.

98. Find the *x*- and *y*-components of **u**, using your calculator.

99. Find the *x*- and *y*-components of **v**, using your calculator.

100. Find the *x*- and *y*-components of **u** + **v** by adding the results you obtained in Exercises 98 and 99.

101. Use your calculator to find the magnitude and direction angle of the vector **u** + **v**.

102. Compare your answers in Exercises 97 and 101.
 (a) What do you notice?
 (b) Which method of solution do you prefer?

7.5 Applications of Vectors

• **The Equilibrant** • **Incline Applications** • **Navigation Applications**

The Equilibrant The previous section covered methods for finding the resultant of two vectors. Sometimes it is necessary to find a vector that will counterbalance the resultant. This opposite vector is called the **equilibrant;** that is, the equilibrant of vector **u** is the vector −**u**.

● ● ● **Example 1** Finding the Magnitude and Direction of an Equilibrant

Find the magnitude of the equilibrant of forces of 48 newtons and 60 newtons acting on a point *A*, if the angle between the forces is 50°. Then find the angle between the equilibrant and the 48-newton force.

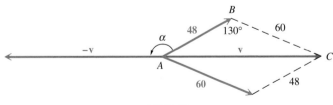

Figure 40

In Figure 40, the equilibrant is −**v**. The magnitude of **v**, and hence of −**v**, is found by using △*ABC* and the law of cosines.

$$|\mathbf{v}|^2 = 48^2 + 60^2 - 2(48)(60)\cos 130°$$
$$|\mathbf{v}|^2 \approx 9606.5$$
$$|\mathbf{v}| \approx 98 \text{ newtons} \quad \text{Two significant digits}$$

The required angle, labeled α in Figure 40, can be found by subtracting angle CAB from $180°$. Use the law of sines to find angle CAB.

$$\frac{98}{\sin 130°} = \frac{60}{\sin CAB}$$

$$\sin CAB \approx .46900680$$

$$CAB \approx 28°$$

Finally, $\qquad\qquad\qquad \alpha \approx 180° - 28° = 152°.$ ● ● ●

Incline Applications The next two examples use vectors to solve incline problems.

● ● ● **Example 2** Finding a Required Force

Find the force required to pull a 50-pound weight up a ramp inclined at $20°$ to the horizontal.

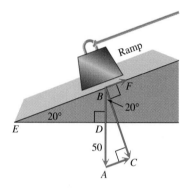

Figure 41

In Figure 41, the vertical 50-pound force **BA** represents the force of gravity. Its components are **BC** and $-$**AC**. The component **BC** represents the force with which the weight pushes against the ramp. The vector **BF** represents the force that would pull the weight up the ramp. Since vectors **BF** and **AC** are equal, $|$**AC**$|$ gives the magnitude of the required force.

Vectors **BF** and **AC** are parallel, so angle EBD equals angle A. Since angle BDE and angle C are right angles, triangles CBA and DEB have two corresponding angles equal and so are similar triangles. Therefore, angle ABC equals angle E, which is $20°$. From right triangle ABC,

$$\sin 20° = \frac{|\mathbf{AC}|}{50}$$

$$|\mathbf{AC}| = 50 \sin 20° \approx 17.$$

To the nearest pound, a 17-pound force will be required to pull the weight up the ramp. ● ● ●

● ● ● **Example 3** Finding an Incline Angle

A force of 16 pounds is required to hold a 40-pound lawn mower on an incline. What angle does the incline make with the horizontal?

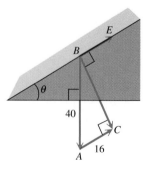

Figure 42

Figure 42 illustrates the situation. Consider right triangle ABC. Angle $B =$ angle θ, the magnitude of vector **BA** represents the weight of the mower, and vector **AC** equals vector **BE**, which represents the force required to hold the mower on the incline. From the figure,

$$\sin B = \frac{16}{40}$$

$$\sin B = .4$$

$$B \approx 23.5782°.$$

Therefore, the hill makes an angle of about $24°$ with the horizontal. ● ● ●

Navigation Applications Problems involving bearing (defined in Section 2.5) can also be worked with vectors.

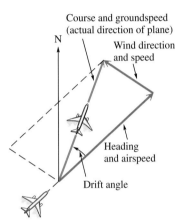

Figure 43

● ● ● **Example 4** Applying Vectors to a Navigation Problem

A ship leaves port on a bearing of 28° and travels 8.2 miles. The ship then turns due east and travels 4.3 miles. How far is the ship from port? What is its bearing from port?

In Figure 43, vectors **PA** and **AE** represent the ship's path. The magnitude and bearing of the resultant **PE** can be found as follows. Triangle *PNA* is a right triangle, so angle $NAP = 90° - 28° = 62°$. Then angle $PAE = 180° - 62° = 118°$. Use the law of cosines to find $|\mathbf{PE}|$, the magnitude of vector **PE**.

$$|\mathbf{PE}|^2 = 8.2^2 + 4.3^2 - 2(8.2)(4.3)\cos 118°$$
$$|\mathbf{PE}|^2 \approx 118.84$$

Therefore, $|\mathbf{PE}| \approx 10.9$,

or 11 miles, rounded to two significant digits.

To find the bearing of the ship from port, first find angle *APE*. Use the law of sines, along with the value of $|\mathbf{PE}|$, before rounding.

$$\frac{\sin APE}{4.3} = \frac{\sin 118°}{10.9}$$

$$\sin APE = \frac{4.3 \sin 118°}{10.9}$$

$$\text{angle } APE \approx 20.4°$$

After rounding, angle *APE* is 20°, so the ship is 11 miles from port on a bearing of $28° + 20° = 48°$.   ● ● ●

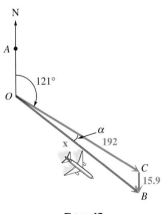

Figure 44

In air navigation, the **airspeed** of a plane is its speed relative to the air, while the **groundspeed** is its speed relative to the ground. Because of wind, these two speeds are usually different. The groundspeed of the plane is represented by the vector sum of the airspeed and windspeed vectors. See Figure 44.

● ● ● **Example 5** Applying Vectors to a Navigation Problem

A plane with an airspeed of 192 mph is headed on a bearing of 121°. A north wind is blowing (from north to south) at 15.9 mph. Find the groundspeed and the actual bearing of the plane.

In Figure 45, the groundspeed is represented by $|\mathbf{x}|$. We must find angle α to determine the bearing, which will be $121° + \alpha$. From Figure 45, angle *BCO* equals angle *AOC*, which equals 121°. Find $|\mathbf{x}|$ by the law of cosines.

$$|\mathbf{x}|^2 = 192^2 + 15.9^2 - 2(192)(15.9)\cos 121°$$
$$|\mathbf{x}|^2 \approx 40{,}261$$

Therefore, $|\mathbf{x}| \approx 200.7$,

or 201 mph. Now find α by using the law of sines. As before, use the value of $|\mathbf{x}|$ before rounding.

$$\frac{\sin \alpha}{15.9} = \frac{\sin 121°}{200.7}$$

$$\sin \alpha \approx .0679$$

$$\alpha \approx 3.89°$$

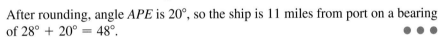

After rounding, α is 3.9°. The groundspeed is about 201 mph, on a bearing of $121° + 3.89° \approx 125°$, to three significant digits.   ● ● ●

7.5 Exercises

Solve each problem. See Examples 1–5.

1. *Angle between Forces* Two forces of 692 newtons and 423 newtons act at a point. The resultant force is 786 newtons. Find the angle between the forces.

2. *Angle between Forces* Two forces of 128 pounds and 253 pounds act at a point. The equilibrant is 320 pounds. Find the angle between the forces.

3. *Angle of a Hill Slope* A force of 25 pounds is required to push an 80-pound crate up a hill. What angle does the hill make with the horizontal?

4. *Force Needed to Keep a Car Parked* Find the force required to keep a 3000-pound car parked on a hill that makes an angle of 15° with the horizontal.

5. *Force Needed to Pull a Monolith* To build the pyramids in Egypt, it is believed that giant causeways were built to transport the building materials to the site. One such causeway is said to have been 3000 feet long, with a slope of about 2.3°. How much force would be required to pull a 60-ton monolith along this causeway?

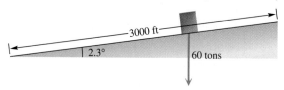

6. *Weight of a Boat* A force of 500 pounds is required to pull a boat up a ramp inclined at 18° with the horizontal. How much does the boat weigh?

7. *Weight of a Box* Two people are carrying a box. One person exerts a force of 150 pounds at an angle of 62.4° with the horizontal. The other person exerts a force of 114 pounds at an angle of 54.9°. Find the weight of the box.

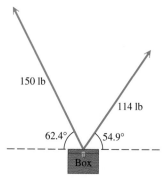

8. *Direction and Magnitude of an Equilibrant* Two tugboats are pulling a disabled speedboat into port with forces of 1240 pounds and 1480 pounds. The angle between these forces is 28.2°. Find the direction and magnitude of the equilibrant.

9. *Weight of a Crate and Tension of a Rope* A crate is supported by two ropes. One rope makes an angle of 46° 20′ with the horizontal and has a tension of 89.6 pounds on it. The other rope is horizontal. Find the weight of the crate and the tension in the horizontal rope.

10. *Angles between Forces* Three forces acting at a point are in equilibrium. The forces are 980 pounds, 760 pounds, and 1220 pounds. Find the angles between the directions of the forces. (*Hint:* Arrange the forces to form the sides of a triangle.)

11. *Magnitudes of Forces* A force of 176 pounds makes an angle of 78° 50′ with a second force. The resultant of the two forces makes an angle of 41° 10′ with the first force. Find the magnitudes of the second force and of the resultant.

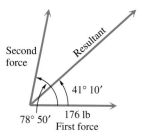

12. *Magnitudes of Forces* A force of 28.7 pounds makes an angle of 42° 10′ with a second force. The resultant of the two forces makes an angle of 32° 40′ with the first force. Find the magnitudes of the second force and of the resultant.

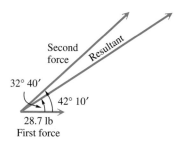

13. *Course of a Plane* A plane flies 650 mph on a bearing of 175.3°. A 25-mph wind, from a direction of 266.6°, blows against the plane. Find the resulting course of the plane.

14. *Airspeed and Groundspeed* A pilot wants to fly on a bearing of 74.9°. By flying due east, he finds that a 42-mph wind, blowing from the south, puts him on course. Find the airspeed and the groundspeed.

15. *Distance of Ship from Its Starting Point* Starting at point *A*, a ship sails 18.5 km on a bearing of 189°, then turns and sails 47.8 km on a bearing of 317°. Find the distance of the ship from point *A*.

16. *Bearing of One Point from Another* Two towns 21 miles apart are separated by a dense forest. To travel from town *A* to town *B*, a person must go 17 miles on a bearing of 325°, then turn and continue for 9 miles to reach town *B*. Find the bearing of *B* from *A*.

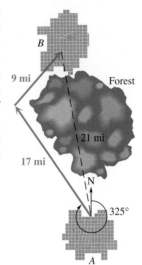

17. *Bearing and Groundspeed of a Plane* An airline route from San Francisco to Honolulu is on a bearing of 233°. A jet flying at 450 mph on that bearing flies into a wind blowing at 39 mph from a direction of 114°. Find the resulting bearing and groundspeed of the plane.

18. *Bearing and Groundspeed of a Plane* A pilot is flying at 168 mph. She wants her flight path to be on a bearing of 57° 40′. A wind is blowing from the south at 27.1 mph. Find the bearing the pilot should fly, and find the plane's groundspeed.

19. *Bearing and Airspeed of a Plane* What bearing and airspeed are required for a plane to fly 400 miles due north in 2.5 hours if the wind is blowing from a direction of 328° at 11 mph?

20. *Groundspeed and Bearing of a Plane* A plane is headed due south with an airspeed of 192 mph. A wind from a direction of 78° is blowing at 23 mph. Find the groundspeed and resulting bearing of the plane.

21. *Groundspeed and Bearing of a Plane* An airplane is headed on a bearing of 174° at an airspeed of 240 km per hour. A 30 km per hour wind is blowing from a direction of 245°. Find the groundspeed and resulting bearing of the plane.

22. *Distance Traveled by a Ship* A ship sailing due east in the North Atlantic has been warned to change course to avoid a group of icebergs. The captain turns and sails on a bearing of 62° for a while, then changes course again to a bearing of 115° until the ship reaches its original course. See the figure. How much farther did the ship have to travel to avoid the icebergs?

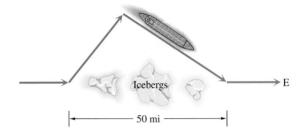

23. *Path Traveled by a Plane* The aircraft carrier *Tallahassee* is traveling at sea on a steady course with a bearing of 30° at 32 mph. Patrol planes on the carrier have enough fuel for 2.6 hours of flight when traveling at a speed of 520 mph. One of the pilots takes off on a bearing of 338° and then turns and heads in a straight line, so as to be able to catch the carrier and land on the deck at the exact instant that his fuel runs out. If the pilot left at 2 P.M., at what time did he turn to head for the carrier?

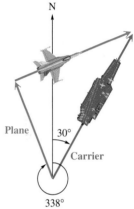

24. *Movement of a Motorboat* Suppose you would like to cross a 132-foot wide river in a motorboat. Assume that the motorboat can travel at 7 mph relative to the water and that the current is flowing west at the rate of 3 mph. The bearing θ is chosen so that the motorboat will land at a point exactly across from the starting point.

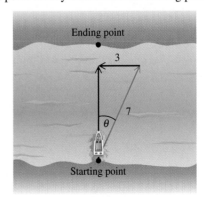

(a) At what speed will the motorboat be traveling relative to the banks?

(b) How long will it take for the motorboat to make the crossing?

(c) What is the measure of angle θ?

25. *Velocity of a Star* The space velocity **v** of a star relative to the sun can be expressed as the resultant vector of two perpendicular vectors—the radial velocity $\mathbf{v}_r$ and the tangential velocity $\mathbf{v}_t$ where $\mathbf{v} = \mathbf{v}_r + \mathbf{v}_t$. If a star is located near the sun and its space velocity is large, then its motion across the sky will also be large. Barnard's Star is a relatively close star with a distance of 35 trillion miles from the sun. It moves across the sky through an angle of $10.34''$ per year, which is the largest motion of any known star. Its radial velocity is $\mathbf{v}_r = 67$ miles per second toward the sun. (*Sources:* Zeilik, M., S. Gregory, and E. Smith, *Introductory Astronomy and Astrophysics,* Second Edition, Saunders College Publishing, 1998; Acker, A. and C. Jaschek, *Astronomical Methods and Calculations,* John Wiley & Sons, 1986.)

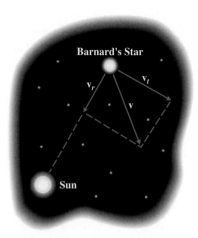

Not to scale

(a) Approximate the tangential velocity $\mathbf{v}_t$ of Barnard's Star. (*Hint:* Use the arc length formula $s = r\theta$.)

(b) Compute the magnitude of **v**.

· · · · · · · · · · · · **Relating Concepts** · · · · · · · · · · · · ·

For individual or collaborative investigation

(Exercises 26–30)

Reading an Electrocardiogram *When reading an electrocardiogram, a cardiologist measures the heights and directions of certain peaks that appear. Also, depending on where the electrodes are attached to the patient, a certain angle is associated with the reading. If we call the measures of the peaks* **a** *and* **b** *and the angle θ, then* **a**, **b**, *and θ are related as shown in the figure. The cardiologist needs to know the length and direction of vector* **v** *in the figure. The following equations give these values.*

$$|\mathbf{v}| = \frac{\sqrt{a^2 + b^2 - 2ab\cos\theta}}{\sin\theta} \qquad \alpha = \cos^{-1}\frac{a}{|\mathbf{v}|}$$

(Do not assume that line L is perpendicular to **b**.*)*

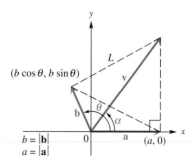

Work Exercises 26–30 in order.

26. As shown in Exercises 78–81 in Section 5.4, the slope of a line equals the tangent of its angle of inclination. Use this fact and the information given in the figure to write the equation of line L.

27. Use the result from Exercise 26 to find the coordinates of the endpoint of **v**.

28. Use the distance formula to find the magnitude of **v**. Rewrite the answer so that all trigonometric functions are expressed in terms of sine and cosine. This result should be the first formula given at the left.

29. What line in the figure corresponds to the quantity under the radical in the numerator of the expression for the magnitude of **v**?

30. Explain how to get the formula given at the left for α.

Chapter 7 **Summary**

Key Terms & Symbols	Key Ideas
7.1 Oblique Triangles and the Law of Sines Side-Angle-Side (SAS) Angle-Side-Angle (ASA) Side-Side-Side (SSS) oblique triangle	**Law of Sines** In any $\triangle ABC$, with sides a, b, and c, $$\frac{a}{\sin A} = \frac{b}{\sin B}, \qquad \frac{a}{\sin A} = \frac{c}{\sin C}, \qquad \text{and} \qquad \frac{b}{\sin B} = \frac{c}{\sin C}.$$ **Area of a Triangle** The area of a triangle is given by half the product of the lengths of two sides and the sine of the angle between the two sides. $$\mathcal{A} = \frac{1}{2}bc \sin A, \qquad \mathcal{A} = \frac{1}{2}ab \sin C, \qquad \mathcal{A} = \frac{1}{2}ac \sin B$$
7.2 The Ambiguous Case of the Law of Sines	Given A, a, and b in $\triangle ABC$. If A is acute, h is the altitude from C, and 1. $a < h$, then there is no triangle. 2. $a = h$, then there is one triangle (a right triangle). 3. $a \geq b$, then there is one triangle. 4. $b > a > h$, then there are two triangles. If A is obtuse and 1. $a \leq b$, then there is no triangle. 2. $a > b$, then there is one triangle.
7.3 The Law of Cosines	**Law of Cosines** In any $\triangle ABC$, with sides a, b, and c, $$a^2 = b^2 + c^2 - 2bc \cos A$$ $$b^2 = a^2 + c^2 - 2ac \cos B$$ $$c^2 = a^2 + b^2 - 2ab \cos C.$$ **Heron's Area Formula** If a triangle has sides of lengths a, b, and c, and if the semiperimeter is $$s = \frac{1}{2}(a + b + c),$$ then the area of the triangle is $$\mathcal{A} = \sqrt{s(s - a)(s - b)(s - c)}.$$
7.4 Vectors and the Dot Product **7.5 Applications of Vectors** scalars vector quantities vector **OP** or $\overrightarrow{OP}$ magnitude $\lvert \mathbf{OP} \rvert$ initial point terminal point parallelogram rule resultant component	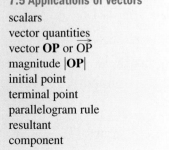 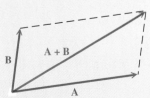 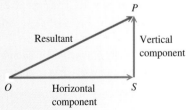 **Magnitude and Direction Angle of a Vector** The magnitude (length) of vector $\mathbf{u} = \langle a, b \rangle$ is given by $\lvert \mathbf{u} \rvert = \sqrt{a^2 + b^2}$. The direction angle θ satisfies $\tan \theta = b/a$, where $a \neq 0$.

Key Terms & Symbols	**Key Ideas**
vertical and horizontal components opposite (of a vector) zero vector scalar product position vector $\langle a, b \rangle$ x- and y-components direction angle unit vectors $\mathbf{i}, \mathbf{j}$ dot product equilibrant airspeed groundspeed	**Vector Operations** For any real numbers a, b, c, d, and k, $$\langle a, b \rangle + \langle c, d \rangle = \langle a + c, b + d \rangle$$ $$k \cdot \langle a, b \rangle = \langle ka, kb \rangle.$$ If $\mathbf{a} = \langle a_1, a_2 \rangle$, then $-\mathbf{a} = \langle -a_1, -a_2 \rangle$. $$\langle a, b \rangle - \langle c, d \rangle = \langle a, b \rangle + -\langle c, d \rangle$$ If $\mathbf{u} = \langle x, y \rangle$ has direction angle θ, then $\mathbf{u} = \langle |\mathbf{u}| \cos \theta, |\mathbf{u}| \sin \theta \rangle$. **i, j Form for Vectors** If $\mathbf{v} = \langle a, b \rangle$, then $\mathbf{v} = a\mathbf{i} + b\mathbf{j}$, where $\mathbf{i} = \langle 1, 0 \rangle$ and $\mathbf{j} = \langle 0, 1 \rangle$. **Dot Product** The dot product of the two vectors $\mathbf{u} = \langle a, b \rangle$ and $\mathbf{v} = \langle c, d \rangle$, denoted $\mathbf{u} \cdot \mathbf{v}$, is given by $$\mathbf{u} \cdot \mathbf{v} = ac + bd.$$ If θ is the angle between $\mathbf{u}$ and $\mathbf{v}$, where $0° \le \theta \le 180°$, then $$\mathbf{u} \cdot \mathbf{v} = |\mathbf{u}| \, |\mathbf{v}| \cos \theta.$$

Chapter 7 Review Exercises

Use the law of sines to find the indicated part of each $\triangle ABC$.

1. $C = 74.2°, c = 96.3$ m, $B = 39.5°$; find b

2. $A = 129.7°, a = 127$ ft, $b = 69.8$ ft; find B

3. $C = 51.3°, c = 68.3$ m, $b = 58.2$ m; find B

4. $a = 165$ m, $A = 100.2°, B = 25.0°$; find b

5. $B = 39° \, 50', b = 268$ m, $a = 340$ m; find A

6. $C = 79° \, 20', c = 97.4$ mm, $a = 75.3$ mm; find A

7. If we are given a, A, and C in a $\triangle ABC$, does the possibility of the ambiguous case exist? If not, explain why.

8. Can $\triangle ABC$ exist if $a = 4.7$, $b = 2.3$, and $c = 7.0$? If not, explain why. Answer this question without using trigonometry.

9. Given $a = 10$ and $B = 30°$, determine the values of b for which A has
    (a) exactly one value   **(b)** two values   **(c)** no value.

10. Given $a = 10$ and $B = 150°$, determine the values of b for which A has
    (a) exactly one value   **(b)** two values   **(c)** no value.

Use the law of cosines to find the indicated part of each $\triangle ABC$.

11. $a = 86.14$ in., $b = 253.2$ in., $c = 241.9$ in.; find A

12. $B = 120.7°, a = 127$ ft, $c = 69.8$ ft; find b

13. $A = 51° \, 20', c = 68.3$ m, $b = 58.2$ m; find a

14. $a = 14.8$ m, $b = 19.7$ m, $c = 31.8$ m; find B

15. $A = 60°, b = 5$ cm, $c = 21$ cm; find a

16. $a = 13$ ft, $b = 17$ ft, $c = 8$ ft; find A

Solve each $\triangle ABC$ having the given information.

17. $A = 25.2°, a = 6.92$ yd, $b = 4.82$ yd

18. $A = 61.7°, a = 78.9$ m, $b = 86.4$ m

19. $a = 27.6$ cm, $b = 19.8$ cm, $C = 42° \, 30'$

20. $a = 94.6$ yd, $b = 123$ yd, $c = 109$ yd

Find the area of each $\triangle ABC$ with the given information.

21. $b = 840.6$ m, $c = 715.9$ m, $A = 149.3°$

22. $a = 6.90$ ft, $b = 10.2$ ft, $C = 35° \, 10'$

23. $a = .913$ km, $b = .816$ km, $c = .582$ km

24. $a = 43$ m, $b = 32$ m, $c = 51$ m

The following identities involve all six parts of a △ABC and are thus useful for checking answers.

Newton's formula $$\dfrac{a+b}{c} = \dfrac{\cos\frac{1}{2}(A-B)}{\sin\frac{1}{2}C}$$

Mollweide's formula $$\dfrac{a-b}{c} = \dfrac{\sin\frac{1}{2}(A-B)}{\cos\frac{1}{2}C}$$

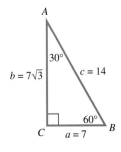

25. Apply Newton's formula to the triangle shown in the figure to verify the accuracy of the information.

26. Apply Mollweide's formula to the triangle shown in the figure to verify the accuracy of the information.

Solve each problem.

27. *Distance across a Canyon*   To measure the distance AB across a canyon for a power line, a surveyor measures angles B and C and the distance BC, as shown in the figure. What is the distance from A to B?

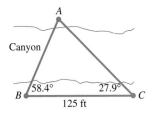

28. *Length of a Brace*   A banner on an 8.0-foot pole is to be mounted on a building at an angle of 115°, as shown in the figure. Find the length of the brace.

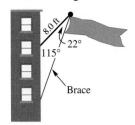

29. *Height of a Tree*   A tree leans at an angle of 8.0° from the vertical. From a point 7.0 meters from the bottom of the tree, the angle of elevation to the top of the tree is 68°. How tall is the tree?

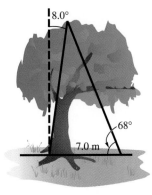

30. *Length of a Tunnel*   To measure the distance through a mountain for a proposed tunnel, a point C is chosen that can be reached from each end of the tunnel. If $AC = 3800$ meters, $BC = 2900$ meters, and angle $C = 110°$, find the length of the tunnel.

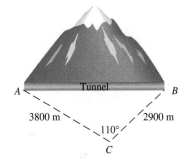

31. *Height of a Tree*   A hill makes an angle of 14.3° with the horizontal. From the base of the hill, the angle of elevation to the top of a tree on top of the hill is 27.2°. The distance along the hill from the base to the tree is 212 feet. Find the height of the tree.

32. *Distance from a Ship to a Rock*   A ship is sailing east. At one point, the bearing of a submerged rock is 45° 20′. After sailing 15.2 miles, the bearing of the rock has become 308° 40′. Find the distance of the ship from the rock at the latter point.

33. *Distance between Two Boats*   Two boats leave a dock together. Each travels in a straight line. The angle between their courses measures 54° 10′. One boat travels 36.2 km per hour, and the other travels 45.6 km per hour. How far apart will they be after 3 hours?

34. *Diagonals of a Parallelogram*   Find the lengths of both diagonals of a parallelogram with adjacent sides of 12 cm and 15 cm if the angle between these sides is 33°.

35. *Distance between a Battleship and a Submarine* From an airplane flying over the ocean, the angle of depression to a submarine lying just under the surface is 24° 10′. At the same moment the angle of depression from the airplane to a battleship is 17° 30′. The distance from the airplane to the battleship is 5120 feet. Find the distance between the battleship and the submarine. (Assume the airplane, submarine, and battleship are in a vertical plane.)

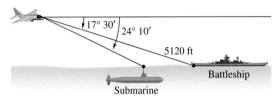

36. *Distance from a Ship to a Lighthouse* A ship sailing parallel to shore sights a lighthouse at an angle of 30° from its direction of travel. After the ship travels 2.0 miles farther, the angle has increased to 55°. At that time, how far is the ship from the lighthouse?

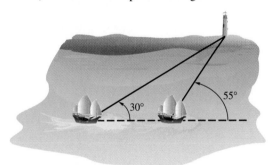

37. *Distances on a Baseball Diamond* A baseball diamond is a square, 90.0 feet on a side, with home plate and the three bases as vertices. The pitcher's rubber is located 60.5 feet from home plate. Find the distance from the pitcher's rubber to each of the bases.

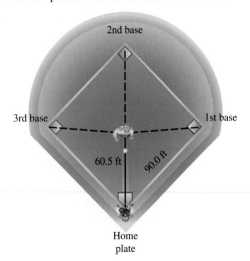

38. *Vietnam Veterans' Memorial* The Vietnam Veterans' Memorial in Washington, D.C., is in the shape of an unenclosed isosceles triangle (that is, V-shaped) with equal sides of length 246.75 feet and the angle between these sides measuring 125° 12′. Find the distance between the ends of the two equal sides. (*Source:* Information pamphlet obtained at the Vietnam Veterans' Memorial.)

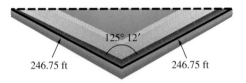

39. *Area of a Triangle* Find the area of the triangle shown in the figure using Heron's area formula.

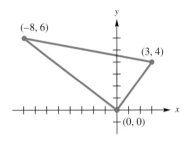

40. *Area of a Quadrilateral* A lot has the shape of the quadrilateral in the figure. What is its area?

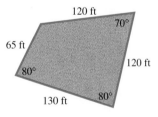

41. *Amount of Paint Required* Raoul plans to paint a triangular wall in his A-frame cabin. Two sides measure 7 meters each, and the third side measures 6 meters. How much paint will he need to buy if a can of paint covers 7.5 square meters?

42. *Concept Check* If angle C of a $\triangle ABC$ measures 90°, what does the law of cosines $c^2 = a^2 + b^2 - 2ab \cos C$ become?

In Exercises 43–45, use the vectors pictured here. Sketch the following.

43. $\mathbf{a} + \mathbf{b}$

44. $\mathbf{a} - \mathbf{b}$

45. $\mathbf{a} + 3\mathbf{c}$

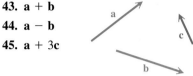

46. *Concept Check* Decide whether each statement is true or false.
  (a) Opposite angles of a parallelogram are equal.
  (b) A diagonal of a parallelogram must bisect two angles of the parallelogram.

Given two forces and the angle between them, find the magnitude of the resultant force.

47. forces of 15 and 23 pounds, forming an angle of 87°

48. forces of 142 and 215 newtons, forming an angle of 112°

49. forces of 475 and 586 pounds, forming an angle of 78° 20′

Vector $\mathbf{v}$ has the given magnitude and direction angle. Find the magnitudes of the horizontal and vertical components of $\mathbf{v}$.

50. $|\mathbf{v}| = 50, \theta = 45°$
(Give exact values.)

51. $|\mathbf{v}| = 69.2, \theta = 75°$

52. $|\mathbf{v}| = 964, \theta = 154° \, 20′$

Find the magnitude and direction angle for $\mathbf{u}$ rounded to the nearest tenth.

53. $\mathbf{u} = \langle 21, -20 \rangle$

54. $\mathbf{u} = \langle -9, 12 \rangle$

Find (a) the dot product and (b) the angle between each pair of vectors.

55. $\mathbf{u} = \langle 6, 2 \rangle, \mathbf{v} = \langle 3, -2 \rangle$

56. $\mathbf{u} = \langle 2\sqrt{3}, 2 \rangle, \mathbf{v} = \langle 5, 5\sqrt{3} \rangle$

Find the vector of magnitude 1 having the same direction angle as the given vector.

57. $\mathbf{u} = \langle -4, 3 \rangle$

58. $\mathbf{u} = \langle 5, 12 \rangle$

Solve each problem.

59. *Force Placed on a Barge* One rope pulls a barge directly east with a force of 100 newtons. Another rope pulls the barge to the northeast with a force of 200 newtons. Find the resultant force acting on the barge and the angle between the resultant and the first rope.

60. *Weight of a Sled and Passenger* Paula and Steve are pulling their daughter Jessie on a sled. Steve pulls with a force of 18 pounds at an angle of 10°. Paula pulls with a force of 12 pounds at an angle of 15°. Find the magnitude of the resultant force on Jessie and the sled.

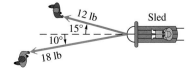

61. *Angle of a Hill* A 186-pound force just keeps a 2800-pound car from rolling down a hill. What angle does the hill make with the horizontal?

62. *Direction and Speed of a Plane* A plane has an airspeed of 520 mph. The pilot wishes to fly on a bearing of 310°. A wind of 37 mph is blowing from a bearing of 212°. What direction should the pilot fly, and what will be her actual speed?

63. *Speed and Direction of a Boat* A boat travels 15 km per hour in still water. The boat is traveling across a large river, on a bearing of 130°. The current in the river, coming from the west, has a speed of 7 km per hour. Find the resulting speed of the boat and its resulting direction of travel.

64. *Car Banking a Curve*   A car going around a banked curve is subject to the forces shown in the figure. If the radius of the curve is 100 feet, what value of θ to the nearest degree would allow an automobile to travel around the curve at a speed of 40 feet per second without depending on friction?

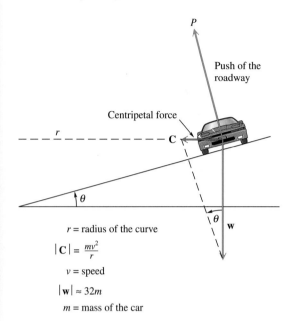

Push of the roadway

Centripetal force

C

θ

θ

w

r = radius of the curve

$|\mathbf{C}| = \dfrac{mv^2}{r}$

v = speed

$|\mathbf{w}| \approx 32m$

m = mass of the car

65. *Control Points*   To obtain accurate aerial photographs, ground control must determine the coordinates of *control points* located on the ground that can be identified in the photographs. Using these known control points, the orientation and scale of each photograph can be found. Then, unknown positions and distances can easily be determined. Before an aerial photograph is taken for highway design, horizontal control points must be located and the distance between them calculated. The figure shows three consecutive control points *A*, *B*, and *C*.

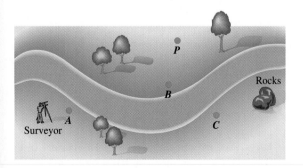

P

B

Rocks

Surveyor   A   C

A surveyor measures a baseline distance of 92.13 feet from *B* to an arbitrary point *P*. Angles *BAP* and *BCP* are found to be 2° 22′ 47″ and 5° 13′ 11″, respectively. Then, angles *APB* and *CPB* are determined to be 63° 4′ 25″ and 74° 19′ 49″, respectively. Determine the distance between control points *A* and *B* and between *B* and *C*. (*Source:* Moffitt, F. and E. Mikhail, *Photogrammetry,* Third Edition, Harper & Row, 1980.)

66. *State Plane Coordinates*   To find the coordinates of control points for aerial photography, ground control must first locate basic control monuments established by the U.S. Coast and Geodetic Survey and the U.S. Geological Survey. These monuments have published *x*- and *y*-coordinates called *state plane coordinates*. Using these monuments and common surveying techniques, coordinates of control points can be determined. Two basic control monuments *A* and *B* have coordinates in feet of $x_A = 2{,}101{,}345.1$, $y_A = 998{,}764.3$ and $x_B = 2{,}131{,}667.8$, $y_B = 923{,}541.7$. The location of an unknown control point *P* is to be determined. If angles *PAB* and *PBA* have measures 37° 41′ 37″ and 57° 52′ 4″, respectively, discuss the steps you would take to determine the state plane coordinates of control point *P*. (*Source:* Moffitt, F. and E. Mikhail, *Photogrammetry,* Third Edition, Harper & Row, 1980.)

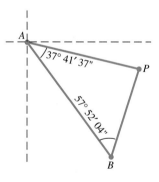

A

37° 41′ 37″

P

57° 52′ 04″

B

Chapter 7   Test

Find the indicated part of each △ABC.

1. $A = 25.2°, a = 6.92$ yd, $b = 4.82$ yd;   find C
2. $C = 118°, b = 130$ km, $a = 75$ km;   find c
3. $a = 17.3$ ft, $b = 22.6$ ft, $c = 29.8$ ft;   find B
4. Find the area of △ABC in Exercise 2.
5. Given $a = 10$ and $B = 150°$ in △ABC, determine the values of b for which A has
   (a) exactly one value
   (b) two values
   (c) no value.
6. Find the area of the triangle having sides of lengths 22, 26, and 40.
7. What conditions determine whether or not three positive numbers can represent the lengths of the sides of a triangle?

8. Find the magnitude and the direction angle for the vector shown in the figure.

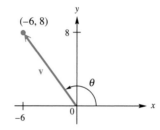

For the vectors $\mathbf{u} = \langle -1, 3 \rangle$ *and* $\mathbf{v} = \langle 2, -6 \rangle$, *find each of the following.*

9. $\mathbf{u} + \mathbf{v}$ 10. $-3\mathbf{v}$ 11. $\mathbf{u} \cdot \mathbf{v}$

Solve each problem.

12. *Height of a Balloon*   The angles of elevation of a balloon from two points A and B on level ground are 24° 50′ and 47° 20′, respectively. As shown in the figure, points A and B are in the same vertical plane and are 8.4 miles apart. Approximate the height of the balloon above the ground to the nearest tenth of a mile.

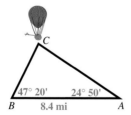

13. *Distances on a Softball Field*   The pitcher's mound on a regulation men's softball field is 46 feet from home plate. The distance between the bases is 60 feet, as shown in the figure. (*Source:* Microsoft Encarta.) How far is the pitcher's mound at point M from third base (point T)? Give your answer to the nearest foot.

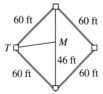

14. *Horizontal and Vertical Components*   Find the horizontal and vertical components of the vector with magnitude 569 that is inclined 127.5° from the horizontal. Give your answer in the form $\langle a, b \rangle$.

15. *Radio Direction Finders*   Radio direction finders are placed at points A and B, which are 3.46 miles apart on an east-west line, with A west of B. From A, the bearing of a certain illegal pirate radio transmitter is 48°, and from B the bearing is 302°. Find the distance between the transmitter and A to the nearest hundredth of a mile.

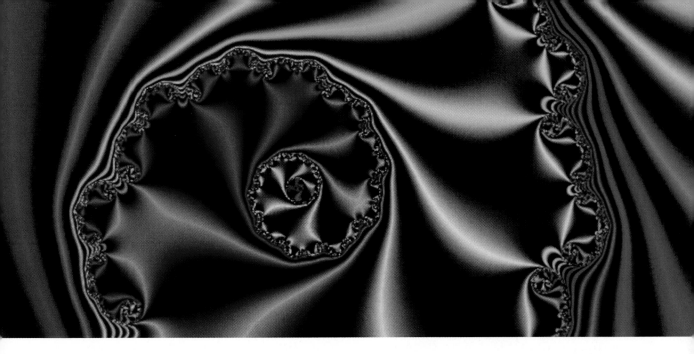

8 Complex Numbers, Polar Equations, and Parametric Equations

8.1   **Complex Numbers**

8.2   **Trigonometric (Polar) Form of Complex Numbers**

8.3   **The Product and Quotient Theorems**

8.4   **Powers and Roots of Complex Numbers**

8.5   **Polar Equations and Graphs**

8.6   **Parametric Equations, Graphs, and Applications**

High-resolution computer graphics and *complex numbers* (introduced in this chapter) make it possible to produce beautiful shapes called *fractals*. Benoit B. Mandelbrot first used the term *fractal* in 1975. At its basic level, a fractal is a unique, enchanting geometric figure with an endless self-similarity property. A fractal image repeats itself infinitely with ever-decreasing dimensions. Although most current applications of fractals are related to creating fascinating images and pictures, fractals do have a tremendous potential in applied science. The example of a fractal shown in the figure on the next page is an amazing graphical solution to a difficult problem first presented by Sir Arthur Cayley in 1879. This fractal, called *Newton's basins of attraction for the cube roots of unity,* is discussed in the exercise for Section 8.4. Other fractals, the theme of this chapter, are presented in the examples and exercises.*

Sources: Crownover, R., *Introduction to Fractals and Chaos,* Jones and Bartlett Publishers, 1995.

Kline, M., *Mathematics: The Loss of Certainty,* Oxford University Press, 1980.

Lauwerier, H., *Fractals,* Princeton University Press, 1991.

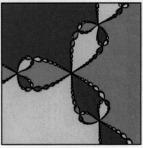

Source: Kincaid, D. and Cheney, W., *Numerical Analysis*, Brooks/Cole Publishing Co., 1991.

8.1 Complex Numbers

| • Basic Terminology and Definitions | • Complex Solutions of Equations | • Operations on Complex Numbers |
| • Powers of i | | |

Basic Terminology and Definitions The complex number system involves a number that is new to us, the imaginary unit i, defined as follows.

> ### The Imaginary Unit i
>
> $$i = \sqrt{-1} \quad \text{or} \quad i^2 = -1$$

Looking Ahead to Calculus
The letters j and k are also used to represent $\sqrt{-1}$ in calculus and some applications (electronics, for example).

A **complex number** is a number that has the form $a + bi$, where a and b are real numbers. The form $a + bi$ is called the **rectangular** (or **standard**) **form** of the complex number. The real number a is called the **real part,** and the real number b is called the **imaginary part.*** If $b \neq 0$, $a + bi$ is also called an **imaginary number.** Each real number is a complex number since a real number a may be thought of as the complex number $a + 0i$. The set of real numbers is a subset of the set of complex numbers. See Figure 1.

Complex Numbers

Imaginary numbers	Rational numbers	Irrational numbers
$8 - i$	$\frac{4}{9}, -\frac{5}{8}, \frac{11}{7}$	$-\sqrt{8}$
$3 - i\sqrt{2}$	Integers $-11, -6, -4$	$\sqrt{15}$
$4i$	Whole numbers	$\sqrt{23}$
$-11i$	0	π
$i\sqrt{7}$	Natural numbers	$\frac{\pi}{4}$
$1 + \pi i$	1, 2, 3, 4, 5, 37, 50	

The figure shows several complex numbers, each with its type and rectangular form. Real numbers are shaded.

Figure 1

---

*Some texts define bi as the imaginary part.

N O T E The form $a + ib$ is often used for symbols such as $i\sqrt{5}$, since $\sqrt{5}i$ could be too easily mistaken for $\sqrt{5i}$.

C O N N E C T I O N S A brief discussion on the development of complex numbers can be found in the *Thirty-first Yearbook of the National Council of Teachers of Mathematics, Historical Topics for the Mathematics Classroom* (1969). Eugene W. Hellmich writes:

Perhaps the earliest encounter with the square root of a negative number is in the expression $\sqrt{81 - 144}$, which appears in the *Stereometrica* of Heron of Alexandria (*c.* A.D. 50); the next known encounter is in Diophantus' attempt to solve the equation $336x^2 + 24 = 172x$ (as we would now write it), in whose solution the quantity $\sqrt{1,849 - 2,016}$ appears (again using modern notation).

The first clear statement of difficulty with the square root of a negative number was given in India by Mahavira (*c.* 850), who wrote: "As in the nature of things, a negative is not a square, it has no square root." Nicolas Chuquet (1484) and Luca Pacioli (1494) in Europe were among those who continued to reject imaginaries.

Girolamo Cardano (1545), who is also known as Jerome Cardan, is credited with some progress in introducing complex numbers in his solution of the cubic equation, even though he regarded them as "fictitious." He is credited also with the first use of the square root of a negative number in solving the now-famous problem, "Divide 10 into two parts such that the product . . . is 40," which Cardano first says is "manifestly impossible"; but then he goes on to say, in a properly adventurous spirit, "Nevertheless, we will operate." (This was due, no doubt, to his medical training!) Thus he found $5 + \sqrt{-15}$ and $5 - \sqrt{-15}$ and showed that they did indeed have a sum of 10 and a product of 40.

Cardano concludes by saying that these quantities are "truly sophisticated" and that to continue working with them would be "as subtle as it would be useless."

While at first there may seem to be no meaningful applications of numbers like $\sqrt{-1}$, they play an important role in many new and exciting fields of applied mathematics and technology. Their development has enabled mathematicians and scientists to solve new problems about the designs of airplane wings, ships, electrical circuits, noise control, and the theme of this chapter, fractals.

For Discussion or Writing

1. Why might a beginning algebra student have difficulty with the concept of $\sqrt{-1}$?
2. Since by definition $i^2 = -1$, predict what i^4 should equal. How about i^3?

The square root of a negative number can be written as the product of a real number and i, using the definition of $\sqrt{-a}$ that follows.

Complex number mode

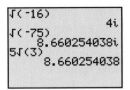

8.660254038 is an
approximation for $5\sqrt{3}$.

Figure 2

$$\sqrt{-a}$$

For positive real numbers a,

$$\sqrt{-a} = i\sqrt{a}.$$

For example, $\sqrt{-16} = i\sqrt{16} = 4i$ and $\sqrt{-75} = i\sqrt{75} = 5i\sqrt{3}$.

A graphing calculator in *complex number mode* can simplify expressions such as $\sqrt{-16}$ and $\sqrt{-75}$ using the imaginary unit i. See Figure 2. (If the calculator were in *real number mode*, it would return an error message for square roots of negative numbers.) ∎

Complex Solutions of Equations

● ● ● **Example 1** Solving a Quadratic Equation for Complex Solutions

Solve $x^2 = -9$ for its complex solutions.

Take the square root on both sides, remembering that we must find both roots, indicated by the $\pm$ sign.

$$x^2 = -9$$
$$x = \pm\sqrt{-9}$$
$$x = \pm i\sqrt{9} \qquad \sqrt{-a} = i\sqrt{a}$$
$$x = \pm 3i \qquad \sqrt{9} = 3$$

The solutions are $-3i$ and $3i$. ● ● ●

● ● ● **Example 2** Solving a Quadratic Equation for Complex Solutions

Find the complex solutions of $9x^2 + 5 = 6x$.

Algebraic Solution

Write the equation in standard form, $9x^2 - 6x + 5 = 0$. Use the quadratic formula, with $a = 9$, $b = -6$, and $c = 5$.

$$x = \frac{-b \pm \sqrt{b^2 - 4ac}}{2a} \qquad \text{Quadratic formula}$$

$$= \frac{-(-6) \pm \sqrt{(-6)^2 - 4(9)(5)}}{2(9)} \qquad a = 9, b = -6, c = 5$$

$$= \frac{6 \pm \sqrt{-144}}{18}$$

$$= \frac{6 \pm 12i}{18} \qquad \sqrt{-144} = 12i$$

$$= \frac{6(1 \pm 2i)}{18} \qquad \text{Factor.}$$

$$x = \frac{1 \pm 2i}{3} \qquad \text{Lowest terms}$$

Graphing Calculator Solution

A program for solving quadratic equations with complex solutions is available on the Web site for this text. Figure 3 supports the results found in the algebraic solution. Notice that $1/3$ and $2/3$ are given in decimal form in Figure 3(b).

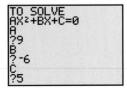

(a)

Figure 3

(continued)

The solutions may be written in standard form as

$$\frac{1}{3} \pm \frac{2}{3}i.$$

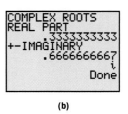

COMPLEX ROOTS
REAL PART
           .3333333333
+-IMAGINARY
           .6666666667
                      i
                   Done

(b)

Figure 3

• • •

N O T E   In the quadratic formula, the expression $b^2 - 4ac$ is called the *discriminant*. If a, b, and c are real numbers, where $a \neq 0$, the quadratic equation $ax^2 + bx + c = 0$ has two real solutions if the discriminant is positive, one real solution if it is 0, and two nonreal, complex solutions if the discriminant is negative. This last case was seen in Example 2.

Operations on Complex Numbers   When working with negative radicands, always use the definition $\sqrt{-a} = i\sqrt{a}$ before using any of the other rules for radicals. In particular, the rule $\sqrt{c} \cdot \sqrt{d} = \sqrt{cd}$ is valid only when c and d are not both negative. For example, multiplying $\sqrt{-2} \cdot \sqrt{-32}$ to get $\sqrt{64}$ gives 8, but this is incorrect. The correct result is

$$\sqrt{-2} \cdot \sqrt{-32} = i\sqrt{2} \cdot i\sqrt{32} \quad \scriptstyle \sqrt{-a} = i\sqrt{a}$$
$$= i^2\sqrt{64}$$
$$= (-1)8 \quad \scriptstyle i^2 = -1$$
$$= -8.$$

• • •   **Example 3**   Simplifying Products and Quotients with Negative Radicands

Express each product or quotient as a real number, or a product of a real number and i.

Algebraic Solution

(a) $\sqrt{-7} \cdot \sqrt{-7} = i\sqrt{7} \cdot i\sqrt{7}$
$$= i^2 \cdot \left(\sqrt{7}\right)^2$$
$$= (-1) \cdot 7 = -7$$

(b) $\sqrt{-6} \cdot \sqrt{-10} = i\sqrt{6} \cdot i\sqrt{10}$
$$= i^2 \cdot \sqrt{6 \cdot 10}$$
$$= -1 \cdot 2\sqrt{15} = -2\sqrt{15}$$

(c) $\dfrac{\sqrt{-50}}{\sqrt{-2}} = \dfrac{i\sqrt{50}}{i\sqrt{2}} = \sqrt{\dfrac{50}{2}} = \sqrt{25} = 5$

(d) $\dfrac{\sqrt{-48}}{\sqrt{24}} = \dfrac{i\sqrt{48}}{\sqrt{24}} = i\sqrt{2}$

Graphing Calculator Solution

The results of parts (a), (c), and (d) are supported in Figure 4.

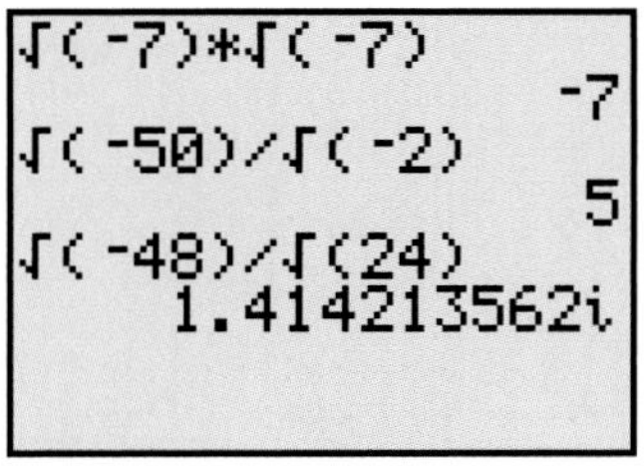

√(-7)*√(-7)
                    -7
√(-50)/√(-2)
                     5
√(-48)/√(24)
          1.414213562i

1.414213562 is an approximation for $\sqrt{2}$.

Figure 4

• • •

Addition and subtraction of complex numbers is defined in a manner similar to these operations on binomials.

> ## Addition and Subtraction of Complex Numbers
> For complex numbers $a + bi$ and $c + di$,
> $$(a + bi) + (c + di) = (a + c) + (b + d)i$$
> and
> $$(a + bi) - (c + di) = (a - c) + (b - d)i.$$

To add complex numbers, add their real parts and add their imaginary parts. Subtraction is accomplished in a similar manner.

● ● ● **Example 4** Adding and Subtracting Complex Numbers

Find each sum or difference.

Algebraic Solution

(a) $(3 - 4i) + (-2 + 6i)$
$$= [3 + (-2)] + [-4 + 6]i$$
$$= 1 + 2i$$

(b) $(-9 + 7i) + (3 - 15i) = -6 - 8i$

(c) $(-4 + 3i) - (6 - 7i)$
$$= (-4 - 6) + [3 - (-7)]i$$
$$= -10 + 10i$$

(d) $(12 - 5i) - (8 - 3i) = 4 - 2i$

Graphing Calculator Solution

The screen in Figure 5 illustrates the operations in parts (a) and (c) of the algebraic solution.

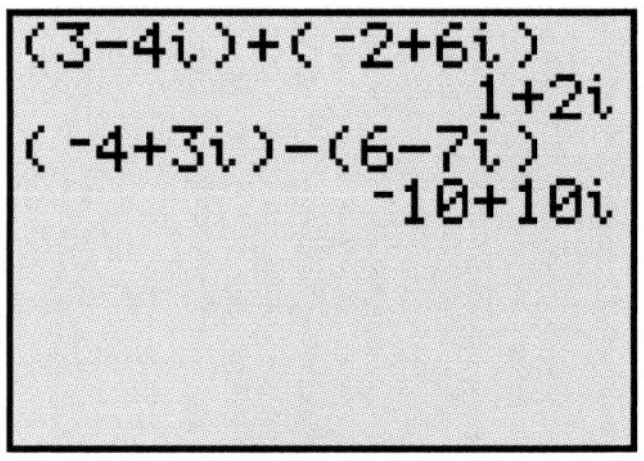

```
(3-4i)+(-2+6i)
              1+2i
(-4+3i)-(6-7i)
           -10+10i
```

Figure 5

● ● ●

The *product* of two complex numbers is found by multiplying as if the numbers were binomials and using the fact that $i^2 = -1$, as follows.

$$(a + bi)(c + di) = ac + adi + bic + bidi$$
$$= ac + adi + bci + bdi^2$$
$$= ac + (ad + bc)i + bd(-1)$$
$$= (ac - bd) + (ad + bc)i$$

> ## Multiplication of Complex Numbers
> For complex numbers $a + bi$ and $c + di$,
> $$(a + bi)(c + di) = (ac - bd) + (ad + bc)i.$$

This formal definition is rarely used when multiplying complex numbers. It is usually easier just to multiply as with binomials, using the FOIL method.

● ● ● **Example 5** **Multiplying Complex Numbers**

Find each product.

Algebraic Solution

(a) $(5 - 4i)(7 - 2i) = 5(7) + 5(-2i) - 4i(7) - 4i(-2i)$
$$= 35 - 10i - 28i + 8i^2$$
$$= 35 - 38i + 8(-1) \quad i^2 = -1$$
$$= 27 - 38i \quad \text{Combine terms.}$$

(b) $(3 - i)(3 + i) = 9 + 3i - 3i - i^2$
$$= 9 - (-1)$$
$$= 10$$

Graphing Calculator Solution

The products found in parts (a) and (b) are illustrated in Figure 6.

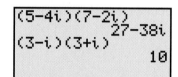

```
(5-4i)(7-2i)
                27-38i
(3-i)(3+i)
                    10
```

Figure 6

● ● ●

The factors in Example 5(b) are called *conjugates*. The **conjugate** of the complex number $a + bi$ is the complex number $a - bi$. Notice that the product of a pair of conjugates is the difference of squares, so Example 5(b) could have been written as

$$(3 - i)(3 + i) = 3^2 - i^2 = 9 - (-1) = 10.$$

The product of conjugates is always a real number. Specifically, the product is the sum of the squares of the real and imaginary parts.

$$(a + bi)(a - bi) = a^2 + b^2$$

Recall that the product of the conjugates $\sqrt{a} + \sqrt{b}$ and $\sqrt{a} - \sqrt{b}$ is also always a real number. We use this fact to rationalize denominators with radicals. Conjugates are used in the same way in division of complex numbers, to express quotients in the form $p + qi$. It can be shown that multiplying both the numerator and denominator of $\dfrac{a + bi}{c + di}$ by $c - di$ gives

$$\underbrace{\frac{ac + bd}{c^2 + d^2}}_{\substack{\text{Real} \\ \text{part}}} + \underbrace{\frac{bc - ad}{c^2 + d^2}}_{\substack{\text{Imaginary} \\ \text{part}}} i.$$

In practice, we do not usually use the above form; rather, we follow this rule.

> ## Division of Complex Numbers
>
> To divide complex numbers, multiply both the numerator and the denominator (divisor) by the conjugate of the denominator.

● ● ●   **Example 6**   Dividing Complex Numbers

Find each quotient.

Algebraic Solution

(a) $\dfrac{3 + 2i}{5 - i}$

Multiply the numerator and denominator by $5 + i$, the conjugate of $5 - i$.

$$\frac{3 + 2i}{5 - i} = \frac{(3 + 2i)(5 + i)}{(5 - i)(5 + i)}$$

$$= \frac{15 + 3i + 10i + 2i^2}{26} \qquad (5 - i)(5 + i) = 5^2 + 1^2 = 26$$

$$= \frac{13 + 13i}{26} = \frac{1}{2} + \frac{1}{2}i$$

To check this answer, show that

$$(5 - i)\left(\frac{1}{2} + \frac{1}{2}i\right) = 3 + 2i.$$

(b) $\dfrac{3}{i} = \dfrac{3(-i)}{i(-i)}$ $-i$ is the conjugate of i.

$$= \frac{-3i}{-i^2}$$

$$= \frac{-3i}{1} \qquad -i^2 = -(-1) = 1$$

$$= -3i$$

Graphing Calculator Solution

The quotients are shown in the screen in Figure 7.

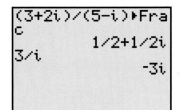

Figure 7

● ● ●

Powers of i   The fact that i^2 is equal to -1 can be used to find higher powers of i.

$$i^0 = 1 \qquad\qquad i^4 = i^2 \cdot i^2 = (-1)(-1) = 1$$

$$i^1 = i \qquad\qquad i^5 = i \cdot i^4 = i \cdot 1 = i$$

$$i^2 = -1 \qquad\qquad i^6 = i^2 \cdot i^4 = (-1) \cdot 1 = -1$$

$$i^3 = i \cdot i^2 = i(-1) = -i \qquad i^7 = i^3 \cdot i^4 = (-i) \cdot 1 = -i$$

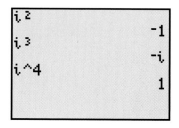

Figure 8

As these examples show, the powers of i rotate through the four numbers 1, i, -1, and $-i$. Larger powers of i can be simplified by using the fact that $i^4 = 1$. For example,

$$i^{75} = (i^4)^{18} \cdot i^3 = 1^{18} \cdot i^3 = 1 \cdot i^3 = i^3 = -i.$$

Figure 8 shows how a graphing calculator computes some small powers of i. ∎

● ● ● **Example 7** Simplifying Powers of i

Find each power of i.

(a) $i^{12} = (i^4)^3 = 1^3 = 1$

(b) $i^{39} = i^{36} \cdot i^3 = (i^4)^9 \cdot i^3 = 1^9 \cdot (-i) = -i$

(c) $i^{-2} = \dfrac{1}{i^2} = \dfrac{1}{-1} = -1$

● ● ●

8.1 Exercises

Concept Check Match the number in Column I with its equivalent in Column II.

	I		**II**
1.	$\sqrt{-5}$	**A.**	$-i\sqrt{5}$
2.	$i\sqrt{-5}$	**B.**	$i\sqrt{5}$
3.	$-i\sqrt{-5}$	**C.**	$-\sqrt{5}$
4.	$-\sqrt{-5}$	**D.**	$\sqrt{5}$

5. *Concept Check* Fill in the blanks with the correct subset.

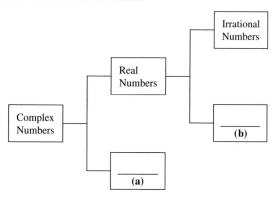

Solve each quadratic equation and express all complex solutions in terms of i. See Examples 1 and 2.

6. $x^2 = -16$ **7.** $y^2 = -36$ **8.** $z^2 + 12 = 0$ **9.** $w^2 + 48 = 0$

10. $3x^2 + 4x + 2 = 0$ **11.** $2k^2 + 3k = -2$ **12.** $m^2 - 6m + 14 = 0$ **13.** $p^2 + 4p + 11 = 0$

14. $4z^2 = 4z - 7$ **15.** $9a^2 + 7 = 6a$ **16.** $m^2 + 1 = -m$ **17.** $y^2 = 2y - 2$

18. (a) Explain how to determine whether a quadratic equation of the form $ax^2 + bx + c = 0$ has nonreal, complex solutions without actually solving the equation.

   (b) Explain how to determine whether a quadratic equation of the form $ax^2 + bx + c = 0$ has nonreal, complex solutions by looking at the graph of $y = ax^2 + bx + c$.

Simplify each of the following. See Example 3.

19. $\sqrt{-3} \cdot \sqrt{-3}$ **20.** $\sqrt{-2} \cdot \sqrt{-2}$ **21.** $\sqrt{-5} \cdot \sqrt{-6}$ **22.** $\sqrt{-27} \cdot \sqrt{-3}$

23. $\dfrac{\sqrt{-12}}{\sqrt{-8}}$ **24.** $\dfrac{\sqrt{-15}}{\sqrt{-3}}$ **25.** $\dfrac{\sqrt{-24}}{\sqrt{72}}$ **26.** $\dfrac{\sqrt{-27}}{\sqrt{9}}$

Decide what answer the calculator will give for each screen shown.

27.

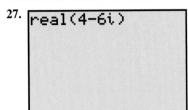

`real(4-6i)`

the real part of $4 - 6i$

28.

`imag(4-6i)`

the imaginary part of $4 - 6i$

29.

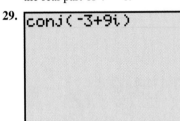

`conj(-3+9i)`

the conjugate of $-3 + 9i$

30.

`conj(-8i)`

the conjugate of $-8i$

Two complex numbers $a + bi$ and $c + di$ are equal if and only if $a = c$ and $b = d$. Use this definition of equality of complex numbers to solve each equation for x and y.

31. $2x + yi = 4 - 3i$ **32.** $x + 3yi = 5 + 2i$ **33.** $7 - 2yi = 14x - 30i$

34. $-5 + yi = x + 6i$ **35.** $x + yi = (2 + 3i)(4 - 2i)$ **36.** $x + yi = (5 - 7i)(1 + i)$

Perform each operation and express all results in rectangular form. See Examples 4–6.

37. $(2 - 5i) + (3 + 2i)$ **38.** $(5 - i) + (3 + 4i)$ **39.** $(-2 + 3i) - (3 + i)$

40. $(4 + 6i) - (-2 - i)$ **41.** $(1 - i) - (5 - 2i)$ **42.** $(-2 + 6i) - (-3 - 8i)$

43. $(2 + i)(3 - 2i)$ **44.** $(-2 + 3i)(4 - 2i)$ **45.** $(2 + 4i)(-1 + 3i)$

46. $(1 + 3i)(2 - 5i)$ **47.** $(2 - i)(2 + i)$ **48.** $(5 + 4i)(5 - 4i)$

49. $\dfrac{5}{i}$ **50.** $\dfrac{-4}{i}$ **51.** $\dfrac{12 - 8i}{5 + i}$ **52.** $\dfrac{-13 - 14i}{8 + 3i}$ **53.** $\dfrac{4 + i}{6 + 2i}$ **54.** $\dfrac{3 - 2i}{5 + 3i}$

55. A student makes the following statement. "I can simplify a large positive power of i by dividing the exponent by 4, and looking at the remainder. If the remainder is 0, it simplifies to 1; if the remainder is 1, it simplifies to i; if the remainder is 2, it simplifies to -1; and if the remainder is 3, it simplifies to $-i$." Explain why this statement is true.

56. Using the procedure of Exercise 55, why don't we have to consider getting a remainder of 4?

Simplify each power of i. See Example 7.

57. i^{12} **58.** i^{9} **59.** i^{18} **60.** i^{99} **61.** i^{-3} **62.** i^{-5} **63.** i^{-10} **64.** i^{-40}

65. Evaluate $1 + i + i^2 + i^3$.   **66.** Evaluate $1 + i + i^2 + \cdots + i^{100}$.

Alternating Current   In work with alternating current, complex numbers are used to describe current, I, voltage, E, and impedance, Z (the opposition to current). These three quantities are related by the equation $E = IZ$. Thus, if any two of these quantities are known, the third can be found. In each of the following problems, solve the equation $E = IZ$ for the missing variable.

67. $I = 8 + 6i$, $Z = 6 + 3i$ **68.** $I = 10 + 6i$, $Z = 8 + 5i$

69. $I = 7 + 5i$, $E = 28 + 54i$ **70.** $E = 35 + 55i$, $Z = 6 + 4i$

Impedance   *Impedance is a measure of the opposition to the flow of alternating electrical current found in common electrical outlets. It consists of two parts called* resistance *and* reactance. *Resistance occurs when a light bulb is turned on, while reactance is produced when electricity passes through a coil of wire like that found in electric motors. Impedance Z in ohms (Ω) can be expressed as a complex number, where the real part represents resistance and the imaginary part represents reactance. For example, if the resistive part is 3 ohms and the reactive part is 4 ohms, then the impedance could be described by the complex number Z = 3 + 4i. In the series circuit shown in the figure, the total impedance will be the sum of the individual impedances.* (*Source:* Wilcox, G. and C. Hesselberth, *Electricity for Engineering Technology,* Allyn & Bacon, 1970.)

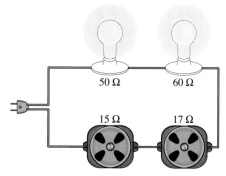

71. The circuit contains two light bulbs and two electric motors. Assuming that the light bulbs are pure resistive and the motors are pure reactive, find the total impedance in this circuit and express it in the form $Z = a + bi$.

72. The phase angle θ measures the phase difference between the voltage and the current in an electrical circuit. θ (in degrees) can be determined by the equation $\tan \theta = b/a$. Find θ for this circuit.

73. Show that $\dfrac{\sqrt{2}}{2} + \dfrac{\sqrt{2}}{2}i$ is a square root of i.   74. Show that $-\dfrac{\sqrt{3}}{2} + \dfrac{1}{2}i$ is a cube root of i.

75. *Concept Check*   What is wrong with the following proof that $-1 = 1$?
$$-1 = i^2 = \sqrt{-1} \cdot \sqrt{-1} = \sqrt{(-1)(-1)} = \sqrt{1} = 1.$$

76. Explain why a real number must be a complex number, but a complex number need not be a real number.

77. *Concept Check*   If the complex number $a + bi$ is real, then what can be said about the value of b?

78. Discuss the similarity between rationalizing the denominator of a fraction such as $\dfrac{a + \sqrt{b}}{c + \sqrt{d}}$ and dividing two complex numbers.

8.2 Trigonometric (Polar) Form of Complex Numbers

> • **The Complex Plane** • **Trigonometric (Polar) Form** • **Converting between Trigonometric and Rectangular Forms**
> • **An Application of Complex Numbers to Fractals**

The Complex Plane   Unlike real numbers, complex numbers cannot be ordered. One way to organize and illustrate them is by using a graph. To graph a complex number such as $2 - 3i$, the familiar coordinate system must be modified. We do this by calling the horizontal axis the **real axis** and the vertical axis the **imaginary axis**. Then complex numbers can be graphed in this **complex plane,** as shown in Figure 9 for the complex number $2 - 3i$.

NOTE   This geometric representation is the reason that $a + bi$ is called the *rectangular form* of a complex number. (*Rectangular form* is also called *standard form.*)

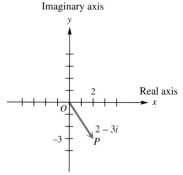

Figure 9

Each nonzero complex number graphed in this way determines a unique directed line segment, the segment from the origin to the point representing the

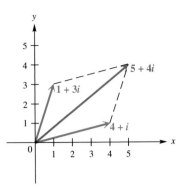

Figure 10

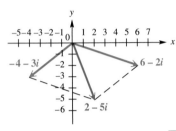

Figure 11

complex number. Recall from Chapter 7 that such directed line segments (like **OP** of Figure 9) are called vectors.

The previous section showed how to find the sum of two complex numbers, such as $4 + i$ and $1 + 3i$.

$$(4 + i) + (1 + 3i) = 5 + 4i$$

Graphically, the sum of two complex numbers is represented by the vector that is the resultant of the vectors corresponding to the two numbers. The vectors representing the complex numbers $4 + i$ and $1 + 3i$ and the resultant vector that represents their sum, $5 + 4i$, are shown in Figure 10.

Example 1 Expressing the Sum of Complex Numbers Graphically

Find the sum of $6 - 2i$ and $-4 - 3i$. Graph both complex numbers and their resultant.

The sum is found by adding the two numbers.

$$(6 - 2i) + (-4 - 3i) = 2 - 5i$$

The graphs are shown in Figure 11.   ● ● ●

C O N N E C T I O N S   In Section 7.4 we saw that the vector **u** with its initial point at the origin and its endpoint at (a, b) could be designated $\langle a, b \rangle$. We then showed how to add and subtract vectors using this new notation. Now we see that the complex number $a + bi$ corresponds to the vector **u** described above. Thus, we have

$$\mathbf{u} = \langle a, b \rangle = a + bi$$

as three ways to designate a complex number or vector.

We can use addition of vectors in the form $\langle a, b \rangle$ to find the sum in Example 1.

$$(6 - 2i) + (-4 - 3i) = \langle 6, -2 \rangle + \langle -4, -3 \rangle$$
$$= \langle 2, -5 \rangle$$
$$= 2 - 5i$$

For Discussion or Writing

1. Find $(6 - 2i) - (-4 - 3i)$ using vectors in the form $\langle a, b \rangle$. Then graph the given complex numbers and the difference.
2. Describe a general method for finding the difference of two vectors.

Trigonometric (Polar) Form   Figure 12 shows the complex number $x + yi$ that corresponds to a vector **OP** with direction angle θ and magnitude r. The following relationships among r, θ, x, and y can be verified from Figure 12.

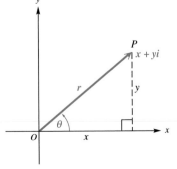

Figure 12

Relationships among x, y, r, and θ	
$x = r \cos \theta$	$r = \sqrt{x^2 + y^2}$
$y = r \sin \theta$	$\tan \theta = \dfrac{y}{x}$, if $x \neq 0$

Substituting $x = r \cos \theta$ and $y = r \sin \theta$ from the relationships given above into $x + yi$ gives

$$x + yi = r \cos \theta + (r \sin \theta)i$$
$$= r(\cos \theta + i \sin \theta).$$

> ### Trigonometric or Polar Form of a Complex Number
> The expression
>
> $$r(\cos \theta + i \sin \theta)$$
>
> is called the **trigonometric form** or **polar form** of the complex number $x + yi$. The expression $\cos \theta + i \sin \theta$ is sometimes abbreviated cis θ. Using this notation,
>
> $$r(\cos \theta + i \sin \theta) \text{ is written } r \text{ cis } \theta.$$

The number r is called the **modulus** or **absolute value** of $x + yi$, while θ is the **argument** of $x + yi$. In this section we will choose the value of θ in the interval $[0°, 360°)$. However, angles coterminal with such angles are also possible; that is, the argument for a particular complex number is not unique.

Converting between Trigonometric and Rectangular Forms

● ● ● **Example 2** Converting from Trigonometric Form to Rectangular Form

Express $2(\cos 300° + i \sin 300°)$ in rectangular form.

Algebraic Solution

Since $\cos 300° = 1/2$ and $\sin 300° = -\sqrt{3}/2$,

$$2(\cos 300° + i \sin 300°) = 2\left(\frac{1}{2} - i\frac{\sqrt{3}}{2}\right)$$
$$= 1 - i\sqrt{3}.$$

Graphing Calculator Solution

The screen in Figure 13 supports the algebraic solution.

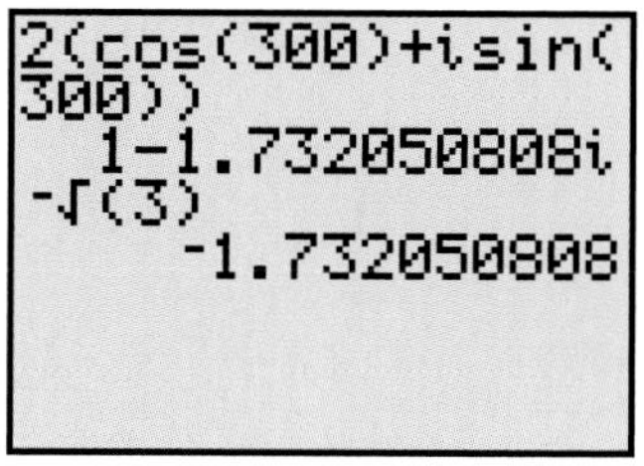

```
2(cos(300)+isin(
300))
    1-1.732050808i
-√(3)
      -1.732050808
```

The imaginary part is an approximation for $-\sqrt{3}$.

Figure 13

● ● ●

In order to convert from rectangular form to trigonometric form, the following procedure is used.

Steps for Converting from Rectangular to Trigonometric Form

Step 1 Sketch a graph of the number in the complex plane.

Step 2 Find r by using the equation $r = \sqrt{x^2 + y^2}$.

Step 3 Find θ by using the equation $\tan \theta = y/x$, $x \neq 0$, choosing the quadrant indicated in Step 1.

C A U T I O N Errors often occur in Step 3 described above. Be sure to choose the correct quadrant for θ by referring to the graph sketched in Step 1.

● ● ● **Example 3** Converting from Rectangular Form to Trigonometric Form

Express each complex number in trigonometric form.

Algebraic Solution

(a) $-\sqrt{3} + i$

Start by sketching the graph of $-\sqrt{3} + i$ in the complex plane, as shown in Figure 14. Next, find r. Since $x = -\sqrt{3}$ and $y = 1$,

$$r = \sqrt{x^2 + y^2} = \sqrt{\left(-\sqrt{3}\right)^2 + 1^2} = \sqrt{3 + 1} = 2.$$

Then find θ.

$$\tan \theta = \frac{y}{x} = \frac{1}{-\sqrt{3}} = -\frac{\sqrt{3}}{3}$$

Since $\tan \theta = -\sqrt{3}/3$, the reference angle for θ is 30°. From the sketch we see that θ is in quadrant II, so $\theta = 180° - 30° = 150°$. Therefore, in trigonometric form,

$$-\sqrt{3} + i = 2(\cos 150° + i \sin 150°)$$
$$= 2 \text{ cis } 150°.$$

Graphing Calculator Solution

We can use choices 4 and 5 in the MATH CPX menu to find the argument (angle) and absolute value of a complex number. (See the first screen in Figure 16.) We then use the results to write the complex number in trigonometric form. The second screen shows that the angle and absolute value, for $-\sqrt{3} + i$, are 150° and 2, respectively, which supports the result in part (a).

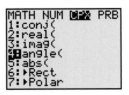

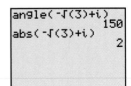

Degree mode

Figure 16

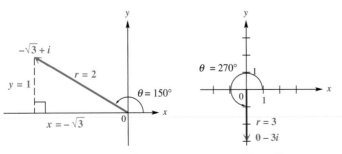

Figure 14   Figure 15

(continued)

(b) $-3i$

The sketch of $-3i$ is shown in Figure 15. Since $-3i = 0 - 3i$, we have $x = 0$ and $y = -3$. Find r as follows.

$$r = \sqrt{0^2 + (-3)^2} = \sqrt{0 + 9} = \sqrt{9} = 3$$

We cannot find θ by using $\tan\theta = y/x$, since $x = 0$. In a case like this, refer to the graph and determine the argument directly from the sketch. A value for θ here is $270°$. In trigonometric form,

$$\begin{aligned}-3i &= 3(\cos 270° + i \sin 270°)\\ &= 3 \text{ cis } 270°.\end{aligned}$$

The result in part (b) can be supported in the same way by entering the rectangular form $0 - 3i$ for choices 4 and 5.

N O T E   In Examples 2 and 3 we gave answers in both forms: $r(\cos\theta + i\sin\theta)$ and r cis θ. We will use these forms interchangeably throughout the rest of this chapter.

●●● **Example 4** Converting between Trigonometric and Rectangular Forms Using Calculator Approximations

Write each complex number in its alternative form, using calculator approximations as necessary.

(a) $6(\cos 115° + i \sin 115°)$

Since $115°$ does not have a special angle as a reference angle, we cannot find exact values for $\cos 115°$ and $\sin 115°$. Use a calculator set in degree mode to find $\cos 115° \approx -.4226182617$ and $\sin 115° \approx .906307787$. Therefore, in rectangular form,

$$\begin{aligned}6(\cos 115° + i \sin 115°) &\approx 6(-.4226182617 + .906307787i)\\ &= -2.53570957 + 5.437846722i.\end{aligned}$$

(b) $5 - 4i$

A sketch of $5 - 4i$ shows that θ must be in quadrant IV. See Figure 17. Here $r = \sqrt{5^2 + (-4)^2} = \sqrt{41}$ and $\tan\theta = -4/5$. Use a calculator to find that one measure of θ is $-38.66°$. In order to express θ in the interval $[0, 360°)$, we find that $\theta = 360° - 38.66° = 321.34°$. Use these results to get

$$5 - 4i = \sqrt{41} \text{ cis } 321.34°.$$ ●●●

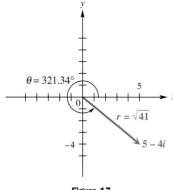

Figure 17

An Application of Complex Numbers to Fractals   We can apply complex numbers to the study of fractals, first discussed in the chapter introduction.

●●● **Example 5** Deciding Whether a Complex Number Is in the Julia Set

The fractal called the **Julia set** is shown in Figure 18 on the next page. It is created by graphing a special set of complex numbers. To determine if a complex number $z = a + bi$ is in this Julia set, perform the following sequence of calculations. Repeatedly compute the values of $z^2 - 1$,

$(z^2 - 1)^2 - 1$, $[(z^2 - 1)^2 - 1]^2 - 1, \ldots$. If the moduli of any of the resulting complex numbers exceeds 2, then the complex number z is not in the Julia set. Otherwise z is part of this set and the point (a, b) should be shaded in the graph. (*Source:* Crownover, R., *Introduction to Fractals and Chaos,* Jones and Bartlett Publishers, 1995.)

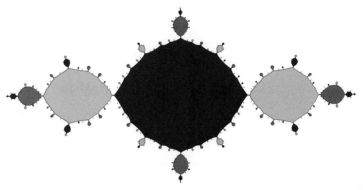

Source: Figure from Crownover, R.: *Introduction to Fractals and Chaos.* Copyright © 1995 Boston: Jones and Bartlett Publishers. Reprinted with permission.

Figure 18

Determine whether the following numbers belong to the Julia set.

(a) $z = 0 + 0i$

Since $z = 0 + 0i = 0$, $z^2 - 1 = 0^2 - 1 = -1$,

$$(z^2 - 1)^2 - 1 = (-1)^2 - 1 = 0,$$
$$[(z^2 - 1)^2 - 1]^2 - 1 = 0^2 - 1 = -1,$$

and so on. We see that the calculations repeat as $0, -1, 0, -1$, and so on. The moduli are either 0 or 1, which do not exceed 2, so $0 + 0i$ is in the Julia set and the point $(0, 0)$ is part of the graph.

(b) $z = 1 + 1i$

We have $z^2 - 1 = (1 + i)^2 - 1 = (1 + 2i + i^2) - 1 = -1 + 2i$. The modulus is $\sqrt{(-1)^2 + 2^2} = \sqrt{5}$. Since $\sqrt{5}$ is greater than 2, $1 + 1i$ is not in the Julia set and $(1, 1)$ is not part of the graph. ● ● ●

8.2 Exercises

1. *Concept Check*   The modulus of a complex number represents the _____ of the vector representing it in the complex plane.

2. *Concept Check*   What is the geometric interpretation of the argument of a complex number?

Graph each complex number. See Example 1.

3. $-2 + 3i$ **4.** $-4 + 5i$ **5.** $8 - 5i$ **6.** $6 - 5i$ **7.** $2 - 2i\sqrt{3}$

8. $4\sqrt{2} + 4i\sqrt{2}$ **9.** $-4i$ **10.** $3i$ **11.** -8 **12.** 2

Give the rectangular form of the complex number represented in each graph.

13.

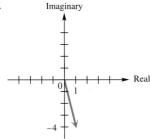

14.

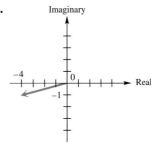

Find the resultant of each pair of complex numbers. See Example 1.

15. $4 - 3i, -1 + 2i$

16. $2 + 3i, -4 - i$

17. $5 - 6i, -2 + 3i$

18. $7 - 3i, -4 + 3i$

19. $-3, 3i$

20. $6, -2i$

21. $2 + 6i, -2i$

22. $4 - 2i, 5$

23. $7 + 6i, 3i$

24. $-5 - 8i, -1$

Write each complex number in rectangular form. See Example 2.

25. $2(\cos 45° + i \sin 45°)$

26. $4(\cos 60° + i \sin 60°)$

27. $10(\cos 90° + i \sin 90°)$

28. $8(\cos 270° + i \sin 270°)$

29. $4(\cos 240° + i \sin 240°)$

30. $2(\cos 330° + i \sin 330°)$

31. cis 30°

32. 3 cis 150°

33. 5 cis 300°

34. 6 cis 135°

35. $\sqrt{2}$ cis 180°

36. $\sqrt{3}$ cis 315°

Write each complex number in trigonometric form $r(\cos θ + i \sin θ)$, with $θ$ in the interval $[0°, 360°)$. See Example 3.

37. $3 - 3i$

38. $-2 + 2i\sqrt{3}$

39. $-3 - 3i\sqrt{3}$

40. $1 + i\sqrt{3}$

41. $\sqrt{3} - i$

42. $4\sqrt{3} + 4i$

43. $-5 - 5i$

44. $-\sqrt{2} + i\sqrt{2}$

45. $2 + 2i$

46. $-\sqrt{3} + i$

47. -4

48. $5i$

49. $-2i$

50. 7

Perform each conversion, using a calculator as necessary. See Example 4.

	Rectangular Form	**Trigonometric Form**
51.	$2 + 3i$	_____
52.	_____	$(\cos 35° + i \sin 35°)$
53.	_____	$3(\cos 250° + i \sin 250°)$
54.	$-4 + i$	_____
55.	$12i$	_____
56.	_____	3 cis 180°
57.	$3 + 5i$	_____
58.	_____	cis 110.5°

Concept Check   *The complex number z, where $z = x + yi$, can be graphed in the plane as (x, y). Describe the graphs of all complex numbers z satisfying the conditions in Exercises 59–62.*

59. The modulus of z is 1.

60. The real and imaginary parts of z are equal.

61. The real part of z is 1.

62. The imaginary part of z is 1.

Julia Set   Refer to Example 5 to solve each problem about fractals.

63. Is $z = -.2i$ in the Julia set?

64. The graph of the Julia set in Figure 18 appears to be symmetric with respect to both the x-axis and y-axis. Complete the following to show that this is true.

(a) Show that complex conjugates have the same modulus.

(b) Compute $z_1^2 - 1$ and $z_2^2 - 1$, where $z_1 = a + bi$ and $z_2 = a - bi$.

(c) Discuss why if (a, b) is in the Julia set then so is $(a, -b)$.

(d) Conclude that the graph of the Julia set must be symmetric with respect to the x-axis.

(e) Using a similar argument, show that the Julia set must also be symmetric with respect to the y-axis.

In Exercises 65 and 66, suppose $z = r(\cos \theta + i \sin \theta)$.

65. Use vectors to show that the conjugate of

$$z = r[\cos(360° - \theta) + i \sin(360° - \theta)]$$

is $r(\cos \theta - i \sin \theta)$.

66. Use vectors to show that

$$-z = r[\cos(\theta + \pi) + i \sin(\theta + \pi)].$$

Concept Check   In Exercises 67–69 identify which geometric condition (A, B, or C) implies the situation.

A. *The corresponding vectors have opposite directions.*

B. *The terminal points of the vectors corresponding to $a + bi$ and $c + di$ lie on a horizontal line.*

C. *The corresponding vectors have the same direction.*

67. The difference between two nonreal complex numbers $a + bi$ and $c + di$ is a real number.

68. The modulus of the sum of two complex numbers $a + bi$ and $c + di$ is equal to the sum of their moduli.

69. The modulus of the difference of two complex numbers $a + bi$ and $c + di$ is equal to the sum of their moduli.

70. Show that z and iz have the same modulus. How are the graphs of these two numbers related?

8.3 The Product and Quotient Theorems

- **The Product of Complex Numbers in Trigonometric Form** • **The Quotient of Complex Numbers in Trigonometric Form**

The Product of Complex Numbers in Trigonometric Form   The product of the two complex numbers $1 + i\sqrt{3}$ and $-2\sqrt{3} + 2i$, which are in rectangular form, can be found by the method shown in Section 8.1.

$$\left(1 + i\sqrt{3}\right)\left(-2\sqrt{3} + 2i\right) = -2\sqrt{3} + 2i - 2i(3) + 2i^2\sqrt{3}$$
$$= -2\sqrt{3} + 2i - 6i - 2\sqrt{3}$$
$$= -4\sqrt{3} - 4i$$

This same product also can be found by first converting the complex numbers $1 + i\sqrt{3}$ and $-2\sqrt{3} + 2i$ to trigonometric form. Using the method explained in the previous section,

$$1 + i\sqrt{3} = 2(\cos 60° + i \sin 60°)$$

and $\qquad\qquad -2\sqrt{3} + 2i = 4(\cos 150° + i \sin 150°).$

If the trigonometric forms are now multiplied together and if the trigonometric identities for cosine and sine of the sum of two angles are used, the result is

$$[2(\cos 60° + i \sin 60°)][4(\cos 150° + i \sin 150°)]$$
$$= 2 \cdot 4(\cos 60° \cdot \cos 150° + i \sin 60° \cdot \cos 150°$$
$$+ i \cos 60° \cdot \sin 150° + i^2 \sin 60° \cdot \sin 150°)$$
$$= 8[(\cos 60° \cdot \cos 150° - \sin 60° \cdot \sin 150°)$$
$$+ i(\sin 60° \cdot \cos 150° + \cos 60° \cdot \sin 150°)]$$
$$= 8[\cos(60° + 150°) + i \sin(60° + 150°)]$$
$$= 8(\cos 210° + i \sin 210°).$$

The modulus of the product, 8, is equal to the product of the moduli of the factors, $2 \cdot 4$, while the argument of the product, $210°$, is the sum of the arguments of the factors, $60° + 150°$.

As we would expect, the product obtained when multiplying by the first method is the rectangular form of the product obtained when multiplying by the second method.

$$8(\cos 210° + i \sin 210°) = 8\left(-\frac{\sqrt{3}}{2} - \frac{1}{2}i\right) = -4\sqrt{3} - 4i$$

Generalizing, the product of the two complex numbers, $r_1(\cos \theta_1 + i \sin \theta_1)$ and $r_2(\cos \theta_2 + i \sin \theta_2)$, is

$$[r_1(\cos \theta_1 + i \sin \theta_1)] \cdot [r_2(\cos \theta_2 + i \sin \theta_2)]$$
$$= r_1 r_2(\cos \theta_1 \cos \theta_2 + i \sin \theta_1 \cos \theta_2 + i \cos \theta_1 \sin \theta_2 + i^2 \sin \theta_1 \sin \theta_2)$$
$$= r_1 r_2[(\cos \theta_1 \cos \theta_2 - \sin \theta_1 \sin \theta_2) + i(\sin \theta_1 \cos \theta_2 + \cos \theta_1 \sin \theta_2)]$$
$$= r_1 r_2[\cos(\theta_1 + \theta_2) + i \sin(\theta_1 + \theta_2)].$$

This work is summarized in the following *product theorem.*

Product Theorem

If $r_1(\cos \theta_1 + i \sin \theta_1)$ and $r_2(\cos \theta_2 + i \sin \theta_2)$ are any two complex numbers, then

$$[r_1(\cos \theta_1 + i \sin \theta_1)] \cdot [r_2(\cos \theta_2 + i \sin \theta_2)]$$
$$= r_1 r_2[\cos(\theta_1 + \theta_2) + i \sin(\theta_1 + \theta_2)].$$

In compact form, this is written

$$(r_1 \text{ cis } \theta_1)(r_2 \text{ cis } \theta_2) = r_1 r_2 \text{ cis}(\theta_1 + \theta_2).$$

● ● ● **Example 1** **Using the Product Theorem**

Find the product of $3(\cos 45° + i \sin 45°)$ and $2(\cos 135° + i \sin 135°)$.

Using the product theorem,

$$[3(\cos 45° + i \sin 45°)][2(\cos 135° + i \sin 135°)]$$
$$= 3 \cdot 2[\cos(45° + 135°) + i \sin(45° + 135°)]$$
$$= 6(\cos 180° + i \sin 180°),$$

which can be expressed as $6(-1 + i \cdot 0) = 6(-1) = -6$. The two complex numbers in this example are complex factors of -6. ● ● ●

The Quotient of Complex Numbers in Trigonometric Form Using the method shown in Section 8.1, the rectangular form of the quotient of the complex numbers $1 + i\sqrt{3}$ and $-2\sqrt{3} + 2i$ is

$$\frac{1 + i\sqrt{3}}{-2\sqrt{3} + 2i} = \frac{\left(1 + i\sqrt{3}\right)\left(-2\sqrt{3} - 2i\right)}{\left(-2\sqrt{3} + 2i\right)\left(-2\sqrt{3} - 2i\right)}$$
$$= \frac{-2\sqrt{3} - 2i - 6i - 2i^2\sqrt{3}}{12 - 4i^2}$$
$$= \frac{-8i}{16} = -\frac{1}{2}i.$$

Writing $1 + i\sqrt{3}$, $-2\sqrt{3} + 2i$, and $-\dfrac{1}{2}i$ in trigonometric form gives

$$1 + i\sqrt{3} = 2(\cos 60° + i \sin 60°),$$

$$-2\sqrt{3} + 2i = 4(\cos 150° + i \sin 150°),$$

$$-\frac{1}{2}i = \frac{1}{2}[\cos(-90°) + i \sin(-90°)].$$

The modulus of the quotient, $1/2$, is the quotient of the two moduli, 2 and 4. The argument of the quotient, $-90°$, is the difference of the two arguments, $60° - 150° = -90°$. It would be easier to find the quotient of these two complex numbers in trigonometric form than in rectangular form. Generalizing from this example leads to another theorem. The proof is similar to the proof of the product theorem, after the numerator and denominator are multiplied by the conjugate of the denominator.

Quotient Theorem

If $r_1(\cos \theta_1 + i \sin \theta_1)$ and $r_2(\cos \theta_2 + i \sin \theta_2)$ are complex numbers, where $r_2(\cos \theta_2 + i \sin \theta_2) \neq 0$, then

$$\frac{r_1(\cos \theta_1 + i \sin \theta_1)}{r_2(\cos \theta_2 + i \sin \theta_2)} = \frac{r_1}{r_2}[\cos(\theta_1 - \theta_2) + i \sin(\theta_1 - \theta_2)].$$

In compact form, this is written

$$\frac{r_1 \text{ cis } \theta_1}{r_2 \text{ cis } \theta_2} = \frac{r_1}{r_2} \text{cis}(\theta_1 - \theta_2).$$

● ● ●   **Example 2**   Using the Quotient Theorem

Find the quotient

$$\frac{10 \text{ cis}(-60°)}{5 \text{ cis } 150°}.$$

Write the result in rectangular form.
  By the quotient theorem,

$$\frac{10 \text{ cis}(-60°)}{5 \text{ cis } 150°} = \frac{10}{5} \text{cis}(-60° - 150°) \qquad \text{Quotient theorem}$$

$$= 2 \text{ cis}(-210°)$$

$$= 2[\cos(-210°) + i \sin(-210°)]$$

$$= 2\left[-\frac{\sqrt{3}}{2} + i\left(\frac{1}{2}\right) \right] \qquad \cos(-210°) = -\frac{\sqrt{3}}{2};$$
$$\sin(-210°) = \frac{1}{2}$$

$$= -\sqrt{3} + i. \qquad \text{Rectangular form} \quad ● ● ●$$

● ● ● **Example 3** Using the Product and Quotient Theorems with a Calculator

Use a calculator to find the following. Write the results in rectangular form.

(a) $(9.3 \text{ cis } 125.2°)(2.7 \text{ cis } 49.8°)$
By the product theorem,

$$(9.3 \text{ cis } 125.2°)(2.7 \text{ cis } 49.8°) = 9.3(2.7) \text{ cis}(125.2° + 49.8°)$$
$$= 25.11 \text{ cis } 175°$$
$$= 25.11(\cos 175° + i \sin 175°)$$
$$\approx 25.11[-.99619470 + i(.08715574)]$$
$$\approx -25.014449 + 2.1884807i.$$

(b)

$$\frac{10.42\left(\cos \dfrac{3\pi}{4} + i \sin \dfrac{3\pi}{4}\right)}{5.21\left(\cos \dfrac{\pi}{5} + i \sin \dfrac{\pi}{5}\right)} = \frac{10.42}{5.21}\left[\cos\left(\dfrac{3\pi}{4} - \dfrac{\pi}{5}\right) + i \sin\left(\dfrac{3\pi}{4} - \dfrac{\pi}{5}\right)\right]$$

Quotient theorem

$$= 2\left(\cos \dfrac{11\pi}{20} + i \sin \dfrac{11\pi}{20}\right)$$
$$\approx -.31286893 + 1.9753767i$$ ● ● ●

8.3 Exercises

Concept Check   Fill in the blanks with the correct responses.

1. When multiplying two complex numbers in trigonometric form, we _____ their absolute values and _____ their arguments.

2. When dividing two complex numbers in trigonometric form, we _____ their absolute values and _____ their arguments.

Find each product and write it in rectangular form. See Example 1.

3. $[3(\cos 60° + i \sin 60°)][2(\cos 90° + i \sin 90°)]$
4. $[4(\cos 30° + i \sin 30°)][5(\cos 120° + i \sin 120°)]$
5. $[2(\cos 45° + i \sin 45°)][2(\cos 225° + i \sin 225°)]$
6. $[8(\cos 300° + i \sin 300°)][5(\cos 120° + i \sin 120°)]$
7. $[4(\cos 60° + i \sin 60°)][6(\cos 330° + i \sin 330°)]$
8. $[8(\cos 210° + i \sin 210°)][2(\cos 330° + i \sin 330°)]$
9. $(5 \text{ cis } 90°)(3 \text{ cis } 45°)$
10. $(6 \text{ cis } 120°)[5 \text{ cis}(-30°)]$
11. $\left(\sqrt{3} \text{ cis } 45°\right)\left(\sqrt{3} \text{ cis } 225°\right)$
12. $\left(\sqrt{2} \text{ cis } 300°\right)\left(\sqrt{2} \text{ cis } 270°\right)$

Find each quotient and write it in rectangular form. In Exercises 19–24, first convert the numerator and the denominator to trigonometric form. See Example 2.

13. $\dfrac{4(\cos 120° + i \sin 120°)}{2(\cos 150° + i \sin 150°)}$
14. $\dfrac{10(\cos 225° + i \sin 225°)}{5(\cos 45° + i \sin 45°)}$
15. $\dfrac{16(\cos 300° + i \sin 300°)}{8(\cos 60° + i \sin 60°)}$

16. $\dfrac{24(\cos 150° + i \sin 150°)}{2(\cos 30° + i \sin 30°)}$
17. $\dfrac{3 \text{ cis } 305°}{9 \text{ cis } 65°}$
18. $\dfrac{12 \text{ cis } 293°}{6 \text{ cis } 23°}$

19. $\dfrac{8}{\sqrt{3} + i}$
20. $\dfrac{2i}{-1 - i\sqrt{3}}$
21. $\dfrac{-i}{1 + i}$

22. $\dfrac{1}{2 - 2i}$
23. $\dfrac{2\sqrt{6} - 2i\sqrt{2}}{\sqrt{2} - i\sqrt{6}}$
24. $\dfrac{4 + 4i}{2 - 2i}$

Use a calculator to perform the indicated operations. Give answers in rectangular form. See Example 3.

25. $[2.5(\cos 35° + i \sin 35°)][3.0(\cos 50° + i \sin 50°)]$ **26.** $[4.6(\cos 12° + i \sin 12°)][2.0(\cos 13° + i \sin 13°)]$

27. $(12 \text{ cis } 18.5°)(3 \text{ cis } 12.5°)$ **28.** $(4 \text{ cis } 19.25°)(7 \text{ cis } 41.75°)$ **29.** $\dfrac{45(\cos 127° + i \sin 127°)}{22.5(\cos 43° + i \sin 43°)}$

30. $\dfrac{30(\cos 130° + i \sin 130°)}{10(\cos 21° + i \sin 21°)}$ **31.** $\left[2 \text{ cis } \dfrac{5\pi}{9}\right]^2$ **32.** $\left[24.3\left(\cos \dfrac{7\pi}{12} + i \sin \dfrac{7\pi}{12}\right)\right]^2$

Relating Concepts

For individual or collaborative investigation
(Exercises 33–39)

Consider the complex numbers

$$w = -1 + i \quad \text{and} \quad z = -1 - i.$$

Work Exercises 33–39 in order.

33. Multiply w and z using their rectangular forms and the FOIL method from Section 8.1. Leave the product in rectangular form.

34. Find the trigonometric forms of w and z.

35. Multiply w and z using their trigonometric forms and the method described in this section.

36. Use the result of Exercise 35 to find the rectangular form of wz. How does this compare to your result in Exercise 33?

37. Find the quotient w/z using their rectangular forms and multiplying both the numerator and the denominator by the conjugate of the denominator. Leave the quotient in rectangular form.

38. Use the trigonometric forms of w and z, found in Exercise 34, to divide w by z using the method described in this section.

39. Use the result of Exercise 38 to find the rectangular form of w/z. How does this compare to your result in Exercise 37?

40. Without actually performing the operations, state why the products

$$[2(\cos 45° + i \sin 45°)] \cdot [5(\cos 90° + i \sin 90°)]$$

and $$\{2[\cos(-315°) + i \sin(-315°)]\} \cdot \{5[\cos(-270°) + i \sin(-270°)]\}$$

are the same.

41. Notice that $(r \text{ cis } \theta)^2 = (r \text{ cis } \theta)(r \text{ cis } \theta) = r^2 \text{ cis}(\theta + \theta) = r^2 \text{ cis } 2\theta$. State in your own words how we can square a complex number in trigonometric form. (In the next section, we will develop this idea more fully.)

42. Show that $1/z = (1/r)(\cos \theta - i \sin \theta)$, where $z = r(\cos \theta + i \sin \theta)$.

Solve each problem.

43. *Electrical Current* The alternating current in an electric inductor is

$$I = \frac{E}{Z}$$

amperes, where E is voltage and $Z = R + X_L i$ is impedance. If $E = 8(\cos 20° + i \sin 20°)$, $R = 6$, and $X_L = 3$, find the current. Give the answer in rectangular form, with real and imaginary parts to the nearest hundredth.

44. *Electrical Current* The current I in a circuit with voltage E, resistance R, capacitive reactance X_c, and inductive reactance X_L is

$$I = \frac{E}{R + (X_L - X_c)i}.$$

Find I if $E = 12(\cos 25° + i \sin 25°)$, $R = 3$, $X_L = 4$, and $X_c = 6$. Give the answer in rectangular form, with real and imaginary parts to the nearest tenth.

Impedance *Refer to Exercises 71 and 72 in Section 8.1. In the parallel electrical circuit shown in the figure here, the impedance Z can be calculated using the equation*

$$Z = \cfrac{1}{\cfrac{1}{Z_1} + \cfrac{1}{Z_2}},$$

where Z_1 and Z_2 are the impedances for the branches of the circuit.

45. If $Z_1 = 50 + 25i$ and $Z_2 = 60 + 20i$, calculate Z.

46. Determine the phase angle θ.

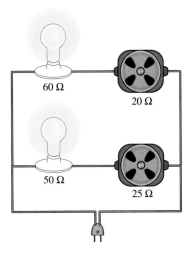

8.4 Powers and Roots of Complex Numbers

• Powers • Roots • **Equations with Complex Solutions**

Powers   In Section 8.3 we studied the product and quotient theorems for complex numbers in trigonometric form. Because raising a number to a positive integer power is a repeated application of the product rule, it would seem likely that a theorem for finding powers of complex numbers exists. This is indeed the case. For example, the square of the complex number $r(\cos \theta + i \sin \theta)$ is

$$\begin{aligned}
[r(\cos \theta + i \sin \theta)]^2 &= [r(\cos \theta + i \sin \theta)][r(\cos \theta + i \sin \theta)] \\
&= r \cdot r[\cos(\theta + \theta) + i \sin(\theta + \theta)] \\
&= r^2(\cos 2\theta + i \sin 2\theta).
\end{aligned}$$

In the same way,

$$[r(\cos \theta + i \sin \theta)]^3 = r^3(\cos 3\theta + i \sin 3\theta).$$

These results suggest the plausibility of the following theorem for positive integer values of n. Although this theorem is stated and can be proved for all n, we will use it only for positive integer values of n and their reciprocals.

De Moivre's Theorem

If $r(\cos \theta + i \sin \theta)$ is a complex number and if n is any real number, then

$$[r(\cos \theta + i \sin \theta)]^n = r^n(\cos n\theta + i \sin n\theta).$$

In compact form, this is written

$$[r \operatorname{cis} \theta]^n = r^n(\operatorname{cis} n\theta).$$

This theorem is named after the French expatriate friend of Issac Newton, Abraham De Moivre (1667–1754), although he never explicitly stated it.

● ● ● **Example 1** Applying De Moivre's Theorem (Finding a Power of a Complex Number)

Find $\left(1 + i\sqrt{3}\right)^8$ and express the result in rectangular form.

To use De Moivre's theorem, first convert $1 + i\sqrt{3}$ into trigonometric form using the methods of Section 8.2.

$$1 + i\sqrt{3} = 2(\cos 60° + i \sin 60°)$$

Now apply De Moivre's theorem.

$$
\begin{aligned}
\left(1 + i\sqrt{3}\right)^8 &= [2(\cos 60° + i \sin 60°)]^8 \\
&= 2^8[\cos(8 \cdot 60°) + i \sin(8 \cdot 60°)] \\
&= 256(\cos 480° + i \sin 480°) \\
&= 256(\cos 120° + i \sin 120°) \qquad \text{\small 480° and 120° are coterminal.} \\
&= 256\left(-\frac{1}{2} + i\frac{\sqrt{3}}{2}\right) \qquad \text{\small $\cos 120° = -\dfrac{1}{2}$;} \\
&\qquad\qquad\qquad\qquad\qquad \text{\small $\sin 120° = \dfrac{\sqrt{3}}{2}$} \\
&= -128 + 128i\sqrt{3} \qquad\quad \text{\small Rectangular form} \qquad ● ● ●
\end{aligned}
$$

Roots In algebra it is shown that every nonzero complex number has exactly n distinct complex nth roots. De Moivre's theorem can be extended to find all nth roots of a complex number. An nth root of a complex number is defined as follows.

---

*n*th Root

For a positive integer n, the complex number $a + bi$ is an ***n*th root** of the complex number $x + yi$ if

$$(a + bi)^n = x + yi.$$

To find the cube roots of the complex number $8(\cos 135° + i \sin 135°)$, for example, look for a complex number, say $r(\cos \alpha + i \sin \alpha)$, that will satisfy

$$[r(\cos \alpha + i \sin \alpha)]^3 = 8(\cos 135° + i \sin 135°).$$

By De Moivre's theorem, this equation becomes

$$r^3(\cos 3\alpha + i \sin 3\alpha) = 8(\cos 135° + i \sin 135°).$$

One way to satisfy this equation is to set $r^3 = 8$ and also $\cos 3\alpha + i \sin 3\alpha = \cos 135° + i \sin 135°$. The first of these conditions implies that $r = 2$, and the second implies that

$$\cos 3\alpha = \cos 135° \qquad \text{and} \qquad \sin 3\alpha = \sin 135°.$$

For these equations to be satisfied, 3α must represent an angle that is coterminal with 135°. Therefore, we must have

$$3\alpha = 135° + 360° \cdot k, \quad k \text{ any integer,}$$

or

$$\alpha = \frac{135° + 360° \cdot k}{3}, \quad k \text{ any integer.}$$

Now let k take on the integer values 0, 1, and 2.

$$\text{If } k = 0, \text{ then} \qquad \alpha = \frac{135° + 0°}{3} = 45°.$$

$$\text{If } k = 1, \text{ then} \qquad \alpha = \frac{135° + 360°}{3} = \frac{495°}{3} = 165°.$$

$$\text{If } k = 2, \text{ then} \qquad \alpha = \frac{135° + 720°}{3} = \frac{855°}{3} = 285°.$$

In the same way, $\alpha = 405°$ when $k = 3$. But note that $\sin 405° = \sin 45°$ and $\cos 405° = \cos 45°$. If $k = 4$, then $\alpha = 525°$, which has the same sine and cosine values as $165°$. To continue with larger values of k would just be repeating solutions already found. Therefore, all of the cube roots (three of them) can be found by letting $k = 0$, 1, and 2.

When $k = 0$, the root is $2(\cos 45° + i \sin 45°)$.

When $k = 1$, the root is $2(\cos 165° + i \sin 165°)$.

When $k = 2$, the root is $2(\cos 285° + i \sin 285°)$.

In conclusion, we see that $2(\cos 45° + i \sin 45°)$, $2(\cos 165° + i \sin 165°)$, and $2(\cos 285° + i \sin 285°)$ are the three cube roots of $8(\cos 135° + i \sin 135°)$.

Notice that the formula for α in the discussion above can be written in the alternative form

$$\alpha = \frac{135°}{3} + \frac{360° \cdot k}{3} = 45° + 120° \cdot k,$$

for $k = 0$, 1, and 2, which is easier to use.

Generalizing the work above leads to the following theorem.

---

nth Root Theorem

If n is any positive integer and r is a positive real number, then the complex number $r(\cos \theta + i \sin \theta)$ has exactly n distinct nth roots, given by

$$\sqrt[n]{r}(\cos \alpha + i \sin \alpha) \qquad \text{or} \qquad \sqrt[n]{r} \text{ cis } \alpha,$$

where

$$\alpha = \frac{\theta + 360° \cdot k}{n} \qquad \text{or} \qquad \alpha = \frac{\theta}{n} + \frac{360° \cdot k}{n},$$

$k = 0, 1, 2, \ldots, n - 1.$

● ● ●   **Example 2**   Finding Roots of a Complex Number

Find and graph all fourth roots of $-8 + 8i\sqrt{3}$. Write the roots in rectangular form.

First write $-8 + 8i\sqrt{3}$ in trigonometric form as

$$-8 + 8i\sqrt{3} = 16 \text{ cis } 120°.$$

Here $r = 16$ and $\theta = 120°$. The fourth roots of this number have modulus $\sqrt[4]{16} = 2$ and arguments given as follows. Using the alternative formula for α,

$$\alpha = \frac{120°}{4} + \frac{360° \cdot k}{4} = 30° + 90° \cdot k.$$

If $k = 0$, then $\quad \alpha = 30° + 90° \cdot 0 = 30°.$

If $k = 1$, then $\quad \alpha = 30° + 90° \cdot 1 = 120°.$

If $k = 2$, then $\quad \alpha = 30° + 90° \cdot 2 = 210°.$

If $k = 3$, then $\quad \alpha = 30° + 90° \cdot 3 = 300°.$

Using these angles, the fourth roots are

$$2 \text{ cis } 30°, \quad 2 \text{ cis } 120°, \quad 2 \text{ cis } 210°, \quad \text{and} \quad 2 \text{ cis } 300°.$$

These four roots can be written in rectangular form as $\sqrt{3} + i$, $-1 + i\sqrt{3}$, $-\sqrt{3} - i$, and $1 - i\sqrt{3}$. The graphs of these roots are all on a circle that has center at the origin and radius 2, as shown in Figure 19. Notice that the roots are equally spaced about the circle 90° apart.

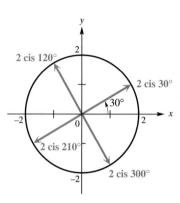

Figure 19

Equations with Complex Solutions

● ● ● **Example 3** Solving an Equation by Finding Complex Roots

Find and graph all complex number solutions of $x^5 - 1 = 0$.

Write the equation as

$$x^5 - 1 = 0$$
$$x^5 = 1.$$

While there is only one real number solution, 1, there are five complex number solutions. To find these solutions, first write 1 in trigonometric form as

$$1 = 1 + 0i = 1(\cos 0° + i \sin 0°).$$

The modulus of the fifth roots is $\sqrt[5]{1} = 1$, and the arguments are given by

$$0° + 72° \cdot k, \qquad k = 0, 1, 2, 3, \text{ or } 4.$$

By using these arguments, the fifth roots are

$$1(\cos 0° + i \sin 0°), \qquad {\scriptstyle k = 0}$$
$$1(\cos 72° + i \sin 72°), \qquad {\scriptstyle k = 1}$$
$$1(\cos 144° + i \sin 144°), \qquad {\scriptstyle k = 2}$$
$$1(\cos 216° + i \sin 216°), \qquad {\scriptstyle k = 3}$$

and $\qquad\qquad\qquad 1(\cos 288° + i \sin 288°). \qquad {\scriptstyle k = 4}$

The first of these roots equals 1; the others cannot easily be expressed in rectangular form. The five fifth roots all lie on a unit circle and are equally spaced around it every 72°, as shown in Figure 20.

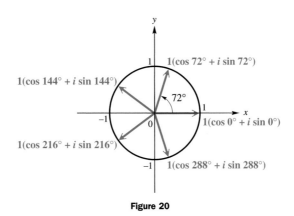

Figure 20

8.4 Exercises

Find each power. Write each answer in rectangular form. See Example 1.

1. $[3(\cos 30° + i \sin 30°)]^3$ **2.** $[2(\cos 135° + i \sin 135°)]^4$ **3.** $(\cos 45° + i \sin 45°)^8$

4. $[2(\cos 120° + i \sin 120°)]^3$ **5.** $[3 \text{ cis } 100°]^3$ **6.** $[3 \text{ cis } 40°]^3$

7. $(\sqrt{3} + i)^5$ **8.** $(2\sqrt{2} - 2i\sqrt{2})^6$ **9.** $(2 - 2i\sqrt{3})^4$

10. $\left(\dfrac{\sqrt{2}}{2} - \dfrac{\sqrt{2}}{2}i\right)^8$ **11.** $(-2 - 2i)^5$ **12.** $(-1 + i)^7$

Find and graph all cube roots of each complex number. Leave answers in trigonometric form. See Example 2.

13. $(\cos 0° + i \sin 0°)$ **14.** $(\cos 90° + i \sin 90°)$ **15.** $8 \text{ cis } 60°$ **16.** $27 \text{ cis } 300°$

17. $-8i$ **18.** $27i$ **19.** -64 **20.** 27

21. $1 + i\sqrt{3}$ **22.** $2 - 2i\sqrt{3}$ **23.** $-2\sqrt{3} + 2i$ **24.** $\sqrt{3} - i$

Find and graph all specified roots of 1.

25. second (square)   **26.** fourth   **27.** sixth   **28.** eighth

Find and graph all specified roots of i.

29. second (square)   **30.** fourth

Find all solutions of each equation. Leave answers in trigonometric form. See Example 3.

31. $x^3 - 1 = 0$ **32.** $x^3 + 1 = 0$ **33.** $x^3 + i = 0$ **34.** $x^4 + i = 0$

35. $x^3 - 8 = 0$ **36.** $x^3 + 27 = 0$ **37.** $x^4 + 1 = 0$ **38.** $x^4 + 16 = 0$

39. $x^4 - i = 0$ **40.** $x^5 - i = 0$ **41.** $x^3 - (4 + 4i\sqrt{3}) = 0$ **42.** $x^4 - (8 + 8i\sqrt{3}) = 0$

43. Solve the equation $x^3 - 1 = 0$ by factoring the left side as the difference of two cubes and setting each factor equal to 0. Apply the quadratic formula as needed. Then compare your solutions to those of Exercise 31.

44. Solve the equation $x^3 + 27 = 0$ by factoring the left side as the sum of two cubes and setting each factor equal to 0. Apply the quadratic formula as needed. Then compare your solutions to those of Exercise 36.

. . . . **Relating Concepts** . . . .
For individual or collaborative investigation
(Exercises 45–48)

In Chapter 5 we derived identities, or formulas, for cos 2θ *and* sin 2θ. *Interestingly, these identities can also be derived using De Moivre's theorem.* **Work Exercises 45–48 in order, to see how this is done.**

45. De Moivre's theorem states that $(\cos\theta + i\sin\theta)^2 =$ _____ .

46. Expand the left side of the equation in Exercise 45 as a binomial and collect terms to write the left side in the form $a + bi$.

47. Use the equality principle stated in the directions to Exercises 31–36 of Section 8.1 to obtain the double-angle formula for the cosine.

48. Repeat Exercise 47, but find the double-angle formula for the sine.

Solve each problem.

49. *Mandelbrot Set* The fractal called the *Mandelbrot set* is shown in the figure. To determine if a complex number $z = a + bi$ is in this set, perform the following sequence of calculations. Repeatedly compute

$$z, \quad z^2 + z, \quad (z^2 + z)^2 + z, \quad [(z^2 + z)^2 + z]^2 + z, \dots.$$

In a manner analogous to the Julia set, the complex number z does not belong to the Mandelbrot set if any of the resulting moduli exceed 2. Otherwise z is in the set and the point (a, b) should be shaded in the graph. Determine whether or not the following numbers belong to the Mandelbrot set. (*Source:* Lauwerier, H., *Fractals,* Princeton University Press, 1991.)
(a) $z = 0 + 0i$ **(b)** $z = 1 - 1i$ **(c)** $z = -.5i$

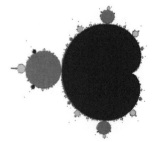

Source: Figure from Crownover, R.: *Introduction to Fractals and Chaos.* Copyright © 1995. Boston: Jones and Bartlett Publishers. Reprinted with permission.

50. *Basins of Attraction* The fractal shown in the figure is the solution to Cayley's problem of determining the basins of attraction for the cube roots of unity. The three cube roots of unity are

$$w_1 = 1, \quad w_2 = -\frac{1}{2} + \frac{\sqrt{3}}{2}i, \quad \text{and} \quad w_3 = -\frac{1}{2} - \frac{\sqrt{3}}{2}i.$$

This fractal can be generated by repeatedly evaluating the function with $f(z) = \dfrac{2z^3 + 1}{3z^2}$, where z is a complex number. One begins by picking $z_1 = a + bi$ and then successively computing $z_2 = f(z_1)$, $z_3 = f(z_2)$, $z_4 = f(z_3), \dots$. If the resulting values of $f(z)$ approach w_1, color the pixel at (a, b) red. If it approaches w_2, color it blue, and if it approaches w_3, color it yellow. If this process continues for a large number of different z_1, the fractal in the figure will appear. Determine the appropriate color of the pixel for each value of z_1. (*Source:* Crownover, R., *Introduction to Fractals and Chaos,* Jones and Bartlett Publishers, 1995.)
(a) $z_1 = i$ **(b)** $z_1 = 2 + i$ **(c)** $z_1 = -1 - i$

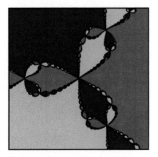

Source: Kincaid, D. and Cheney, W., *Numerical Analysis,* Brooks/Cole Publishing Co., 1991.

51. The screens here illustrate how a pentagon can be graphed using a graphing calculator. Note that a pentagon has five sides, and the T-step is $360/5 = 72$. The display at the bottom of the graph screen indicates that one fifth root of 1 is $1 + 0i = 1$. Use this technique to find all fifth roots of 1, and express the real and imaginary parts in decimal form.

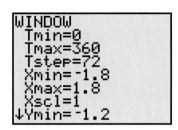

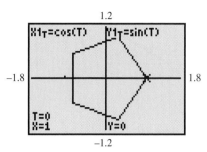

This is a continuation of the previous screen.

The calculator is in parametric, degree, and connected graph modes.

52. Use the method of Exercise 51 to find the first three of the ten 10th roots of 1.

53. One of the three cube roots of a complex number is $2 + 2\sqrt{3}i$. Determine the rectangular form of its other two cube roots.

Use a calculator to find all solutions of each equation in rectangular form.

54. $x^3 + 4 - 5i = 0$ **55.** $x^5 + 2 + 3i = 0$

56. *Concept Check*   How many complex 64th roots does 1 have? How many are real? How many are not?

57. *Concept Check*   True or false: Every real number must have two distinct real square roots.

58. *Concept Check*   True or false: Some real numbers have three real cube roots.

59. Show that if z is an nth root of 1, then so is $1/z$.

60. Explain why a real number can have only one real cube root.

61. Explain why the n nth roots of 1 are equally spaced around the unit circle.

62. Refer to Figure 20. A regular pentagon can be created by joining the tips of the arrows. Explain how you can use this principle to create a regular octagon.

8.5   Polar Equations and Graphs

• **Polar Coordinates** • **Graphs of Polar Equations** • **Summary of Polar Graphs** • **Converting between Equation Forms**

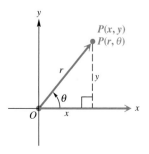

Figure 21

Figure 22

Polar Coordinates   Throughout this text we have been using the Cartesian coordinate system to graph equations. Another coordinate system that is particularly useful for graphing many relations is the **polar coordinate system.** The system is based on a point, called the **pole,** and a ray, called the **polar axis.** The polar axis is usually drawn in the direction of the positive x-axis, as shown in Figure 21.

In Figure 22 the pole has been placed at the origin of a Cartesian coordinate system, so the polar axis coincides with the positive x-axis. Point P has coordinates (x, y) in the Cartesian coordinate system. Point P can also be located by giving the directed angle θ from the positive x-axis to ray OP and the directed distance r from the pole to point P. The ordered pair (r, θ) gives the **polar coordinates** of point P.

The use of polar coordinates was first suggested by Sir Isaac Newton in about 1671. His work was expanded upon by Jakob Bernoulli in 1691. In later years, Jacob Hermann and Leonhard Euler provided further development of the polar coordinate system.

● ● ● **Example 1** Graphing Points with Polar Coordinates

Plot each point, given its polar coordinates.

(a) $P(2, 30°)$

In this case, $r = 2$ and $\theta = 30°$, so the point P is located 2 units from the pole in the positive direction on a ray making a $30°$ angle with the polar axis, as shown in Figure 23.

(b) $Q(-4, 120°)$

Since r is negative, Q is 4 units in the negative direction from the pole on an extension of the $120°$ ray. See Figure 24.

(c) $R(5, -45°)$

Point R is shown in Figure 25. Since θ is negative, the angle is measured in the clockwise direction.

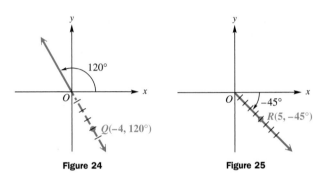

Figure 24   **Figure 25**   ● ● ●

One important difference between Cartesian coordinates and polar coordinates is that while a given point in the plane can have only one pair of Cartesian coordinates, this same point can have an infinite number of pairs of polar coordinates. For example, $(2, 30°)$ locates the same point as $(2, 390°)$ or $(2, -330°)$ or $(-2, 210°)$.

● ● ● **Example 2** Giving Alternative Forms for Coordinates of a Point

(a) Give three other pairs of polar coordinates for the point $P(3, 140°)$.

Three pairs that could be used for the point are $(3, -220°)$, $(-3, 320°)$, and $(-3, -40°)$. See Figure 26.

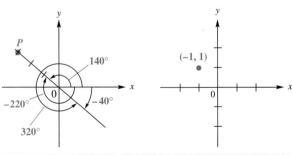

Figure 26   **Figure 27**

(b) Determine two pairs of polar coordinates for the point with rectangular coordinates $(-1, 1)$.

As shown in Figure 27, the point $(-1, 1)$ lies in the second quadrant. Since $\tan \theta = \dfrac{1}{-1} = -1$, one possible value for θ is $135°$. Also,

$$r = \sqrt{x^2 + y^2} = \sqrt{(-1)^2 + 1^2} = \sqrt{2}.$$

Therefore, two pairs of polar coordinates are $\left(\sqrt{2}, 135°\right)$ and $\left(\sqrt{2}, -225°\right)$. (Any angle coterminal with $135°$ could have been used for the second angle.)

● ● ●

Graphs of Polar Equations   An equation like $r = 3 \sin \theta$, where r and θ are the variables, is a **polar equation.** (Equations in x and y are called **rectangular** or **Cartesian equations.**) The simplest equation for many useful curves turns out to be a polar equation.

> **C O N N E C T I O N S**   Lines and circles are two of the most common types of graphs studied in rectangular coordinates. While their rectangular forms are the ones most often encountered, they can also be defined in terms of polar coordinates. It can be shown that the line $ax + by = c$ has an equivalent polar equation
>
> $$r = \frac{c}{a \cos \theta + b \sin \theta}.$$
>
> The following two screens show the same line; the one on the left was graphed in function graphing mode and the one on the right in polar graphing mode. (Notice the defining equation at the top left of each screen.)
>
>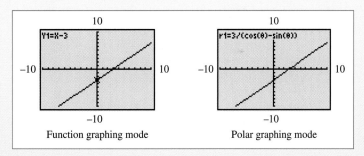
>
> Function graphing mode   Polar graphing mode
>
> The circle $x^2 + y^2 = a^2$ has polar form $r = a$. The circle centered at the origin with radius 2 is shown in both screens that follow. Again, the one on the left was graphed in function mode (as the union of two functions) and the one on the right in polar mode. The defining equations are shown on the screens.
>
>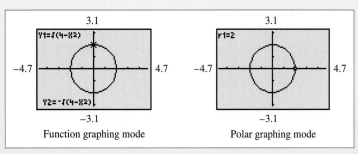
>
> Function graphing mode   Polar graphing mode
>
> *(continued)*

For Discussion or Writing

1. Give the $ax + by = c$ form of the equation of the line shown in the first two screens.
2. Give the $x^2 + y^2 = a^2$ form of the equation of the circle shown in the second two screens.

Graphing a polar equation in the traditional manner is much the same as graphing a Cartesian equation. Find some representative ordered pairs, (r, θ), satisfying the equation, and then sketch the graph.

A graphing calculator can be used to graph an equation in the form $r = f(\theta)$. Refer to your owner's manual to see how your model handles polar graphs. As always, it is necessary to set the window appropriately and choose the correct angle mode (radians or degrees). You will need to decide on maximum and minimum values of θ. Keep in mind the periods of the functions, so a complete set of ordered pairs is generated. ∎

● ● ● **Example 3** Graphing a Polar Equation (Cardioid)

Graph $r = 1 + \cos \theta$.

Traditional Approach

To graph this equation, find some ordered pairs (as in the table) and then connect the points in order — from $(2, 0°)$ to $(1.9, 30°)$ to $(1.7, 45°)$ and so on. The graph is shown in Figure 28. This curve is called a **cardioid** because of its heart shape.

θ	$\cos \theta$	$r = 1 + \cos \theta$	θ	$\cos \theta$	$r = 1 + \cos \theta$
0°	1	2	135°	−.7	.3
30°	.9	1.9	150°	−.9	.1
45°	.7	1.7	180°	−1	0
60°	.5	1.5	270°	0	1
90°	0	1	315°	.7	1.7
120°	−.5	.5			

Once the pattern of values of r becomes clear, it is not necessary to find more ordered pairs. That is why we stopped with the ordered pair $(1.7, 315°)$ in the table above. From the pattern, the pair $(1.9, 330°)$ also would satisfy the relation.

Graphing Calculator Approach

For this equation, we will choose degree mode and graph it for values of θ in the interval $[0°, 360°]$. The screens in Figure 29(a) and (b) show the choices needed to generate the graph shown in Figure 29(c).

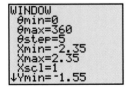

(a)

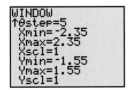

This is a continuation of the previous screen.

(b)

Figure 29

(continued)

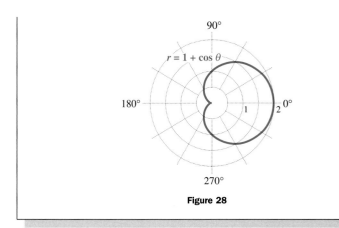

Figure 28

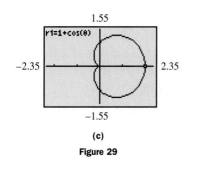

(c)
Figure 29

• • •

• • • **Example 4** Graphing a Polar Equation (Lemniscate)

Graph $r^2 = \cos 2\theta$.

Traditional Approach

First complete a table of ordered pairs as shown, and then sketch the graph, as in Figure 30. The point $(-1, 0°)$, with r negative, may be plotted as $(1, 180°)$. Also, $(-.7, 30°)$ may be plotted as $(.7, 210°)$, and so on. This curve is called a **lemniscate.**

θ	0°	30°	45°	135°	150°	180°
2θ	0°	60°	90°	270°	300°	360°
$\cos 2\theta$	1	.5	0	0	.5	1
$r = \pm\sqrt{\cos 2\theta}$	± 1	$\pm .7$	0	0	$\pm .7$	± 1

Graphing Calculator Approach

To graph $r^2 = \cos 2\theta$ with a graphing calculator, define r_1 as $\sqrt{\cos 2\theta}$ and r_2 as $-\sqrt{\cos 2\theta}$. See Figure 31.

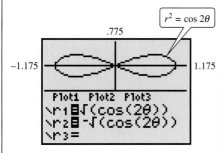

Figure 31

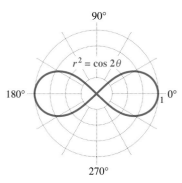

Figure 30

(continued)

Values of θ for $45° < \theta < 135°$ are not included in the table because the corresponding values of cos 2θ are negative (quadrants II and III) and so do not have real square roots. Values of θ larger than 180° give 2θ larger than 360° and would repeat the points already found.

● ● ● **Example 5** Graphing a Polar Equation (Rose)

Graph $r = 3 \cos 2\theta$.

Traditional Approach

Because of the argument 2θ, the graph requires a large number of points. A few ordered pairs are given below. You should complete the table similarly through the first 180°, so that 2θ has values up to 360°.

θ	0°	15°	30°	45°	60°	75°	90°
2θ	0°	30°	60°	90°	120°	150°	180°
cos 2θ	1	.9	.5	0	−.5	−.9	−1
$r = 3 \cos 2\theta$	3	2.7	1.5	0	−1.5	−2.7	−3

Plotting these points in order gives the graph, called a **four-leaved rose.** Notice in Figure 32 how the graph is developed with a continuous curve, beginning with the upper half of the right horizontal leaf and ending with the lower half of that leaf. As the graph is traced, the curve goes through the pole four times.

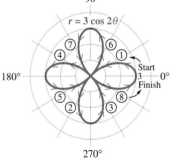

Figure 32

Graphing Calculator Approach

The screen in Figure 33 shows the graph of $r = 3 \cos 2\theta$. You can duplicate this screen and watch how the graph takes shape, comparing it to the description in Figure 32.

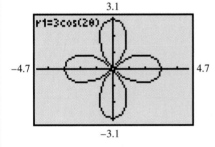

Figure 33

The equation in Example 5 has a graph that belongs to a family of curves called **roses.** The graphs of $r = \sin n\theta$ and $r = \cos n\theta$ are roses, with n petals if n is odd, and $2n$ petals if n is even.

● ● ●   **Example 6**   Graphing a Polar Equation (Spiral of Archimedes)

Graph $r = 2\theta$ (θ measured in radians).

Traditional Approach

Some ordered pairs are shown below. Since $r = 2\theta$, rather than a trigonometric function of θ, it is also necessary to consider negative values of θ. The radian measures have been rounded for simplicity.

θ (degrees)	θ (radians)	$r = 2\theta$
-180	-3.1	-6.2
-90	-1.6	-3.2
-45	$-.8$	-1.6
0	0	0
30	$.5$	1
60	1	2
90	1.6	3.2
180	3.1	6.2
270	4.7	9.4
360	6.3	12.6

Graphing Calculator Approach

Figure 35 shows much more of the spiral than is seen in the traditional graph.

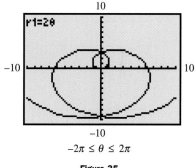

$-2\pi \le \theta \le 2\pi$

Figure 35

Figure 34 shows this graph, called a **spiral of Archimedes.**

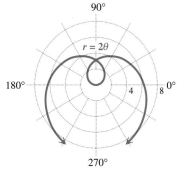

Figure 34

● ● ●

Summary of Polar Graphs The following chart summarizes some of the more common polar graphs and forms of their equations. (In addition to circles, lemniscates, and roses just presented, we include *limaçons*. Cardioids are a special case of limaçons, where $|a/b| \geq 1$.)

Circles and Lemniscates

Circles		Lemniscates	
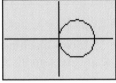	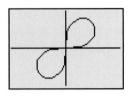	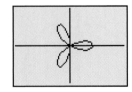	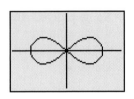
$r = a \cos \theta$	$r = a \sin \theta$	$r^2 = a^2 \sin 2\theta$	$r^2 = a^2 \cos 2\theta$

Limaçons

$$r = a \pm b \sin \theta \quad \text{or} \quad r = a \pm b \cos \theta$$

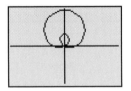	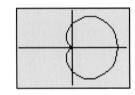	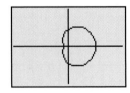	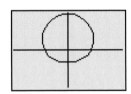
$\dfrac{a}{b} < 1$	$\dfrac{a}{b} = 1$	$1 < \dfrac{a}{b} < 2$	$\dfrac{a}{b} \geq 2$

Rose Curves

$2n$ petals if n is even, $n \geq 2$		n petals if n is odd	
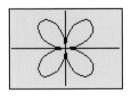	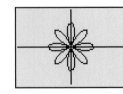		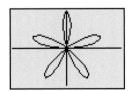
$n = 2$ $r = a \sin n\theta$	$n = 4$ $r = a \cos n\theta$	$n = 3$ $r = a \cos n\theta$	$n = 5$ $r = a \sin n\theta$

Converting between Equation Forms Sometimes an equation given in polar form is easier to graph in rectangular (Cartesian) form. To convert a polar equation to a rectangular equation, we use the following relationships, which were introduced in Section 8.2. See triangle *POQ* in Figure 36.

> ## Converting between Polar and Rectangular Coordinates
>
> $$x = r \cos \theta \qquad r = \sqrt{x^2 + y^2}$$
>
> $$y = r \sin \theta \qquad \tan \theta = \frac{y}{x}, \quad \text{if } x \neq 0$$

● ● ●   **Example 7**   **Converting a Polar Equation to a Rectangular Equation**

Convert the equation to rectangular coordinates, and graph.

$$r = \frac{4}{1 + \sin \theta}$$

Multiply both sides of the equation by the denominator on the right, to clear the fraction.

$$r = \frac{4}{1 + \sin \theta}$$

$$r + r \sin \theta = 4$$

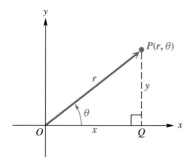

Figure 36

Now substitute $\sqrt{x^2 + y^2}$ for r and y for $r \sin \theta$.

$$\sqrt{x^2 + y^2} + y = 4$$

$$\sqrt{x^2 + y^2} = 4 - y$$

Square both sides to eliminate the radical.

$$x^2 + y^2 = (4 - y)^2$$

$$x^2 + y^2 = 16 - 8y + y^2$$

$$x^2 = -8y + 16$$

$$x^2 = -8(y - 2)$$

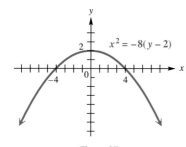

Figure 37

The final equation represents a parabola and can be graphed using rectangular coordinates. See Figure 37.   ● ● ●

● ● ●   **Example 8**   **Converting a Rectangular Equation to a Polar Equation**

Convert the equation $3x + 2y = 4$ to a polar equation.
   Use $x = r \cos \theta$ and $y = r \sin \theta$ to get

$$3x + 2y = 4$$

$$3r \cos \theta + 2r \sin \theta = 4.$$

Now solve for r. First factor out r on the left.

$$r(3 \cos \theta + 2 \sin \theta) = 4$$

$$r = \frac{4}{3 \cos \theta + 2 \sin \theta}$$

The polar equation of the line $3x + 2y = 4$ is

$$r = \frac{4}{3 \cos \theta + 2 \sin \theta}.$$ ● ● ●

In Examples 7 and 8, we presented methods of equation conversion. To support our results, see Figures 38 and 39, which show how a graphing calculator graphs the polar equations directly.

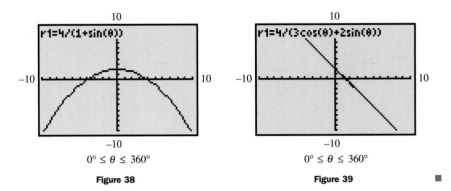

$0° \leq \theta \leq 360°$

Figure 38

$0° \leq \theta \leq 360°$

Figure 39

8.5 Exercises

1. *Concept Check* For each point given in polar coordinates, state the quadrant in which the point lies if it is graphed in a rectangular coordinate system.
   (a) $(5, 135°)$ **(b)** $(2, 60°)$ **(c)** $(6, -30°)$ **(d)** $(4.6, 213°)$

2. *Concept Check* For each point given in polar coordinates, state the axis on which the point lies if it is graphed in a rectangular coordinate system. Also state whether it is on the positive portion or the negative portion of the axis. (For example, $(5, 0°)$ lies on the positive x-axis.)
   (a) $(7, 360°)$ **(b)** $(4, 180°)$ **(c)** $(2, -90°)$ **(d)** $(8, 450°)$

Plot each point, given its polar coordinates. Give two other pairs of polar coordinates for each point. See Examples 1 and 2(a).

3. $(1, 45°)$ **4.** $(3, 120°)$ **5.** $(-2, 135°)$ **6.** $(-4, 27°)$ **7.** $(5, -60°)$

8. $(2, -45°)$ **9.** $(-3, -210°)$ **10.** $(-1, -120°)$ **11.** $(3, 300°)$ **12.** $(4, 270°)$

Plot the point whose rectangular coordinates are given. Then determine two pairs of polar coordinates for the point with $0° \leq \theta < 360°$. See Example 2(b).

13. $(-1, 1)$ **14.** $(1, 1)$ **15.** $(0, 3)$ **16.** $(0, -3)$ **17.** $\left(\sqrt{2}, \sqrt{2}\right)$

18. $\left(-\sqrt{2}, \sqrt{2}\right)$ **19.** $\left(\dfrac{\sqrt{3}}{2}, \dfrac{3}{2}\right)$ **20.** $\left(-\dfrac{\sqrt{3}}{2}, -\dfrac{1}{2}\right)$ **21.** $(3, 0)$

22. *Concept Check* Match the polar graphs below to their corresponding equations in choices A–D.

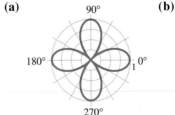 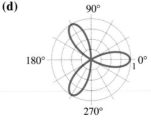

A. $r = 3$ **B.** $r = \cos 3\theta$ **C.** $r = \cos 2\theta$ **D.** $r = \dfrac{2}{\cos \theta + \sin \theta}$

Graph each polar equation for θ in $[0°, 360°)$. *In Exercises 23–32, also identify the type of polar graph. Use traditional methods or a graphing calculator, as directed by your instructor. See Examples 3–6.*

23. $r = 2 + 2 \cos \theta$

24. $r = 2(4 + 3 \cos \theta)$

25. $r = 3 + \cos \theta$

26. $r = 2 - \cos \theta$

27. $r = 4 \cos 2\theta$

28. $r = 3 \cos 5\theta$

29. $r^2 = 4 \cos 2\theta$

30. $r^2 = 4 \sin 2\theta$

31. $r = 4(1 - \cos \theta)$

32. $r = 3(2 - \cos \theta)$

33. $r = 2 \sin \theta \tan \theta$
(This is a *cissoid*.)

34. $r = \dfrac{\cos 2\theta}{\cos \theta}$
(This is a *cissoid* with a loop.)

· · · · · · · · · · · · **Relating Concepts** · · · · · · · · · · · · · · ·

For individual or collaborative investigation
(Exercises 35–42)

You have probably observed symmetry in the polar graphs in this section. Visualize an xy-plane superimposed on the polar coordinate system, with the pole at the origin and the polar axis on the positive x-axis. Then a polar graph may be symmetric with respect to the x-axis (the polar axis), the y-axis (the line $\theta = \pi/2$), *or the origin (the pole).* **Work Exercises 35–42 in order.**

35. Complete the missing ordered pairs in the graphs below.

(a)

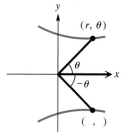

(b)

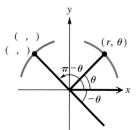

(c)

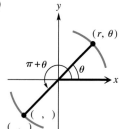

Use your answers for Exercise 35 to complete the sentences in Exercises 36–42.

36. The graph of $r = f(\theta)$ is symmetric with respect to the polar axis if substitution of _____ for θ leads to an equivalent equation.

37. The graph of $r = f(\theta)$ is symmetric with respect to the vertical line $\theta = \pi/2$ if substitution of _____ for θ leads to an equivalent equation.

38. Alternatively, the graph of $r = f(\theta)$ is symmetric with respect to the vertical line $\theta = \pi/2$ if substitution of _____ for r and _____ for θ leads to an equivalent equation.

39. The graph of $r = f(\theta)$ is symmetric with respect to the pole if substitution of _____ for r leads to an equivalent equation.

40. Alternatively, the graph of $r = f(\theta)$ is symmetric with respect to the pole if substitution of _____ for θ leads to an equivalent equation.

41. In general, the completed statements in Exercises 36–40 mean that the graphs of polar equations of the form $r = a \pm b \cos \theta$ (where a may be 0) are symmetric with respect to _____.

42. In general, the completed statements in Exercises 36–40 mean that the graphs of polar equations of the form $r = a \pm b \sin \theta$ (where a may be 0) are symmetric with respect to _____.

In Exercises 43 and 44, find the greatest value of $|r|$ of any point on the graph. Also, find all values of θ for which $r = 0$.

43. $r = 4 \cos 2\theta, 0° \leq \theta < 360°$

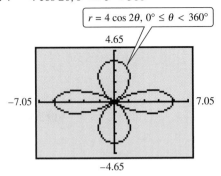

$r = 4 \cos 2\theta, 0° \leq \theta < 360°$

4.65

−7.05   7.05

−4.65

44. $r = 5 \sin 3\theta, 0° \leq \theta < 180°$

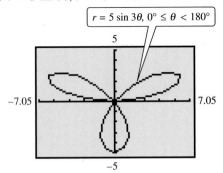

$r = 5 \sin 3\theta, 0° \leq \theta < 180°$

5

−7.05   7.05

−5

45. The screens below indicate the same point. Verify algebraically that the polar coordinates shown in the left screen and the rectangular coordinates shown in the right screen are equivalent.

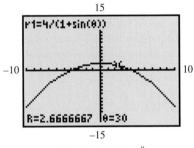

15

r1=4/(1+sin(θ))

−10   10

R=2.6666667 θ=30

−15

When $\theta = 30°$, R $= \frac{8}{3}$.

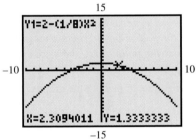

15

Y1=2−(1/8)X²

−10   10

X=2.3094011 Y=1.3333333

−15

When X $= \frac{4\sqrt{3}}{3}$, Y $= \frac{4}{3}$.

Find the polar coordinates of the points of intersection of the given curves for the specified interval of θ.

46. $r = 4 \sin \theta, r = 1 + 2 \sin \theta; \quad 0 \leq \theta < 2\pi$

47. $r = 2 + \sin \theta, r = 2 + \cos \theta; \quad 0 \leq \theta < 2\pi$

48. $r = \sin 2\theta, r = \sqrt{2} \cos \theta; \quad 0 \leq \theta < \pi$

49. Explain the method used to plot a point (r, θ) in polar coordinates, if $r < 0$.

50. Refer to Example 8. Would you find it easier to graph the equation using the Cartesian form or the polar form? Why?

For each equation, find an equivalent equation in rectangular coordinates and graph. Give a traditional or a calculator graph, as directed by your instructor. See Example 7.

51. $r = 2 \sin \theta$

52. $r = 2 \cos \theta$

53. $r = \dfrac{2}{1 - \cos \theta}$

54. $r = \dfrac{3}{1 - \sin \theta}$

55. $r + 2 \cos \theta = -2 \sin \theta$

56. $r = \dfrac{3}{4 \cos \theta - \sin \theta}$

57. $r = 2 \sec \theta$

58. $r = -5 \csc \theta$

59. $r(\cos \theta + \sin \theta) = 2$

60. $r(2 \cos \theta + \sin \theta) = 2$

For each equation, find an equivalent equation in polar coordinates. See Example 8.

61. $x + y = 4$

62. $2x - y = 5$

63. $x^2 + y^2 = 16$

64. $x^2 + y^2 = 9$

65. $y = 2$

66. $x = 4$

67. Graph $r = \theta$, a spiral of Archimedes. (See Example 6.) Use both positive and nonpositive values for θ.

68. Use a graphing calculator to graph a great deal more of $r = 2\theta$ (a spiral of Archimedes) than what is shown in Figure 34.

69. Find the polar equation of the line that passes through the points $(1, 0°)$ and $(2, 90°)$.

70. *Orbits of Planets*   The polar equation

$$r = \frac{a(1 - e^2)}{1 + e \cos \theta}$$

can be used to graph the orbits of the planets, where a is the average distance in astronomical units from the sun and e is a constant called *eccentricity*. The sun will be located at the pole. The table lists a and e for the planets.

(a) Graph the orbits of the four planets closest to the sun on the same polar axis. Choose a viewing window that results in a graph with nearly circular orbits.

(b) Plot the orbits of Earth, Jupiter, Uranus, and Pluto on the same polar axis. How does Earth's distance from the sun compare to these planets?

(c) Use graphing to determine whether or not Pluto is always the farthest planet from the sun.

Planet	a	e	Planet	a	e
Mercury	.39	.206	Saturn	9.54	.056
Venus	.78	.007	Uranus	19.2	.047
Earth	1.00	.017	Neptune	30.1	.009
Mars	1.52	.093	Pluto	39.4	.249
Jupiter	5.20	.048			

Sources: Karttunen, H., P. Kröger, H. Oja, M. Putannen, and K. Donners (editors), *Fundamental Astronomy,* Springer-Verlag, 1994; Zeilik, M., S. Gregory, and E. Smith, *Introductory Astronomy and Astrophysics,* Fourth Edition, Saunders College Publishers, 1998.

8.6   Parametric Equations, Graphs, and Applications

• **Basic Concepts**   • **Parametric Graphs and Their Rectangular Equivalents**   • **The Cycloid**   • **Applications of Parametric Equations**

Basic Concepts   Throughout this text, we have graphed sets of ordered pairs of real numbers that correspond to a function of the form $y = f(x)$ or $r = g(\theta)$. Another way to determine a set of ordered pairs involves two functions f and g defined by $x = f(t)$ and $y = g(t)$, where t is a real number in some interval I. Each value of t leads to a corresponding x-value and a corresponding y-value, and thus to an ordered pair (x, y).

> **Parametric Equations of a Plane Curve**
>
> A **plane curve** is a set of points (x, y) such that $x = f(t)$, $y = g(t)$, and f and g are both defined on an interval I. The equations $x = f(t)$ and $y = g(t)$ are **parametric equations** with **parameter t.**

In addition to graphing rectangular and polar equations, graphing calculators are capable of graphing plane curves defined by parametric equations. The calculator must be set in parametric mode, and the window requires intervals for the parameter t, as well as for x and y. ∎

Parametric Graphs and Their Rectangular Equivalents

● ● ● **Example 1** Graphing a Plane Curve Defined Parametrically

Let $x = t^2$ and $y = 2t + 3$ for t in $[-3, 3]$. Graph the set of ordered pairs (x, y).

Traditional Approach

Begin by making a table of values.

t	-3	-2	-1	0	1	2	3
x	9	4	1	0	1	4	9
y	-3	-1	1	3	5	7	9

Now graph the points (x, y) from the table of values and connect them with a smooth curve as in Figure 40. Since the domain of t is a closed interval, the graph has endpoints at $(9, -3)$ and $(9, 9)$.

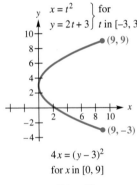

$$4x = (y - 3)^2$$
for x in $[0, 9]$

Figure 40

Graphing Calculator Approach

For this equation, we make the choices seen in the first two screens in Figure 41. The actual graph is shown in the final screen.

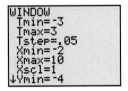

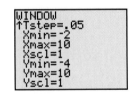

This is a continuation of the previous screen.

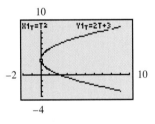

Figure 41

● ● ●

Sometimes it is possible to eliminate the parameter from a pair of parametric equations to get a rectangular equation, relating x and y.

● ● ● **Example 2** Finding an Equivalent Rectangular Equation

Find a rectangular equation for the plane curve of Example 1 defined as follows.

$$x = t^2, \, y = 2t + 3, \quad \text{for } t \text{ in } [-3, 3]$$

To eliminate the parameter t, solve either equation for t. Here, only the second equation, $y = 2t + 3$, leads to a unique solution for t, so choose it.

$$y = 2t + 3$$

$$2t = y - 3$$

$$t = \frac{y - 3}{2}$$

Now substitute this result in the first equation to get

$$x = t^2 = \left(\frac{y - 3}{2}\right)^2 = \frac{(y - 3)^2}{4}$$

or $4x = (y - 3)^2.$

This is the equation of a horizontal parabola opening to the right, which agrees with the graph given in Figure 40. Because t is in $[-3, 3]$, x is in $[0, 9]$ and y is in $[-3, 9]$. The rectangular equation must be given with its restricted domain as

$$4x = (y - 3)^2, \qquad \text{for } x \text{ in } [0, 9]. \qquad \bullet\ \bullet\ \bullet$$

Trigonometric functions are often used to define a plane curve parametrically.

$\bullet\ \bullet\ \bullet$ **Example 3** **Graphing a Plane Curve Defined Parametrically with Trigonometric Functions**

Graph the plane curve defined by $x = 2 \sin t$, $y = 3 \cos t$, for t in $[0, 2\pi]$.

Traditional Approach

To convert to a rectangular equation, it is not productive here to solve either equation for t. Instead, we use the fact that $\sin^2 t + \cos^2 t = 1$ to apply another approach. Square both sides of each equation; solve one for $\sin^2 t$, the other for $\cos^2 t$.

$$x = 2 \sin t \qquad y = 3 \cos t$$

$$x^2 = 4 \sin^2 t \qquad y^2 = 9 \cos^2 t$$

$$\frac{x^2}{4} = \sin^2 t \qquad \frac{y^2}{9} = \cos^2 t$$

Now add corresponding sides of the two equations.

$$\frac{x^2}{4} + \frac{y^2}{9} = \sin^2 t + \cos^2 t$$

$$\frac{x^2}{4} + \frac{y^2}{9} = 1$$

Graphing Calculator Approach

The ellipse can be graphed directly with the calculator in parametric mode. See Figure 43.

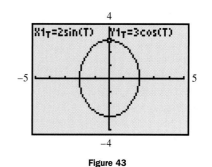

Figure 43

(continued)

This is the equation of an ellipse with vertical major axis, as shown in Figure 42.

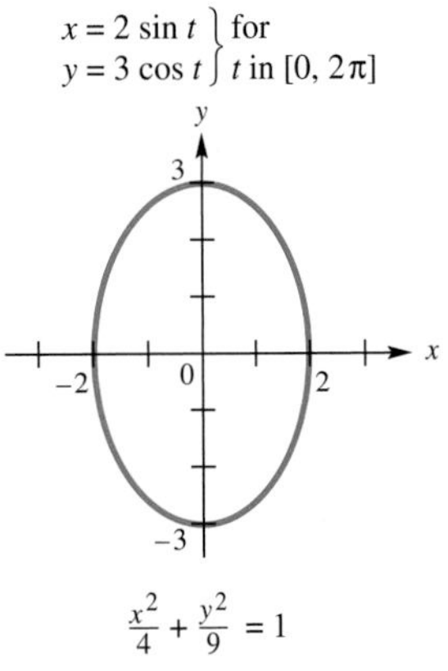

$$x = 2 \sin t \ \} \ \text{for}$$
$$y = 3 \cos t \ \} \ t \text{ in } [0, 2\pi]$$

$$\frac{x^2}{4} + \frac{y^2}{9} = 1$$

Figure 42

Parametric representations of a curve are not unique. In fact, there are infinitely many parametric representations of a given curve. If the curve can be described by a rectangular equation $y = f(x)$, with domain X, then one simple parametric representation is

$$x = t, y = f(t), \qquad \text{for } t \text{ in } X.$$

Example 4 Finding Alternative Parametric Equation Forms

Give three parametric representations for the parabola

$$y = (x - 2)^2 + 1.$$

The simplest choice is to let

$$x = t, y = (t - 2)^2 + 1, \qquad \text{for } t \text{ in } (-\infty, \infty).$$

Another choice, which leads to a simpler equation for y, is

$$x = t + 2, y = t^2 + 1, \qquad \text{for } t \text{ in } (-\infty, \infty).$$

Sometimes trigonometric functions are desirable; one choice here might be

$$x = 2 + \tan t, y = \sec^2 t, \qquad \text{for } t \text{ in } \left(-\frac{\pi}{2}, \frac{\pi}{2}\right).$$

Looking Ahead to Calculus

The cycloid is a special case of a curve traced out by a point at a given distance from the center of a circle as the circle rolls along a straight line. Such a curve is called a *trochoid*. It is just one of several parametrically defined curves studied in calculus. *Bezier curves* are used in manufacturing, and *Conchoids of Nicodemes* are so named because the shape of their outer branches resembles a conch shell. Other examples are *hypocycloids, epicycloids, the witch of Agnesi, swallowtail catastrophe curves,* and *Lissajou figures. (Source: Stewart, J., Calculus, Third Edition, Brooks/Cole Publishing Co., 1995.)*

The Cycloid The path traced by a fixed point on the circumference of a circle rolling along a line is called a *cycloid*. A **cycloid** is defined by

$$x = at - a \sin t, y = a - a \cos t, \qquad \text{for } t \text{ in } (-\infty, \infty).$$

• • •   **Example 5**   Graphing a Cycloid

Graph the cycloid with $a = 1$ for t in $[0, 2\pi]$.

Traditional Approach

There is no simple way to find a rectangular equation for the cycloid from its parametric equations. Instead, begin with a table of values.

t	0	$\dfrac{\pi}{4}$	$\dfrac{\pi}{2}$	π	$\dfrac{3\pi}{2}$	2π
x	0	.08	.6	π	5.7	2π
y	0	.3	1	2	1	0

Plotting the ordered pairs (x, y) from the table of values leads to the portion of the graph in Figure 44 from 0 to 2π.

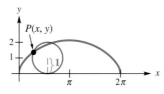

Figure 44

Graphing Calculator Approach

It is much easier to graph a cycloid with a graphing calculator in parametric mode than with traditional methods. See Figure 45.

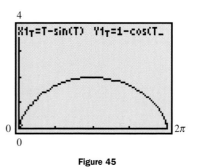

Figure 45

• • •

Figure 46

The cycloid has an interesting physical property. If a flexible cord or wire goes through points P and Q as in Figure 46, and a bead is allowed to slide due to the force of gravity without friction along this path from P to Q, the path that requires the shortest time takes the shape of the graph of an inverted cycloid.

Applications of Parametric Equations   An important application of parametric equations is to determine the path of a moving object whose position is given by the functions $x = f(t)$, $y = g(t)$, where t represents time. The parametric equations give the position of the object at any time t.

• • •   **Example 6**   Examining Parametric Equations Defining the Position of an Object in Motion

The motion of a projectile (neglecting air resistance) is given by

$$x = (v_0 \cos \theta)t, \quad y = (v_0 \sin \theta)t - 16t^2, \qquad \text{for } t \text{ in } [0, k],$$

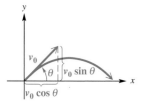

Figure 47

where t is time in seconds, v_0 is the initial speed of the projectile in the direction θ with the horizontal, x and y are in feet, and k is a positive real number. See Figure 47. Find the rectangular form of the downward-opening parabola seen in the figure.

Begin by solving the first equation for t to get

$$t = \frac{x}{v_0 \cos \theta}.$$

Now, substitute this expression for t into the second equation.

$$y = (v_0 \sin \theta)\left(\frac{x}{v_0 \cos \theta}\right) - 16\left(\frac{x}{v_0 \cos \theta}\right)^2$$

$$y = (\tan \theta)x - \frac{16}{v_0^2 \cos^2 \theta}x^2$$

● ● ●

If the projectile in Example 6 is fired with initial velocity $v_0 = 48$ feet per second and at an angle $\theta = 45°$ with respect to the horizontal, then the rectangular equation

$$y = (\tan \theta)x - \frac{16}{v_0^2 \cos^2 \theta}x^2$$

becomes

$$y = (\tan 45°)x - \frac{16}{48^2 \cos^2 45°}x^2.$$

Simplifying this last equation, we get

$$y = x - \frac{1}{72}x^2. \quad \text{Recall that } x \text{ and } y \text{ are in feet.}$$

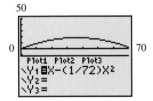

Figure 48

Figure 48 shows a calculator graph of this rectangular equation. Here we see the parabolic path of the projectile. In other words, we see *where* the projectile has been. However, the rectangular equation does not tell us *when* the projectile was at a particular point (x, y). The next example illustrates how we can use the parameter t, representing time, to determine when the projectile was at any point (x, y).

NOTE In this section we will assume that the only force acting on any projectile is gravity.

● ● ● **Example 7** Analyzing the Path of a Projectile

Use the parametric equations given in Example 6 with $v_0 = 48$ feet per second and $\theta = 45°$ to find the coordinates of the points where the projectile is located after 0 seconds and after 1 second.

Algebraic Solution

With $\theta = 45°$ and $v_0 = 48$ feet per second, the equations become

$$x = (48 \cos 45°)t$$
$$x = 24\sqrt{2}\,t$$

Graphing Calculator Solution

Figure 49 supports both of the algebraic results. The first is supported with the display at the bottom of the upper half of the screen, while the

(continued)

and

$$y = -16t^2 + (48 \sin 45°)t$$
$$y = -16t^2 + 24\sqrt{2}\,t.$$

Letting $t = 0$, we get $x = 0$ and $y = 0$, so at time $t = 0$, the coordinates of the location of the projectile are $(0, 0)$. Letting $t = 1$, we find that the coordinates are

$$\left(24\sqrt{2}, -16 + 24\sqrt{2}\right) \approx (33.94, 17.94).$$

second is supported in the bottom entry of the table in the lower half of the screen.

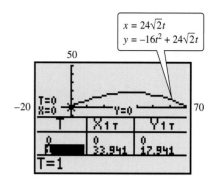

Figure 49

Notice that the graphs in Figures 48 and 49 are the same. They were, however, generated differently: in Figure 48, function mode was used, while in Figure 49, parametric mode was used.

• • •

So far we have analyzed the path of a projectile that has been launched from ground level. In general, the path of a projectile fired with an initial velocity v_0 feet per second, from a height h feet, and with an angle θ degrees from the horizontal is modeled by the parametric equations

$$x = (v_0 \cos \theta)t \qquad \text{and} \qquad y = h - 16t^2 + (v_0 \sin \theta)t,$$

where x and y are in feet and t is in seconds.

• • • **Example 8** Examining Parametric Equations Defining the Position of an Object in Motion

A small rocket is launched from a table that is 3.36 feet above the ground. Its initial velocity is 64 feet per second, and it is launched at an angle of 30° with respect to the ground. Its path is defined by the parametric equations

$$x = (64 \cos 30°)t \qquad \text{and} \qquad y = 3.36 - 16t^2 + (64 \sin 30°)t$$

or, equivalently,

$$x = 32\sqrt{3}\,t \qquad \text{and} \qquad y = -16t^2 + 32t + 3.36.$$

Find the rectangular equation that models this path.

We know that the parametric equations are

$$x = 32\sqrt{3}\,t \qquad \text{and} \qquad y = -16t^2 + 32t + 3.36.$$

From $x = 32\sqrt{3}\,t$, we get

$$t = \frac{x}{32\sqrt{3}}.$$

Substituting into the other parametric equation for t yields

$$y = -16\left(\frac{x}{32\sqrt{3}}\right)^2 + 32\left(\frac{x}{32\sqrt{3}}\right) + 3.36.$$

Simplifying, we find that the rectangular equation is

$$y = -\frac{1}{192}x^2 + \frac{\sqrt{3}}{3}x + 3.36.$$

• • •

• • • **Example 9** Analyzing the Path of a Projectile

Determine the total flight time and the horizontal distance traveled by the rocket in Example 8.

Algebraic Solution

To determine the total time the rocket is in the air, use the equation

$$y = -16t^2 + 32t + 3.36$$

since it tells the vertical position of the rocket for any time t. We need to determine those values of t for which $y = 0$ since this corresponds to the rocket at ground level. This yields

$$0 = -16t^2 + 32t + 3.36.$$

Using the quadratic formula to solve for t, we determine that $t = -.1$ or $t = 2.1$. Since t represents time, $t = -.1$ is an unacceptable answer. Therefore, the flight time is 2.1 seconds.

Since we know that the rocket was in the air for 2.1 seconds, we can use $t = 2.1$ and the parametric equation that models the horizontal position, $x = 32\sqrt{3}\,t$, to get

$$x = 32\sqrt{3}\,(2.1) \approx 116.4 \text{ feet.}$$

Graphing Calculator Solution

Figure 50 shows that when $t = 2.1$, the horizontal distance covered is approximately 116.4 feet, which supports the algebraic solution.

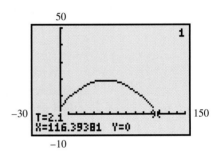

Figure 50

• • •

8.6 Exercises

Concept Check Match the ordered pair from Column II with the pair of parametric equations in Column I on whose graph the point lies. In each case, consider the given value of t.

I	II
1. $x = 3t + 6, y = -2t + 4$; $t = 2$	**A.** $(5, 25)$
2. $x = \cos t, y = \sin t$; $t = \pi/4$	**B.** $(7, 2)$
3. $x = t, y = t^2$; $t = 5$	**C.** $(12, 0)$
4. $x = t^2 + 3, y = t^2 - 2$; $t = 2$	**D.** $\left(\dfrac{\sqrt{2}}{2}, \dfrac{\sqrt{2}}{2}\right)$

Graph each plane curve defined by the given parametric equations. Give a traditional or a calculator graph, as directed by your instructor. Then find a rectangular equation for each curve. See Examples 1 and 2.

5. $x = 2t, y = t + 1$, for t in $[-2, 3]$

6. $x = t + 2, y = t^2$, for t in $[-1, 1]$

7. $x = \sqrt{t}, y = 3t - 4$, for t in $[0, 4]$

8. $x = t^2, y = \sqrt{t}$, for t in $[0, 4]$

9. $x = t^3 + 1, y = t^3 - 1$, for t in $(-\infty, \infty)$

10. $x = 2t - 1, y = t^2 + 2$, for t in $(-\infty, \infty)$

11. $x = 2 \sin t, y = 2 \cos t$, for t in $[0, 2\pi]$

12. $x = \sqrt{5} \sin t, y = \sqrt{3} \cos t$, for t in $[0, 2\pi]$

13. $x = 3 \tan t, y = 2 \sec t$, for t in $\left(-\dfrac{\pi}{2}, \dfrac{\pi}{2}\right)$

14. $x = \cot t, y = \csc t$, for t in $(0, \pi)$

Find a rectangular equation for each curve defined as follows and graph the curve. Give a traditional or a calculator graph, as directed by your instructor. See Examples 1 and 2.

15. $x = \sin t, y = \csc t$, for t in $(0, \pi)$

16. $x = \tan t, y = \cot t$, for t in $\left(0, \dfrac{\pi}{2}\right)$

17. $x = t, y = \sqrt{t^2 + 2}$, for t in $(-\infty, \infty)$

18. $x = \sqrt{t}, y = t^2 - 1$, for t in $[0, \infty)$

19. $x = 2 + \sin t, y = 1 + \cos t$, for t in $[0, 2\pi]$

20. $x = 1 + 2 \sin t, y = 2 + 3 \cos t$, for t in $[0, 2\pi]$

21. $x = t + 2, y = \dfrac{1}{t + 2}$, for $t \neq -2$

22. $x = t - 3, y = \dfrac{2}{t - 3}$, for $t \neq 3$

Graph each plane curve defined by the parametric equations, using a traditional or a calculator graph, as directed by your instructor. Assume that the interval for t is the set of all real numbers for which $x = f(t)$ and $y = g(t)$ are both defined. See Examples 2 and 3.

23. $x = \sin t, y = \cos t$

24. $x = t, y = \dfrac{\sqrt{4 - 4t^2}}{2}$

25. $x = t + 2, y = t - 4$

26. $x = t^2 + 2, y = t^2 - 4$

Graph each cycloid defined by the given equations for t in the specified interval. Use a traditional or a calculator graph, as directed by your instructor. See Example 5.

27. $x = t - \sin t, y = 1 - \cos t$, for t in $[0, 4\pi]$

28. $x = 2t - 2 \sin t, y = 2 - 2 \cos t$, for t in $[0, 8\pi]$

In Exercises 29–32, do each of the following.
(a) *Determine the parametric equations that model the path of the projectile.*
(b) *Determine the rectangular equation that models the path of the projectile.*
(c) *Determine how long the projectile is in flight and the horizontal distance covered.*
See Examples 6–9.

29. *(Modeling) Flight of a Model Rocket*   A model rocket is launched from the ground with a velocity of 48 feet per second at an angle of 60° with respect to the ground.

30. *(Modeling) Flight of a Golf Ball*   Tiger is playing golf. He hit a golf ball from the ground at an angle of 60° with respect to the ground at a velocity of 150 feet per second.

31. *(Modeling) Flight of a Softball*   Sally hits a softball when it is 2 feet above the ground. The ball leaves her bat at an angle of 20° with respect to the ground at a velocity of 88 feet per second.

32. *(Modeling) Flight of a Baseball*   Mark hits a baseball when it is 2.5 feet above the ground. The ball leaves his bat at an angle of 29° from the horizontal with a velocity of 136 feet per second.

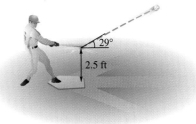

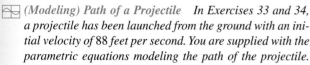

(Modeling) Path of a Projectile *In Exercises 33 and 34, a projectile has been launched from the ground with an initial velocity of 88 feet per second. You are supplied with the parametric equations modeling the path of the projectile.*
(a) *Graph the parametric equations.*
(b) *Approximate θ, the angle the projectile makes with the horizontal at launch, to the nearest tenth of a degree.*
(c) *Based on your answer to part (b), write parametric equations for the projectile using the cosine and sine functions.*

33. $x = 82.69265063t, y = -16t^2 + 30.09777261t$
34. $x = 56.56530965t, y = -16t^2 + 67.41191099t$

For Exercises 35–38, see Example 4.

35. Give two parametric representations of the line through the point (x_1, y_1) with slope m.

36. Give two parametric representations of the parabola

$$y = a(x - h)^2 + k.$$

37. Give a parametric representation of the hyperbola

$$\frac{x^2}{a^2} - \frac{y^2}{b^2} = 1.$$

38. Give a parametric representation of the ellipse

$$\frac{x^2}{a^2} + \frac{y^2}{b^2} = 1.$$

39. The spiral of Archimedes has polar equation $r = a\theta$, where $r^2 = x^2 + y^2$. Show that a parametric representation of the spiral of Archimedes is

$$x = a\theta \cos \theta, y = a\theta \sin \theta, \qquad \text{for } \theta \text{ in } (-\infty, \infty).$$

40. Show that the hyperbolic spiral $r\theta = a$, where $r^2 = x^2 + y^2$, is given parametrically by

$$x = \frac{a \cos \theta}{\theta}, y = \frac{a \sin \theta}{\theta}, \qquad \text{for } \theta \text{ in } (-\infty, 0) \cup (0, \infty).$$

41. The parametric equations $x = \cos t$, $y = \sin t$, for t in $[0, 2\pi]$ and the parametric equations $x = \cos t$, $y = -\sin t$, for t in $[0, 2\pi]$ both have the unit circle as their graph. However, in one case the circle is traced out clockwise (as t moves from 0 to 2π) and in the other case the circle is traced out counterclockwise. For which equations is the circle traced out in the clockwise direction?

42. *Concept Check* Consider the parametric equations $x = f(t)$, $y = g(t)$, for t in $[a, b]$.
(a) How is the graph affected if the equation $x = f(t)$ is replaced by $x = c + f(t)$?
(b) How is the graph affected if the equation $y = g(t)$ is replaced by $y = d + g(t)$?

Chapter 8 Summary

Key Terms & Symbols	Key Ideas
8.1 Complex Numbers complex number $a + bi$ rectangular (standard) form real part imaginary part imaginary number conjugate	**The Imaginary Unit i** $$i = \sqrt{-1} \qquad \text{or} \qquad i^2 = -1$$ For positive real numbers a, $\sqrt{-a} = i\sqrt{a}$. The **conjugate** of $a + bi$ is $a - bi$. For complex numbers $a + bi$ and $c + di$, follow these procedures to perform the four arithmetic operations. **Addition of Complex Numbers** $$(a + bi) + (c + di) = (a + c) + (b + d)i$$ **Subtraction of Complex Numbers** $$(a + bi) - (c + di) = (a - c) + (b - d)i.$$ **Multiplication of Complex Numbers** To find the product $(a + bi)(c + di)$, use FOIL and the definition of i^2. **Division of Complex Numbers** To find the quotient $(a + bi)/(c + di)$, multiply both the numerator and the denominator by the conjugate of the denominator, $c - di$.

Key Terms & Symbols	Key Ideas
8.2 Trigonometric (Polar) Form of Complex Numbers real axis imaginary axis complex plane trigonometric (polar) form modulus (absolute value) argument	**Trigonometric (Polar) Form of Complex Numbers** If the complex number $x + yi$ corresponds to the vector with direction angle θ and magnitude r, then $$x = r \cos \theta \qquad r = \sqrt{x^2 + y^2}$$ $$y = r \sin \theta \qquad \tan \theta = \frac{y}{x}, \quad \text{if } x \neq 0.$$ The expression $$r(\cos \theta + i \sin \theta) \quad \text{or} \quad r \operatorname{cis} \theta$$ is the trigonometric form (or polar form) of $x + yi$.
8.3 The Product and Quotient Theorems	**The Product and Quotient Theorems** For any two complex numbers $r_1(\cos \theta_1 + i \sin \theta_1)$ and $r_2(\cos \theta_2 + i \sin \theta_2)$, $$[r_1(\cos \theta_1 + i \sin \theta_1)] \cdot [r_2(\cos \theta_2 + i \sin \theta_2)]$$ $$= r_1 r_2[\cos(\theta_1 + \theta_2) + i \sin(\theta_1 + \theta_2)]$$ and $\quad \dfrac{r_1(\cos \theta_1 + i \sin \theta_1)}{r_2(\cos \theta_2 + i \sin \theta_2)} = \dfrac{r_1}{r_2}[\cos(\theta_1 - \theta_2) + i \sin(\theta_1 - \theta_2)],$ where $r_2 \operatorname{cis} \theta_2 \neq 0$.
8.4 Powers and Roots of Complex Numbers nth root of a complex number	**De Moivre's Theorem** $$[r(\cos \theta + i \sin \theta)]^n = r^n(\cos n\theta + i \sin n\theta)$$ **nth Root Theorem** If n is any positive integer and r is a positive real number, then the nonzero complex number $r(\cos \theta + i \sin \theta)$ has exactly n distinct nth roots, given by $$\sqrt[n]{r}\,(\cos \alpha + i \sin \alpha),$$ where $$\alpha = \frac{\theta + 360°k}{n} \quad \text{or} \quad \alpha = \frac{\theta}{n} + \frac{360°k}{n},$$ $k = 0, 1, 2, \ldots, n - 1.$
8.5 Polar Equations and Graphs polar coordinate cardioid system lemniscate pole rose curve polar axis (four-leaved rose) polar coordinates spiral of Archimedes polar equation rectangular (Cartesian) equation	**Polar Graphs** Polar coordinates determine a point by locating it θ degrees from the polar axis (the positive x-axis) and r units from the origin. Polar equations are graphed in the same way as Cartesian equations, by point plotting or with a graphing calculator.
8.6 Parametric Equations, Graphs, and Applications parametric equations of a plane curve parameter cycloid	**Plane Curve** A plane curve is a set of points (x, y) such that $x = f(t)$, $y = g(t)$, and f and g are both defined on an interval I. The equations $x = f(t)$ and $y = g(t)$ are parametric equations with parameter t.

Chapter 8 Review Exercises

Write as a multiple of i.

1. $\sqrt{-9}$

2. $\sqrt{-12}$

Solve each quadratic equation.

3. $x^2 = -81$

4. $x(2x + 3) = -4$

Perform each operation. Write answers in rectangular form.

5. $(1 - i) - (3 + 4i) + 2i$

6. $(2 - 5i) + (9 - 10i) - 3$

7. $(6 - 5i) + (2 + 7i) - (3 - 2i)$

8. $(4 - 2i) - (6 + 5i) - (3 - i)$

9. $(3 + 5i)(8 - i)$

10. $(4 - i)(5 + 2i)$

11. $(2 + 6i)^2$

12. $(6 - 3i)^2$

13. $(1 - i)^3$

14. $(2 + i)^3$

15. $\dfrac{6 + 2i}{3 - i}$

16. $\dfrac{2 - 5i}{1 + i}$

17. $\dfrac{2 + i}{1 - 5i}$

18. $\dfrac{3 + 2i}{i}$

19. i^{53}

20. i^{-41}

21. $1 \cdot i \cdot i^2 \cdot i^3$

22. $1 \cdot i \cdot i^2 \cdot \ldots \cdot i^{100}$

23. $[5(\cos 90° + i \sin 90°)][6(\cos 180° + i \sin 180°)]$

24. $[3 \text{ cis } 135°][2 \text{ cis } 105°]$

25. $\dfrac{2(\cos 60° + i \sin 60°)}{8(\cos 300° + i \sin 300°)}$

26. $\dfrac{4 \text{ cis } 270°}{2 \text{ cis } 90°}$

27. $\left(\sqrt{3} + i\right)^3$

28. $(2 - 2i)^5$

29. $(\cos 100° + i \sin 100°)^6$

30. *Concept Check* The vector representing a real number will lie on the _____-axis in the complex plane.

31. Explain the geometric similarity between the absolute value of a real number and the absolute value (or modulus) of a complex number. (*Hint:* Think in terms of distance.)

Graph each complex number.

32. $5i$

33. $-4 + 2i$

34. $3 - 3i\sqrt{3}$

35. Find and graph the resultant of $7 + 3i$ and $-2 + i$.

Complete the chart in Exercises 36–42.

Rectangular Form	Trigonometric Form
36. $-2 + 2i$	_____
37. _____	$3(\cos 90° + i \sin 90°)$
38. _____	$2(\cos 225° + i \sin 225°)$
39. $-4 + 4i\sqrt{3}$	_____
40. $1 - i$	_____
41. _____	$4 \text{ cis } 240°$
42. $-4i$	_____

Concept Check The complex number z, where $z = x + yi$, can be graphed in the plane as (x, y). Describe the graphs of all complex numbers z satisfying the conditions in Exercises 43 and 44.

43. The modulus of z is 2.

44. The imaginary part of z is the negative of the real part of z.

Find all roots as indicated. Express them in trigonometric form.

45. the fifth roots of $-2 + 2i$

46. the cube roots of $1 - i$

47. How many real fifth roots does -32 have?

48. How many real sixth roots does -64 have?

Solve each equation. Leave answers in trigonometric form.

49. $x^3 + 125 = 0$

50. $x^4 + 16 = 0$

51. $x^2 + i = 0$

52. Convert $\left(-1, \sqrt{3}\right)$ to polar coordinates, with $0° \le \theta < 360°$.

53. Convert $(5, 315°)$ to rectangular coordinates.

54. If a point lies on an axis in the rectangular plane, then what kind of angle must θ be if (r, θ) represents the point in polar coordinates?

55. What will the graph of $r = k$ be, for $k > 0$?

Identify and graph each polar equation for θ in $[0°, 360°)$. Use a traditional or a calculator graph, as directed by your instructor.

56. $r = 4 \cos \theta$ **57.** $r = -1 + \cos \theta$ **58.** $r = 2 \sin 4\theta$

*In Exercises 59–62 identify the geometric symmetry (**A**, **B**, or **C**) that the graph will possess.*
    A. *symmetry with respect to the origin*
    B. *symmetry with respect to the y-axis*
    C. *symmetry with respect to the x-axis*

59. Whenever (r, θ) is on the graph, then so is $(-r, -\theta)$.

60. Whenever (r, θ) is on the graph, then so is $(-r, \theta)$.

61. Whenever (r, θ) is on the graph, then so is $(r, -\theta)$.

62. Whenever (r, θ) is on the graph, then so is $(r, \pi - \theta)$.

Find an equivalent equation in rectangular coordinates.

63. $r = \dfrac{3}{1 + \cos \theta}$ **64.** $r = \sin \theta + \cos \theta$ **65.** $r = 2$

Find an equivalent equation in polar coordinates.

66. $y = x$ **67.** $y = x^2$

In Exercises 68–70, find a polar equation having the given graph. (Note: The values of Xscl and Yscl are 1.)

68.

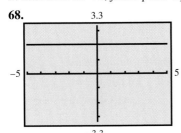

69.

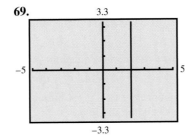

70.

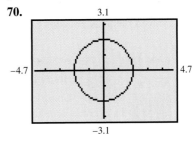

71. Graph the plane curve defined by the parametric equations $x = t + \cos t$, $y = \sin t$, for t in $[0, 2\pi]$. Use a traditional or a calculator graph, as directed by your instructor.

Find a rectangular equation for each plane curve with the given parametric equations.

72. $x = 3t + 2, y = t - 1$, for t in $[-5, 5]$ **73.** $x = \sqrt{t - 1}, y = \sqrt{t}$, for t in $[1, \infty)$

74. $x = t^2 + 5, y = \dfrac{1}{t^2 + 1}$, for t in $(-\infty, \infty)$ **75.** $x = 5 \tan t, y = 3 \sec t$, for t in $(-\pi/2, \pi/2)$

76. $x = \cos 2t, y = \sin t$, for t in $(-\pi, \pi)$

77. Find the vertex of the parabola given by the rectangular equation $y = (\tan \theta)x - \dfrac{16}{v_0^2 \cos^2 \theta} x^2$.

78. Find a pair of parametric equations whose graph is the circle with center $(3, 4)$ that contains the origin.

79. *Mandelbrot Set*   Follow the steps in Exercise 64 of Section 8.2 to show that the graph of the Mandelbrot set in Exercise 49 of Section 8.4 is symmetric with respect to the x-axis.

80. *Flight of a Baseball*   A baseball is hit when it is 3.2 feet above the ground. It leaves the bat with a velocity of 118 feet per second at an angle of 27° with respect to the ground. Follow the directions for Exercises 29–32 in Section 8.6.

Chapter 8 Test

1. For the complex numbers $w = 2 - 4i$ and $z = 5 + i$, find each of the following in rectangular form.

  (a) $w + z$ (and give a geometric representation)   **(b)** $w - z$ **(c)** wz **(d)** $\dfrac{w}{z}$

2. Express each of the following in rectangular form.
  (a) i^{15} **(b)** $(1 + i)^2$

3. Find all complex solutions of $2x^2 - x + 4 = 0$.

4. Write each complex number in trigonometric (polar) form, where $0° \le \theta < 360°$.
  (a) $3i$ **(b)** $1 + 2i$ **(c)** $-1 - \sqrt{3}i$

5. Write each complex number in rectangular form.
  (a) $3(\cos 30° + i \sin 30°)$ **(b)** $4 \operatorname{cis} 40°$ **(c)** $3(\cos 90° + i \sin 90°)$

6. For the complex numbers $w = 8(\cos 40° + i \sin 40°)$ and $z = 2(\cos 10° + i \sin 10°)$, find each of the following in the form specified.
  (a) wz (trigonometric form)   **(b)** $\dfrac{w}{z}$ (rectangular form)   **(c)** z^3 (rectangular form)

7. Find the four complex fourth roots of $-16i$. Express them in trigonometric form.

8. Convert the given rectangular coordinates to polar coordinates. Give two pairs of polar coordinates for each point.
  (a) $(0, 5)$ **(b)** $(-2, -2)$

9. Convert the given polar coordinates to rectangular coordinates.
  (a) $(3, 315°)$ **(b)** $(-4, 90°)$

Identify and graph each polar equation for θ in $[0°, 360°]$. Use a traditional or a calculator graph, as directed by your instructor.

10. $r = 1 - \cos \theta$ **11.** $r = 3 \cos 3\theta$

12. Convert the polar equation $r = \dfrac{4}{2 \sin \theta - \cos \theta}$ to a rectangular equation, and sketch its graph.

Graph each pair of parametric equations. Use a traditional or a calculator graph, as directed by your instructor.

13. $x = 4t - 3, y = t^2$, for t in $[-3, 4]$ **14.** $x = 2 \cos 2t, y = 2 \sin 2t$, for t in $[0, 2\pi]$

15.   *Julia Set*   Consider the complex number $z = -1 + i$. Compute the value of $z^2 - 1$, and show that its modulus exceeds 2, indicating that $-1 + i$ is not in the Julia set.

Chapter 8 Internet Project

The Art of Undersampling

When we draw a line or circle with pencil and paper, the pencil point is drawing an infinite number of points on a continuous curve. A graphing calculator plots a series of points and connects the points with line segments. This means that the polar circle $r = 8$, shown in the figure graphed in degree mode with default setting θ-step $(7.5°)$, is not truly a circle but a polygon with vertices plotted every $7.5°$ around the origin and connected by line segments. Since $360/7.5 = 48$, the "circle" is really a polygon with 48 vertices (a 48-gon).

The Internet project for this chapter illustrates the importance of plotting enough points to accurately represent a curve. With "selected sampling," we can distort expected results. This strategy is sometimes used by statisticians to manipulate conclusions drawn from a set of data. The Web site for this book, found at www.awl.com/lhs, explores the art of *undersampling*.

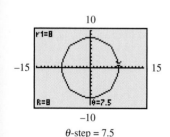

r1=8

−15   15

R=8   θ=7.5

−10

θ-step = 7.5
θmax = 360

Exponential and
Logarithmic Functions

9

9.1 **Exponential Functions**

9.2 **Logarithmic Functions**

9.3 **Evaluating Logarithms and the Change-of-Base Theorem**

9.4 **Exponential and Logarithmic Equations**

The burning of fossil fuels, deforestation, and changes in land use from 1850–1986 put approximately 312 billion tons of carbon into the atmosphere, mostly in the form of carbon dioxide. Burning fossil fuels produces 5.4 billion tons of carbon each year, which is absorbed by both the atmosphere and the oceans. A critical aspect of the accumulation of carbon dioxide in the atmosphere is that it is irreversible and its effect requires hundreds of years to disappear. In 1990 the International Panel of Climate Change (IPCC) reported that if current trends of burning of fossil fuels and deforestation continue, then future amounts of atmospheric carbon dioxide in parts per million (ppm) would increase as shown in the table.

Atmospheric Carbon Dioxide

Year	Carbon Dioxide (ppm)
1990	353
2000	375
2075	590
2175	1090
2275	2000

How can these data be used to predict when the amount of carbon dioxide will double? What will be the resulting global warming? How are carbon dioxide levels and global temperature increases related? In this chapter, we will attempt to answer these and other questions related to ecology, the chapter theme.*

· ·

9.1 Exponential Functions

- Exponents and Properties • Graphs of Exponential Functions • Exponential Equations • Compound Interest
- The Number e • Exponential Growth and Decay • Curve Fitting

Exponents and Properties Recall the definition of a^x, where x is a rational number: if $x = m/n$, then for appropriate values of m and n,

$$a^{m/n} = \left(\sqrt[n]{a}\right)^m.$$

For example,

$$16^{3/4} = \left(\sqrt[4]{16}\right)^3 = 2^3 = 8,$$

$$27^{-1/3} = \frac{1}{27^{1/3}} = \frac{1}{\sqrt[3]{27}} = \frac{1}{3},$$

and

$$64^{-1/2} = \frac{1}{64^{1/2}} = \frac{1}{\sqrt{64}} = \frac{1}{8}.$$

The definition of a^x is extended here to include all real (not just rational) values of the exponent. For example, $2^{\sqrt{3}}$ might be evaluated by approximating the exponent $\sqrt{3}$ with the numbers 1.7, 1.73, 1.732, and so on. Since these decimals approach the value of $\sqrt{3}$ more and more closely, it seems reasonable that $2^{\sqrt{3}}$ should be approximated more and more closely by the numbers $2^{1.7}$, $2^{1.73}$, $2^{1.732}$, and so on. (Recall, for example, that $2^{1.7} = 2^{17/10} = \sqrt[10]{2^{17}}$.) In fact, this is exactly how $2^{\sqrt{3}}$ is defined (in a more advanced course). To show that this assumption is reasonable, Figure 1 gives graphs of the function $f(x) = 2^x$ with three different domains.

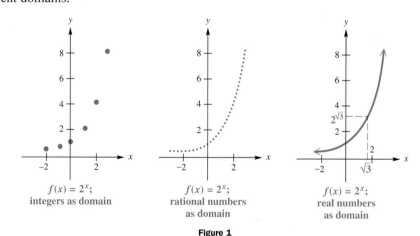

$f(x) = 2^x$;
integers as domain

$f(x) = 2^x$;
rational numbers
as domain

$f(x) = 2^x$;
real numbers
as domain

Figure 1

Sources: Clime, W., *The Economics of Global Warming,* Institute for International Economics, Washington, D.C., 1992.

Kraljic, M. (Editor), *The Greenhouse Effect,* The H. W. Wilson Company, New York, 1992.

International Panel on Climate Change (IPCC), 1990.

Wuebbles, D. and J. Edmonds, *Primer of Greenhouse Gases,* Lewis Publishers, Inc., Chelsea, Michigan, 1991.

Using this interpretation of real exponents, all rules and theorems for exponents are valid for real number exponents as well as rational ones. In addition to the rules for exponents presented earlier, we use several new properties in this chapter. For example, if $y = 2^x$, then each real value of x leads to exactly one value of y, and therefore, $y = 2^x$ defines a function. Furthermore,

$$\text{if } 2^x = 2^4, \text{ then } x = 4,$$

and for $p > 0$,

$$\text{if } p^2 = 3^2, \text{ then } p = 3.$$

Also,
$$4^2 < 4^3 \quad \text{but} \quad \left(\frac{1}{2}\right)^2 > \left(\frac{1}{2}\right)^3,$$

so when $a > 1$, increasing the exponent on a leads to a *larger* number, but when $0 < a < 1$, increasing the exponent on a leads to a *smaller* number.

These properties are generalized below. Proofs of the properties are not given here, as they require more advanced mathematics.

Additional Properties of Exponents

For any real number $a > 0$, $a \neq 1$, and any real number x, the following statements are true.

(a) a^x is a unique real number.

(b) $a^b = a^c$ if and only if $b = c$.

(c) If $a > 1$ and $m < n$, then $a^m < a^n$.

(d) If $0 < a < 1$ and $m < n$, then $a^m > a^n$.

These properties require $a > 0$ so that a^x is always defined. For example, $(-6)^x$ is not a real number if $x = 1/2$. This means that a^x will always be positive, since a must be positive. In Property (a), $a \neq 1$ because $1^x = 1$ for every real number value of x, so each value of x leads to the *same* real number, 1. For Property (b) to hold, a must not equal 1 since, for example, $1^4 = 1^5$, even though $4 \neq 5$.

● ● ●   **Example 1**   Evaluating an Exponential Expression

If $f(x) = 2^x$, find each of the following.

Algebraic Solution

(a) $f(-1)$

Replace x with -1.

$$f(-1) = 2^{-1} = \frac{1}{2}$$

(b) $f(3) = 2^3 = 8$

(c) $f(5/2) = 2^{5/2} = (2^5)^{1/2} = 32^{1/2} = \sqrt{32} = 4\sqrt{2}$

(d) $f(4.92) \approx 30.2738447$ Use a calculator.

Graphing Calculator Solution

Figures 2 and 3 illustrate how a graphing calculator supports the algebraic results. Here, $Y_1 = f(x) = 2^x$.

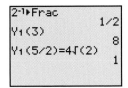

Figure 2

(continued)

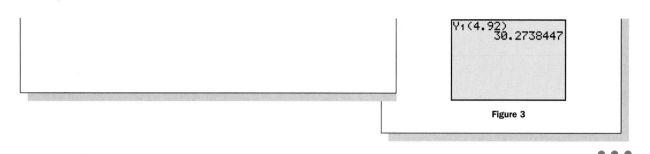

Figure 3

• • •

Graphs of Exponential Functions Figure 4 shows the graph of $f(x) = 2^x$ from Example 1. The base of this exponential function is 2. The y-intercept is

$$y = 2^0 = 1.$$

Since $2^x > 0$ for all x and $2^x \to 0$ as $x \to -\infty$, the x-axis is a horizontal asymptote. The table to the left of Figure 4 gives several points on the graph of the function. Plotting these points and then drawing a smooth curve through them gives the graph in Figure 4. As the graph suggests, the domain of the function is $(-\infty, \infty)$ and the range is $(0, \infty)$. The function is increasing on its entire domain, and it is one-to-one by the horizontal line test. Figure 5 shows a graph and table generated by a graphing calculator.

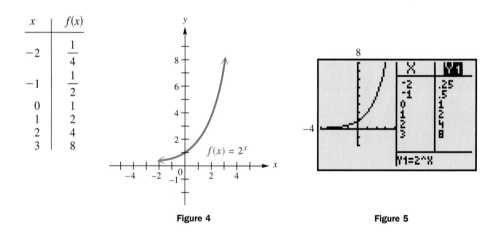

x	$f(x)$
-2	$\frac{1}{4}$
-1	$\frac{1}{2}$
0	1
1	2
2	4
3	8

$f(x) = 2^x$

Figure 4   **Figure 5**

We can now define a function $f(x) = a^x$ whose domain is the set of all real numbers (and not just the rationals).

Exponential Function

If $a > 0$ and $a \neq 1$, then

$$f(x) = a^x$$

defines the **exponential function** with base a.

N O T E   If $a = 1$, the function becomes the constant function with $f(x) = 1$, not an exponential function.

● ● ●   **Example 2**   **Graphing an Exponential Function**

Graph $f(x) = \left(\dfrac{1}{2}\right)^x$.

Algebraic Solution

The y-intercept is 1, and the x-axis is a horizontal asymptote. Plot a few ordered pairs, and draw a smooth curve through them. For example, several points are shown in the table to the left of Figure 6. Like the function $f(x) = 2^x$, this function also has domain $(-\infty, \infty)$ and range $(0, \infty)$ and is one-to-one. The graph is decreasing on its entire domain.

x	$f(x)$
-3	8
-2	4
-1	2
0	1
1	$\dfrac{1}{2}$
2	$\dfrac{1}{4}$

$f(x) = \left(\tfrac{1}{2}\right)^x$

Figure 6

Graphing Calculator Solution

The graph and table shown in Figure 7, as generated by a graphing calculator, support the algebraic results.

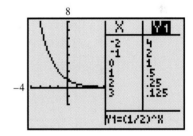

Figure 7

● ● ●

Starting with $f(x) = 2^x$ and replacing x with $-x$ gives $f(-x) = 2^{-x} = (2^{-1})^x = (1/2)^x$. For this reason, the graph of $f(x) = 2^x$ and $f(x) = (1/2)^x$ are reflections of each other across the y-axis. This is supported by the graphs in Figures 4–7.

The graph of $f(x) = 2^x$ is typical of graphs of $f(x) = a^x$ where $a > 1$. For larger values of a, the graphs rise more steeply, but the general shape is similar to the graph in Figure 4. When $0 < a < 1$, the graph decreases in a manner similar to the graph of $f(x) = (1/2)^x$. In Figure 8 on the next page, the graphs of several typical exponential functions illustrate these facts.

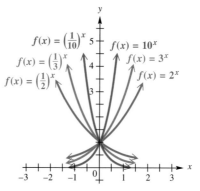

$f(x) = a^x$
Domain: $(-\infty, \infty)$
Range: $(0, \infty)$
When $a > 1$, the function is increasing.
When $0 < a < 1$, the function is decreasing.

Figure 8

In summary, the graph of a function of the form $f(x) = a^x$ has the following features.

Characteristics of the Graph of $f(x) = a^x$

1. The points $(0, 1)$ and $(1, a)$ are on the graph.
2. If $a > 1$, f is an increasing function; if $0 < a < 1$, f is a decreasing function.
3. The x-axis is a horizontal asymptote.
4. The domain is $(-\infty, \infty)$, and the range is $(0, \infty)$.

● ● ● **Example 3**   Graphing Reflections and Translations

Graph each function.

Algebraic Solution

(a) $f(x) = -2^x$

The graph is that of $f(x) = 2^x$ reflected across the x-axis. The domain is $(-\infty, \infty)$, and the range is $(-\infty, 0)$. See Figure 9.

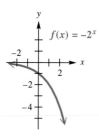

Figure 9

Graphing Calculator Solution

Figure 12 shows how a graphing calculator can be directed to graph the three functions of this example. Y_1 is defined as 2^x, and Y_2, Y_3, and Y_4 are defined as reflections and/or translations of Y_1.

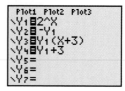

Figure 12

(continued)

(b) $f(x) = 2^{x+3}$

The graph is the graph of $f(x) = 2^x$ translated 3 units to the left, as shown in Figure 10.

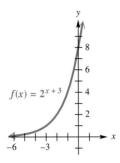

Figure 10

(c) $f(x) = 2^x + 3$

This graph is that of $f(x) = 2^x$ translated 3 units upward. See Figure 11.

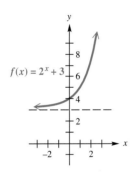

Figure 11

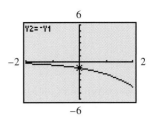

Figure 13

Compare the graph with Figure 9.

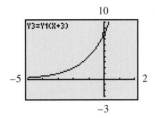

Figure 14

Compare the graph with Figure 10.

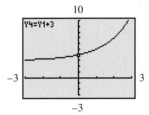

Figure 15

Compare the graph with Figure 11.

• • •

Exponential Equations Property (b) given at the beginning of this section is useful in solving equations, as shown in the next examples.

• • • **Example 4** Using a Property of Exponents to Solve an Equation

Solve $\left(\dfrac{1}{3}\right)^x = 81$.

Algebraic Solution

Write each side of the equation using a common base. First, write $1/3$ as 3^{-1}, so $(1/3)^x = (3^{-1})^x = 3^{-x}$. Since $81 = 3^4$,

$$\left(\frac{1}{3}\right)^x = 81$$

Graphing Calculator Solution

The screen shown in Figure 16 on the next page supports the algebraic result, using the x-intercept method of solution first introduced in Section 6.2.

(continued)

becomes $3^{-x} = 3^4.$

By Property (b),

$$-x = 4 \quad \text{or} \quad x = -4.$$

The solution of the original equation is -4.

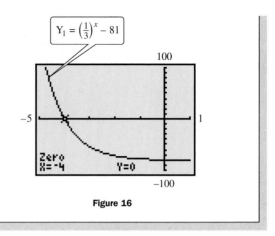

Figure 16

● ● ●

Example 5 Using a Property of Exponents to Solve an Equation

Solve $2^{x+4} = 8^{x-6}$.

Write each side of the equation using a common base.

$$2^{x+4} = 8^{x-6}$$
$$2^{x+4} = (2^3)^{x-6} \qquad \text{Write 8 as a power of 2.}$$
$$2^{x+4} = 2^{3x-18} \qquad (x^a)^b = x^{ab}$$
$$x + 4 = 3x - 18 \qquad \text{Set exponents equal to each other.}$$
$$-2x = -22$$
$$x = 11$$

The solution is 11.   ● ● ●

Later in this chapter, we describe a more general method for solving exponential equations where the approach used in Examples 4 and 5 is not possible. For instance, the above method could not be used to solve an equation like $7^x = 12$ since it is not easy to express both sides as exponential expressions with the same base.

● ● ● **Example 6** Using a Property of Exponents to Solve an Equation

Solve $81 = b^{4/3}$.

Begin by writing $b^{4/3}$ as $\left(\sqrt[3]{b}\right)^4$.

$$81 = \left(\sqrt[3]{b}\right)^4 \qquad \text{Definition of rational exponent}$$
$$\pm 3 = \sqrt[3]{b} \qquad \text{Take fourth roots on both sides.}$$
$$\pm 27 = b \qquad \text{Cube both sides.}$$

Check *both* solutions in the original equation. Both check, so the solutions are -27 and 27.   ● ● ●

Compound Interest The formula for *compound interest* (interest paid on both principal and interest) is an important application of exponential functions. Recall the formula for simple interest, $I = Prt$, where P is principal (amount left at interest), r is annual rate of interest expressed as a decimal, and t is time in years that the principal earns interest. Suppose $t = 1$ year. Then at the end of the year the amount has grown to

$$P + Pr = P(1 + r),$$

the original principal plus interest. If this amount is left at the same interest rate for another year, the total amount becomes

$$[P(1 + r)] + [P(1 + r)]r = [P(1 + r)](1 + r)$$
$$= P(1 + r)^2.$$

After the third year, this will grow to

$$[P(1 + r)^2] + [P(1 + r)^2]r = [P(1 + r)^2](1 + r)$$
$$= P(1 + r)^3.$$

Continuing in this way produces the following formula for compound interest.

> **Compound Interest**
>
> If P dollars is deposited in an account paying an annual rate of interest r compounded (paid) m times per year, then after t years the account will contain A dollars, where
>
> $$A = P\left(1 + \frac{r}{m}\right)^{tm}.$$

For example, suppose $1000 is deposited in an account paying 8% per year compounded quarterly, or four times per year. After 10 years the account will contain

$$P\left(1 + \frac{r}{m}\right)^{tm} = 1000\left(1 + \frac{.08}{4}\right)^{10(4)}$$
$$= 1000(1 + .02)^{40}$$
$$= 1000(1.02)^{40}$$

dollars. Using a calculator, $(1.02)^{40} = 2.20804$, to five decimal places. The amount on deposit after 10 years is

$$1000(1.02)^{40} = 1000(2.20804) = 2208.04 \quad \text{or} \quad \$2208.04.$$

In the formula for compound interest, A is sometimes called the **future value** and P the **present value.**

Example 7 Finding Present Value

An accountant wants to buy a new computer in three years that will cost $20,000.

(a) How much should be deposited now, at 6% interest compounded annually, to give the required $20,000 in three years?

Since the money deposited should amount to $20,000 in three years, $20,000 is the future value of the money. To find the present value P of $20,000 (the amount to deposit now), use the compound interest formula with $A = 20{,}000$, $r = .06$, $m = 1$, and $t = 3$.

$$A = P\left(1 + \frac{r}{m}\right)^{tm}$$

$$20{,}000 = P\left(1 + \frac{.06}{1}\right)^{3(1)} = P(1.06)^3$$

$$\frac{20{,}000}{(1.06)^3} = P$$

$$P \approx 16{,}792.38566 \qquad \text{Use a calculator to approximate.}$$

The accountant must deposit $16,792.39.

(b) If only $15,000 is available to deposit now, what annual interest rate is required for it to increase to $20,000 in three years?

Here $P = 15{,}000$, $A = 20{,}000$, $m = 1$, $t = 3$, and r is unknown. Substitute the known values into the compound interest formula and solve for r.

$$A = P\left(1 + \frac{r}{m}\right)^{tm}$$

$$20{,}000 = 15{,}000\left(1 + \frac{r}{1}\right)^3$$

$$\frac{4}{3} = (1 + r)^3 \qquad \text{Divide both sides by 15,000.}$$

$$\left(\frac{4}{3}\right)^{1/3} = 1 + r \qquad \text{Take cube roots on both sides.}$$

$$\left(\frac{4}{3}\right)^{1/3} - 1 = r \qquad \text{Subtract 1 on both sides.}$$

$$r \approx .10 \qquad \text{Use a calculator to approximate.}$$

An interest rate of 10% will produce enough interest to increase the $15,000 deposit to the $20,000 needed at the end of three years. ● ● ●

The Number e Perhaps the single most useful base for an exponential function is the irrational number e. Base e exponential functions provide a good model for many natural, as well as economic, phenomena. The letter e was chosen to represent this number in honor of the Swiss mathematician Leonhard Euler (pronounced "oiler") (1707–1783). Applications of exponential functions with base e are given later in this chapter.

The number e occurs naturally when using the formula for compound interest. Suppose a lucky investment produces an annual interest rate of 100%, so $r = 1.00$, or $r = 1$. Suppose also that only $1 can be deposited at this rate, and for only one year. Then $P = 1$ and $t = 1$. Substitute into the formula for compound interest:

$$P\left(1 + \frac{r}{m}\right)^{tm} = 1\left(1 + \frac{1}{m}\right)^{1(m)} = \left(1 + \frac{1}{m}\right)^m.$$

If interest is compounded annually, making $m = 1$, the total amount on deposit is

$$\left(1 + \frac{1}{m}\right)^m = \left(1 + \frac{1}{1}\right)^1 = 2^1 = 2,$$

m	$\left(1 + \dfrac{1}{m}\right)^m$ (rounded)
1	2
2	2.25
5	2.48832
10	2.59374
25	2.66584
50	2.69159
100	2.70481
500	2.71557
1000	2.71692
10,000	2.71815
1,000,000	2.71828

so an investment of $1 becomes $2 in one year. As interest is compounded more and more often, the value of this expression will increase.

A calculator was used to get the results in the table at the left. The table suggests that as m increases, the value of $(1 + 1/m)^m$ gets closer and closer to some fixed number. It turns out that this is indeed the case. This fixed number is called e. Figure 17 shows how the table feature of a graphing calculator does this as well, for selected values of x, where $Y_1 = (1 + 1/x)^x$.

Value of e

To nine decimal places,

$$e \approx 2.718281828.$$

Figure 18 shows how a graphing calculator is used to find e and selected powers of e. Figure 19 shows the functions defined by $y = 2^x$, $y = 3^x$, and $y = e^x$. Notice that because $2 < e < 3$, the graph of $y = e^x$ lies "between" the other two graphs.

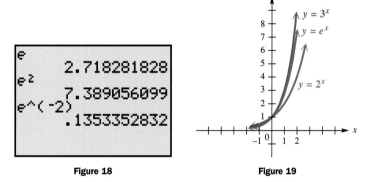

Figure 17

Figure 18

Figure 19

Looking Ahead to Calculus

In calculus the derivative of a function is a limit that allows us to determine the slope of a tangent line to the graph of the function. For the function $f(x) = e^x$, the derivative is the function f itself: $f'(x) = e^x$. Geometrically this means that the slope of a tangent line to its graph is the same as the y-coordinate of the point of tangency. Will this slope ever be negative? Will it ever be 0?

C O N N E C T I O N S   In calculus, it is shown that

$$e^x = 1 + x + \frac{x^2}{2 \cdot 1} + \frac{x^3}{3 \cdot 2 \cdot 1} + \frac{x^4}{4 \cdot 3 \cdot 2 \cdot 1} + \frac{x^5}{5 \cdot 4 \cdot 3 \cdot 2 \cdot 1} + \cdots.$$

By using more and more terms, a more and more accurate approximation may be obtained for e^x.

For Discussion or Writing

1. Use the terms shown here and replace x with 1 to approximate $e^1 = e$ to three decimal places. Check your results with a calculator.
2. Use the terms shown here and replace x with $-.05$ to approximate $e^{-.05}$ to four decimal places. Check your results with a calculator.
3. Give the next term in the sum for e^x.

Exponential Growth and Decay As mentioned above, the number e is important as the base of an exponential function because many practical applications require an exponential function with base e. For example, it can be shown that in situations involving growth or decay of a quantity, the amount or number present at time t often can be closely modeled by a function defined by

$$y = y_0 e^{kt},$$

where y_0 is the amount or number present at time $t = 0$ and k is a constant.

The next example, which refers to the problem stated at the beginning of this chapter, illustrates exponential growth.

● ● ● **Example 8** Using Data to Model Exponential Growth

If current trends of burning fossil fuels and deforestation continue, then future amounts of atmospheric carbon dioxide in parts per million (ppm) will increase as shown in the table.

Year	Carbon Dioxide (ppm)
1990	353
2000	375
2075	590
2175	1090
2275	2000

Source: International Panel on Climate Change (IPCC), 1990.

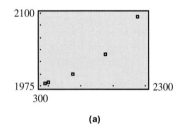

(a)

(a) Make a scatter diagram of the data. Do the carbon dioxide levels appear to grow exponentially?

We show a calculator-generated graph for the data in Figure 20(a). The data do appear to have the shape of the graph of an increasing exponential function.

(b) The function defined by

$$y = 353e^{.0060857(t-1990)}$$

is a good model for the data.

(Later in this chapter we will show how this expression for y was obtained.) A graph of this function in Figure 20(b) shows that it is very close to the data points. From the graph, estimate when future levels of carbon dioxide will double and triple over the preindustrial level of 280 ppm.

In Figure 21, we graph $y = 2 \cdot 280 = 560$ and $y = 3 \cdot 280 = 840$ on the same coordinate axes as the function and use the calculator to find the intersection points.

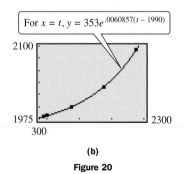

For $x = t$, $y = 353e^{.0060857(t-1990)}$

(b)

Figure 20

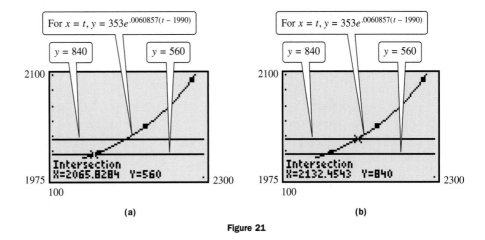

Figure 21

The graph of the function intersects the horizontal lines at approximately 2065.8 and 2132.5. According to this model, carbon dioxide levels will double by 2065 and triple by 2132. ● ● ●

Curve Fitting Graphing calculators are capable of fitting exponential curves to scatter diagrams like the one found in Example 8. Figure 22(a) shows how the TI-83 displays another (different) equation for the atmospheric carbon dioxide example: $y = .0019 \cdot 1.0061^x$. (Coefficients are rounded here.) Notice that this calculator-generated form differs from the model in Example 8. Figure 22(b) shows the data points and the graph of this exponential regression equation.

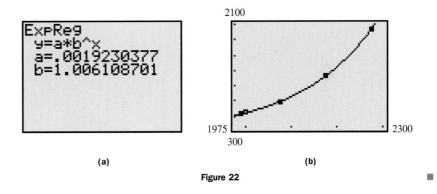

Figure 22

9.1 Exercises

If $f(x) = 3^x$ and $g(x) = (1/4)^x$, find each of the following. If a result is irrational, give the answer to as many decimal places as your calculator shows. See Example 1.

1. $f(2)$ **2.** $f(3)$ **3.** $f(-2)$ **4.** $f(-3)$

5. $g(2)$ **6.** $g(3)$ **7.** $g(-2)$ **8.** $g(-3)$

9. $f(1.5)$ **10.** $g(1.5)$ **11.** $g(2.34)$ **12.** $f(1.68)$

· · · · · · · · · · · · · · **Relating Concepts** · · · · · · · · · · · · · ·

For individual or collaborative investigation
(Exercises 13–18)

In Exercises 13–18, assume $f(x) = a^x$, where $a > 1$. **Work these exercises in order.**

13. Is f a one-to-one function? If so, based on Section 6.1, what kind of related function exists for f?

14. If f has an inverse function f^{-1}, sketch f and f^{-1} on the same set of axes.

15. If f^{-1} exists, find an equation for $y = f^{-1}(x)$ using the method described in Section 6.1. You need not solve for y.

16. If $a = 10$, what is the equation for $y = f^{-1}(x)$? (You need not solve for y.)

17. If $a = e$, what is the equation for $y = f^{-1}(x)$? (You need not solve for y.)

18. If the point (p, q) is on the graph of f, then the point _____ is on the graph of f^{-1}.

Graph each function. Give a traditional graph or a calculator graph, as directed by your instructor. See Examples 2 and 3.

19. $f(x) = 3^x$

20. $f(x) = 4^x$

21. $f(x) = \left(\dfrac{1}{3}\right)^x$

22. $f(x) = \left(\dfrac{1}{4}\right)^x$

23. $f(x) = \left(\dfrac{3}{2}\right)^x$

24. $f(x) = \left(\dfrac{2}{3}\right)^x$

25. $f(x) = e^x$

26. $f(x) = 10^x$

27. $f(x) = e^{-x}$

28. $f(x) = 10^{-x}$

29. $f(x) = 2^{|x|}$

30. $f(x) = 2^{-|x|}$

Sketch the graph of $f(x) = 2^x$. Then refer to it and use the techniques of Example 3 to graph each function defined.

31. $f(x) = 2^x + 1$

32. $f(x) = 2^x - 4$

33. $f(x) = 2^{x+1}$

34. $f(x) = 2^{x-4}$

Sketch the graph of $f(x) = (1/3)^x$. Then refer to it and use the techniques of Example 3 to graph each function defined.

35. $f(x) = \left(\dfrac{1}{3}\right)^x - 2$

36. $f(x) = \left(\dfrac{1}{3}\right)^x + 4$

37. $f(x) = \left(\dfrac{1}{3}\right)^{x+2}$

38. $f(x) = \left(\dfrac{1}{3}\right)^{x-4}$

Concept Check The graphs of $y = a^x$ for $a = 1.8, 2.3, 3.2, .4, .75,$ and $.31$ are given in the figure. They are identified by letter, but not necessarily in the same order as the values of a just given. Use your knowledge of how the exponential function behaves for various values of a to identify each lettered graph.

39. A

40. B

41. C

42. D

43. E

44. F

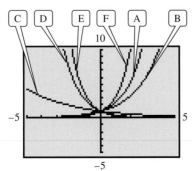

⊞ *Use a graphing calculator to graph each function defined. See Examples 2 and 3.*

45. $f(x) = \dfrac{e^x - e^{-x}}{2}$

46. $f(x) = \dfrac{e^x + e^{-x}}{2}$

47. $f(x) = x \cdot 2^x$

48. $f(x) = x^2 \cdot 2^{-x}$

Solve each equation. See Examples 4–6.

49. $4^x = 2$

50. $125^r = 5$

51. $\left(\dfrac{1}{2}\right)^k = 4$

52. $\left(\dfrac{2}{3}\right)^x = \dfrac{9}{4}$

53. $2^{3-y} = 8$

54. $5^{2p+1} = 25$

55. $\dfrac{1}{27} = b^{-3}$

56. $\dfrac{1}{81} = k^{-4}$

57. $4 = r^{2/3}$

58. $z^{5/2} = 32$

59. $27^{4z} = 9^{z+1}$

60. $32^t = 16^{1-t}$

61. $\left(\dfrac{1}{2}\right)^{-x} = \left(\dfrac{1}{4}\right)^{x+1}$

62. $\left(\dfrac{2}{3}\right)^{k-1} = \left(\dfrac{81}{16}\right)^{k+1}$

Solve each problem involving compound interest. See Example 7.

63. *Future Value*   Find the future value of $8906.54 at 5% compounded semiannually for 9 years.

64. *Future Value*   Find the future value of $56,780 at 5.3% compounded quarterly for 23 quarters.

65. *Present Value*   Find the present value for a future value of $25,000 if interest is 6% compounded quarterly for 11 quarters.

66. *Present Value*   Find the present value for a future value of $45,678.93 if interest is 9.6% compounded monthly for 11 months.

67. *Interest Rate*   Find the required annual interest rate to the nearest tenth for $65,000 to grow to $65,325, if interest is compounded monthly for 6 months.

68. *Interest Rate*   Find the required annual interest rate to the nearest tenth for $1200 to grow to $1780 if interest is compounded quarterly for 5 years.

Solve each problem. See Example 8.

69. *(Modeling) Atmospheric Pressure*   The atmospheric pressure (in millibars) at a given altitude (in meters) is shown in the table at the right.

⊞ **(a)** Use a graphing calculator to make a scatter diagram of the data for atmospheric pressure P at altitude x.

(b) Would a linear or exponential function fit the data better?

⊞ **(c)** The function defined by

$$P(x) = 1013e^{-.0001341x}$$

approximates the data. Use a graphing calculator to graph P and the data on the same coordinate axes.

(d) Use P to predict the pressure at 1500 m and 11,000 m and compare it to the actual values of 846 millibars and 227 millibars, respectively.

Altitude	Pressure	Altitude	Pressure
0	1013	6000	472
1000	899	7000	411
2000	795	8000	357
3000	701	9000	308
4000	617	10,000	265
5000	541		

Source: Miller, A. and J. Thompson, *Elements of Meteorology,* Fourth Edition, Charles E. Merrill Publishing Company, Columbus, Ohio, 1993.

70. *(Modeling) Radiative Forcing*   Carbon dioxide in the atmosphere traps heat from the sun. Presently, the net incoming solar radiation reaching the earth's surface is 240 watts per square meter (w/m²). The relationship between additional watts per square meter of heat trapped by the increased carbon dioxide R and the average rise in global temperature T (in °F) is shown in the graph on the next page. This additional solar radiation trapped by carbon dioxide is called *radiative forcing*. It is measured in watts per square meter.

(a) Is T a linear or exponential function of R?

(b) Let T represent the temperature increase resulting from an additional radiative forcing of R w/m². Use the graph to write T as a function of R.

(c) Find the global temperature increase when $R = 5$ w/m².

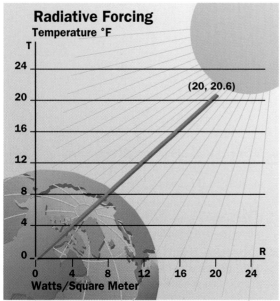

Radiative Forcing

Source: Clime, W., *The Economics of Global Warming,* Institute for International Economics, Washington, D.C., 1992.

71. *(Modeling) World Population Growth* Since 1980, world population in millions closely fits the exponential function defined by

$$y = 4481e^{.0156x},$$

where x is the number of years since 1980.

(a) The world population was about 5320 million in 1990. How closely does the function approximate this value?

(b) Use this model to approximate the population in 1995.

(c) Use this model to predict the population in 2005.

(d) Explain why this model may not be accurate for 2005.

72. *(Modeling) Deer Population* The exponential growth of the deer population in Massachusetts can be calculated using the model

$$T = 50,000(1 + .06)^n,$$

where 50,000 is the initial deer population and .06 is the rate of growth. T is the total population after n years have passed.

(a) Predict the total population after 4 years.

(b) If the initial population was 30,000 and the growth rate was .12, approximately how many deer would be present after 3 years?

(c) How many additional deer can we expect in 5 years if the initial population is 45,000 and the current growth rate is .08?

73. *(Modeling) Employee Training* A person learning certain skills involving repetition tends to learn quickly at first. Then learning tapers off and approaches some upper limit. Suppose the number of symbols per minute that a person using a word processor can type is given by

$$p(t) = 250 - 120(2.8)^{-.5t},$$

where t is the number of months the operator has been in training. Find each of the following.

(a) $p(2)$ **(b)** $p(4)$ **(c)** $p(10)$

(d) What happens to the number of symbols per minute after several months of training?

74. *(Modeling) Median Home Cost in the U.S.* The median cost, in dollars, of homes in the United States since 1990 can be modeled by the exponential function defined by

$$C(x) = 130,700e^{.027x},$$

where x is the number of years since 1990. Use this function to approximate the median cost for the following years. Give answers to the nearest hundred dollars. (*Source: Statistical Abstract of the United States,* 1998).

(a) 1990 **(b)** 1993 **(c)** 1998

(d) Graph $C(x) = 130,700e^{.027x}$ for $0 \le x \le 10$.

9.2 Logarithmic Functions

● Meaning of Logarithm ● Logarithmic Equations ● Logarithmic Functions ● Properties of Logarithms

Meaning of Logarithm The previous section dealt with exponential functions of the form $y = a^x$ for all positive values of a, where $a \ne 1$. The horizontal line test shows that exponential functions are one-to-one, and thus have inverse functions. In this section we discuss inverses of exponential functions. The equation defining the inverse of a function is found by interchanging x and y in the equation that defines the function. Doing so with $y = a^x$ gives

$$x = a^y$$

as the equation of the inverse function of the exponential function defined by $y = a^x$. This equation can be solved for y by using the following definition.

Logarithm

For all real numbers y, and all positive numbers a and x, where $a \neq 1$:

$$y = \log_a x \qquad \text{if and only if} \qquad x = a^y.$$

The "log" in the definition above is an abbreviation for *logarithm.* Read $\log_a x$ as "the logarithm to the base a of x."

Consider the following fill-in-the-box problems.

$$4^3 = \boxed{} \qquad 5^{\boxed{}} = 25$$

The answers, of course, are

$$4^3 = \boxed{64} \qquad 5^{\boxed{2}} = 25\,.$$

When we solve the problem on the left, we are "doing" exponents. When we solve the one on the right, we are "doing" logarithms. That is, we are finding the power to which 5 must be raised in order to get 25. Therefore, $2 = \log_5 25$. In a certain sense, logarithms are just exponents.

By the definition of logarithm, if $y = \log_a x$, then the power to which a must be raised to obtain x is y, or $x = a^y$. It is important to remember the location of the base and the exponent in each form.

Logarithmic form: $y = \log_a x$
(Exponent points to y; Base points to a)

Exponential form: $a^y = x$
(Exponent points to y; Base points to a)

The chart on the next page shows several pairs of equivalent statements, written in both exponential and logarithmic forms.

Exponential Form	Logarithmic Form
$2^3 = 8$	$\log_2 8 = 3$
$\left(\dfrac{1}{2}\right)^{-4} = 16$	$\log_{1/2} 16 = -4$
$10^5 = 100{,}000$	$\log_{10} 100{,}000 = 5$
$3^{-4} = \dfrac{1}{81}$	$\log_3\left(\dfrac{1}{81}\right) = -4$
$5^1 = 5$	$\log_5 5 = 1$
$\left(\dfrac{3}{4}\right)^0 = 1$	$\log_{3/4} 1 = 0$

Logarithmic Equations The definition of logarithm can be used to solve logarithmic equations, as shown in the next example.

● ● ● **Example 1** Solving Logarithmic Equations

Solve each equation.

(a) $\log_x \dfrac{8}{27} = 3$

First, write the expression in exponential form.

$$x^3 = \frac{8}{27}$$

$$x^3 = \left(\frac{2}{3}\right)^3 \qquad \frac{8}{27} = \left(\frac{2}{3}\right)^3$$

$$x = \frac{2}{3} \qquad \text{Property (b) of exponents}$$

The solution is 2/3.

(b) $\log_4 x = 5/2$

$$4^{5/2} = x \quad \text{Write in exponential form.}$$

$$(4^{1/2})^5 = x$$

$$2^5 = x$$

$$32 = x$$

The solution is 32. ● ● ●

Logarithmic Functions The logarithmic function with base a is defined as follows.

> ### Logarithmic Function
>
> If $a > 0$, $a \neq 1$, and $x > 0$, then
>
> $$f(x) = \log_a x$$
>
> defines the **logarithmic function** with base a.

Exponential and logarithmic functions are inverses of each other. Since the domain of an exponential function is the set of all real numbers, the range of a logarithmic function also will be the set of all real numbers. In the same way, both the range of an exponential function and the domain of a logarithmic function are the set of all positive real numbers, so logarithms can be found for positive numbers only.

The graph of $y = 2^x$ is shown in red in Figure 23(a). The graph of its inverse is found by reflecting the graph of $y = 2^x$ across the line $y = x$. The graph of the inverse function, defined by $y = \log_2 x$, shown in blue, has the y-axis as a vertical asymptote. Figure 23(b) shows a calculator-generated graph of the two functions.

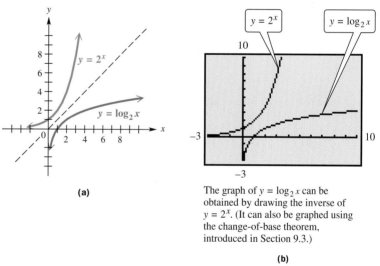

(a)

The graph of $y = \log_2 x$ can be obtained by drawing the inverse of $y = 2^x$. (It can also be graphed using the change-of-base theorem, introduced in Section 9.3.)

(b)

Figure 23

The graph of $y = (1/2)^x$ is shown in red in Figure 24(a) on the next page. The graph of its inverse, defined by $y = \log_{1/2} x$, in blue, is found by reflecting the graph of $y = (1/2)^x$ across the line $y = x$. As the figure suggests, the graph of $y = \log_{1/2} x$ also has the y-axis as a vertical asymptote. Figure 24(b) shows a graphing calculator version.

Calculator-generated graphs of logarithmic functions do not, in general, give an accurate picture of the behavior of the graphs near the vertical asymptotes. While it may seem as if the graph has an endpoint, this is not the case. The resolution of the calculator screen is not precise enough to indicate that the graph approaches the vertical asymptote as the value of x gets closer to it. Do not draw incorrect conclusions just because the calculator does not show this behavior. ∎

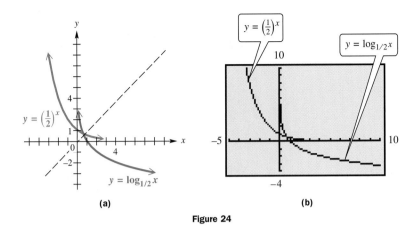

(a)   **(b)**

Figure 24

The graphs of $y = \log_2 x$ in Figure 23 and $y = \log_{1/2} x$ in Figure 24 suggest the following generalizations about the graphs of logarithmic functions of the form $f(x) = \log_a x$.

Characteristics of the Graph of $f(x) = \log_a x$

1. The points $(1, 0)$ and $(a, 1)$ are on the graph.
2. If $a > 1$, f is an increasing function; if $0 < a < 1$, f is a decreasing function.
3. The y-axis is a vertical asymptote.
4. The domain is $(0, \infty)$, and the range is $(-\infty, \infty)$.

In writing logarithmic functions, it is important to use parentheses and brackets to make the intent clear. Just as we put parentheses around $x - 2$ in $f(x - 2)$, we put parentheses around $x - 2$ in $\log_a(x - 2)$. Similarly, we write $\log_a(xy)$ to avoid the misinterpretation $(\log_a x)y$. Also, we write $\log_a(x^2)$ to distinguish it from $(\log_a x)^2$. However, we will continue to write $\log_a x$ without parentheses, because the meaning is clear in this case.

● ● ● **Example 2**   Graphing a Translated Logarithmic Function

Graph each function.

(a) $f(x) = \log_2(x - 1)$

The graph of $f(x) = \log_2(x - 1)$ is the graph of $f(x) = \log_2 x$ translated 1 unit to the right. The vertical asymptote is $x = 1$. The domain of the function defined by $f(x) = \log_2(x - 1)$ is $(1, \infty)$ since logarithms can be found only for positive numbers. To find some ordered pairs to plot, use the equivalent equations in exponential form,

$$x - 1 = 2^y \quad \text{or} \quad x = 2^y + 1,$$

choosing values for y and then calculating each of the corresponding x-values. See Figure 25(a).

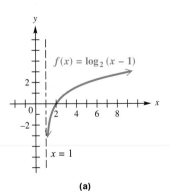

(a)

Figure 25

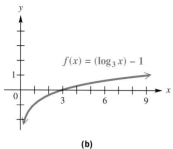

$f(x) = (\log_3 x) - 1$

(b)

Figure 25

(b) $f(x) = (\log_3 x) - 1$

This function has the same graph as $g(x) = \log_3 x$ translated 1 unit downward. Ordered pairs to plot can be found by writing $y = (\log_3 x) - 1$ in exponential form.

$$y = (\log_3 x) - 1$$
$$y + 1 = \log_3 x$$
$$x = 3^{y+1}$$

Again, it is easier to choose y-values and calculate the corresponding x-values. The graph is shown in Figure 25(b). • • •

CAUTION If you write a logarithmic function in exponential form, choosing y-values to calculate x-values as we did in Example 2, be careful to get the ordered pairs in the correct order.

Properties of Logarithms Since a logarithmic statement can be written as an exponential statement, it is not surprising that there are properties of logarithms based on the properties of exponents. The properties of logarithms allow us to change the form of logarithmic statements so that products can be converted to sums, quotients can be converted to differences, and powers can be converted to products.

Properties of Logarithms

If x and y are any positive real numbers, r is any real number, and a is any positive real number, $a \neq 1$, then the following properties are true.

(a) $\log_a xy = \log_a x + \log_a y$ **(b)** $\log_a \dfrac{x}{y} = \log_a x - \log_a y$

(c) $\log_a x^r = r \log_a x$ **(d)** $\log_a a = 1$

(e) $\log_a 1 = 0$

Looking Ahead to Calculus

The product rule for logarithms (as well as the quotient and power rules) is proved in calculus using a different method than the one labeled "Proof" here. The typical calculus proof involves the derivative of the base e logarithmic function.

Proof To prove Property (a), let

$$m = \log_a x \quad \text{and} \quad n = \log_a y.$$

$a^m = x$ and $a^n = y$	Change to exponential form.
$a^m \cdot a^n = xy$	Multiply.
$a^{m+n} = xy$	Property of exponents
$\log_a xy = m + n$	Definition of logarithm

Since $m = \log_a x$ and $n = \log_a y$,

$$\log_a xy = \log_a x + \log_a y.$$

Properties (b) and (c) are proven in a similar way. Properties (d) and (e) follow directly from the definition of logarithm since $a^1 = a$ and $a^0 = 1$.

• • • **Example 3** Using the Properties of Logarithms

Assume that all variables represent positive real numbers. Rewrite each expression.

(a) $\log_6(7 \cdot 9) = \log_6 7 + \log_6 9$ Property (a)

(b) $\log_9\left(\dfrac{15}{7}\right) = \log_9 15 - \log_9 7$ Property (b)

(c) $\log_5\sqrt{8} = \log_5(8^{1/2}) = \dfrac{1}{2}\log_5 8$ Property (c)

(d) $\log_a\left(\dfrac{mnq}{p^2}\right) = \log_a m + \log_a n + \log_a q - 2\log_a p$

(e) $\log_a\sqrt[3]{m^2} = \dfrac{2}{3}\log_a m$

(f) $\log_b\sqrt[n]{\dfrac{x^3 y^5}{z^m}} = \dfrac{1}{n}\log_b\left(\dfrac{x^3 y^5}{z^m}\right)$

$$= \dfrac{1}{n}(\log_b(x^3) + \log_b(y^5) - \log_b(z^m))$$

$$= \dfrac{1}{n}(3\log_b x + 5\log_b y - m\log_b z)$$

$$= \dfrac{3}{n}\log_b x + \dfrac{5}{n}\log_b y - \dfrac{m}{n}\log_b z \quad \text{Distributive property}$$

Notice the use of parentheses in the second step. The factor $1/n$ applies to each term. • • •

• • • **Example 4** Using the Properties of Logarithms

Write each expression as a single logarithm with coefficient 1. Assume that all variables represent positive real numbers.

(a) $\log_3(x + 2) + \log_3 x - \log_3 2$
Using Properties (a) and (b),

$$\log_3(x + 2) + \log_3 x - \log_3 2 = \log_3\left[\dfrac{(x + 2)x}{2}\right].$$

(b) $2\log_a m - 3\log_a n = \log_a(m^2) - \log_a(n^3)$ Property (c)

$$= \log_a\left(\dfrac{m^2}{n^3}\right) \quad\quad\quad \text{Property (b)}$$

(c) $\dfrac{1}{2}\log_b m + \dfrac{3}{2}\log_b(2n) - \log_b(m^2 n)$

$$= \log_b(m^{1/2}) + \log_b[(2n)^{3/2}] - \log_b(m^2 n) \quad \text{Property (c)}$$

$$= \log_b\left(\dfrac{m^{1/2}(2n)^{3/2}}{m^2 n}\right) \quad\quad\quad \text{Properties (a) and (b)}$$

$$= \log_b\left(\dfrac{2^{3/2} n^{1/2}}{m^{3/2}}\right) \quad\quad\quad \text{Rules for exponents}$$

$$= \log_b\left[\left(\frac{2^3 n}{m^3}\right)^{1/2}\right]$$

Rules for exponents

$$= \log_b\sqrt{\frac{8n}{m^3}}$$

Definition of $a^{1/n}$

● ● ●

C A U T I O N   There is no property of logarithms to rewrite a logarithm of a *sum* or *difference*. That is why, in Example 4(a), $\log_3(x + 2)$ was not written as $\log_3 x + \log_3 2$. Remember, $\log_3 x + \log_3 2 = \log_3(x \cdot 2)$.

The distributive property does not apply here, because $\log(x + y)$ is one term; "log" is not a factor.

● ● ●   **Example 5**   Using the Properties of Logarithms with Numerical Values

Assume that $\log_{10} 2 = .3010$. Find the base 10 logarithms of 4 and 5.

By the properties of logarithms,

$$\log_{10} 4 = \log_{10}(2^2) = 2\log_{10} 2 = 2(.3010) = .6020$$

$$\log_{10} 5 = \log_{10}\left(\frac{10}{2}\right) = \log_{10} 10 - \log_{10} 2 = 1 - .3010 = .6990.$$

We used Property (d) to replace $\log_{10} 10$ with 1.   ● ● ●

If $f(x)$ and $g(x)$ are inverse functions, then

$$f[g(x)] = x \qquad \text{and} \qquad g[f(x)] = x.$$

Since $f(x) = a^x$ and $g(x) = \log_a x$ are inverses, the next theorem follows.

Theorem on Inverses

For $a > 0, a \neq 1$:

$$a^{\log_a x} = x \qquad \text{and} \qquad \log_a(a^x) = x.$$

By the results of this theorem,

$$\log_5 5^3 = 3, \qquad 7^{\log_7 10} = 10, \qquad \text{and} \qquad \log_r r^{k+1} = k + 1.$$

The second statement in the theorem will be useful in Section 9.4 when we solve other logarithmic and exponential equations.

C O N N E C T I O N S   Long before the days of calculators and computers, the search for making calculations easier was an ongoing process. Machines built by Charles Babbage and Blaise Pascal, a system of "rods" used by John Napier, and slide rules were the forerunners of today's electronic marvels. The invention of logarithms by John Napier in the sixteenth century was a great breakthrough in the search for easier methods of calculation.

(continued)

Since logarithms are exponents, their properties allowed users of tables of common logarithms to multiply by adding, divide by subtracting, raise to powers by multiplying, and take roots by dividing. Although logarithms are no longer used for computations, they play an important part in higher mathematics.

For Discussion or Writing

1. To multiply 458.3 by 294.6 using logarithms, we add $\log_{10} 458.3$ and $\log_{10} 294.6$, then find 10 to the sum. Perform this multiplication using the log* key and the 10^x key on your calculator. Check your answer by multiplying directly with your calculator.
2. Try division, raising to a power, and taking a root by this method.

9.2 Exercises

In Exercises 1–8, match the logarithm in Column I with its value in Column II. Remember that $\log_a x$ *is the exponent to which a must be raised in order to obtain x.*

I	**II**
1. $\log_2 16$	**A.** 0
2. $\log_3 1$	**B.** $\dfrac{1}{2}$
3. $\log_{10} .1$	**C.** 5
4. $\log_2 \sqrt{2}$	**D.** not a real number
5. $\log_{10} 10^5$	**E.** 4
6. $\log_e\left(\dfrac{1}{e^2}\right)$	**F.** -3
7. $\log_{1/2} 8$	**G.** -1
8. $\log_5(-1)$	**H.** -2

For each statement, write an equivalent statement in logarithmic form.

9. $3^4 = 81$ **10.** $2^5 = 32$ **11.** $(2/3)^{-3} = 27/8$ **12.** $10^{-4} = .0001$

For each statement, write an equivalent statement in exponential form.

13. $\log_6 36 = 2$ **14.** $\log_5 5 = 1$ **15.** $\log_{\sqrt{3}} 81 = 8$ **16.** $\log_4\left(\dfrac{1}{64}\right) = -3$

17. Explain why logarithms of negative numbers are not defined.

18. Why does $\log_a 1$ always equal 0 for any valid base a?

Solve each logarithmic equation. See Example 1.

19. $x = \log_5\left(\dfrac{1}{625}\right)$ **20.** $x = \log_3\left(\dfrac{1}{81}\right)$ **21.** $x = \log_{10} .001$ **22.** $x = \log_6\left(\dfrac{1}{216}\right)$

---

*In this text, the notation log x is used to mean $\log_{10} x$. This is also the meaning of the log key on calculators.

23. $x = 2^{\log_2 9}$

24. $x = 8^{\log_8 11}$

25. $\log_x 25 = -2$

26. $\log_x\left(\dfrac{1}{16}\right) = -2$

27. $\log_4 x = 3$

28. $\log_2 x = -1$

29. $x = \log_4 \sqrt[3]{16}$

30. $x = \log_5 \sqrt[4]{25}$

31. Compare the summary of characteristics of the graph of $f(x) = \log_a x$ with the similar summary about the graph of $f(x) = a^x$ in Section 9.1. Make a list of characteristics that reinforce the idea that these are inverse functions.

32. *Concept Check* A calculator-generated graph of $y = \log_2 x$ shows the values of the ordered pair with $x = 5$. What does the value of y represent?

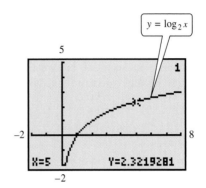

Sketch the graph of $f(x) = \log_2 x$. Then refer to it and use the techniques of Example 2 to graph each function.

33. $f(x) = (\log_2 x) + 3$

34. $f(x) = \log_2(x + 3)$

35. $f(x) = |\log_2(x + 3)|$

Sketch the graph of $f(x) = \log_{1/2} x$. Then refer to it and use the techniques of Example 2 to graph each function.

36. $f(x) = (\log_{1/2} x) - 2$

37. $f(x) = \log_{1/2}(x - 2)$

38. $f(x) = |\log_{1/2}(x - 2)|$

Graph each function. See Example 2.

39. $f(x) = \log_3 x$

40. $f(x) = \log_{10} x$

41. $f(x) = \log_{1/2}(1 - x)$

42. $f(x) = \log_{1/3}(3 - x)$

43. $f(x) = \log_3(x - 1)$

44. $f(x) = \log_2(x^2)$

Concept Check In Exercises 45–50, match the function with its graph from choices A–F.

45. $f(x) = \log_2 x$

46. $f(x) = \log_2(2x)$

47. $f(x) = \log_2\left(\dfrac{1}{x}\right)$

48. $f(x) = \log_2\left(\dfrac{x}{2}\right)$

49. $f(x) = \log_2(x - 1)$

50. $f(x) = \log_2(-x)$

A.

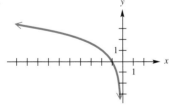

B.

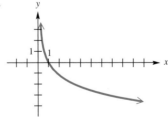

C.

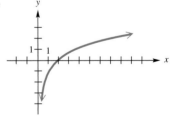

D.

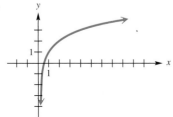

E.

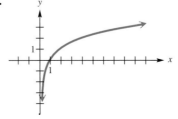

F.

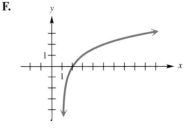

Use the log *key on your graphing calculator (for* $\log_{10} x$*) to graph each function.*

51. $f(x) = x \log_{10} x$

52. $f(x) = x^2 \log_{10} x$

· · · · · · · · · · · · **Relating Concepts** · · · · · · · · · ·

For individual or collaborative investigation
(Exercises 53–56)

Exercises 53–56 show how the quotient property for logarithms is related to the vertical translation of graphs. **Work these exercises in order.**

53. Complete the following statement of the quotient property for logarithms: If x and y are positive numbers, then $\log_a \dfrac{x}{y} =$ _____.

54. Use the quotient property to explain how the graph of $f(x) = \log_2\left(\dfrac{x}{4}\right)$ can be obtained from the graph of $g(x) = \log_2 x$ by a vertical translation.

55. Graph f and g on the same axes and explain how these graphs support your answer in Exercise 54.

56. If $x = 4$, $\log_2\left(\dfrac{x}{4}\right) =$ _____; since $\log_2 x =$ _____ and $\log_2 4 =$ _____, $\log_2 x - \log_2 4 =$ _____. How does this support the quotient property stated in Exercise 53?

Write each expression as a sum, difference, or product of logarithms. Simplify the result if possible. Assume that all variables represent positive real numbers. See Example 3.

57. $\log_2\left(\dfrac{6x}{y}\right)$

58. $\log_3\left(\dfrac{4p}{q}\right)$

59. $\log_5\left(\dfrac{5\sqrt{7}}{3}\right)$

60. $\log_2\left(\dfrac{2\sqrt{3}}{5}\right)$

61. $\log_4(2x + 5y)$

62. $\log_6(7m + 3q)$

63. $\log_m \sqrt{\dfrac{5r^3}{z^5}}$

64. $\log_p \sqrt[3]{\dfrac{m^5 n^4}{t^2}}$

Write each expression as a single logarithm with coefficient 1. Assume that all variables represent positive real numbers. See Example 4.

65. $\log_a x + \log_a y - \log_a m$

66. $(\log_b k - \log_b m) - \log_b a$

67. $2 \log_m a - 3 \log_m(b^2)$

68. $\dfrac{1}{2} \log_y(p^3 q^4) - \dfrac{2}{3} \log_y(p^4 q^3)$

69. $2 \log_a(z - 1) + \log_a(3z + 2), \quad z > 1$

70. $\log_b(2y + 5) - \dfrac{1}{2} \log_b(y + 3)$

Given $\log_{10} 2 = .3010$ *and* $\log_{10} 3 = .4771$*, find each logarithm without using a calculator. See Example 5.*

71. $\log_{10} 6$

72. $\log_{10} 12$

73. $\log_{10}\left(\dfrac{9}{4}\right)$

74. $\log_{10}\left(\dfrac{20}{27}\right)$

Solve each problem.

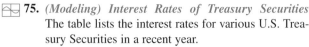

75. *(Modeling) Interest Rates of Treasury Securities*
The table lists the interest rates for various U.S. Treasury Securities in a recent year.
  (a) Make a scatter diagram of the data.
  (b) Discuss which type of function will model these data best: linear, exponential, or logarithmic.

Time	Yield	Time	Yield
3-month	5.71%	3-year	7.52%
6-month	6.37%	5-year	7.63%
1-year	6.87%	10-year	7.68%
2-year	7.34%	30-year	7.79%

Source: Reuters.

76. *Concept Check* Use the graph to estimate each logarithm.

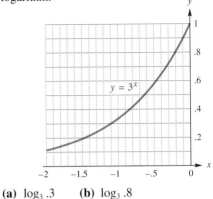

(a) $\log_3 .3$ **(b)** $\log_3 .8$

77. *Concept Check* Suppose $f(x) = \log_a x$ and $f(3) = 2$. Determine each function value.

(a) $f\left(\dfrac{1}{9}\right)$ **(b)** $f(27)$

78. Use properties of logarithms to evaluate each expression.
(a) $100^{\log_{10} 3}$ **(b)** $\log_{10} .01^3$

9.3 Evaluating Logarithms and the Change-of-Base Theorem

- **Common Logarithms** • **Applications and Modeling with Common Logarithms** • **Natural Logarithms**
- **Applications and Modeling with Natural Logarithms** • **Logarithms to Other Bases** • **Curve Fitting**

```
log(1000)
               3
log(142)
      2.152288344
log(.005832)
     -2.234182485
```
Figure 26

```
10^(log(1000))
            1000
10^(log(142))
             142
10^(log(.005832))
        .005832
```
Figure 27

Common Logarithms The bases 10 and e are so important for logarithms that scientific and graphing calculators have keys for these bases. Base 10 logarithms are called **common logarithms.** The common logarithm of the number x, or $\log_{10} x$, is often abbreviated as log x, and we will use that convention from now on. A calculator with a log key can be used to find base 10 logarithms of any positive number. Consult your owner's manual for the keystrokes needed to find common logarithms.

Figure 26 shows how a graphing calculator displays common logarithms. A common logarithm of a power of 10, such as 1000, will be given as an integer (in this case, 3). Most common logarithms used in applications, such as log 142 and log .005832, are irrational numbers.

Figure 27 reinforces the concept presented in the previous section: log x is the exponent to which 10 must be raised in order to obtain x. ∎

NOTE Base a, $a > 1$, logarithms of numbers less than 1 are always negative, as suggested by the graphs in Section 9.2.

Applications and Modeling with Common Logarithms In chemistry, the pH of a solution is defined as

$$\text{pH} = -\log[\text{H}_3\text{O}^+],$$

where $[\text{H}_3\text{O}^+]$ is the hydronium ion concentration in moles* per liter. The pH value is a measure of the acidity or alkalinity of solutions. Pure water has a pH of 7.0, substances with pH values greater than 7.0 are alkaline, and substances with pH values less than 7.0 are acidic.

---

*A *mole* is the amount of a substance that contains the same number of molecules as the number of atoms in exactly 12 grams of carbon 12.

● ● ●　**Example 1**　Finding pH

(a)　Find the pH of a solution with $[H_3O^+] = 2.5 \times 10^{-4}$.

Algebraic Solution

$$
\begin{aligned}
pH &= -\log[H_3O^+] \\
&= -\log(2.5 \times 10^{-4}) \qquad \text{Substitute.} \\
&= -(\log 2.5 + \log 10^{-4}) \qquad \text{Property (a) of logarithms}
\end{aligned}
$$

Evaluate log 2.5 with a calculator.

$$
\begin{aligned}
&= -(.3979 - 4) \qquad \log 10^{-4} = -4 \\
&= -.3979 + 4 \\
pH &\approx 3.6
\end{aligned}
$$

It is customary to round pH values to the nearest tenth.

Graphing Calculator Solution

A graphing calculator can determine $-\log(2.5 \times 10^{-4})$ directly, as shown in Figure 28.

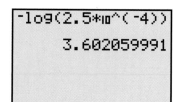

Figure 28

(b)　Find the hydronium ion concentration of a solution with pH $= 7.1$.

Algebraic Solution

$$
\begin{aligned}
pH &= -\log[H_3O^+] \\
7.1 &= -\log[H_3O^+] \qquad \text{Substitute.} \\
-7.1 &= \log[H_3O^+] \qquad \text{Multiply by } -1. \\
[H_3O^+] &= 10^{-7.1} \qquad \text{Write in exponential form.}
\end{aligned}
$$

Evaluate $10^{-7.1}$ with a calculator to get

$$
[H_3O^+] \approx 7.9 \times 10^{-8}.
$$

Graphing Calculator Solution

Use the exponent key with base 10 to find the concentration. See Figure 29. (Note that E^{-8} means "times 10^{-8}".)

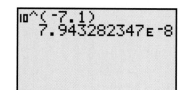

Figure 29

● ● ●

● ● ●　**Example 2**　Using pH in an Application

Wetlands are classified as *bogs, fens, marshes,* and *swamps.* These classifications are based on pH values. A pH value between 6.0 and 7.5, such as that of Summerby Swamp in Michigan's Hiawatha National Forest, indicates that the wetland is a "rich fen." When the pH is between 4.0 and 6.0, it is a "poor fen," and if the pH falls to 3.0 or less, the wetland is a "bog."

(*Source:* R. Mohlenbrock, "Summerby Swamp, Michigan," *Natural History,* March 1994.) Suppose that the hydronium ion concentration of a sample of water from a wetland is 6.3×10^{-5}. How would this wetland be classified?

Use the definition of pH.

$$
\begin{aligned}
\text{pH} &= -\log[\text{H}_3\text{O}^+] \\
&= -\log(6.3 \times 10^{-5}) \\
&= -(\log 6.3 + \log 10^{-5}) \qquad \text{\small Property (a) of logarithms} \\
&= -\log 6.3 - (-5) \qquad \text{\small Distributive property; } \log 10^n = n \\
&= -\log 6.3 + 5 \\
\text{pH} &\approx 4.2 \qquad\qquad\qquad \text{\small Use a calculator.}
\end{aligned}
$$

Since the pH is between 4.0 and 6.0, the wetland is a poor fen. ● ● ●

● ● ● **Example 3** Measuring the Loudness of Sound

The loudness of sounds is measured in a unit called a *decibel*. To measure with this unit, we first assign an intensity of I_0 to a very faint sound, called the *threshold sound*. If a particular sound has intensity I, then the decibel rating of this louder sound is

$$
d = 10 \log \frac{I}{I_0}.
$$

Find the decibel rating of a sound with intensity $10,000I_0$.

Let $I = 10,000I_0$ and find d.

$$
\begin{aligned}
d &= 10 \log \frac{10,000I_0}{I_0} \\
&= 10 \log 10,000 \\
&= 10(4) \qquad \text{\small } \log 10,000 = 4 \\
&= 40
\end{aligned}
$$

The sound has a decibel rating of 40. ● ● ●

Natural Logarithms In Section 9.1, we introduced the irrational number e. In most practical applications of logarithms, e is used as base. Logarithms to base e are called **natural logarithms,** since they occur in the life sciences and economics in natural situations that involve growth and decay. The base e logarithm of x is written $\ln x$ (read "el-en x"). A graph of the natural logarithmic function defined by $f(x) = \ln x$ is given in Figure 30, in both traditional and graphing calculator forms.

Looking Ahead to Calculus

The natural logarithmic function $f(x) = \ln x$ and the reciprocal function $g(x) = \frac{1}{x}$ have an important relationship in calculus. The derivative of the natural logarithmic function is the reciprocal function. Using *Leibniz notation* (named after one of the co-inventors of calculus), this fact is written $\frac{d}{dx}(\ln x) = \frac{1}{x}$.

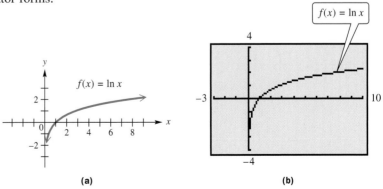

(a) (b)

Figure 30

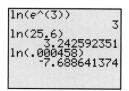

Figure 31

Figure 32

⧅ Natural logarithms can be found using a calculator. (Consult your owner's manual.) As in the case of common logarithms, when used in applications natural logarithms are usually irrational numbers. Figure 31 shows how three natural logarithms are evaluated with a graphing calculator.

Figure 32 reinforces the fact that ln x is the exponent to which e must be raised in order to obtain x. ■

Applications and Modeling with Natural Logarithms

Example 4 Measuring the Age of Rocks

Geologists sometimes measure the age of rocks by using "atomic clocks." By measuring the amounts of potassium 40 and argon 40 in a rock, the age t of the specimen in years is found with the formula

$$t = (1.26 \times 10^9)\frac{\ln[1 + 8.33(A/K)]}{\ln 2}.$$

A and K are respectively the numbers of atoms of argon 40 and potassium 40 in the specimen.

(a) How old is a rock in which $A = 0$ and $K > 0$?

If $A = 0$, $A/K = 0$ and the equation becomes

$$t = (1.26 \times 10^9)\frac{\ln 1}{\ln 2} = (1.26 \times 10^9)(0) = 0.$$

The rock is 0 years old or new.

(b) The ratio A/K for a sample of granite from New Hampshire is .212. How old is the sample?

Since A/K is .212, we have

$$t = (1.26 \times 10^9)\frac{\ln[1 + 8.33(.212)]}{\ln 2} \approx 1.85 \times 10^9.$$

The granite is about 1.85 billion years old. ●●●

Example 5 Modeling Global Temperature Increase

Carbon dioxide in the atmosphere traps heat from the sun. The additional solar radiation trapped by carbon dioxide is called *radiative forcing*. It is measured in watts per square meter. In 1896 the Swedish scientist Svante Arrhenius modeled radiative forcing R caused by additional atmospheric carbon dioxide using the logarithmic equation

$$R = k \ln(C/C_0),$$

where C_0 is the preindustrial amount of carbon dioxide, C is the current carbon dioxide level, and k is a constant. Arrhenius determined that $10 \le k \le 16$ when $C = 2C_0$. (*Source:* Clime, W., *The Economics of Global Warming,* Institute for International Economics, Washington, D.C., 1992.)

(a) Let $C = 2C_0$. Is the relationship between R and k linear or logarithmic?

If $C = 2C_0$, then $C/C_0 = 2$, so $R = k \ln 2$ is a linear relation, because $\ln 2$ is a constant.

(b) The average global temperature increase T (in °F) is given by $T(R) = 1.03R$. (See Section 9.1, Exercise 70.) Write T as a function of k. Use the expression for R given above.

$$T(R) = 1.03R$$

$$T(k) = 1.03k \ln(C/C_0)$$ • • •

Logarithms to Other Bases A calculator can be used to find the values of either natural logarithms (base e) or common logarithms (base 10). However, sometimes it is convenient to use logarithms to other bases. For example, base 2 logarithms are important in computer science. The following theorem can be used to convert logarithms from one base to another.

Change-of-Base Theorem

For any positive real numbers x, a, and b, where $a \neq 1$ and $b \neq 1$:

$$\log_a x = \frac{\log_b x}{\log_b a}.$$

This theorem is proved by using the definition of logarithm to write $y = \log_a x$ in exponential form.

Proof Let

$$y = \log_a x.$$

$$a^y = x \qquad \text{Change to exponential form.}$$

$$\log_b a^y = \log_b x \qquad \text{Take logarithms on both sides.}$$

$$y \log_b a = \log_b x \qquad \text{Property (c) of logarithms}$$

$$y = \frac{\log_b x}{\log_b a} \qquad \text{Divide both sides by } \log_b a.$$

$$\log_a x = \frac{\log_b x}{\log_b a} \qquad \text{Substitute } \log_a x \text{ for } y.$$

Any positive number other than 1 can be used for base b in the change-of-base theorem, but usually the only practical bases are e and 10 since calculators give logarithms only for these two bases.

Refer to Figures 23(b) and 24(b) in Section 9.2. We obtained the graphs of $y = \log_2 x$ and $y = \log_{1/2} x$ by directing the calculator to "draw" the inverses of $y = 2^x$ and $y = \left(\dfrac{1}{2}\right)^x$. With the change-of-base theorem, we can now graph them by directing the calculator to graph $y = \dfrac{\log x}{\log 2}$ and $y = \dfrac{\log x}{\log(1/2)}$, or equivalently, $y = \dfrac{\ln x}{\ln 2}$ and $y = \dfrac{\ln x}{\ln(1/2)}$. ∎

• • • **Example 6** Using the Change-of-Base Theorem

Use logarithms and the change-of-base theorem to find each of the following. Round to four decimal places.

Algebraic Solution

(a) $\log_5 17$

In this example, we use natural logarithms.

$$\log_5 17 = \frac{\ln 17}{\ln 5} \approx \frac{2.8332}{1.6094} \approx 1.7604$$

(b) $\log_2 .1$

Here, we use common logarithms.

$$\log_2 .1 = \frac{\log .1}{\log 2} \approx \frac{-1.0000}{.3010} \approx -3.3219$$

Graphing Calculator Solution

Figure 33 shows how the result of part (a) can be found with a graphing calculator using *common* logarithms, and how the result of part (b) can be found using *natural* logarithms. Notice that the results are the same.

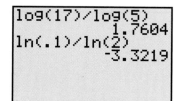

```
log(17)/log(5)
            1.7604
ln(.1)/ln(2)
           -3.3219
```

Figure 33

• • •

NOTE In the algebraic solution of Example 6, logarithms evaluated in the intermediate steps, such as ln 17 and ln 5, were shown to four decimal places. However, the final answers were obtained *without* rounding off these intermediate values, using all the digits obtained with the calculator. In general, it is best to wait until the final step to round off the answer; otherwise, a build-up of round-off errors may cause the final answer to have an incorrect final decimal place digit.

• • • **Example 7** Modeling Diversity of Species

 One measure of the diversity of the species in an ecological community is modeled by the formula

$$H = -[P_1 \log_2 P_1 + P_2 \log_2 P_2 + \cdots + P_n \log_2 P_n],$$

where $P_1, P_2, \ldots, P_n$ are the proportions of a sample belonging to each of n species found in the sample. For example, in a community with two species, where there are 90 of one species and 10 of the other, $P_1 = 90/100 = .9$ and $P_2 = 10/100 = .1$. Thus,

$$H = -[.9 \log_2 .9 + .1 \log_2 .1].$$

In Example 6(b), $\log_2 .1$ was found to be approximately -3.32. Now find $\log_2 .9$.

$$\log_2 .9 = \frac{\ln .9}{\ln 2} \approx \frac{-.1054}{.6931} \approx -.152$$

Therefore,

$$H \approx -[.9(-.152) + .1(-3.32)] \approx .469.$$ • • •

Curve Fitting   At the end of Section 9.1, we saw that graphing calculators are capable of fitting exponential curves to data that suggest such behavior. The same is true for logarithmic curves. Figure 34(a) shows the data of Exercise 61 in this section, and Figure 34(b) shows how the calculator gives the best-fitting natural logarithmic curve: $y \approx -273 + 74 \ln x$. Figure 34(c) shows the data points, along with the graph of the curve.

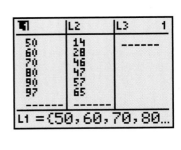

(a)

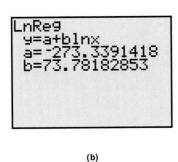

(b)

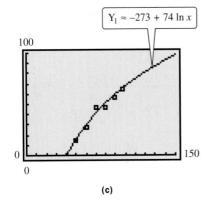

(c)

Figure 34

9.3 Exercises

Concept Check   To check your understanding of the concepts presented so far in this chapter, answer each of the following.

1. For the exponential function defined by $f(x) = a^x$, where $a > 1$, is the function increasing or decreasing over its entire domain?

2. For the logarithmic function defined by $g(x) = \log_a x$, where $a > 1$, is the function increasing or decreasing over its entire domain?

3. If $f(x) = 5^x$, what is the rule for $f^{-1}(x)$?

4. What is the name given to the exponent to which 4 must be raised in order to obtain 11?

5. A base e logarithm is called a(n) _____ logarithm; a base 10 logarithm is called a(n) _____ logarithm.

6. How is $\log_3 12$ written in terms of natural logarithms?

7. Why is $\log_2 0$ undefined?

8. Between what two consecutive integers must $\log_2 12$ lie?

Use a calculator with logarithm keys to find an approximation for each expression. Give answers to four decimal places.

9. $\log 36$

10. $\log 72$

11. $\log .042$

12. $\log .319$

13. $\log(2 \times 10^4)$

14. $\log(2 \times 10^{-4})$

15. $\ln 36$

16. $\ln 72$

17. $\ln .042$

18. $\ln .319$

19. $\ln(2 \times e^4)$

20. $\ln(2 \times e^{-4})$

For each substance, find the pH from the given hydronium ion concentration. See Example 1(a).

21. grapefruit, 6.3×10^{-4}

22. crackers, 3.9×10^{-9}

23. limes, 1.6×10^{-2}

24. sodium hydroxide (lye), 3.2×10^{-14}

Find the $[H_3O^+]$ for each substance with the given pH. See Example 1(b).

25. soda pop, 2.7

26. wine, 3.4

27. beer, 4.8

28. drinking water, 6.5

In Exercises 29–31, suppose that water from a wetland area is sampled and found to have the given hydronium ion concentration. Determine whether the wetland is a rich fen, poor fen, or bog. See Example 2.

29. 2.49×10^{-5} **30.** 2.49×10^{-2} **31.** 2.49×10^{-7}

32. Use your calculator to find approximations of each logarithm.
   (a) log 398.4   **(b)** log 39.84   **(c)** log 3.984
   (d) From your answers to parts (a)–(c), make a conjecture concerning the decimal values in the approximations of common logarithms of numbers greater than 1 that have the same digits.

Use the change-of-base theorem to find an approximation for each logarithm. Give answers to four decimal places. See Example 6.

33. $\log_2 5$ **34.** $\log_2 9$ **35.** $\log_8 .59$ **36.** $\log_8 .71$

37. $\log_{\sqrt{13}} 12$ **38.** $\log_{\sqrt{19}} 5$ **39.** $\log_{.32} 5$ **40.** $\log_{.91} 8$

41. *Concept Check*   Which of the following is the same as $2\ln(3x)$ for $x > 0$?
   A. $\ln 9 + \ln x$ **B.** $\ln(6x)$ **C.** $\ln 6 + \ln x$ **D.** $\ln(9x^2)$

42. *Concept Check*   Which of the following is the same as $\ln(4x) - \ln(2x)$ for $x > 0$?

   A. $2\ln x$ **B.** $\ln(2x)$ **C.** $\dfrac{\ln(4x)}{\ln(2x)}$ **D.** $\ln 2$

Let $u = \ln a$ and $v = \ln b$. Write the following expressions in terms of u and v without using the $\ln$ function.

43. $\ln\left(b^4\sqrt{a}\right)$ **44.** $\ln\dfrac{a^3}{b^2}$ **45.** $\ln\sqrt{\dfrac{a^3}{b^5}}$ **46.** $\ln\left(\sqrt[3]{a} \cdot b^4\right)$

47. Given $g(x) = e^x$, evaluate the following.

   (a) $g(\ln 3)$ **(b)** $g[\ln(5^2)]$ **(c)** $g\left[\ln\left(\dfrac{1}{e}\right)\right]$

48. Given $f(x) = 3^x$, evaluate the following.
   (a) $f(\log_3 7)$ **(b)** $f[\log_3(\ln 3)]$
   (c) $f[\log_3(2\ln 3)]$

49. Given $f(x) = \ln x$, evaluate the following.
   (a) $f(e^5)$ **(b)** $f(e^{\ln 3})$ **(c)** $f(e^{2\ln 3})$

50. Given $f(x) = \log_2 x$, evaluate the following.
   (a) $f(2^3)$ **(b)** $f(2^{\log_2 2})$ **(c)** $f(2^{2\log_2 2})$

51. The function defined by $f(x) = \ln|x|$ plays a prominent role in calculus. Find its domain, range, and symmetries.

52. Consider the function defined by $f(x) = \log_3|x|$.
   (a) What is the domain of this function?
   (b) Use a graphing calculator to graph $f(x) = \log_3|x|$ in the window $[-4, 4]$ by $[-4, 4]$.
   (c) How might one easily misinterpret the domain of the function simply by observing the calculator-generated graph?

53. The table is for $Y_1 = \log_3(4 - x)$. Why do the values of Y_1 show ERROR for $x \geq 4$?

X	Y1	
1	1	
2	.63093	
3	0	
4	ERROR	
5	ERROR	
6	ERROR	
7	ERROR	
X=1		

54. The function defined by Y_1 is of the form $\log_a x$. What is the value of a?

X	Y1	
1	0	
4	1	
8	1.5	
16	2	
32	2.5	
64	3	
128	3.5	
X=1		

Solve each application of logarithms. See Examples 3–5.

55. *Decibel Levels* Find the decibel ratings of sounds having the following intensities.

(a) $100I_0$ (b) $1000I_0$

(c) $100,000I_0$ (d) $1,000,000I_0$

(e) If the intensity of a sound is doubled, by how much is the decibel rating increased?

56. *Decibel Levels* Find the decibel ratings of the following sounds, having intensities as given. Round each answer to the nearest whole number.

(a) whisper, $115I_0$

(b) busy street, $9,500,000I_0$

(c) heavy truck, 20 meters away, $1,200,000,000I_0$

(d) rock music, $895,000,000,000I_0$

(e) jetliner at takeoff, $109,000,000,000,000I_0$

57. *Earthquake Intensity* The magnitude of an earthquake, measured on the Richter scale, is $\log_{10}(I/I_0)$, where I is the amplitude registered on a seismograph 100 km from the epicenter of the earthquake, and I_0 is the amplitude of an earthquake of a certain (small) size. Find the Richter scale ratings for earthquakes having the following amplitudes.

(a) $1000I_0$ (b) $1,000,000I_0$ (c) $100,000,000I_0$

58. *Earthquake Intensity* On June 16, 1999, the city of Puebla in central Mexico was shaken by an earthquake that measured 6.7 on the Richter scale. Express this reading in terms of I_0. See Exercise 57. (*Source: Times Picayune.*)

59. *Earthquake Intensity* On September 19, 1985, Mexico's largest recent earthquake, measuring 8.1 on the Richter scale, killed about 9500 people. Express the magnitude of an 8.1 reading in terms of I_0. (*Source: Times Picayune.*)

60. Compare your answers to Exercises 58 and 59. How much greater was the force of the 1985 earthquake than the 1999 earthquake?

61. *(Modeling) Visitors to U.S. National Parks* The heights of the bars in the graph represent the number of visitors (in millions) to U.S. National Parks from 1950–1997. Suppose x represents the number of years since 1900—thus, 1950 is represented by 50, 1960 is represented by 60, and so on. The logarithmic function defined by

$$f(x) = -273 + 74 \ln x$$

closely models the data. Use this function to estimate the number of visitors in the year 2000. What assumption must we make to estimate the number of visitors in years beyond 1997?

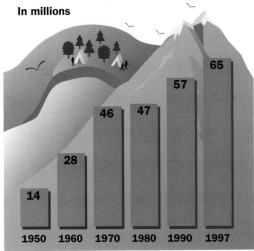

Source: Statistical Abstract of the United States, 1998.

62. *(Modeling) Volunteerism among College Freshmen* The growth in the percentage of college freshmen who reported involvement in volunteer work during their last year of high school is shown in the bar graph. Connecting the tops of the bars with a continuous curve would give a graph that indicates logarithmic growth. The function defined by

$$f(t) = -608.5 + 149 \ln t, \quad t \geq 90,$$

where t represents the number of years since 1900 and $f(t)$ is the percent, approximates the curve reasonably well.

(a) What does the function predict for the percent of freshmen entering college in 1999 who performed volunteer work during their last year of high school? How does this compare to the percent shown in the graph? The actual percent is 75.3.

(b) Explain why an exponential function would *not* provide a good model for these data.

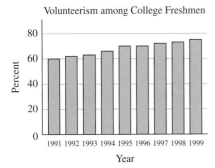

Source: The American Freshman: National Norms for Fall 1998, American Council on Education, UCLA.

63. *(Modeling) Diversity of Species* The number of species in a sample is given by

$$S(n) = a \ln\left(1 + \frac{n}{a}\right).$$

Here n is the number of individuals in the sample, and a is a constant that indicates the diversity of species in the community. If $a = .36$, find $S(n)$ for the following values of n. (*Hint: S(n)* must be a whole number.)
(a) 100   (b) 200   (c) 150   (d) 10

64. *(Modeling) Diversity of Species* In Exercise 63, find $S(n)$ if a changes to .88. Use the following values of n. (*Hint: S(n)* must be a whole number.)
(a) 50   (b) 100   (c) 250

65. *(Modeling) Diversity of Species* Suppose a sample of a small community shows two species with 50 individuals each. Find the index of diversity H. (See Example 7.)

66. *(Modeling) Diversity of Species* A virgin forest in northwestern Pennsylvania has 4 species of large trees with the following proportions of each: hemlock, .521; beech, .324; birch, .081; maple, .074. Find the index of diversity H. (See Example 7.)

67. *(Modeling) Global Temperature Increase* In Example 5, we expressed the average global temperature increase T (in °F) as

$$T(k) = 1.03k \ln(C/C_0),$$

where C_0 is the preindustrial amount of carbon dioxide, C is the current carbon dioxide level, and k is a constant. Arrhenius determined that $10 \le k \le 16$ when C was double the value C_0. Use $T(k)$ to find the range of the rise in global temperature T (rounded to the nearest degree) that Arrhenius predicted. (*Source:* Clime, W., *The Economics of Global Warming,* Institute for International Economics, Washington, D.C., 1992.)

68.   *(Modeling) Global Temperature Increase* (Refer to Exercise 67.) According to the IPCC, if present trends continue, future increases in average global temperatures (in °F) can be modeled by

$$T(x) = 6.489 \ln(C/280),$$

where C is the concentration of atmospheric carbon dioxide (in ppm). C can be modeled by the function with

$$C(x) = 353(1.006)^{x-1990},$$

where x is the year. (*Source:* International Panel on Climate Change (IPCC), 1990.)
(a) Write T as a function of x.
(b) Using a graphing calculator, graph $C(x)$ and $T(x)$ on the interval $[1990, 2275]$ using different coordinate axes. Describe the graph of each function. How are C and T related?
(c) Approximate the slope of the graph of T. What does this slope represent?
(d) Use graphing to estimate x and $C(x)$ when $T(x) = 10°F$.

69. *(Modeling) Planets' Distances from the Sun and Period of Revolution* The following table contains the planets' average distances D from the sun and their periods P of revolution around the sun in years. The distances have been normalized so that Earth is one unit away from the sun. For example, since Jupiter's distance is 5.2, its distance from the sun is 5.2 times farther than Earth's.

Planet	D	P
Mercury	.39	.24
Venus	.72	.62
Earth	1	1
Mars	1.52	1.89
Jupiter	5.2	11.9
Saturn	9.54	29.5
Uranus	19.2	84.0
Neptune	30.1	164.8

Source: Ronan, C., *The Natural History of the Universe,* MacMillan Publishing Co., New York, 1991.

(a) Make a scatter diagram by plotting the point $(\ln D, \ln P)$ for each planet on the xy-coordinate axes using a graphing calculator. Do the data points appear to be linear?
(b) Determine a linear equation that models the data points. Graph your line and the data on the same coordinate axes.
(c) Use this linear model to predict the period of the planet Pluto if its distance is 39.5. Compare your answer to the actual value of 248.5 years.

9.4 Exponential and Logarithmic Equations

> • A Property of Logarithms • Exponential Equations • Logarithmic Equations • Modeling

A Property of Logarithms Some simple equations were solved in earlier sections of this chapter. More general methods for solving these equations depend on the property below. This property follows from the fact that logarithmic functions are one-to-one.

Property of Logarithms

(f) If $x > 0$, $y > 0$, $a > 0$, and $a \neq 1$, then

$$x = y \quad \text{if and only if} \quad \log_a x = \log_a y.$$

(Properties (a)–(e) of logarithms were stated in Section 9.2.)

Exponential Equations The first examples illustrate a general method, using the new property in the algebraic solution, for solving exponential equations.

● ● ● **Example 1** Solving an Exponential Equation

Solve $7^x = 12$. Give the solution to four decimal places.

Algebraic Solution

The properties of exponents given in Section 9.1 cannot be used to solve this equation, so we apply Property (f). While any appropriate base b can be used, the best practical base is base 10 or base e. Taking base e (natural) logarithms of both sides gives

$$7^x = 12$$

$$\ln 7^x = \ln 12 \qquad \text{Property (f) of logarithms}$$

$$x \ln 7 = \ln 12 \qquad \text{Property (c) of logarithms}$$

$$x = \frac{\ln 12}{\ln 7} \qquad \text{Divide by } \ln 7.$$

$$x \approx 1.2770. \qquad \text{Use a calculator.}$$

The solution is 1.2770.

Graphing Calculator Solution

Graph $Y_1 = 7^x$ and $Y_2 = 12$. Use the intersection-of-graphs method to find the x-coordinate of their point of intersection. As seen in the display at the bottom of Figure 35, when rounded to four decimal places, the solution agrees with that found algebraically.

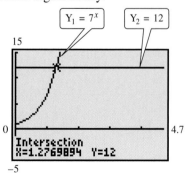

Figure 35

CAUTION Be careful when evaluating a quotient like $\dfrac{\ln 12}{\ln 7}$ in Example 1. Do not confuse this quotient with $\ln\left(\dfrac{12}{7}\right)$, which can be written as $\ln 12 - \ln 7$. You *cannot* change the quotient of *two logarithms* to a difference of logarithms.

$$\frac{\ln 12}{\ln 7} \neq \ln\left(\frac{12}{7}\right)$$

● ● ● **Example 2** Solving an Exponential Equation

Solve $3^{2x-1} = .4^{x+2}$. Give the solution to four decimal places.

Algebraic Solution

$$\ln 3^{2x-1} = \ln .4^{x+2} \qquad \text{Take natural logarithms on both sides.}$$

$$(2x - 1)\ln 3 = (x + 2)\ln .4 \qquad \text{Property (c) of logarithms}$$

$$2x \ln 3 - \ln 3 = x \ln .4 + 2 \ln .4 \qquad \text{Distributive property}$$

$$2x \ln 3 - x \ln .4 = 2 \ln .4 + \ln 3$$

$$x(2 \ln 3 - \ln .4) = 2 \ln .4 + \ln 3 \qquad \text{Factor out } x.$$

$$x = \frac{2 \ln .4 + \ln 3}{2 \ln 3 - \ln .4}$$

$$x = \frac{\ln .16 + \ln 3}{\ln 9 - \ln .4} \qquad \text{Property (c) of logarithms}$$

$$x = \frac{\ln .48}{\ln\left(\dfrac{9}{.4}\right)}$$

$$x \approx -.2357 \qquad \text{Use a calculator.}$$

The solution is $-.2357$.

Graphing Calculator Solution

Use the intersection-of-graphs method as in Example 1. In the display at the bottom of Figure 36, the calculator is set to show the coordinates correct to four decimal places, which supports the algebraic result.

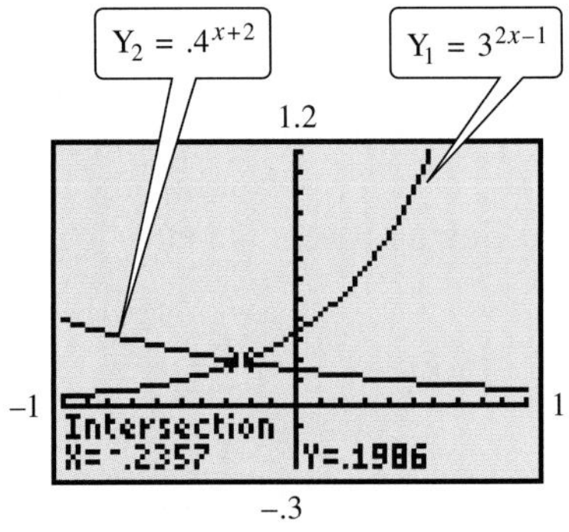

$Y_2 = .4^{x+2}$ $Y_1 = 3^{2x-1}$

Intersection
X=-.2357   Y=.1986

Figure 36

● ● ●

● ● ● **Example 3** Solving a Base e Exponential Equation

(a) Solve $e^{x^2} = 200$, and give solutions to four decimal places.

Take natural logarithms on both sides; then use properties of logarithms.

$$\ln e^{x^2} = \ln 200$$

$$x^2 = \ln 200 \qquad \ln e^{x^2} = x^2$$

$$x = \pm\sqrt{\ln 200}$$

$$x \approx \pm 2.3018 \qquad \text{Use a calculator.}$$

The solutions are ± 2.3018.

(b) Solve $\dfrac{e^x - 1}{e^{-x} - 1} = -3$, and give the exact solution(s).

Begin by multiplying both sides by the denominator.

$$e^x - 1 = -3(e^{-x} - 1)$$
$$e^x - 1 = -3e^{-x} + 3 \qquad \text{Distributive property}$$
$$e^x - 4 + 3e^{-x} = 0 \qquad \text{Write with 0 alone on one side.}$$
$$e^{2x} - 4e^x + 3 = 0 \qquad \text{Multiply both sides by } e^x.$$
$$(e^x)^2 - 4e^x + 3 = 0 \qquad \text{Rewrite as a quadratic equation in } e^x.$$

Let $u = e^x$ and solve first for u, factoring to get the solutions 1 and 3. Then solve for x.

$$u = 1 \quad \text{or} \quad u = 3$$
$$e^x = 1 \quad \text{or} \quad e^x = 3 \qquad \text{Replace } u \text{ with } e^x.$$
$$x = 0 \quad \text{or} \quad x = \ln 3$$

Because we solved by multiplying both sides by an expression involving a variable $(e^{-x} - 1)$, we must check for extraneous solutions. Letting $x = 0$ leads to 0 in the denominator, so 0 must be rejected. The solution $\ln 3$ leads to a true statement, so the only solution is $\ln 3$. ● ● ●

Logarithmic Equations   The next examples show some ways to solve logarithmic equations. The properties of logarithms given in Section 9.2 are useful here, as is Property (f).

● ● ●   **Example 4**   Solving a Logarithmic Equation

Solve $\log_a(x + 6) - \log_a(x + 2) = \log_a x$.

Rewrite the equation as

$$\log_a \frac{x + 6}{x + 2} = \log_a x. \qquad \text{Property (b) of logarithms}$$

Now the equation is in the proper form to use Property (f).

$$\frac{x + 6}{x + 2} = x \qquad \text{Property (f) of logarithms}$$
$$x + 6 = x(x + 2) \qquad \text{Multiply by } x + 2.$$
$$x + 6 = x^2 + 2x \qquad \text{Distributive property}$$
$$x^2 + x - 6 = 0 \qquad \text{Write with 0 alone on one side.}$$
$$(x + 3)(x - 2) = 0 \qquad \text{Use the zero-factor property.}$$
$$x = -3 \quad \text{or} \quad x = 2$$

The negative solution $(x = -3)$ cannot be used since it is not in the domain of $\log_a x$ in the original equation. For this reason, the only valid solution is the positive number 2. ● ● ●

CAUTION   Recall that the domain of $y = \log_b x$ is $(0, \infty)$. For this reason, it is always necessary to check that the apparent solution of a logarithmic equation results in logarithms of positive numbers in the original equation.

● ● ●   **Example 5**   Solving a Logarithmic Equation

Solve $\log(3x + 2) + \log(x - 1) = 1$.

Algebraic Solution

Since $\log x$ is an abbreviation for $\log_{10} x$, and $1 = \log_{10} 10$, the properties of logarithms give

$$\log[(3x + 2)(x - 1)] = \log 10 \qquad \text{Property (a) of logarithms}$$

$$(3x + 2)(x - 1) = 10 \qquad \text{Property (f) of logarithms}$$

$$3x^2 - x - 2 = 10$$

$$3x^2 - x - 12 = 0.$$

Now use the quadratic formula to get

$$x = \frac{1 \pm \sqrt{1 + 144}}{6}.$$

The number $(1 - \sqrt{145})/6$ is negative, so $x - 1$ is negative. Therefore, $\log(x - 1)$ is not defined and this proposed solution must be discarded. Since $(1 + \sqrt{145})/6 > 1$, both $3x + 2$ and $x - 1$ are positive and the solution is

$$\frac{1 + \sqrt{145}}{6}.$$

Graphing Calculator Solution

Figure 37 shows how the equation-solving feature of a graphing calculator supports the result found algebraically. Recall that the equation must first be written with 0 on one side.

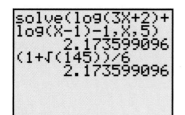

Figure 37

● ● ●

N O T E   The definition of logarithm could have been used in Example 5 by first writing

$$\log(3x + 2) + \log(x - 1) = 1$$

$$\log_{10}[(3x + 2)(x - 1)] = 1 \qquad \text{Property (a) of logarithms}$$

$$(3x + 2)(x - 1) = 10^1, \qquad \text{Definition of logarithm}$$

then continuing as shown above.

● ● ●   **Example 6**   Solving a Logarithmic Equation

Solve $\ln e^{\ln x} - \ln(x - 3) = \ln 2$.

On the left side of the equation, $\ln e^{\ln x}$ can be written as $\ln x$ using the theorem on inverses at the end of Section 9.2. The equation becomes

$$\ln x - \ln(x - 3) = \ln 2$$

$$\ln \frac{x}{x - 3} = \ln 2 \qquad \text{Property (b) of logarithms}$$

$$\frac{x}{x - 3} = 2 \qquad \text{Property (f) of logarithms}$$

$$x = 2x - 6 \qquad \text{Multiply by } x - 3.$$

$$6 = x.$$

Verify that the solution is 6. ● ● ●

A summary of the methods used for solving equations in this section follows.

Solving Exponential or Logarithmic Equations

To solve an exponential or logarithmic equation, first use the properties of algebra to change the given equation into one of the following forms, where a and b are real numbers with appropriate restrictions.

1. $a^{f(x)} = b$
   To solve, take logarithms on both sides.

2. $\log_a f(x) = b$
   Solve by changing to exponential form $a^b = f(x)$.

3. $\log_a f(x) = \log_a g(x)$
   From the given equation, obtain the equation $f(x) = g(x)$, then solve algebraically.

4. In a more complicated equation, such as the one in Example 3(b), it may be necessary to first solve for $e^{f(x)}$ or $\log_a f(x)$ and then solve the resulting equation using one of the methods given above.

Modeling

● ● ● **Example 7** Modeling Coal Consumption in the U.S.

The use of coal as an energy source continues to increase due to its low cost and plentiful supply. However, coal takes a toll on the environment. The burning of coal releases harmful greenhouse gases, and the mining of coal (especially strip and open-pit mining) can negatively alter the ecology of a region. The table gives the U. S. coal consumption (in quadrillions of British thermal units, or *quads*) for several years.

Year	Coal Consumption (in quads)
1975	12.66
1980	15.42
1985	17.48
1990	19.10
1995	20.08

Source: Statistical Abstract of the United States, 1998.

The data can be modeled with the function defined by

$$f(t) = 31.52 \ln t - 122.98, \qquad t \geq 75,$$

where t is the number of years after 1900.

(a) Approximately what amount of coal was consumed in the United States in 1993?

The year 1993 is represented by $1993 - 1900 = 93$. Use a calculator to find $f(93)$.

$$f(93) = 31.52 \ln 93 - 122.98$$
$$\approx 19.89$$

Based on this model, 19.89 quads were used in 1993.

(b) If this trend continues, approximately when will the annual consumption reach 25 quads?

Let $f(t) = 25$, and solve for t.

$$25 = 31.52 \ln t - 122.98$$
$$147.98 = 31.52 \ln t$$
$$\ln t = \frac{147.98}{31.52}$$
$$t = e^{147.98/31.52}$$
$$t \approx 109$$

Add 109 to 1900 to get 2009. The annual consumption will reach 25 quads in approximately 2009. ● ● ●

9.4 Exercises

Concept Check   An exponential equation such as $5^x = 9$ can be solved for its exact solution using the meaning of logarithm and the change-of-base theorem. Since x is the exponent to which 5 must be raised in order to obtain 9, the solution is $\log_5 9$, or $\dfrac{\log 9}{\log 5}$ or $\dfrac{\ln 9}{\ln 5}$. For the following equations, give the exact solution in three forms similar to the forms explained here.

1. $7^x = 19$ **2.** $3^x = 10$ **3.** $\left(\dfrac{1}{2}\right)^x = 12$ **4.** $\left(\dfrac{1}{3}\right)^x = 4$

Solve each equation. When solutions are irrational, give them as decimals correct to four decimal places. See Examples 1–6.

5. $3^x = 6$ **6.** $4^x = 12$ **7.** $6^{1-2x} = 8$ **8.** $3^{2x-5} = 13$

9. $2^{x+3} = 5^x$ **10.** $6^{x+3} = 4^x$ **11.** $e^{x-1} = 4$ **12.** $e^{2-x} = 12$

13. $2e^{5x+2} = 8$ **14.** $10e^{3x-7} = 5$ **15.** $2^x = -3$ **16.** $3^x = -6$

17. $e^{8x} \cdot e^{2x} = e^{20}$ **18.** $e^{6x} \cdot e^x = e^{21}$ **19.** $100(1.02)^{x/4} = 200$ **20.** $500(1.05)^{x/4} = 200$

21. $\log x + \log(x - 21) = 2$ **22.** $\log x + \log(3x - 13) = 1$ **23.** $\ln(5 + 4x) - \ln(3 + x) = \ln 3$

24. $\ln(2x + 5) + \ln x = \ln 7$ **25.** $\log_6 4x - \log_6(x - 3) = \log_6 12$ **26.** $\log_2 3x + \log_2 3 = \log_2(2x + 15)$

27. $5^{x+2} = 2^{2x-1}$ **28.** $6^{x-3} = 3^{4x+1}$ **29.** $\ln e^x - \ln e^3 = \ln e^5$

30. $\ln e^x - 2 \ln e = \ln e^4$ **31.** $\log_2(\log_2 x) = 1$ **32.** $\log x = \sqrt{\log x}$

33. $\log x^2 = (\log x)^2$ **34.** $\log_2 \sqrt{2x^2} = \dfrac{3}{2}$

35. Suppose you overhear the following statement: "I must reject any negative answer when I solve an equation involving logarithms." Is this correct? Write an explanation of why it is or is not correct.

Find $f^{-1}(x)$, and give the domain and the range.

37. $f(x) = e^{x+1} - 4$

36. What values of x could not possibly be solutions of the following equation?

$$\log_a(4x - 7) + \log_a(x^2 + 4) = 0$$

38. $f(x) = 2 \ln 3x$

Use a graphing calculator to solve each equation. Give irrational solutions correct to the nearest hundredth. See Examples 1, 2, and 5.

39. $e^x + \ln x = 5$

40. $e^x - \ln(x + 1) = 3$

41. $2e^x + 1 = 3e^{-x}$

42. $e^x + 6e^{-x} = 5$

43. $\log x = x^2 - 8x + 14$

44. $\ln x = -\sqrt[3]{x + 3}$

Solve each application. See Example 7.

45. *(Modeling) Personal Computer Shipments* The table shows worldwide shipments of personal computers (in millions) from 1990–1998.

Year	Millions of Computers
1990	23.7
1991	27.0
1992	32.4
1993	38.9
1994	47.9
1995	60.2
1996	71.1
1997	82.4
1998	97.3

Source: The Wall Street Journal Almanac, 1999.

Letting y represent the number of personal computers (in millions) and x represent the number of years since 1990, we find that the function defined by

$$f(x) = 23 \cdot 1.2^x$$

models the data quite well. According to this function, when will worldwide shipments double their 1998 value?

46. *(Modeling) Race Speed* At the World Championship races held at Rome's Olympic Stadium in 1987, American sprinter Carl Lewis ran the 100-meter race in 9.86 seconds. His speed in meters per second after t seconds is closely modeled by the function defined by

$$f(t) = 11.65(1 - e^{-t/1.27}).$$

(Source: Banks, Robert B., Towing Icebergs, Falling Dominoes, and Other Adventures in Applied Mathematics, Princeton University Press, 1998.)

(a) How fast was he running as he crossed the finish line?

(b) After how many seconds was he running at the rate of 10 meters per second?

47. *(Modeling) Fatherless Children* The percent of U.S. children growing up without a father has increased rapidly since 1950. If x represents the number of years since 1900, the function defined by

$$f(x) = \frac{25}{1 + 1364.3e^{-x/9.316}}$$

models the percent fairly well. *(Sources: National Longitudinal Survey of Youth; U.S. Department of Commerce; U.S. Bureau of the Census.)* **What percent of U.S. children lived in a home without a father in 1997?**

48. *(Modeling) Height of the Eiffel Tower* Paris's Eiffel Tower was constructed in 1889 to commemorate the one hundredth anniversary of the French Revolution. The right side of the Eiffel Tower has a shape that can be approximated by the graph of the function defined by

$$f(x) = -301 \ln(x/207).$$

See the figure. *(Source: Banks, Robert B., Towing Icebergs, Falling Dominoes, and Other Adventures in Applied Mathematics, Princeton University Press, 1998.)*

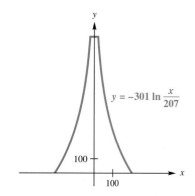

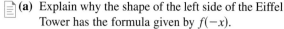

(a) Explain why the shape of the left side of the Eiffel Tower has the formula given by $f(-x)$.

(b) The short horizontal line at the top of the figure has length 15.7488 feet. Approximately how tall is the Eiffel Tower?

(c) Approximately how far from the center of the tower is the point on the right side that is 500 feet above the ground?

49. 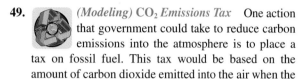 (*Modeling*) CO_2 *Emissions Tax*   One action that government could take to reduce carbon emissions into the atmosphere is to place a tax on fossil fuel. This tax would be based on the amount of carbon dioxide emitted into the air when the fuel is burned. The *cost-benefit* equation

$$\ln(1 - P) = -.0034 - .0053T$$

models the approximate relationship between a tax of T dollars per ton of carbon and the corresponding percent reduction P (in decimal form) of emissions of carbon dioxide. (*Source:* Nordhause, W., "To Slow or Not to Slow: The Economics of the Greenhouse Effect," Yale University, New Haven, Connecticut.)

(a) Write P as a function of T.

(b) Graph P for $0 \le T \le 1000$. Discuss the benefit of continuing to raise taxes on carbon.

(c) Determine P when $T = \$60$, and interpret this result.

(d) What value of T will give a 50% reduction in carbon emissions?

50. 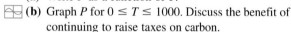 (*Modeling*) *Radiative Forcing*   (Refer to Example 5 in Section 9.3 and Exercise 70 in Section 9.1.) Using computer models, the International Panel on Climate Change (IPCC) in 1990 estimated k to be 6.3 in the radiative forcing equation

$$R = k \ln(C/C_0),$$

where C_0 is the preindustrial amount of carbon dioxide and C is the current level. (*Source:* Clime, W., *The Economics of Global Warming,* Institute for International Economics, Washington, D.C., 1992.)

(a) What radiative forcing R (in w/m²) is expected by the IPCC if the carbon dioxide level in the atmosphere doubles from its preindustrial level?

(b) Determine the global temperature increase predicted by the IPCC if the carbon dioxide levels were to double.

For Exercises 51–54, refer to the formula for compound interest given in Section 9.1.

$$A = P\left(1 + \frac{r}{m}\right)^{tm}$$

51. *Interest on an Account*   Tom Tupper wants to buy a \$30,000 car. He has saved \$27,000. Find the number of years (to the nearest tenth) it will take for his \$27,000 to grow to \$30,000 at 6% interest compounded quarterly.

52. *Investment Time*   Find t to the nearest hundredth if \$1786 becomes \$2063.40 at 11.6%, with interest compounded monthly.

53. *Interest Rate*   Find the interest rate that will produce \$2500 if \$2000 is left at interest compounded semiannually for 3.5 years.

54. *Interest Rate*   At what interest rate will \$16,000 grow to \$20,000 if invested for 5.25 years and interest is compounded quarterly?

Quantitative Reasoning

55. *Financial Planning for Retirement—What are your options?*   The IRA (Individual Retirement Account) is the most common tax-deferred savings plan in the United States. Earned income deposited into an IRA is not taxed in the current year, and no taxes are incurred on the interest paid in subsequent years. However, when you withdraw the money from the account after age $59\frac{1}{2}$, you pay taxes on the entire amount. Suppose you deposit \$2000 of earned income into an IRA, you can earn an annual interest rate of 8%, and you are in a 40% tax bracket. (*Note:* We recognize that interest rates and tax brackets are subject to change over a long period of time, but some assumptions must be made to evaluate the investment.) Also, suppose you deposit the \$2000 at age 25 and withdraw it at age 60.

(a) How much money will remain after you pay the taxes at age 60?

(b) Suppose that instead of depositing the money into an IRA, you pay taxes on the money and the annual interest. How much money will you have at age 60? (*Note:* You effectively start with \$1200 (60% of \$2000), and the money earns 4.8% (60% of 8%) interest after taxes.)

(c) How much additional money will you earn with the IRA?

(d) Suppose you pay taxes on the original \$2000, but are then able to earn 8% in a tax-free investment. Compare your balance at age 60 with the IRA balance.

Chapter 9   Summary

Key Terms & Symbols	Key Ideas
9.1 Exponential Functions exponential function compound interest future value present value e	**Additional Properties of Exponents** **(a)** If $a > 0$ and $a \neq 1$, then a^x is a unique real number for all real numbers x. **(b)** If $a > 0$ and $a \neq 1$, then $a^b = a^c$ if and only if $b = c$. **(c)** If $a > 1$ and $m < n$, then $a^m < a^n$. **(d)** If $0 < a < 1$ and $m < n$, then $a^m > a^n$.
9.2 Logarithmic Functions logarithm $\log_a x$ logarithmic function	**Properties of Logarithms** For any positive real numbers x and y, real number r, and positive real number a, $a \neq 1$: **(a)** $\log_a xy = \log_a x + \log_a y$ **(b)** $\log_a \dfrac{x}{y} = \log_a x - \log_a y$ **(c)** $\log_a x^r = r \log_a x$ **(d)** $\log_a a = 1$ **(e)** $\log_a 1 = 0$.
9.3 Evaluating Logarithms and the Change-of-Base Theorem common logarithm $\log x$ natural logarithm $\ln x$	**Change-of-Base Theorem** For any positive real numbers x, a, and b, where $a \neq 1$ and $b \neq 1$: $$\log_a x = \frac{\log_b x}{\log_b a}.$$
9.4 Exponential and Logarithmic Equations	**Property of Logarithms** **(f)** If $x > 0$, $y > 0$, $a > 0$, and $a \neq 1$, then $$x = y \quad \text{if and only if} \quad \log_a x = \log_a y.$$

Chapter 9   Review Exercises

Concept Check   *Match each equation with one of the graphs below.*

1. $y = \log_{.3} x$

2. $y = e^x$

3. $y = \ln x$

4. $y = (.3)^x$

A.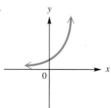

B.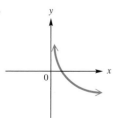

C.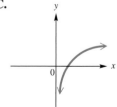

D.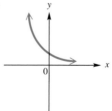

Write each equation in logarithmic form.

5. $2^5 = 32$

6. $100^{1/2} = 10$

7. $(3/4)^{-1} = 4/3$

Graph each function.

8. $f(x) = 1.5^x$

9. $f(x) = \log_3 x$

10. Is the logarithm to the base 3 of 4 written as $\log_4 3$ or $\log_3 4$?

. . . . **Relating Concepts** . . . .
For individual or collaborative investigation
(Exercises 11–16)

Work Exercises 11–16 in order.

11. What is the exact value of $\log_3 9$?

12. What is the exact value of $\log_3 27$?

13. Between what two consecutive integers must $\log_3 16$ lie?

14. Use the change-of-base theorem to support your answer for Exercise 13.

15. Repeat Exercises 11 and 12 for $\log_5(1/5)$ and $\log_5 1$.

16. Repeat Exercises 13 and 14 for $\log_5 .68$.

Write each equation in exponential form.

17. $\log_9 27 = \dfrac{3}{2}$

18. $\log 3.45 \approx .5378$

19. $\ln 45 \approx 3.8067$

20. What is the base of the logarithmic function whose graph contains the point $(81, 4)$?

21. What is the base of the exponential function whose graph contains the point $(-4, 1/16)$?

Use properties of logarithms to write, if possible, each logarithm as a sum, difference, or product of logarithms. Assume all variables represent positive numbers.

22. $\log_3\left(\dfrac{mn}{5r}\right)$

23. $\log_5\left(x^2 y^4 \sqrt[5]{m^3 p}\right)$

24. $\log_7(7k + 5r^2)$

Find each logarithm. Round to four decimal places.

25. $\log 45.6$

26. $\log .0411$

27. $\ln 470$

28. $\ln 144,000$

29. $\log_3 769$

30. $\log_{2/3}(5/8)$

Solve each equation. Round irrational solutions to four decimal places.

31. $8^x = 32$

32. $x^{-3} = \dfrac{8}{27}$

33. $10^{2x-3} = 17$

34. $4^{x+3} = 5^{2-x}$

35. $e^{x+1} = 10$

36. $\log_{64} x = \dfrac{1}{3}$

37. $\ln(6x) - \ln(x + 1) = \ln 4$

38. $\log_{16} \sqrt{x + 1} = \dfrac{1}{4}$

39. $\ln x + 3 \ln 2 = \ln\left(\dfrac{2}{x}\right)$

40. $\log x + \log(x - 3) = 1$

41. $\ln[\ln(e^{-x})] = \ln 3$

42. $S = a \ln\left(1 + \dfrac{n}{a}\right)$ for n

Solve each problem.

43. *Earthquake Intensity* On July 14, 1991, Peshawar, Pakistan, was shaken by an earthquake that measured 6.6 on the Richter scale.

(a) Express this reading in terms of I_0. (See Section 9.3 Exercises.)

(b) In February of the same year a quake measuring 6.5 on the Richter scale killed about 900 people in the mountains of Pakistan and Afghanistan. Express the magnitude of a 6.5 reading in terms of I_0.

(c) How much greater was the force of the earthquake with a measure of 6.6?

44. *Earthquake Intensity*

(a) The San Francisco earthquake of 1906 had a Richter scale rating of 8.3. Express the magnitude of this earthquake as a multiple of I_0.

(b) In 1989, the San Francisco region experienced an earthquake with a Richter scale rating of 7.1. Express the magnitude of this earthquake as a multiple of I_0.

(c) Compare the magnitudes of the two San Francisco earthquakes discussed in parts (a) and (b).

45. *(Modeling) Decibel Levels*   Recall from Section 9.3 that the model for the decibel rating of the loudness of a sound is

$$d = 10 \log \frac{I}{I_0}.$$

A few years ago, there was a controversy about a proposed government limit on factory noise. One group wanted a maximum of 89 decibels, while another group wanted 86. This difference seemed very small to many people. Find the percent by which the 89-decibel intensity exceeds that for 86 decibels.

46. *Interest Rate*   What annual interest rate, to the nearest tenth, will produce $8780 if $3500 is left at interest (compounded annually) for 10 years?

47. *Growth of an Account*   Find the number of years (to the nearest tenth) needed for $48,000 to become $58,344 at 5% interest compounded semiannually.

48. *Growth of an Account*   Manuel deposits $10,000 for 12 years in an account paying 12% compounded annually. He then puts this total amount on deposit in another account paying 10% compounded semiannually for another 9 years. Find the total amount on deposit after the entire 21-year period.

49. *Growth of an Account*   Anne Kelly deposits $12,000 for 8 years in an account paying 5% compounded annually. She then leaves the money alone with no further deposits at 6% compounded annually for an additional 6 years. Find the total amount on deposit after the entire 14-year period.

50. *Cost from Inflation*   If the inflation rate were 10%, use the formula for continuous compounding to find the number of years, to the nearest tenth, for a $1 item to cost $2.

51. *(Modeling) Software Exports*   India has become an important exporter of software to the United States. The table shows India's software exports (in millions of U.S. dollars) in selected years since 1985.

Year (x)	Million $ (y)	Year (x)	Million $ (y)
1985	6	1993	225
1987	39	1995	483
1989	67	1997	1000
1991	128		

Sources: NIIT, NASSCOM. From *Scientific American,* September, 1994, page 95.

Letting y represent software (in millions of dollars) and x represent the number of years since 1900, we find that the function defined by

$$f(x) = 6.2(10)^{-12}(1.4)^x$$

models the data reasonably well. According to this function, when did software exports double their 1997 value?

52. *Racial Mix in the United States*   In 1995, the U.S. racial mix was 75.3% white, 9.0% Hispanic, 12.0% African American, 2.9% Asian/Pacific Islanders, and .8% Native American. Their percentages of executives, managers, and administrators in private-industry communication firms were 84.0%, 9.5%, 3.4%, 2.0%, and 0%, respectively. Assume these percents remain constant. If the number of African American executives, managers, and administrators increases at a rate of 5% per year, determine the year when their representation will reach 12.0% of the total number of executives, managers, and administrators in communications. (*Source:* U.S. Labor Department's Glass Ceiling Commission.)

53. *Atmospheric Pressure*   (Refer to Exercise 69 in Section 9.1.) The atmospheric pressure (in millibars) at a given altitude (in meters) is listed in the table.

Altitude (x)	Pressure (P)
0	1013
1000	899
2000	795
3000	701
4000	617
5000	541
6000	472
7000	411
8000	357
9000	308
10,000	265

Source: Miller, A. and J. Thompson, *Elements of Meteorology,* Fourth Edition, Charles E. Merrill Publishing Company, Columbus, Ohio, 1993.

(a) Make a scatter diagram of the points $(x, \ln P)$. Is there a linear relationship between x and $\ln P$?

(b) If $P = Ce^{kx}$ with constants C and k, explain why there is a linear relationship between x and $\ln P$.

54. *(Modeling) Transistors on Computer Chips* Computing power of personal computers has increased dramatically as a result of the ability to place an increasing number of transistors on a single processor chip. The table lists the number of transistors on some popular computer chips made by Intel.

Year	Chip	Transistors
1971	4004	2300
1986	386DX	275,000
1989	486DX	1,200,000
1993	Pentium	3,300,000
1995	P6	5,500,000
1997	Pentium III	9,500,000

Source: Intel.

(a) Make a scatter diagram of the data. Let the x-axis represent the year, where $x = 0$ corresponds to 1971, and let the y-axis represent the number of transistors.

(b) Discuss which type of function $y = f(x)$ describes the data best, where a and b are constants.
  (i) $f(x) = ax + b$ (linear)
  (ii) $f(x) = a \ln b(x + 1)$ (logarithmic)
  (iii) $f(x) = ab^x$ (exponential)

(c) Determine a function f that approximates these data. Plot f and the data on the same coordinate axes.

(d) Assuming that the present trend continues, use f to predict the number of transistors on a chip in the year 2002.

55. *(Modeling) Drug Level in the Bloodstream* After a medical drug is injected directly into the bloodstream it is gradually eliminated from the body. Graph the following functions on the interval $[0, 10]$. Use $[0, 500]$ for the range of $A(t)$. Determine the function that best models the amount $A(t)$ (in milligrams) of a drug remaining in the body after t hours if 350 milligrams were initially injected.

(a) $A(t) = t^2 - t + 350$
(b) $A(t) = 350 \log(t + 1)$
(c) $A(t) = 350(.75)^t$
(d) $A(t) = 100(.95)^t$

56. The graphs of $y = x^2$ and $y = 2^x$ have the points $(2, 4)$ and $(4, 16)$ in common. There is a third point common to the graphs whose coordinates can be approximated by using a graphing calculator. Find the coordinates, giving as many decimal places as your calculator displays.

57. Consider $f(x) = \log_4(2x^2 - x)$.

(a) Use the change-of-base theorem with base e to write $\log_4(2x^2 - x)$ in a suitable form to graph with a calculator.

(b) Graph the function using a graphing calculator. Use the window $[-2.5, 2.5]$ by $[-5, 2.5]$.

(c) What are the x-intercepts?

(d) Give the equations of the vertical asymptotes.

(e) Explain why there is no y-intercept.

Chapter 9 Test

1. Consider the functions defined by $f(x) = a^x$ and $g(x) = \log_a x$.
  (a) How are f and g related?
  (b) How are the graphs of f and g related?
  (c) If $(2, 9)$ belongs to f, give an ordered pair that belongs to g.

2. Match each equation with its graph.

  (a) $y = \log_{1/3} x$ **(b)** $y = e^x$ **(c)** $y = \ln x$ **(d)** $y = \left(\dfrac{1}{3}\right)^x$

A.
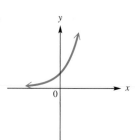

B.
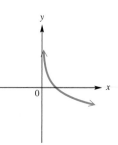

C.
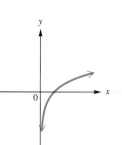

D.
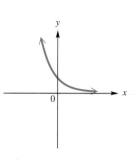

3. Solve $\left(\dfrac{1}{8}\right)^{2x-3} = 16^{x+1}$.

4. (a) Write $4^{3/2} = 8$ in logarithmic form.

   (b) Write $\log_8 4 = \dfrac{2}{3}$ in exponential form.

5. Graph $f(x) = \left(\dfrac{1}{2}\right)^x$ and $g(x) = \log_{1/2} x$ on the same axes. What is their relationship?

6. Use properties of logarithms to write the following as a sum, difference, or product of logarithms. Assume all variables represent positive numbers.

$$\log_7\left(\frac{x^2\sqrt[4]{y}}{z^3}\right)$$

Use a calculator to find an approximation for each logarithm. Express answers to four decimal places.

7. $\log 237.4$

8. $\ln .0467$

9. $\log_9 13$

10. $\log(2.49 \times 10^{-3})$

Use properties of logarithms to solve each equation. Express the solutions as directed.

11. $\log_x 25 = 2$ (exact)

12. $\log_4 32 = x$ (exact)

13. $\log_2 x + \log_2(x + 2) = 3$ (exact)

14. $5^{x+1} = 7^x$ (approximate, to four decimal places)

15. $\ln x - 4\ln 3 = \ln\left(\dfrac{5}{x}\right)$ (approximate, to four decimal places)

16. One of your friends is taking another mathematics course and tells you, "I have no idea what an expression like $\log_5 27$ really means." Write an explanation of what it means, and tell how you can find an approximation for it with a calculator.

Solve each problem.

17. *(Modeling) Skydiver Fall Speed* A skydiver in free fall travels at a speed modeled by

$$v(t) = 176(1 - e^{-.18t})$$

feet per second after t seconds. How long will it take for the skydiver to attain the speed of 147 feet per second (100 mph)?

18. *Growth of an Account* How many years, to the nearest tenth, will be needed for $5000 to increase to $18,000 at 6.8% compounded **(a)** monthly **(b)** continuously?

19. *(Modeling) Radioactive Decay* The amount of radioactive material, in grams, present after t days is modeled by

$$A(t) = 600e^{-.05t}.$$

Find the amount present after 12 days.

20. *(Modeling) Population Growth* In the year 2000, the population of New York state was 18.15 million and increasing exponentially with growth constant .0021. The population of Florida was 15.23 million and increasing exponentially with growth constant .0131. Assuming these trends were to continue, in what year would the population of Florida equal the population of New York?

Chapter 9  Internet Project

Modeling Growth of Internet Hosts

A model is valid only within the domain of the given data. *Extrapolating* too far into the future can lead to erroneous and even impossible results. (See, for example, the world population growth model in Section 9.1, Exercise 71.)

The Internet project for this chapter, found at www.awl.com/lhs, involves finding a model for the growth of Internet hosts and illustrates how one can test the model to see how accurately it might predict growth in future years.

Answers to Selected Exercises

To The Student

If you need further help with trigonometry, you may want to obtain a copy of the *Student's Solution Manual* that goes with this book. It contains solutions to all the odd-numbered section and chapter review exercises and all the chapter test exercises. Your college bookstore either has the *Manual* or can order it for you.

In this section we provide the answers that we think most students will obtain when they work the exercises using the methods explained in the text. If your answer does not look exactly like the one given here, it is not necessarily wrong. In many cases there are equivalent forms of the answer. For example, if the answer section shows $\frac{3}{4}$ and your answer is .75, you have obtained the correct answer but written it in a different (yet equivalent) form. Unless the directions specify otherwise, .75 is just as valid an answer as $\frac{3}{4}$. In general, if your answer does not agree with the one given in the text, see whether it can be transformed into the other form. If it can, then it is the correct answer. If you still have doubts, talk with your instructor.

CHAPTER 1 THE TRIGONOMETRIC FUNCTIONS

Connections *(page 5)*

1. $\left(-\dfrac{1}{2}, 1\right)$ **2.** $(-4, 2)$

1.1 Exercises *(page 11)*

1.–7.

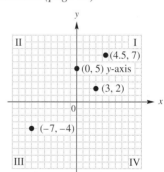

9. II **11.** III **15.** I and III **17.** $\sqrt{34}$ **19.** $\sqrt{34}$ **21.** $\sqrt{133}$

23. 6 **25.** $(-2, 1)$ and $(3, -\pi)$ **29.** yes **31.** no

33. The statement is true. For example, let $a = 3$, $b = 4$, and $c = 5$. Then $a^2 + b^2 = c^2$, or $3^2 + 4^2 = 5^2$ is true, but $3 + 4 \neq 5$.

35. yes **37.** no **39.** $5, -1$ **41.** $9 + \sqrt{119}, 9 - \sqrt{119}$

43. $x^2 + y^2 = 25$

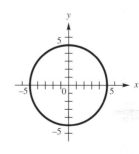

45. 58.02 ft **47.** 31.6 ft **49.** 231.3 m **51.** 65.28 ft **53.** $[10, \infty)$

55. $[8, 13]$ **57.** $(-\infty, 2)$ **61.** 6 **63.** 0 **65.** $-2a^2 + 8$

67. (a) 1.2 (b) 5 (c) $(0, 3.6)$ (d) $(3, 0)$ **69.** (a) no; yes (b) 1.88 yr; Mars takes 1.88 yr to complete its orbit around the sun. (c) 7.165 AU

71. $(-\infty, \infty); (-\infty, \infty)$; function **73.** $(-\infty, \infty); [-5, \infty)$; function

75. $(-\infty, \infty); [-5, \infty)$; function **77.** $[-4, \infty); [0, \infty)$; function

79. $[-1, 1]; [0, 1]$; function **81.** $[-5, \infty); [0, \infty)$; function **83.** $[-4, 4]; [-3, 3]$; not a function **85.** $(-\infty, -1) \cup (-1, \infty)$ **87.** $a + b; a + b; a^2 + 2ab + b^2$

88. $\dfrac{1}{2}ab; 2ab; c^2$ **89.** $2ab + c^2$ **90.** $a^2 + 2ab + b^2; 2ab + c^2$ **91.** $a^2 + b^2; c^2$

1.2 Exercises *(page 21)*

3. 45°   **5.** 150°   **7.** 70°; 110°   **9.** 55°; 35°   **11.** 80°; 100°   **13.** $(90 - x)°$ **15.** $(x - 360)°$ **17.** 83° 59′

19. 23° 49′   **21.** 38° 32′   **23.** 17° 1′ 49″   **25.** 20.900°   **27.** 91.598°   **29.** 274.316°   **31.** 31° 25′ 47″

33. 89° 54′ 1″   **35.** 178° 35′ 58″   **39.** 320°   **41.** 235°   **43.** 179°   **45.** 130°   **47.** $30° + n \cdot 360°$

49. $135° + n \cdot 360°$ **51.** $-90° + n \cdot 360°$ **55.** 320°

Angles other than those given are possible in Exercises 57–63.

57.

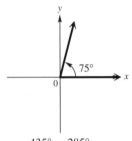

435°; −285°;
quadrant I

59.

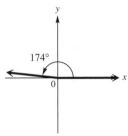

534°; −186°;
quadrant II

61.

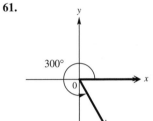

660°; −60°;
quadrant IV

63.

299°; −421°;
quadrant IV

65. $3\sqrt{2}$

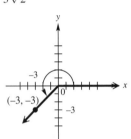

67. $\sqrt{34}$

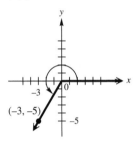

69. 4

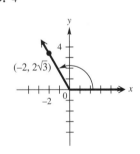

71. 1.5

73. 1800°

75. 12.5 rotations per hr

77. 89° 59′ 59.237″

79. 4 sec

1.3 Exercises *(page 29)*

1. vertical angles   **3.** 51°; 51°   **5.** 50°; 60°; 70°   **7.** 60°; 60°; 60°   **9.** 65°; 115°   **11.** 49°; 49°   **13.** 48°; 132°

15. 91°   **17.** 2° 29′   **19.** 25.4°   **23.** Answers are given in numerical order: 55°, 65°, 60°, 65°, 60°, 120°, 60°, 60°, 55°, 55°

25. right; scalene   **27.** acute; equilateral   **29.** right; scalene   **31.** right; isosceles   **33.** obtuse; scalene

35. acute; isosceles   **41.** A and P; B and Q; C and R; AC and PR; BC and QR; AB and PQ **43.** A and C; E and D;

ABE and CBD; EB and DB; AB and CB; AE and CD **45.** $P = 90°$; $Q = 42°$; $B = R = 48°$ **47.** $B = 106°$; $A = M = 44°$

49. $X = M = 52°$ **51.** $a = 20$; $b = 15$ **53.** $a = 6$; $b = 7\frac{1}{2}$ **55.** $x = 6$ **57.** 30 m   **59.** 500 m, 700 m

61. 112.5 ft   **63.** $x = 110$ **65.** $c = 111.1$ **67.** The unknown side in the first quadrilateral is 40 cm; The unknown

sides in the second quadrilateral are 27 cm and 36 cm.   **69.** $x = 10$; $y = 2$ **71.** **(a)** about $\frac{1}{4}$ **(b)** about 30 arc degrees

73. **(a)** about 1,830,000 mi   **(b)** yes   **74.** **(a)–(b)** Corresponding sides of similar triangles AHD and ACG are proportional.

(c) $BD = EF$; substitution   (d) Substitute 1 (pace) for EF.

Connections *(page 38)*

1. $\sin\theta = \dfrac{y}{1} = y = PQ$; $\cos\theta = \dfrac{x}{1} = x = OQ$; $\tan\theta = \dfrac{y}{x} = \dfrac{PQ}{OQ} = \dfrac{BA}{1}$, so $BA = \tan\theta$

2.

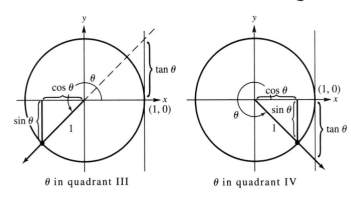

θ in quadrant III θ in quadrant IV

1.4 Exercises *(page 41)*

1.

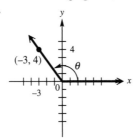

3.

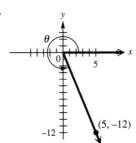

In Exercises 5–11 and 25–29 we give, in order, sine, cosine, tangent, cotangent, secant, and cosecant.

5. $\dfrac{4}{5}$; $-\dfrac{3}{5}$; $-\dfrac{4}{3}$; $-\dfrac{3}{4}$; $\dfrac{5}{3}$; $\dfrac{5}{4}$ **7.** 1; 0; undefined; 0; undefined; 1   **9.** $\dfrac{\sqrt{3}}{2}$; $\dfrac{1}{2}$; $\sqrt{3}$; $\dfrac{\sqrt{3}}{3}$; 2; $\dfrac{2\sqrt{3}}{3}$ **11.** $-.34727$; $.93777$;

$-.37031$; -2.7004; 1.0664; -2.8796 **15.** 0   **17.** positive   **19.** negative   **21.** positive   **23.** negative

25.

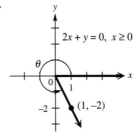

27.

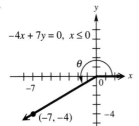

29.

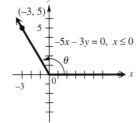

$-\dfrac{2\sqrt{5}}{5}$; $\dfrac{\sqrt{5}}{5}$; -2; $-\dfrac{1}{2}$; $\sqrt{5}$; $-\dfrac{\sqrt{5}}{2}$ $-\dfrac{4\sqrt{65}}{65}$; $-\dfrac{7\sqrt{65}}{65}$; $\dfrac{4}{7}$; $\dfrac{7}{4}$; $-\dfrac{\sqrt{65}}{7}$; $-\dfrac{\sqrt{65}}{4}$ $\dfrac{5\sqrt{34}}{34}$; $-\dfrac{3\sqrt{34}}{34}$; $-\dfrac{5}{3}$; $-\dfrac{3}{5}$; $-\dfrac{\sqrt{34}}{3}$; $\dfrac{\sqrt{34}}{5}$

33. -3 **35.** -3 **37.** 5   **39.** 1   **41.** -1 **43.** 0   **45.** undefined   **47.** They are equal.   **49.** They are negatives of each other.   **53.** $45°$, $225°$ **55.** $\theta = 143.13°$ or $\theta = 216.87°$ (Other answers are possible.)   **57.** about .940; about .342   **59.** $35°$ **61.** decrease; increase   **63. (a)** $\tan\theta = \dfrac{y}{x}$ **(b)** $x = \dfrac{y}{\tan\theta}$

1.5 Exercises *(page 50)*

1. $1; \theta = 90°$ **3.** $\dfrac{1}{3}$ **5.** -5 **7.** $-\sqrt{7}$ **9.** .70069071   **15.** $\dfrac{1}{2}$ **17.** $\sqrt{3}$ **19.** -100 **21.** $2°$ **23.** $1°$

27. II   **29.** I or III   **31.** II or III   **33.** $+; +; +$ **35.** $-; -; +$ **37.** $+; +; +$ **39.** $-; +; -$ **41.** $\tan 30°$

43. $\sec 33°$ **45.** $\cos 26°$ **47.** impossible   **49.** possible   **51.** possible   **53.** possible   **55.** possible

57. $-2\sqrt{2}$ **59.** $-\dfrac{\sqrt{5}}{2}$ **61.** $-\dfrac{\sqrt{15}}{4}$ **63.** 3.44701905   **65.** .36   **67.** yes

In Exercises 69–75 we give, in order, sine, cosine, tangent, cotangent, secant, and cosecant.

69. $\dfrac{15}{17}, -\dfrac{8}{17}, -\dfrac{15}{8}, -\dfrac{8}{15}, -\dfrac{17}{8}, \dfrac{17}{15}$ **71.** $-\dfrac{\sqrt{3}}{2}, -\dfrac{1}{2}, \sqrt{3}; \dfrac{\sqrt{3}}{3}; -2; -\dfrac{2\sqrt{3}}{3}$ **73.** $\dfrac{\sqrt{5}}{7}, \dfrac{2\sqrt{11}}{7}, \dfrac{\sqrt{55}}{22}, \dfrac{2\sqrt{55}}{5}, \dfrac{7\sqrt{11}}{22}, \dfrac{7\sqrt{5}}{5}$

75. $-.555762; .831342; -.668512; -1.49586; 1.20287; -1.79933$

79. False; For example, $\sin 30° + \cos 30° \approx .5 + .8660 = 1.3660 \neq 1$.   **81.** 146 ft   **85.** positive   **87.** negative

Chapter 1 Review Exercises *(page 55)*

3. $(-\infty, -4]$ **5. (a)** 12.25 mi   **(b)** 12.25 mi   **7.** $4\sqrt{5}$ **11.** $-x^2 + x + 4$ **13.** $(-\infty, \infty); (-\infty, \infty)$; function

15. $[0, \infty); [0, \infty)$; function   **17.** $[-5, 5]; [-3, 3]$; not a function   **19.** $309°$ **21.** $72°$ **23.** $1280°$ **25.** $47.420°$

27. $-61° \, 30' \, 12''$ **29.** $58°; 58°$ **31.** $\theta = \beta - \alpha$ **33.** $V = 41°; Z = 32°; Y = U = 107°$ **35.** $m = 45; n = 60$

37. $r = \dfrac{108}{7}$ **39.** proportional; equal

In Exercises 41–51, we give, in order, sine, cosine, tangent, cotangent, secant, and cosecant.

41. $-\dfrac{\sqrt{2}}{2}; -\dfrac{\sqrt{2}}{2}; 1; 1; -\sqrt{2}; -\sqrt{2}$ **43.** $0; -1; 0$; undefined; -1; undefined   **45.** $\dfrac{15}{17}, -\dfrac{8}{17}, -\dfrac{15}{8}, -\dfrac{8}{15}, -\dfrac{17}{8}, \dfrac{17}{15}$

47. $\dfrac{5\sqrt{26}}{26}; \dfrac{\sqrt{26}}{26}; -5; -\dfrac{1}{5}; \sqrt{26}; -\dfrac{\sqrt{26}}{5}$ **49.** $-\dfrac{1}{2}; \dfrac{\sqrt{3}}{2}; -\dfrac{\sqrt{3}}{3}; -\sqrt{3}; \dfrac{2\sqrt{3}}{3}; -2$ **51.** $\dfrac{5\sqrt{34}}{34}; \dfrac{3\sqrt{34}}{34}; \dfrac{5}{3}; \dfrac{3}{5}; \dfrac{\sqrt{34}}{3}; \dfrac{\sqrt{34}}{5}$

53.

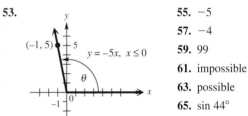

$y = -5x, \ x \le 0$

55. -5

57. -4

59. 99

61. impossible

63. possible

65. $\sin 44°$

In Exercises 67–71, we give, in order, sine, cosine, tangent, cotangent, secant, and cosecant.

67. $-\dfrac{\sqrt{39}}{8}; -\dfrac{5}{8}; \dfrac{\sqrt{39}}{5}; \dfrac{5\sqrt{39}}{39}; -\dfrac{8}{5}; -\dfrac{8\sqrt{39}}{39}$ **69.** $\dfrac{2\sqrt{5}}{5}; -\dfrac{\sqrt{5}}{5}; -2; -\dfrac{1}{2}; -\sqrt{5}; \dfrac{\sqrt{5}}{2}$ **71.** $-\dfrac{3}{5}; \dfrac{4}{5}; -\dfrac{3}{4}; -\dfrac{4}{3}; \dfrac{5}{4}; -\dfrac{5}{3}$

75. approximately 9500 ft

Chapter 1 Test *(page 59)*

1. $3\sqrt{2}$ **2.** 1400 km   **3. (a)** $(-\infty, \infty)$ **(b)** $[-2, \infty)$ **(c)** 0   **4.** $74.2983°$ **5.** $203°$ **6.** $2700°$

7. $x = 30°; y = 30°$ **8.** 40 yd   **9.** $\sin \theta = -\dfrac{5\sqrt{29}}{29}; \cos \theta = \dfrac{2\sqrt{29}}{29}; \tan \theta = -\dfrac{5}{2}$ **10.** III

11. $\sin \theta = -\dfrac{3}{5}; \tan \theta = -\dfrac{3}{4}; \cot \theta = -\dfrac{4}{3}; \sec \theta = \dfrac{5}{4}; \csc \theta = -\dfrac{5}{3}$

CHAPTER 2   ACUTE ANGLES AND RIGHT TRIANGLES

2.1 Exercises *(page 68)*

1. C   **3.** B   **5.** E

In Exercises 7 and 9, we give, in order, sine, cosine, tangent, cotangent, secant, and cosecant.

7. $\dfrac{21}{29}, \dfrac{20}{29}, \dfrac{21}{20}, \dfrac{20}{21}, \dfrac{29}{20}, \dfrac{29}{21}$ **9.** $\dfrac{n}{p}, \dfrac{m}{p}, \dfrac{n}{m}, \dfrac{m}{n}, \dfrac{p}{m}, \dfrac{p}{n}$

In Exercises 11 and 13, we give, in order, the unknown side, sine, cosine, tangent, cotangent, secant, and cosecant.

11. $c = 13; \dfrac{12}{13}, \dfrac{5}{13}, \dfrac{12}{5}, \dfrac{5}{12}, \dfrac{13}{5}, \dfrac{13}{12}$ **13.** $b = \sqrt{13}; \dfrac{\sqrt{13}}{7}, \dfrac{6}{7}, \dfrac{\sqrt{13}}{6}, \dfrac{6\sqrt{13}}{13}, \dfrac{7}{6}, \dfrac{7\sqrt{13}}{13}$

15. $\sin A = \cos(90° - A)$; $\cos A = \sin(90° - A)$; $\tan A = \cot(90° - A)$; $\cot A = \tan(90° - A)$; $\sec A = \csc(90° - A)$; $\csc A = \sec(90° - A)$ **17.** $\csc 51°$ **19.** $\tan(100° - \beta)$ **21.** $\cos 51.3°$ **23.** $40°$ **25.** $20°$ **27.** $12°$

29. true   **31.** false   **33.** true   **35.** $\dfrac{\sqrt{3}}{3}$ **37.** $\dfrac{1}{2}$ **39.** $\sqrt{2}$ **41.** $\dfrac{\sqrt{3}}{2}$

43. 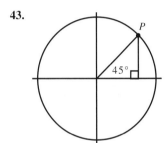   **44.** 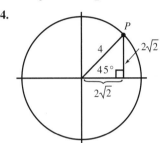   **45.** the legs; $(2\sqrt{2}, 2\sqrt{2})$ **46.** $(1, \sqrt{3})$

47. $\sin x, \tan x$ **49.** $60°$ **51.** $\left(\dfrac{\sqrt{2}}{2}, \dfrac{\sqrt{2}}{2}\right)$; $45°$

53. $y = \dfrac{\sqrt{3}}{3}x$ **55.** $60°$

57. (a) $60°$ (b) k (c) $k\sqrt{3}$ (d) $2; \sqrt{3}; 30°; 60°$

59. $a = 12; b = 12\sqrt{3}; d = 12\sqrt{3}; c = 12\sqrt{6}$ **61.** $m = \dfrac{7\sqrt{3}}{3}; a = \dfrac{14\sqrt{3}}{3}; n = \dfrac{14\sqrt{3}}{3}; q = \dfrac{14\sqrt{6}}{3}$ **63.** $A = \dfrac{s^2\sqrt{3}}{4}$

65. yes   **69.** approximately 78 mph

2.2 Exercises (page 75)

1. C   **3.** A   **5.** D   **11.** $\dfrac{\sqrt{3}}{3}; \sqrt{3}$ **13.** $\dfrac{\sqrt{3}}{2}; \dfrac{\sqrt{3}}{3}; \dfrac{2\sqrt{3}}{3}$ **15.** $-1; -1$ **17.** $-\dfrac{\sqrt{3}}{2}; -\dfrac{2\sqrt{3}}{3}$

In Exercises 19–33, we give, in order, sine, cosine, tangent, cotangent, secant, and cosecant.

19. $-\dfrac{\sqrt{2}}{2}; -\dfrac{\sqrt{2}}{2}; 1; 1; -\sqrt{2}; -\sqrt{2}$ **21.** $-\dfrac{\sqrt{2}}{2}; \dfrac{\sqrt{2}}{2}; -1; -1; \sqrt{2}; -\sqrt{2}$ **23.** $\dfrac{\sqrt{3}}{2}; \dfrac{1}{2}; \sqrt{3}; \dfrac{\sqrt{3}}{3}; 2; \dfrac{2\sqrt{3}}{3}$

25. $\dfrac{\sqrt{2}}{2}; -\dfrac{\sqrt{2}}{2}; -1; -1; -\sqrt{2}; \sqrt{2}$ **27.** $\dfrac{1}{2}; \dfrac{\sqrt{3}}{2}; \dfrac{\sqrt{3}}{3}; \sqrt{3}; \dfrac{2\sqrt{3}}{3}; 2$ **29.** $\dfrac{\sqrt{3}}{2}; \dfrac{1}{2}; \sqrt{3}; \dfrac{\sqrt{3}}{3}; 2; \dfrac{2\sqrt{3}}{3}$

31. $-\dfrac{1}{2}; \dfrac{\sqrt{3}}{2}; -\dfrac{\sqrt{3}}{3}; -\sqrt{3}; \dfrac{2\sqrt{3}}{3}; -2$ **33.** $\dfrac{\sqrt{3}}{2}; \dfrac{1}{2}; \sqrt{3}; \dfrac{\sqrt{3}}{3}; 2; \dfrac{2\sqrt{3}}{3}$ **35.** $(-5\sqrt{2}, -5\sqrt{2})$ **37.** yes   **39.** positive

41. negative   **43.** negative   **47.** $-.4$ **49.** $135°, 315°$ **51.** false; $\dfrac{1 + \sqrt{3}}{2} \neq 1$ **53.** true   **55.** false; $\dfrac{\sqrt{3}}{2} \neq 0$

57. true   **59.** (a) approximately 550 ft (b) approximately 369 ft

2.3 Exercises (page 80)

1. sin; 1   **3.** reciprocal; reciprocal   **5.** .5657728   **7.** 1.1342773   **9.** 1.4267182   **11.** 15.055723
13. 1.4887142   **15.** .6743024   **17.** .9999905   **19.** .4320857   **21.** .6494076   **23.** .0643581   **25.** 1.2162701
29. 57.997172°   **31.** 81.168073°   **33.** 30.502748°   **35.** 46.173582°   **37.** 56°   **39.** 1   **41.** 0
43. 2×10^8 m per sec   **45.** 19°   **47.** 48.7°   **49.** false   **51.** true   **53.** false   **55.** false   **57.** -100.5 lb
59. 2.866°   **61.** 2771 lb   **63.** 2200-lb car on 2° uphill grade   **65.** (a) 703 ft (b) 1701 ft (c) R would decrease.
67. (a) 67.00 ft; 67.14 ft; 66.84 ft; D increases and then decreases. (b) 64.40 ft; 67.14 ft; 69.93 ft; D increases.
(c) v; The shotputter should concentrate on achieving as large a value of v as possible.

Connections *(page 87)*

Our Steps 1 and 2 correspond to Polya's Steps 1 and 2. Our Step 3 corresponds to Polya's Steps 3 and 4.

2.4 Exercises *(page 87)*

1. 16,454.5 to 16,455.5   **3.** 546.5 to 547.5   **9.** .05   **11.** $B = 53° \, 40'$; $a = 571$ m; $b = 777$ m

13. $M = 38.8°$; $n = 154$ m; $p = 198$ m   **15.** $A = 47.9108°$; $c = 84.816$ cm; $a = 62.942$ cm

17.

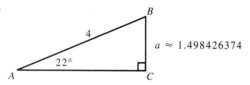

$a \approx 1.498426374$

23. $B = 62.00°$; $a = 8.17$ ft; $b = 15.4$ ft   **25.** $A = 17.00°$; $a = 39.1$ in.; $c = 134$ in.   **27.** $c = 85.9$ yd; $A = 62° \, 50'$; $B = 27° \, 10'$ **29.** $b = 42.3$ cm; $A = 24° \, 10'$; $B = 65° \, 50'$ **31.** $B = 36° \, 36'$; $a = 310.8$ ft, $b = 230.8$ ft

33. $A = 50° \, 51'$; $a = .4832$ m; $b = .3934$ m   **35.** The angle of elevation from X to Y is 90° whenever Y is directly above X.

39. 9.35 m   **41.** 88.3 m   **43.** 26.92 in.   **45.** 583 ft   **47.** 28.0 m   **49.** 469 m   **51.** 146 m   **53.** 37° 35'

55. 3.342 mm   **57. (a)** 29,008 ft **(b)** shorter   **59. (a)** 23.4 ft **(b)** 48.3 ft **(c)** The faster the speed, the more land needs to be cleared inside the curve.

2.5 Exercises *(page 95)*

1. It should be shown as an angle measured from due north.   **3.** A sketch is important to show the relationships among the given data and the unknowns.   **5.** 270°; N 90° W or S 90° W   **7.** 315°; N 45° W   **9.** $y = \dfrac{\sqrt{3}}{3}x, \, x \le 0$ **11.** 220 mi

13. 47 nautical mi   **15.** 120 mi   **17.** 148 mi   **19.** $x = \dfrac{b}{a - c}$ **21.** $y = (\tan 35°)(x - 25)$ **23.** 433 ft

25. 114 ft   **27.** 5.18 m   **29. (a)** $d = \dfrac{b}{2}\left(\cot\dfrac{\alpha}{2} + \cot\dfrac{\beta}{2}\right)$ **(b)** 345.3951 m   **31.** 10.8 ft   **33.** 84.7 m

Chapter 2 Review Exercises *(page 99)*

In Exercises 1, 13, and 15, we give, in order, sine, cosine, tangent, cotangent, secant, and cosecant.

1. $\dfrac{60}{61}, \dfrac{11}{61}, \dfrac{60}{11}, \dfrac{11}{60}, \dfrac{61}{11}, \dfrac{61}{60}$ **3.** 10°   **5.** 7°   **7.** true   **9.** true   **13.** $-\dfrac{\sqrt{3}}{2}, \dfrac{1}{2}; -\sqrt{3}; -\dfrac{\sqrt{3}}{3}; 2; -\dfrac{2\sqrt{3}}{3}$

15. $-\dfrac{1}{2}; \dfrac{\sqrt{3}}{2}; -\dfrac{\sqrt{3}}{3}; -\sqrt{3}; \dfrac{2\sqrt{3}}{3}; -2$ **17.** 120°; 240°   **19.** 150°; 210°   **21.** $3 - \dfrac{2\sqrt{3}}{3}$ **25.** -1.3563417

27. 1.0210339   **29.** .20834446   **31.** 55.673870°   **33.** 12.733938°   **35.** 63.008286°   **37.** 47.1°; 132.9°

39. false; $1.4088321 \ne 1$ **41.** true   **43.** no   **45.** III   **47.** $B = 31° \, 30'$; $a = 638$; $b = 391$

49. $B = 50.28°$; $a = 32.38$ m; $c = 50.66$ m   **51.** 73.7 ft   **53.** 18.75 cm   **55.** 1200 m   **57.** 140 mi

61. AB **63.** OB **65. (a)** $x_Q = x_P + d \sin \theta$, $y_Q = y_P + d \cos \theta$ **(b)** (181.34, 523.02)

Chapter 2 Test *(page 102)*

1. $\sin A = \dfrac{12}{13}$, $\cos A = \dfrac{5}{13}$, $\tan A = \dfrac{12}{5}$, $\cot A = \dfrac{5}{12}$, $\sec A = \dfrac{13}{5}$, $\csc A = \dfrac{13}{12}$ **2.** $x = 4$, $y = 4\sqrt{3}$, $z = 4\sqrt{2}$, $w = 8$

3. 15°   **4.** 16.16664145°   **5.** 135°, 225°   **6.** Take the reciprocal of $\tan \theta$ to get $\cot \theta = .5960011896$.

7. (a) true **(b)** false; For $0° \le \theta \le 90°$, cosine is decreasing. **(c)** true   **8.** $-\sqrt{3}$

9. (a) .97939940 **(b)** .20834446 **(c)** 1.9362132   **10.** $B = 31° \, 30'$, $a = 638$, $b = 391$ **11.** 15.5 ft   **12.** 110 km

13. $\dfrac{k}{4\sqrt{h}}$ or $\dfrac{k\sqrt{h}}{4h}$ sec   **14.** $8\sqrt{h}$ ft per sec; The velocity depends only on the height h; the length k does not matter.

CHAPTER 3 RADIAN MEASURE AND THE CIRCULAR FUNCTIONS

3.1 Exercises (page 107)

1. 1 **3.** 3 **5.** $\dfrac{\pi}{3}$ **7.** $\dfrac{\pi}{2}$ **9.** $\dfrac{5\pi}{6}$ **11.** $\dfrac{5\pi}{3}$ **13.** $\dfrac{5\pi}{2}$ **21.** 60° **23.** 315° **25.** 330° **27.** −30°

29. 126° **31.** 48° **33.** 153° **35.** .68 **37.** .742 **39.** 2.43 **41.** 1.122 **43.** .9847 **45.** .832391

47. 114° 35′ **49.** 99° 42′ **51.** 19° 35′ **53.** 287° 6′ **57.** $\dfrac{\sqrt{3}}{2}$ **59.** 1 **61.** $\dfrac{2\sqrt{3}}{3}$ **63.** 1 **65.** $-\sqrt{3}$

67. $\dfrac{1}{2}$ **69.** −1 **71.** $-\dfrac{\sqrt{3}}{2}$ **73.** $\dfrac{1}{2}$ **75.** We begin the answers with the blank next to 30°, and then proceed

counterclockwise from there: $\dfrac{\pi}{6}$; 45; $\dfrac{\pi}{3}$; 120; 135; $\dfrac{5\pi}{6}$; π; $\dfrac{7\pi}{6}$; $\dfrac{5\pi}{4}$; 240; 300; $\dfrac{7\pi}{4}$; $\dfrac{11\pi}{6}$. **77. (a)** $\dfrac{\pi}{200}$ **(b)** 3.15°; .055 radian

79. (a) 16π **(b)** 60π **81. (a)** 5π **(b)** $\dfrac{8\pi}{3}$

Connections (page 112)

Answers will vary. The longitude at Greenwich is 0°.

3.2 Exercises (page 113)

1. 2 radians **3.** 2π **5.** 8 **7.** 1 **9.** 25.8 cm **11.** 5.05 m **13.** 2.5 **15.** $\dfrac{20}{\pi}$ in.

17. The length is doubled. **19.** 3500 km **21.** 5900 km **23.** 44° N **25. (a)** 11.6 in. **(b)** 37° 5′ **27.** 38.5°

29. 146 in. **31.** .20 km **33.** 2100 mi **35.** 6π **37.** 1.5 **39.** 1120 m² **41.** 1300 cm² **43.** 114 cm²

45. 3.6 **47.** 16 m **49.** The area of a circle of radius r is πr^2. **51.** $A = \dfrac{\pi r^2 \theta}{360}$ **53. (a)** $13\tfrac{1}{3}°$; $\dfrac{2\pi}{27}$ **(b)** 480 ft

(c) $\dfrac{160}{9} \approx 17.8$ ft **(d)** approximately 672 ft² **55.** 75.4 in.² **57.** 7800 mi **59.** $V = \dfrac{1}{2}\theta(r_1^2 - r_2^2)h$ (θ in radians)

61. $r = \dfrac{L}{\theta}$ **62.** $h = r\cos\dfrac{\theta}{2}$ **63.** $d = r\left(1 - \cos\dfrac{\theta}{2}\right)$ **64.** $d = \dfrac{L}{\theta}\left(1 - \cos\dfrac{\theta}{2}\right)$ **65. (a)** Triangle RQP is similar to

triangle RMO because angle R = angle R and angle Q = angle M. **(b)** $\dfrac{r}{c} = \dfrac{c/2}{b}$; $r = \dfrac{c^2}{2b}$ **(c)** $a^2 + b^2 = c^2$; $r = \dfrac{a^2 + b^2}{2b}$

(d) The radius is 5 in.

3.3 Exercises (page 124)

1. circular **3.** trigonometric **5.** circular **7.** circular **9.** .7 **11.** .2 **13.** 4 **15. (a)** negative

(b) negative **(c)** negative **(d)** positive **17.** $\dfrac{1}{2}$ **19.** $-\dfrac{\sqrt{3}}{2}$ **21.** −2 **23.** $-\sqrt{3}$ **25.** $-\dfrac{1}{2}$ **27.** $\dfrac{2\sqrt{3}}{3}$

29. $-\dfrac{\sqrt{2}}{2}$ **31.** .80036052 **33.** .99813420 **35.** 1.0170372 **37.** .92607765 **39.** −.44357977

41. −.75469733 **43.** −.99668945 **45.** .67180620 **47.** 1.2797997 **49.** 1.2043741 **51.** 1.4139143

53. 1.3631380 **55.** $\dfrac{2\pi}{3}$ **57.** $\dfrac{7\pi}{6}$ **59.** $\dfrac{11\pi}{6}$ **61.** .9846 **63.** (−.80114362, .59847214)

65. (.43854733, −.89870810) **67.** I **69.** II **71. (a)** 30° **(b)** 60° **(c)** 75° **(d)** 86° **(e)** 86° **(f)** 60°

73. $\left\{\dfrac{\pi}{2} - 1, \dfrac{3\pi}{2} - 1\right\}$ **75.** 8.6 hr; 15.4 hr

3.4 Exercises (page 130)

1. 2π sec **3.** $\dfrac{5\pi}{4}$ radians **5.** $\dfrac{\pi}{25}$ radian per sec **7.** 9 min **9.** 10.768 radians **13.** 8π m per sec

15. $\frac{9}{5}$ radians per sec   **17.** 1.83333 radians per sec   **19.** 18π cm   **21.** 12 sec   **23.** $\frac{3\pi}{32}$ radian per sec

27. $\frac{\pi}{6}$ radian per hr   **29.** $\frac{\pi}{30}$ radian per sec   **31.** $\frac{7\pi}{30}$ cm per min   **33.** 168π m per min   **35.** 1500π m per min

37. 15.5 mph   **39. (a)** $\frac{2\pi}{365}$ radian   **(b)** $\frac{\pi}{4380}$ radian per hr   **(c)** 66,700 mph   **41.** larger pulley: $\frac{25\pi}{18}$ radians per sec;

smaller pulley: $\frac{125\pi}{48}$ radians per sec   **43.** 3.73 cm   **45.** about 29 sec

Chapter 3 Review Exercises *(page 133)*

3. Three of many possible answers are $1 + 2\pi$, $1 + 4\pi$, and $1 + 6\pi$.

5. $\frac{\pi}{4}$ **7.** $\frac{35\pi}{36}$ **9.** $\frac{40\pi}{9}$ **11.** 225°   **13.** 480°   **15.** −110°   **17.** π in.   **19.** 12π in.   **21.** 35.8 cm

23. 7.683 cm   **25.** 273 m²   **27.** 4500 km   **29.** $\frac{3}{4}$; 1.5 sq units   **31. (a)** $\frac{\pi}{3}$ radians   **(b)** 2π in.   **33.** $\sqrt{3}$

35. $-\frac{1}{2}$ **37.** 2   **39.** tan 1   **41.** sin 2   **43.** .86602663   **45.** .97030688   **47.** 1.6755332   **49.** .38974894

51. .51489440   **53.** 1.1053762   **55.** $\frac{\pi}{4}$ **57.** $\frac{7\pi}{6}$ **59.** −.4   **61.** 60°; $\frac{\pi}{3}$ radians   **63.** $\frac{15}{32}$ sec

65. $\frac{\pi}{20}$ radian per sec   **67.** 285.3 cm   **69. (b)** $\frac{\pi}{6}$ **(c)** less ultraviolet light when $\theta = \frac{\pi}{3}$

Chapter 3 Test *(page 135)*

2. $\frac{2\pi}{3}$ **3.** 162°   **4.** $\frac{4}{3}$ **5.** 15,000 cm²   **6.** $\sin \frac{7\pi}{6} = -\frac{1}{2}$; $\cos \frac{7\pi}{6} = -\frac{\sqrt{3}}{2}$; $\tan \frac{7\pi}{6} = \frac{\sqrt{3}}{3}$

7. $\left\{ s \mid s \neq \frac{\pi}{2} + n\pi \right\}$ **8.** .97169234   **9.** approximately 8.3 mi per sec   **10.** 46.65 ft   **11.** $\frac{\pi}{36}$ radian per sec

12. (a) $y = 2.625 \sin \theta$ **(b)** $u = 2.625 \cos \theta$ **(c)** $s = 10.5 \cos \alpha$ **(d)** $x = 2.625 \cos \theta + 10.5 \cos \alpha$

(e) The maximum velocity of 21.575 mph occurs when $\theta = 4.94415$ radians.

CHAPTER 4   GRAPHS OF THE CIRCULAR FUNCTIONS

Connections *(page 141)*

1. One example is $f(x) = x^2$.   **2.** One example is $f(x) = x^3$.   **3. (a)** even   **(b)** odd   **(c)** neither

Connections *(page 148)*

1. X = −.4161468, Y = .90929743; X is cos 2 and Y is sin 2.   **2.** X = 1.9, Y = .94630009; sin 1.9 = .94630009
3. X = 1.9, Y = −.3232896; cos 1.9 = −.3232896

4.1 Exercises *(page 150)*

1. G   **3.** E   **5.** B   **7.** F

9. 2   **11.** $\frac{2}{3}$ **13.** 1

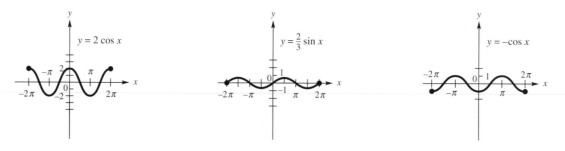

15. 2

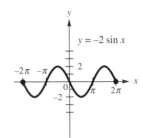

17. 4π; 1

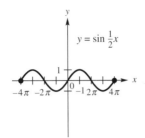

19. 6π; 1

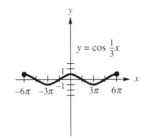

21. $\dfrac{2\pi}{3}$; 1

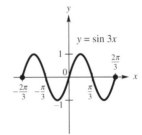

23. 8π; 2

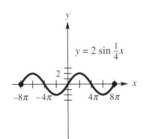

25. $\dfrac{2\pi}{3}$; 2

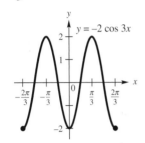

27. 2; 1

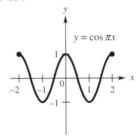

There are other correct answers in Exercises 29 and 31.

29. $y = 4 \sin \dfrac{1}{2}x$ **31.** $y = 4 \cos\left(\dfrac{1}{2}x - \dfrac{\pi}{2}\right)$

33. 24 hr   **35.** approximately 6:00 P.M.; approximately .2 ft

37. approximately 2:00 A.M.; approximately 2.6 ft

39. (a) 20 **(b)** 75   **41. (a)** 80°; 50° **(b)** 15° **(c)** about 35,000 yr

(d) downward   **43.** 1; 240° or $\dfrac{4\pi}{3}$

45. (a) 5; $\dfrac{1}{60}$ **(b)** 60

(c) 5; 1.545; −4.045; −4.045; 1.545

(d)

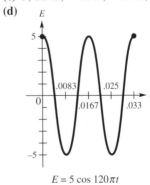

$E = 5 \cos 120\pi t$

47. (a)

$$L(x) = .022x^2 + .55x + 316 + 3.5 \sin(2\pi x)$$

365

15
325

35

(b) maximums: $x = \dfrac{1}{4}, \dfrac{5}{4}, \dfrac{9}{4}, \ldots;$

minimums: $x = \dfrac{3}{4}, \dfrac{7}{4}, \dfrac{11}{4}, \ldots$

53. 1   **55.** $\dfrac{\pi}{2}$ and $\dfrac{3\pi}{2}$ **57. (a)** 5 in.   **(b)** 2 cycles per sec; $\dfrac{1}{2}$ sec   **(c)** after $\dfrac{1}{4}$ sec   **(d)** approximately 4; After 1.3 seconds,

the weight is about 4 inches above the equilibrium position.   **59. (a)** $y = -3 \cos 12t$ **(b)** $\dfrac{\pi}{6}$ sec

4.2 Exercises *(page 162)*

1. B   **3.** C   **5.** D   **7.** H   **9.** B   **11.** F   **13.** right   **15.** 2; 2π; none; π to the right

17. 4; 4π; none; π to the left   **19.** 3; π; none; $\dfrac{\pi}{4}$ to the right   **21.** 1; $\dfrac{2\pi}{3}$; up 2; $\dfrac{\pi}{15}$ to the right

23.

$y = \cos\left(x - \dfrac{\pi}{2}\right)$

25.

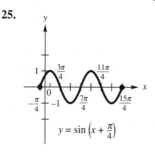

$y = \sin\left(x + \dfrac{\pi}{4}\right)$

27.

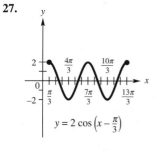

$y = 2 \cos\left(x - \dfrac{\pi}{3}\right)$

29.

$y = \dfrac{3}{2}\sin 2\left(x + \dfrac{\pi}{4}\right)$

31.

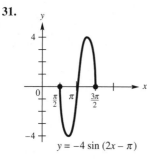

$y = -4 \sin(2x - \pi)$

33.

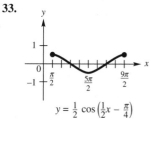

$y = \dfrac{1}{2}\cos\left(\dfrac{1}{2}x - \dfrac{\pi}{4}\right)$

35.

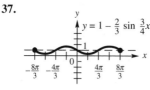

$y = -3 + 2 \sin x$

37.

$y = 1 - \dfrac{2}{3}\sin\dfrac{3}{4}x$

39.

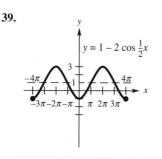

$y = 1 - 2\cos\dfrac{1}{2}x$

41.

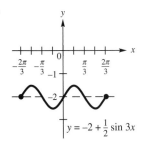

$$y = -2 + \frac{1}{2}\sin 3x$$

43.

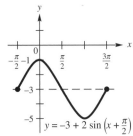

$$y = -3 + 2\sin\left(x + \frac{\pi}{2}\right)$$

45.

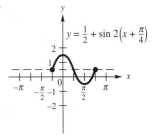

$$y = \frac{1}{2} + \sin 2\left(x + \frac{\pi}{4}\right)$$

There are other correct answers in Exercises 49 and 51.

49. $y = 3\sin 2\left(x - \dfrac{\pi}{4}\right)$ **51.** $y = 3\cos 2\left(x - \dfrac{\pi}{2}\right)$

53. (a) yes

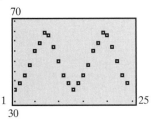

(b) It represents the average yearly temperature.

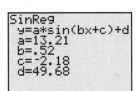

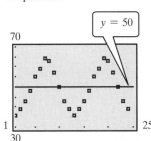

(c) 14; 12; 4.2

(d) $f(x) = 14\sin\left[\dfrac{\pi}{6}(x - 4.2)\right] + 50$

(e) The function gives an excellent model for the given data.

$$f(x) = 14\sin\left[\dfrac{\pi}{6}(x - 4.2)\right] + 50$$

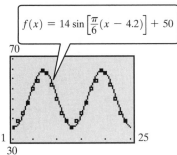

(f)
```
SinReg
 y=a*sin(bx+c)+d
 a=13.21
 b=.52
 c=-2.18
 d=49.68
```

TI-83 fixed to the
nearest hundredth

55. (a)

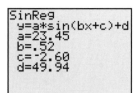 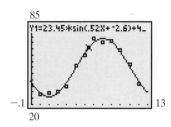

TI-83 fixed to the
nearest hundredth

(b)

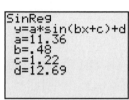 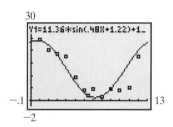

TI-83 fixed to the
nearest hundredth

4.3 Exercises *(page 177)*

1. true   **3.** true   **5.** false; $\tan(-x) = -\tan x$ for all x in the domain.   **7.** B   **9.** E   **11.** D

13.

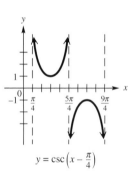

$y = \csc\left(x - \frac{\pi}{4}\right)$

15.

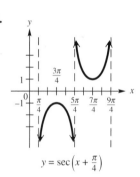

$y = \sec\left(x + \frac{\pi}{4}\right)$

17.

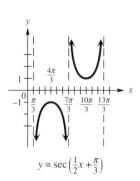

$y = \sec\left(\frac{1}{2}x + \frac{\pi}{3}\right)$

19.

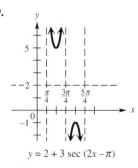

$y = 2 + 3\sec(2x - \pi)$

21.

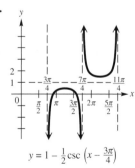

$y = 1 - \frac{1}{2}\csc\left(x - \frac{3\pi}{4}\right)$

23.

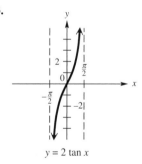

$y = 2\tan x$

25.

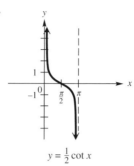

$y = \frac{1}{2} \cot x$

27.

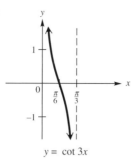

$y = \cot 3x$

29.

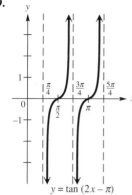

$y = \tan (2x - \pi)$

31.

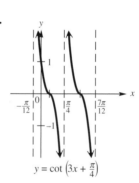

$y = \cot \left(3x + \frac{\pi}{4}\right)$

33.

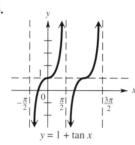

$y = 1 + \tan x$

35.

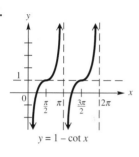

$y = 1 - \cot x$

37.

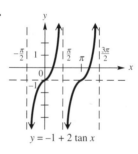

$y = -1 + 2 \tan x$

39.

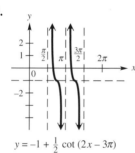

$y = -1 + \frac{1}{2} \cot (2x - 3\pi)$

41.

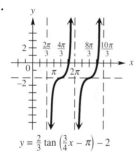

$y = \frac{2}{3} \tan \left(\frac{3}{4}x - \pi\right) - 2$

43. domain: $\left\{x \mid x \neq (2n + 1)\dfrac{\pi}{4}, \text{ where } n \text{ is an integer}\right\}$; range $(-\infty, \infty)$ **45.** four   **47. (a)** 0 m   **(b)** -2.9 m   **(c)** -12.3 m

(d) 12.3 m   **(e)** It leads to $\tan \dfrac{\pi}{2}$, which is undefined.   **51.** We show the display for $Y_1 + Y_2$ at $x = \dfrac{\pi}{6}$.   **53.** π **54.** $\dfrac{5\pi}{4}$

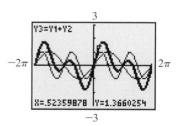

55. $y = \dfrac{5\pi}{4} + n\pi$ **56.** approximately .3217505544   **57.** approximately 3.463343208   **58.** .3217505544 $+ n\pi$

Chapter 4 Review Exercises *(page 180)*

1. B   **3.** sine, cosine, tangent, and cotangent   **5.** 2; 2π; none; none   **7.** $\frac{1}{2}$; $\frac{2\pi}{3}$; none; none   **9.** 2; 8π; 1 up; none

11. 3; 2π; none; $\frac{\pi}{2}$ to the left   **13.** not applicable; π; none; $\frac{\pi}{8}$ to the right   **15.** not applicable; $\frac{\pi}{3}$; none; $\frac{\pi}{9}$ to the right

17. tangent   **19.** cosine   **21.** cotangent

25.

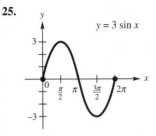

$y = 3 \sin x$

27.

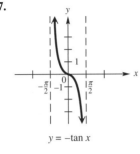

$y = -\tan x$

29.

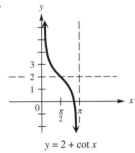

$y = 2 + \cot x$

31.

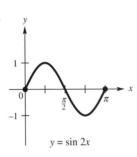

$y = \sin 2x$

33.

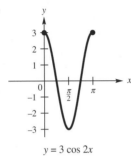

$y = 3 \cos 2x$

35.

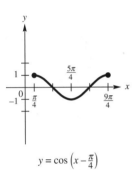

$y = \cos\left(x - \frac{\pi}{4}\right)$

37.

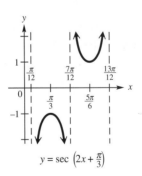

$y = \sec\left(2x + \frac{\pi}{3}\right)$

39.

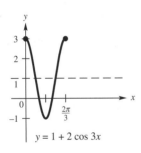

$y = 1 + 2 \cos 3x$

41.

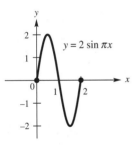

$y = 2 \sin \pi x$

45. (a)

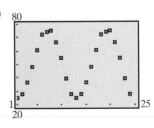

(b) $f(x) = 25 \sin\left[\frac{\pi}{6}(x - 4.2)\right] + 50$

(d) The function gives an excellent model for the data.

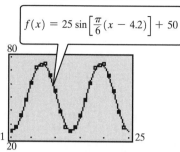

$$f(x) = 25 \sin\left[\frac{\pi}{6}(x - 4.2)\right] + 50$$

(e)

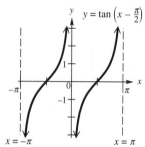

SinReg
y=a*sin(bx+c)+d
a=25.77
b=.52
c=⁻2.19
d=50.57

47. (a) 100 **(b)** 258 **(c)** 122 **(d)** 296

Chapter 4 Test *(page 182)*

1. (a) π **(b)** 6 **(c)** $[-3, 9]$ **(d)** -3 **(e)** $\frac{\pi}{4}$ to the left $\left(\text{that is, } -\frac{\pi}{4}\right)$

2.

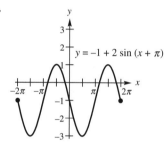

$y = -1 + 2\sin(x + \pi)$

3.

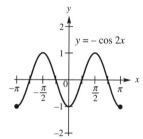

$y = -\cos 2x$

4.

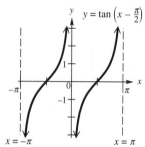

$y = \tan\left(x - \frac{\pi}{2}\right)$

5.

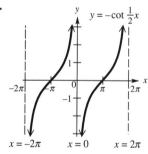

$y = -\cot\frac{1}{2}x$

6.

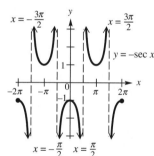

$y = -\sec x$

7.

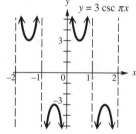

$y = 3\csc \pi x$

8. (a)

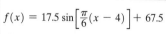

$$f(x) = 17.5 \sin\left[\frac{\pi}{6}(x - 4)\right] + 67.5$$

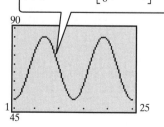

(b) 17.5; 12; 4 to the right; 67.5 up

(c) approximately 52°F

(d) 50°F in January; 85°F in July

(e) approximately 67.5°; This is the vertical translation.

10. C

CHAPTER 5   TRIGONOMETRIC IDENTITIES

5.1 Exercises (page 191)

1. -2.6 **3.** $.625$ **5.** $\dfrac{\sqrt{7}}{4}$ **7.** $-\dfrac{2\sqrt{5}}{5}$ **9.** $-\dfrac{\sqrt{105}}{11}$ **13.** $f(-x) = -f(x)$ **15.** $\cos\theta = -\dfrac{\sqrt{5}}{3}$; $\tan\theta = -\dfrac{2\sqrt{5}}{5}$;

$\cot\theta = -\dfrac{\sqrt{5}}{2}$; $\sec\theta = -\dfrac{3\sqrt{5}}{5}$; $\csc\theta = \dfrac{3}{2}$ **17.** $\sin\theta = -\dfrac{\sqrt{17}}{17}$; $\cos\theta = \dfrac{4\sqrt{17}}{17}$; $\cot\theta = -4$; $\sec\theta = \dfrac{\sqrt{17}}{4}$; $\csc\theta = -\sqrt{17}$

19. $\sin\theta = \dfrac{3}{5}$; $\cos\theta = \dfrac{4}{5}$; $\tan\theta = \dfrac{3}{4}$; $\sec\theta = \dfrac{5}{4}$; $\csc\theta = \dfrac{5}{3}$ **21.** $\sin\theta = -\dfrac{\sqrt{7}}{4}$; $\cos\theta = \dfrac{3}{4}$; $\tan\theta = -\dfrac{\sqrt{7}}{3}$; $\cot\theta = -\dfrac{3\sqrt{7}}{7}$;

$\csc\theta = -\dfrac{4\sqrt{7}}{7}$ **23.** B   **25.** E   **27.** A   **29.** A   **31.** D   **35.** $\sin\theta = \dfrac{\pm\sqrt{2x+1}}{x+1}$ **37.** $\cos\theta$ **39.** $\cot\theta$

41. $\cos^2\theta$ **43.** $\sec\theta - \cos\theta$ **45.** $\cot\theta - \tan\theta$ **47.** $\sec\theta\csc\theta$ **49.** $\cos^2\theta$ **51.** $\sec^2\theta$

53. $\dfrac{\pm\sqrt{1+\cot^2\theta}}{1+\cot^2\theta}$; $\dfrac{\pm\sqrt{\sec^2\theta-1}}{\sec\theta}$ **55.** $\dfrac{\pm\sin\theta\sqrt{1-\sin^2\theta}}{1-\sin^2\theta}$; $\dfrac{\pm\sqrt{1-\cos^2\theta}}{\cos\theta}$; $\pm\sqrt{\sec^2\theta-1}$; $\dfrac{\pm\sqrt{\csc^2\theta-1}}{\csc^2\theta-1}$

57. $\dfrac{\pm\sqrt{1-\sin^2\theta}}{1-\sin^2\theta}$; $\pm\sqrt{\tan^2\theta+1}$; $\dfrac{\pm\sqrt{1+\cot^2\theta}}{\cot\theta}$; $\dfrac{\pm\csc\theta\sqrt{\csc^2\theta-1}}{\csc^2\theta-1}$ **59.** $\dfrac{25\sqrt{6}-60}{12}$; $\dfrac{-25\sqrt{6}-60}{12}$ **61.** $-\sin(2x)$

62. It is the negative of $\sin(2x)$.   **63.** $\cos(4x)$ **64.** It is the same function.   **65. (a)** $-\sin(4x)$ **(b)** $\cos(2x)$

(c) $5\sin(3x)$ **67.** The graph of $y = \csc(-x)$ is a reflection across the x-axis of the graph of $y = \csc x$.

69. The graph of $y = \cot(-x)$ is a reflection across the x-axis of the graph of $y = \cot x$.   **71.** not an identity

73. identity   **75.** not an identity

Connections (page 199)

$\sqrt{(1-x^2)^3} = \sin^3\theta$

5.2 Exercises (page 200)

1. $\csc\theta\sec\theta$ or $\dfrac{1}{\sin\theta\cos\theta}$ **3.** $1 + \sec s$ **5.** 1 **7.** 1 **9.** $2 + 2\sin t$ **11.** $-\dfrac{2\cos x}{\sin^2 x}$ or $-2\cot x\csc x$

13. $(\sin\gamma + 1)(\sin\gamma - 1)$ **15.** $4\sin x$ **17.** $(2\sin x + 1)(\sin x + 1)$ **19.** $(\cos^2 x + 1)^2$

21. $(\sin x - \cos x)(1 + \sin x\cos x)$ **23.** $\sin\theta$ **25.** 1 **27.** $\tan^2\beta$ **29.** $\tan^2 x$ **31.** $\sec^2 x$

71. $(\sec\theta + \tan\theta)(1 - \sin\theta) = \cos\theta$ **73.** $\dfrac{\cos\theta + 1}{\sin\theta + \tan\theta} = \cot\theta$ **75.** identity   **77.** not an identity

79. not an identity   **81.** not an identity   **83.** identity

5.3 Exercises (page 207)

3. $\dfrac{\sqrt{6}-\sqrt{2}}{4}$ **5.** $\dfrac{\sqrt{2}-\sqrt{6}}{4}$ **7.** $\dfrac{\sqrt{2}-\sqrt{6}}{4}$ **9.** 0 **11.** 0 **13.** The calculator gives a value of 0 for the

expression.   **15.** $\cot 3°$ **17.** $\sin\dfrac{5\pi}{12}$ **19.** $\sec 104° 24'$ **21.** $\cos\left(-\dfrac{\pi}{8}\right)$ **23.** $\csc(-56° 42')$

25. $\tan(-86.9814°)$ **27.** $\tan$ **29.** $\cos$ **31.** $\csc$ **33.** $15°$ **35.** $\dfrac{140°}{3}$ **37.** $20°$ **39.** $\cos\theta$ **41.** $-\cos\theta$

43. $\cos\theta$ **45.** $-\cos\theta$ **47.** $\dfrac{4-6\sqrt{6}}{25}$; $\dfrac{4+6\sqrt{6}}{25}$ **49.** $\dfrac{16}{65}$; $-\dfrac{56}{65}$ **51.** $\dfrac{2\sqrt{638}-\sqrt{30}}{56}$; $\dfrac{2\sqrt{638}+\sqrt{30}}{56}$

53. true   **55.** false   **57.** true   **59.** true   **61.** false   **74.** $-\dfrac{\sqrt{6}+\sqrt{2}}{4}$ **75.** $-\dfrac{\sqrt{6}+\sqrt{2}}{4}$

76. (a) $\dfrac{\sqrt{2}-\sqrt{6}}{4}$ **(b)** $-\dfrac{\sqrt{6}+\sqrt{2}}{4}$ **81.** 3

83. (a)

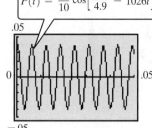

For $x = t$,
$$P(t) = \frac{4}{10} \cos\left[\frac{20\pi}{4.9} - 1026t\right]$$

The pressure P is oscillating.

(b)

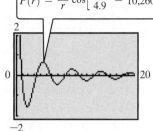

For $x = r$,
$$P(r) = \frac{3}{r} \cos\left[\frac{2\pi r}{4.9} - 10{,}260\right]$$

The pressure oscillates and amplitude decreases as r increases.

(c) $P = \dfrac{a}{n\lambda} \cos(ct)$

5.4 Exercises *(page 214)*

3. C   **5.** E   **7.** B   **9.** $\dfrac{\sqrt{6} + \sqrt{2}}{4}$ **11.** $2 - \sqrt{3}$ **13.** $\dfrac{-\sqrt{6} - \sqrt{2}}{4}$ **15.** $\dfrac{\sqrt{2}}{2}$ **17.** -1 **19.** 0

21. 1   **23.** $\dfrac{\sqrt{3}\cos\theta - \sin\theta}{2}$ **25.** $\dfrac{\cos\theta - \sqrt{3}\sin\theta}{2}$ **27.** $\dfrac{\sqrt{2}(\sin x - \cos x)}{2}$ **29.** $\dfrac{\sqrt{3}\tan\theta + 1}{\sqrt{3} - \tan\theta}$

31. $\dfrac{\sqrt{2}(\cos x + \sin x)}{2}$ **33.** $-\cos\theta$ **35.** $-\tan\theta$ **37.** $-\tan\theta$ **41.** $\dfrac{63}{65}, \dfrac{33}{65}, \dfrac{63}{16}, \dfrac{33}{56}$; I; I   **43.** $\dfrac{4\sqrt{2} + \sqrt{5}}{9}$;

$\dfrac{4\sqrt{2} - \sqrt{5}}{9}; \dfrac{-8\sqrt{5} - 5\sqrt{2}}{20 - 2\sqrt{10}}; \dfrac{-8\sqrt{5} + 5\sqrt{2}}{20 + 2\sqrt{10}}$; II; II   **45.** $\dfrac{77}{85}, \dfrac{13}{85}; -\dfrac{77}{36}, \dfrac{13}{84}$; II; I   **47.** $-\dfrac{33}{65}; -\dfrac{63}{65}, \dfrac{33}{56}, \dfrac{63}{16}$; III; III

49. $\dfrac{-(3\sqrt{22} + \sqrt{21})}{20}; \dfrac{-3\sqrt{22} + \sqrt{21}}{20}; \dfrac{-(66\sqrt{7} + 7\sqrt{66})}{154 - 3\sqrt{462}}; \dfrac{-66\sqrt{7} + 7\sqrt{66}}{154 + 3\sqrt{462}}$; IV; IV   **51.** $\sin\left(\dfrac{\pi}{2} + x\right) = \cos x$

53. $\tan\left(\dfrac{\pi}{2} + x\right) = -\cot x$ **65.** $\dfrac{\sqrt{6} - \sqrt{2}}{4}$ **67.** $2 + \sqrt{3}$ **69.** $\dfrac{-\sqrt{6} - \sqrt{2}}{4}$ **71.** $\dfrac{\sqrt{6} - \sqrt{2}}{4}$ **73.** $-2 + \sqrt{3}$

78. $180° - \beta$ **79.** $\theta = \beta - \alpha$ **80.** $\tan\theta = \dfrac{\tan\beta - \tan\alpha}{1 + \tan\beta\tan\alpha}$ **82.** $18.4°$ **83.** $80.8°$

85. (a)

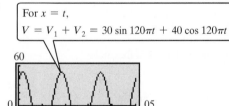

For $x = t$,
$$V = V_1 + V_2 = 30\sin 120\pi t + 40\cos 120\pi t$$

(b) $a = 50, \phi \approx -5.353$

5.5 Exercises *(page 223)*

1. C   **3.** B   **5.** C   **7.** $\cos\theta = \dfrac{2\sqrt{5}}{5}$; $\sin\theta = \dfrac{\sqrt{5}}{5}$; $\tan\theta = \dfrac{1}{2}$; $\sec\theta = \dfrac{\sqrt{5}}{2}$; $\csc\theta = \sqrt{5}$; $\cot\theta = 2$

9. $\cos x = -\dfrac{\sqrt{42}}{12}$; $\sin x = \dfrac{\sqrt{102}}{12}$; $\tan x = -\dfrac{\sqrt{119}}{7}$; $\cot x = -\dfrac{\sqrt{119}}{17}$; $\sec x = -\dfrac{2\sqrt{42}}{7}$; $\csc x = \dfrac{2\sqrt{102}}{17}$ **11.** $\cos 2\theta = \dfrac{17}{25}$;

$\sin 2\theta = -\dfrac{4\sqrt{21}}{25}$; $\tan 2\theta = -\dfrac{4\sqrt{21}}{17}$; $\sec 2\theta = \dfrac{25}{17}$; $\csc 2\theta = -\dfrac{25\sqrt{21}}{84}$; $\cot 2\theta = -\dfrac{17\sqrt{21}}{84}$ **13.** $\tan 2x = -\dfrac{4}{3}$; $\sec 2x = -\dfrac{5}{3}$;

$\cos 2x = -\dfrac{3}{5}$; $\cot 2x = -\dfrac{3}{4}$; $\sin 2x = \dfrac{4}{5}$; $\csc 2x = \dfrac{5}{4}$ **15.** $\sin 2\alpha = -\dfrac{4\sqrt{55}}{49}$; $\cos 2\alpha = \dfrac{39}{49}$; $\tan 2\alpha = -\dfrac{4\sqrt{55}}{39}$;

$\cot 2\alpha = -\dfrac{39\sqrt{55}}{220}$; $\sec 2\alpha = \dfrac{49}{39}$; $\csc 2\alpha = -\dfrac{49\sqrt{55}}{220}$ **17.** .2   **21.** $\dfrac{\sqrt{3}}{2}$ **23.** $\dfrac{\sqrt{3}}{2}$ **25.** $-\dfrac{\sqrt{2}}{2}$ **27.** $\dfrac{1}{2}\tan 102°$

29. $\dfrac{1}{4}\cos 94.2°$ **31.** $-\cos\dfrac{4\pi}{5}$ **33.** $\sin 10x$ **35.** 1   **37.** $-\dfrac{1}{2}$ **39.** $-\dfrac{1}{2}$ **41.** $\sqrt{3}$ **43.** $\sqrt{3}$ **45.** 0

47. $\cos^4 x - \sin^4 x = \cos 2x$ **49.** $\dfrac{\cot^2 x - 1}{2\cot x} = \cot 2x$ **69.** $\cos 3x = 4\cos^3 x - 3\cos x$ **71.** $\tan 3x = \dfrac{3\tan x - \tan^3 x}{1 - 3\tan^2 x}$

73. 980.799 cm per sec^2 **75.** $\dfrac{1}{2}(\sin 70° - \sin 20°)$ **77.** $\dfrac{3}{2}(\cos 8x + \cos 2x)$

79. $\dfrac{1}{2}[\cos 2\theta - \cos(-4\theta)] = \dfrac{1}{2}(\cos 2\theta - \cos 4\theta)$ **81.** $-4[\cos 9y + \cos(-y)] = -4(\cos 9y + \cos y)$

83. (a) $D = \dfrac{v^2 \sin(2\theta)}{32}$ (b) approximately 35 ft

85. (a)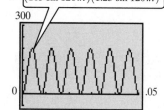

(b) maximum: 200.49 watts; minimum: 0 watts

(c) $a = -100.245$, $\omega = 240\pi$, $c = 100.245$

(e) 100.245 watts

5.6 Exercises *(page 230)*

1. −   **3.** +   **5.** C   **7.** D   **9.** F   **11.** $\dfrac{\sqrt{2 + \sqrt{2}}}{2}$ **13.** $\dfrac{-\sqrt{2 + \sqrt{3}}}{2}$ **15.** $\dfrac{-\sqrt{2 + \sqrt{3}}}{2}$

17. .1270166538   **21.** $\dfrac{\sqrt{10}}{4}$ **23.** 3   **25.** $\dfrac{\sqrt{50 - 10\sqrt{5}}}{10}$ **27.** $-\sqrt{7}$ **29.** $\dfrac{\sqrt{5}}{5}$ **31.** $-\dfrac{\sqrt{42}}{12}$ **33.** $\sin 20°$

35. $\tan 73.5°$ **37.** $\tan 29.87°$ **39.** $\cos 9x$ **41.** $\tan 4\theta$ **43.** $\cos\dfrac{x}{8}$ **45.** $\dfrac{\sin x}{1 + \cos x} = \tan\dfrac{x}{2}$

47. $\dfrac{\tan\dfrac{x}{2} + \cot\dfrac{x}{2}}{\cot\dfrac{x}{2} - \tan\dfrac{x}{2}} = \sec x$ **59.** 106°   **61.** 47°   **63.** 2   **64.** They are both radii of the circle.

65. It is the supplement of a 30° angle.   **66.** Their sum is $180° - 150° = 30°$, and they are equal.   **67.** $2 + \sqrt{3}$

69. $\dfrac{\sqrt{6} + \sqrt{2}}{4}$ **70.** $\dfrac{\sqrt{6} - \sqrt{2}}{4}$ **71.** $2 - \sqrt{3}$

Chapter 5 Review Exercises *(page 234)*

1. B   **3.** C   **5.** H   **7.** D   **9.** $\dfrac{\cos^2\theta}{\sin\theta}$ **11.** $\dfrac{1 + \cos\theta}{\sin\theta}$ **13.** $-\dfrac{\cos\theta}{\sin\theta}$ **15.** cosine, secant

17. $\sec x = -\dfrac{\sqrt{41}}{4}$; $\cos x = -\dfrac{4\sqrt{41}}{41}$; $\cot x = -\dfrac{4}{5}$; $\sin x = \dfrac{5\sqrt{41}}{41}$; $\csc x = \dfrac{\sqrt{41}}{5}$ **19.** $\sin 165° = \dfrac{\sqrt{6} - \sqrt{2}}{4}$;

$\cos 165° = \dfrac{-\sqrt{6} - \sqrt{2}}{4}$; $\tan 165° = -2 + \sqrt{3}$; $\csc 165° = \sqrt{6} + \sqrt{2}$; $\sec 165° = -\sqrt{6} + \sqrt{2}$; $\cot 165° = -2 - \sqrt{3}$

21. B   **23.** A   **25.** C   **27.** D   **29.** .5   **31.** $\dfrac{4 + 3\sqrt{15}}{20}$; $\dfrac{4\sqrt{15} + 3}{20}$; $\dfrac{4 + 3\sqrt{15}}{4\sqrt{15} - 3}$; I

33. $\dfrac{4 - 9\sqrt{11}}{50}$; $\dfrac{12\sqrt{11} - 3}{50}$; $\dfrac{4 - 9\sqrt{11}}{12\sqrt{11} + 3}$; IV   **35.** $\sin\theta = \dfrac{\sqrt{14}}{4}$; $\cos\theta = \dfrac{\sqrt{2}}{4}$ **37.** $\sin 2x = \dfrac{3}{5}$; $\cos 2x = -\dfrac{4}{5}$

39. $\dfrac{1}{2}$ **41.** $\dfrac{\sqrt{5}-1}{2}$ **43.** $-\dfrac{\sin 2x + \sin x}{\cos 2x - \cos x} = \cot\dfrac{x}{2}$ **45.** $\dfrac{\sin x}{1 - \cos x} = \cot\dfrac{x}{2}$ **47.** $\dfrac{2(\sin x - \sin^3 x)}{\cos x} = \sin 2x$

Chapter 5 Test *(page 236)*

1. $\sin x = -\dfrac{5\sqrt{61}}{61};\ \cos x = \dfrac{6\sqrt{61}}{61}$ **2.** -1 **3.** $\sin(x+y) = \dfrac{2 - 2\sqrt{42}}{15};\ \cos(x-y) = \dfrac{4\sqrt{2} - \sqrt{21}}{15};$

$\tan(x+y) = \dfrac{2\sqrt{2} - 4\sqrt{21}}{8 + \sqrt{42}}$ **4.** $\dfrac{-\sqrt{2 - \sqrt{2}}}{2}$ **5.** $\sec x - \sin x \tan x = \cos x$ **6.** $\cot\dfrac{x}{2} - \cot x = \csc x$

9. The cofunction identities in Chapter 2 applied only to acute angles. The cofunction identities in this chapter apply to any angles and to any real numbers.

10. (a) $V = 163 \cos\left(\dfrac{\pi}{2} - \omega t\right)$ **(b)** 163 volts; $\dfrac{1}{240}$ sec

CHAPTER 6 INVERSE TRIGONOMETRIC FUNCTIONS AND TRIGONOMETRIC EQUATIONS

Connections *(page 240)*

$f^{-1}(x) = \dfrac{x-5}{2}$; HAPPINESS IS STAYING OUT OF HOSPITALS AND COURTROOMS.

6.1 Exercises *(page 249)*

1. (a) $[-1, 1]$ **(b)** $\left[-\dfrac{\pi}{2}, \dfrac{\pi}{2}\right]$ **(c)** increasing   **(d)** -2 is not in the domain.   **3. (a)** $(-\infty, \infty)$ **(b)** $\left(-\dfrac{\pi}{2}, \dfrac{\pi}{2}\right)$

(c) increasing   **(d)** no   **5.** $-\dfrac{\pi}{6}$ **7.** $\dfrac{\pi}{4}$ **9.** π **11.** $-\dfrac{\pi}{2}$ **13.** 0   **15.** $\dfrac{\pi}{2}$ **17.** $\dfrac{\pi}{4}$ **19.** $\dfrac{5\pi}{6}$ **21.** $\dfrac{3\pi}{4}$

23. $-\dfrac{\pi}{6}$ **25.** $\dfrac{\pi}{6}$ **27.** $\dfrac{\pi}{3}$ **29.** $-45°$ **31.** $-60°$ **33.** $120°$ **35.** $-30°$ **37.** $-7.6713835°$

39. $113.500970°$ **41.** $30.987961°$ **43.** $.83798122$ **45.** 2.3154725 **47.** 1.1900238

49. $(-\infty, \infty);\ (0, 2\pi)$ **51.** $(-\infty, -2] \cup [2, \infty);\ \left[0, \dfrac{\pi}{2}\right) \cup \left(\dfrac{\pi}{2}, \pi\right]$

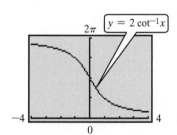

$y = 2\cot^{-1}x$

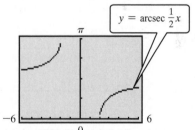

$y = \operatorname{arcsec}\dfrac{1}{2}x$

52. In both cases, the result is x. In each case, the graph is a straight line bisecting quadrants I and III (i.e., the line $y = x$).

53. It is the graph of $y = x$.

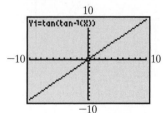

54. It does not agree because the range of the inverse tangent function is $\left(-\dfrac{\pi}{2}, \dfrac{\pi}{2}\right)$, not $(-\infty, \infty)$, as was the case in Exercise 53.

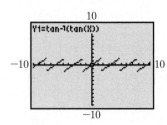

55. The first screen supports parts (b) and (c) and the second supports part (d).

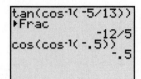 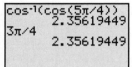

57. $\dfrac{\sqrt{7}}{3}$ **59.** $\dfrac{\sqrt{5}}{5}$ **61.** $-\dfrac{\sqrt{5}}{2}$ **63.** 2   **65.** $\dfrac{\pi}{4}$ **67.** $\dfrac{\pi}{3}$ **69.** $\dfrac{120}{169}$ **71.** $-\dfrac{7}{25}$ **73.** $\dfrac{4\sqrt{6}}{25}$ **75.** $-\dfrac{24}{7}$

77. $\dfrac{\sqrt{10} - 3\sqrt{30}}{20}$ **79.** $-\dfrac{16}{65}$ **81.** .894427191   **83.** .1234399811   **85.** $\sqrt{1 - u^2}$ **87.** $\dfrac{\sqrt{1 - u^2}}{u}$

89. $\dfrac{\sqrt{u^2 - 4}}{|u|}$ **91.** $\dfrac{u\sqrt{2}}{2}$ **93.** all values in the interval $[0, \pi]$ **95.** (a) 45° (b) $\theta = 45$ **97.** (a) 113° (b) 84°

(c) 60° **(d)** 47°   **99.** about 44.7%   **100.** $20[9 \arctan(\sqrt{8}) - \sqrt{8}] \approx 165$ cu ft

6.2 Exercises *(page 259)*

1. D   **3.** A   **5.** $360° \cdot n,\ 120° + 360° \cdot n,\ 240° + 360° \cdot n$, where n is an integer   **7.** $\dfrac{3\pi}{4}, \dfrac{7\pi}{4}$ **9.** $\dfrac{\pi}{6}, \dfrac{5\pi}{6}$

11. no solution   **13.** $\dfrac{\pi}{4}, \dfrac{2\pi}{3}, \dfrac{5\pi}{4}, \dfrac{5\pi}{3}$ **15.** π **17.** $\dfrac{7\pi}{6}, \dfrac{3\pi}{2}, \dfrac{11\pi}{6}$ **19.** $\dfrac{\pi}{4}, \dfrac{\pi}{2}, \dfrac{3\pi}{4}, \dfrac{5\pi}{4}, \dfrac{3\pi}{2}, \dfrac{7\pi}{4}$

21. 30°, 210°, 240°, 300°   **23.** 90°, 210°, 330°   **25.** 45°, 135°, 225°, 315°   **27.** 45°, 225°   **29.** 0°, 30°, 150°, 180°

31. 0°, 45°, 135°, 180°, 225°, 315°   **33.** 0°, 90°   **35.** 90°, 221.8°, 318.2°   **37.** 135°, 315°, 71.6°, 251.6°

39. 71.6°, 90°, 251.6°, 270°   **41.** 53.6°, 126.4°, 187.9°, 352.1°   **43.** 149.6°, 329.6°, 106.3°, 286.3°   **45.** no solution

47. 57.7°, 159.2°   **49.** $\dfrac{\pi}{2} + 2n\pi, \dfrac{7\pi}{6} + 2n\pi, \dfrac{11\pi}{6} + 2n\pi$, where n is an integer

51. $\dfrac{\pi}{3} + 2n\pi, \dfrac{2\pi}{3} + 2n\pi, \dfrac{4\pi}{3} + 2n\pi, \dfrac{5\pi}{3} + 2n\pi$, where n is an integer   **53.** no solution   **55.** .68058878, 1.4158828

57. $\dfrac{\pi}{3}, \dfrac{5\pi}{3}$ **59. (a)**

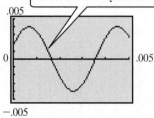

For $x = t$, $P(t) = .004 \sin\left[2\pi(261.63)t + \dfrac{\pi}{7}\right]$

(b) .00164 and .00355   **(c)** [.00164, .00355]   **(d)** outward   **61.** 14°   **63. (a)** 2 sec **(b)** $3\frac{1}{3}$ sec

6.3 Exercises *(page 266)*

1. $\dfrac{\pi}{3}, \pi, \dfrac{4\pi}{3}$ **3.** 60°, 210°, 240°, 310°   **5.** $\dfrac{\pi}{12}, \dfrac{11\pi}{12}, \dfrac{13\pi}{12}, \dfrac{23\pi}{12}$ **7.** $\dfrac{\pi}{2}, \dfrac{7\pi}{6}, \dfrac{11\pi}{6}$ **9.** $\dfrac{\pi}{18}, \dfrac{7\pi}{18}, \dfrac{13\pi}{18}, \dfrac{19\pi}{18}, \dfrac{25\pi}{18}, \dfrac{31\pi}{18}$

11. $\dfrac{3\pi}{8}, \dfrac{5\pi}{8}, \dfrac{11\pi}{8}, \dfrac{13\pi}{8}$ **13.** $\dfrac{\pi}{2}, \dfrac{3\pi}{2}$ **15.** $0, \dfrac{\pi}{4}, \dfrac{\pi}{2}, \dfrac{3\pi}{4}, \pi, \dfrac{5\pi}{4}, \dfrac{3\pi}{2}, \dfrac{7\pi}{4}$ **17.** no solution   **19.** $\dfrac{\pi}{2}$ **21.** $\dfrac{\pi}{3}, \pi, \dfrac{5\pi}{3}$

23. 15°, 45°, 135°, 165°, 255°, 285°   **25.** 0°   **27.** 120°, 240°   **29.** 30°, 150°, 270°   **31.** 0°, 30°, 150°, 180°

33. 60°, 300°   **35.** 11.8°, 78.2°, 191.8°, 258.2°   **37.** 30°, 90°, 150°, 210°, 270°, 330°   **39.** 1.2801888

41.

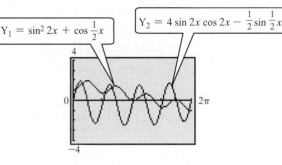

$$Y_1 = \sin^2 2x + \cos \frac{1}{2}x$$

$$Y_2 = 4 \sin 2x \cos 2x - \frac{1}{2} \sin \frac{1}{2}x$$

42. In both cases, the value is approximately .7621.

43. (a)

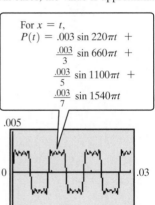

For $x = t$,
$$P(t) = .003 \sin 220\pi t + $$
$$\frac{.003}{3} \sin 660\pi t + $$
$$\frac{.003}{5} \sin 1100\pi t + $$
$$\frac{.003}{7} \sin 1540\pi t$$

(b) The graph is periodic, and the wave has "jagged square" tops and bottoms.

(c) This will occur when t is in one of these intervals: (.0045, .0091), (.0136, .0182), (.0227, .0273).

45. (a)

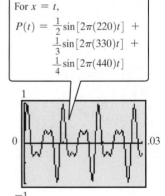

For $x = t$,
$$P(t) = \frac{1}{2}\sin\left[2\pi(220)t\right] + $$
$$\frac{1}{3}\sin\left[2\pi(330)t\right] + $$
$$\frac{1}{4}\sin\left[2\pi(440)t\right]$$

(b) .0007576, .009847, .01894, .02803
(c) 110 Hz

(d)

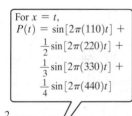
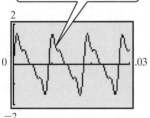

For $x = t$,
$$P(t) = \sin\left[2\pi(110)t\right] + $$
$$\frac{1}{2}\sin\left[2\pi(220)t\right] + $$
$$\frac{1}{3}\sin\left[2\pi(330)t\right] + $$
$$\frac{1}{4}\sin\left[2\pi(440)t\right]$$

47. .001 sec   **49.** .004 sec   **51.** 420° and 660° are not in the interval [0°, 360°].

6.4 Exercises *(page 272)*

1. C   **3.** C   **5.** $x = \arccos \frac{y}{5}$ **7.** $x = \frac{1}{3} \operatorname{arccot} 2y$ **9.** $x = \frac{1}{2} \arctan \frac{y}{3}$ **11.** $x = 4 \arccos \frac{y}{6}$

13. $x = \frac{1}{5} \arccos \left(-\frac{y}{2}\right)$ **15.** $x = -3 + \arccos y$ **17.** $x = \arcsin (y + 2)$ **19.** $x = \arcsin \left(\frac{y + 4}{2}\right)$ **23.** $-2\sqrt{2}$

25. $\pi - 3$ **27.** $\dfrac{3}{5}$ **29.** $\dfrac{4}{5}$ **31.** 0   **33.** $\dfrac{1}{2}$ **35.** $-\dfrac{1}{2}$ **37.** 0   **39.**

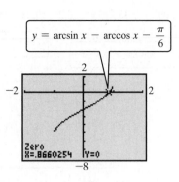

$$y = \arcsin x - \arccos x - \dfrac{\pi}{6}$$

41. 4.4622037   **43. (a)** $A \approx .00506, \phi \approx .484; P = .00506 \sin(440\pi t + .484)$

(b) The two graphs are the same.

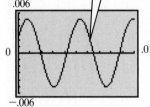

For $x = t$,
$P(t) = .00506 \sin(440\pi t + .484)$
$P_1(t) + P_2(t) = .0012 \sin(440\pi t + .052) +$
$.004 \sin(440\pi t + .61)$

45. (a) $\tan \alpha = \dfrac{x}{z}$; $\tan \beta = \dfrac{x + y}{z}$ **(b)** $\dfrac{x}{\tan \alpha} = \dfrac{x + y}{\tan \beta}$ **(c)** $\alpha = \arctan\left(\dfrac{x \tan \beta}{x + y}\right)$ **(d)** $\beta = \arctan\left(\dfrac{(x + y) \tan \alpha}{x}\right)$

47. (a) $t = \dfrac{1}{2\pi f} \arcsin \dfrac{e}{E_{\max}}$ **(b)** .00068 sec   **49. (a)** $t = \dfrac{3}{4\pi} \arcsin 3y$ **(b)** .27 sec

Chapter 6 Review Exercises *(page 276)*

1. false; The range of the inverse sine function is $\left[-\dfrac{\pi}{2}, \dfrac{\pi}{2}\right]$, while that of the inverse cosine is $[0, \pi]$.

3. false; $\arcsin\left(-\dfrac{1}{2}\right) = -\dfrac{\pi}{6}$, not $\dfrac{11\pi}{6}$.   **5.** $\dfrac{\pi}{4}$ **7.** $-\dfrac{\pi}{3}$ **9.** $\dfrac{3\pi}{4}$ **11.** $\dfrac{2\pi}{3}$ **13.** $\dfrac{3\pi}{4}$ **15.** $-60°$

17. 60.67924514°   **19.** 36.4895081°   **21.** 73.26220613°   **25.** $(-\infty, \infty)$ **27.** $\dfrac{1}{2}$ **29.** -1 **31.** $\dfrac{3\pi}{4}$

33. $\dfrac{\pi}{4}$ **35.** $\dfrac{\sqrt{7}}{4}$ **37.** $\dfrac{\sqrt{3}}{2}$ **39.** $\dfrac{294 + 125\sqrt{6}}{92}$ **41.** $\sqrt{1 - u^2}$

43. $[-1, 1]; \left[-\dfrac{\pi}{2}, \dfrac{\pi}{2}\right]$ **45.** $(-\infty, \infty); (0, \pi)$

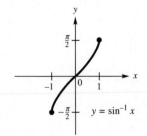

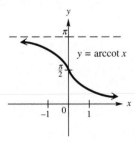

47. .463647609, 3.605240263   **49.** $\dfrac{\pi}{4}, \dfrac{3\pi}{4}, \dfrac{5\pi}{4}, \dfrac{7\pi}{4}$ **51.** $\dfrac{\pi}{8}, \dfrac{3\pi}{8}, \dfrac{5\pi}{8}, \dfrac{7\pi}{8}, \dfrac{9\pi}{8}, \dfrac{11\pi}{8}, \dfrac{13\pi}{8}, \dfrac{15\pi}{8}$ **53.** $\dfrac{\pi}{3}, \pi, \dfrac{5\pi}{3}$ **55.** $270°$

57. $45°, 90°, 225°, 270°$ **59.** $70.5°, 180°, 289.5°$ **61.** $0°, 60°, 90°, 120°, 180°, 240°, 270°, 300°$ **63.** $x = \arcsin 2y$

65. $x = \left(\dfrac{1}{3}\arctan 2y\right) - \dfrac{2}{3}$ **67.** no solution   **69.** $-\dfrac{1}{2}$ **71.** $t = \dfrac{50}{\pi}\arccos\left(\dfrac{d - 550}{450}\right)$

73. (b) 8.6602567 ft; There may be a discrepancy in the final digits.   **(c)** $5\sqrt{3}$ **75.** The light beam is completely underwater.

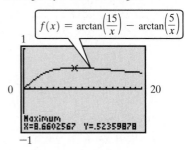

77.

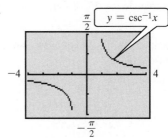

Radian mode

Chapter 6 Test *(page 279)*

1. $[-1, 1]$; $\left[-\dfrac{\pi}{2}, \dfrac{\pi}{2}\right]$ **2. (a)** $\dfrac{2\pi}{3}$ **(b)** $-\dfrac{\pi}{3}$ **(c)** 0 **(d)** $\dfrac{2\pi}{3}$ **(e)** $\dfrac{\pi}{3}$ **(f)** $\dfrac{\pi}{6}$ **3. (a)** $\dfrac{\sqrt{5}}{3}$ **(b)** $\dfrac{4\sqrt{2}}{9}$

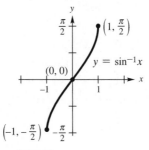

4. $\dfrac{u\sqrt{1 - u^2}}{1 - u^2}$ **5.** $90°, 270°$ **6.** $18.4°, 135°, 198.4°, 315°$ **7.** $0, \dfrac{2\pi}{3}, \dfrac{4\pi}{3}$ **8.** $120°, 240°$ **9. (a)** $x = \dfrac{1}{3}\arccos y$

(b) $\dfrac{4}{5}$ **10.** $\dfrac{5}{6}$ sec, $\dfrac{11}{6}$ sec, $\dfrac{17}{6}$ sec

CHAPTER 7 APPLICATIONS OF TRIGONOMETRY AND VECTORS

Connections *(page 288)*
1. House: $X_H = 1131.8$ ft, $Y_H = 4390.2$ ft; Fire: $X_F = 2277.5$ ft, $Y_F = -2596.2$ ft   **2.** 7079.7 ft

7.1 Exercises *(page 289)*
1. C   **3.** $\sqrt{3}$ **5.** $C = 95°, b = 13$ m, $a = 11$ m   **7.** $B = 37.3°, a = 38.5$ ft, $b = 51.0$ ft   **9.** $C = 57.36°,$

$b = 11.13$ ft, $c = 11.55$ ft   **11.** $B = 18.5°$, $a = 239$ yd, $c = 230$ yd   **13.** $A = 56°\ 00'$, $AB = 361$ ft, $BC = 308$ ft

15. $B = 110.0°$, $a = 27.01$ m, $c = 21.36$ m   **17.** $A = 34.72°$, $a = 3326$ ft, $c = 5704$ ft

19. $C = 97°\ 34'$, $b = 283.2$ m, $c = 415.2$ m   **25.** 118 m   **27.** 1.93 mi   **29.** 10.4 in.   **31.** 111°

33. first location: 5.1 mi; second location: 7.2 mi   **35.** approximately 419,000 km   **37.** appoximately 6600 ft

39. $\dfrac{\sqrt{3}}{2}$ **41.** $\dfrac{\sqrt{2}}{2}$ **43.** 46.4 m^2 **45.** 356 cm^2 **47.** 722.9 in.^{2} **49.** 100 m^2 **51.** increasing

53. $b = \dfrac{a \sin B}{\sin A}$ **54.** $b = \dfrac{a \sin B}{\sin A} = a \cdot \dfrac{\sin B}{\sin A}$. Since $\dfrac{\sin B}{\sin A} < 1$, $b = a \cdot \dfrac{\sin B}{\sin A} < a \cdot 1 = a$, so $b < a$.

57. $x = \dfrac{d \sin \alpha \sin \beta}{\sin(\beta - \alpha)}$ **58. (b)** $1.12257R^2$ **(c) (i)** 8.77 in.^{2} **(ii)** 5.32 in.^{2} **(iii)** red

7.2 Exercises *(page 298)*

1. A   **3. (a)** $4 < h < 5$ **(b)** $h = 4$ or $h > 5$ **(c)** $h < 4$ **5.** 1   **7.** 2   **9.** 0   **11.** 45°   **13.** $B_1 = 49.1°$,
$C_1 = 101.2°$, $B_2 = 130.9°$, $C_2 = 19.4°$ **15.** $B = 26°\ 30'$, $A = 112°\ 10'$ **17.** no such triangle   **19.** $B = 27.19°$,
$C = 10.68°$ **21.** $B = 20.6°$, $C = 116.9°$, $c = 20.6$ ft   **23.** no such triangle   **25.** $B_1 = 49°\ 20'$, $C_1 = 92°\ 00'$,
$c_1 = 15.5$ km, $B_2 = 130°\ 40'$, $C_2 = 10°\ 40'$, $c_2 = 2.88$ km   **27.** $B = 37.77°$, $C = 45.43°$, $c = 4.174$ ft   **29.** $A_1 = 53.23°$,
$C_1 = 87.09°$, $c_1 = 37.16$ m, $A_2 = 126.77°$, $C_2 = 13.55°$, $c_2 = 8.719$ m   **31.** 1; 90°; a right triangle   **35.** It cannot exist.

Connections *(page 303)*

1. In a triangle, the ratio of the tangent of half the difference between two angles to the tangent of half the sum of those angles is equal to the ratio of the difference between the sides opposite those angles and the sum of those sides.

2. In a triangle, the sum of two sides divided by the third side is equal to the cosine of half the difference between the angles opposite those two sides, divided by the sine of half the angle opposite the third side.

Connections *(page 305)*

1.–3. All three methods give the area as 9.5 square units.

7.3 Exercises *(page 305)*

1. (a) law of cosines; $C = 112.5°$ **(b)** law of cosines; $c = 4.52$ **(c)** law of sines; $b = 20.54$ **(d)** Neither is applicable.
3. 7   **5.** 30°   **7.** $c = 2.83$ in., $A = 44.9°$, $B = 106.8°$ **9.** $c = 6.46$ m, $A = 53.1°$, $B = 81.3°$ **11.** $a = 156$ cm,
$B = 64°\ 50'$, $C = 34°\ 30'$ **13.** $b = 9.529$ in., $A = 64.59°$, $C = 40.61°$ **15.** $a = 15.7$ m, $B = 21.6°$, $C = 45.6°$
17. $A = 30°$, $B = 56°$, $C = 94°$ **19.** $A = 82°$, $B = 37°$, $C = 61°$ **21.** $A = 42.0°$, $B = 35.9°$, $C = 102.1°$
23. $A = 47.7°$, $B = 44.9°$, $C = 87.4°$ **25.** 257 m   **27.** 22 ft   **29.** 281 km   **31.** 18 ft   **33.** 2000 km
35. 1470 m   **37.** 16.26°   **39.** $24\sqrt{3}$ **41.** 78 m^2 **43.** 12,600 cm^2 **45.** 3650 ft^2 **47.** 25.24983 mi
49. 33 cans   **51.** Area and perimeter are both 36.   **53.** Any attempt leads to finding the inverse cosine of a number that is
not in the domain of the inverse cosine function.   **55. (a)** 87.8° and 92.2° both appear possible.   **(b)** 92.2°   **(c)** With the law
of cosines we are required to find the inverse cosine of a negative number. Therefore, we know that angle C is greater than 90°.
57. 24.2 ft, 4.14 ft   **61.** Since A is obtuse, $90° < A < 180°$. The cosine of a quadrant II angle is negative.
62. In $a^2 = b^2 + c^2 - 2bc \cos A$, $\cos A$ is negative, so $a^2 = b^2 + c^2$ plus a positive quantity. Thus $a^2 > b^2 + c^2$.
63. $b^2 + c^2 > b^2$ and $b^2 + c^2 > c^2$. If $a^2 > b^2 + c^2$, then $a^2 > b^2$ and $a^2 > c^2$ from which $a > b$ and $a > c$ because a, b, and c
are nonnegative.   **64.** Because A is obtuse it is the largest angle, so the longest side should be a, not c.

7.4 Exercises *(page 317)*

3. **m** and **p**; **n** and **r**   **5.** **m** and **p** equal 2**t**, or **t** is one half **m** or **p**; also **m** = 1**p** and **n** = 1**r**

7. 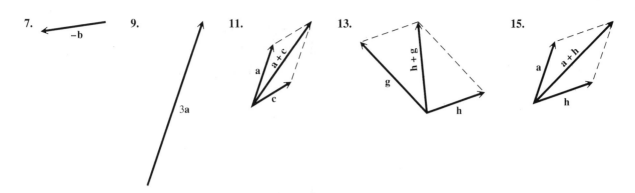   **9.**   **11.**   **13.**   **15.**

17.   **19.**   **21.**   **23.**

25. Vector addition is associative.   **27.** 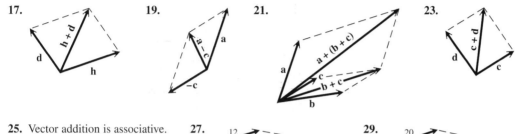   **29.**

31. 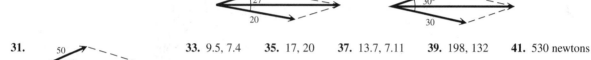   **33.** 9.5, 7.4   **35.** 17, 20   **37.** 13.7, 7.11   **39.** 198, 132   **41.** 530 newtons

43. 27.2 lb   **45.** 88.2 lb   **47.** $\sqrt{2}$; 45°   **49.** 16; 315°   **51.** 17; 331.9°   **53.** 6; 180°   **55.** $\langle 10\sqrt{2}, 10\sqrt{2} \rangle$

57. $\langle -123.258, 154.956 \rangle$ **59.** $\langle -22.116, -65.465 \rangle$ **61.** $\langle 2, 8 \rangle$ **63.** $\langle 6, -2 \rangle$ **65.** $\langle -20, -15 \rangle$ **67.** $-5\mathbf{i} + 8\mathbf{j}$

69. $2\mathbf{i}$ **71.** $4\sqrt{2}\mathbf{i} + 4\sqrt{2}\mathbf{j}$ **73.** $-.254\mathbf{i} + .544\mathbf{j}$ **75.** 7   **77.** -3 **79.** 20   **81.** 135°   **83.** 90°

85. 36.87°   **87.** -6 **89.** -24 **91.** orthogonal   **93.** not orthogonal   **95.** not orthogonal

97. magnitude: 9.52082827; direction angle: 119.0646784°   **98.** $\langle -4.10424172, 11.27631145 \rangle$

99. $\langle -.520944533, -2.954423259 \rangle$ **100.** $\langle -4.625186253, 8.321888191 \rangle$

101. magnitude: 9.52082827; direction angle: 119.0646784°   **102.** **(a)** They are the same.

7.5 Exercises *(page 322)*

1. 93.9°   **3.** 18°   **5.** 2.4 tons   **7.** 226 lb   **9.** weight: 64.8 lb; tension: 61.9 lb   **11.** 190 lb and 283 lb, respectively

13. 173.1°   **15.** 39.2 km   **17.** 237°; 470 mph   **19.** 358°; 170 mph   **21.** 230 km per hr; 167°   **23.** 3:21 P.M.

25. **(a)** approximately 56 mi per sec   **(b)** approximately 87   **26.** $y - b \sin \theta = (-\cot \theta)(x - b \cos \theta)$ (Other answers are

possible.)   **27.** $(a, -a \cot \theta + b \cos \theta \cot \theta + b \sin \theta)$ **28.** $|\mathbf{v}| = \dfrac{\sqrt{a^2 + b^2 - 2ab \cos \theta}}{\sin \theta}$

29. the line joining the endpoints of **a** and **b**

Chapter 7 Review Exercises *(page 326)*

1. 63.7 m   **3.** 41.7°   **5.** 54° 20′ or 125° 40′   **9.** **(a)** $b = 5, b \ge 10$ **(b)** $5 < b < 10$ **(c)** $b < 5$

11. 19.87° or 19° 52′   **13.** 55.5 m   **15.** 19 cm   **17.** $B = 17.3°, C = 137.5°, c = 11.0$ yd   **19.** $c = 18.7$ cm,

$A = 91° 40′, B = 45° 50′$ **21.** 153,600 m²   **23.** .234 km²   **25.** Each expression is equal to $\dfrac{1 + \sqrt{3}}{2}$.   **27.** 58.6 ft

29. 13 m   **31.** 53.2 ft   **33.** 115 km   **35.** 1450 ft   **37.** distance to both first and third bases: 63.7 ft; distance to second base: 66.8 ft   **39.** 25   **41.** He needs about 2.5 cans, so he must buy 3 cans.   **43.**

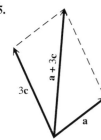

45.   **47.** 28 lb   **49.** 826 lb   **51.** 17.9, 66.8   **53.** 29, 316.4°   **55. (a)** 14   **(b)** 52.13°

57. $\left\langle -\dfrac{4}{5}, \dfrac{3}{5} \right\rangle$ **59.** 280 newtons, 30.4°   **61.** 3° 50′

63. speed: 21 km per hr; bearing: 118°   **65.** $AB = 1978.28$ ft; $BC = 975.05$ ft

Chapter 7 Test *(page 331)*

1. 137.5°   **2.** 180 km   **3.** 49.0°   **4.** 4300 km²   **5. (a)** $b > 10$ **(b)** none   **(c)** $b \le 10$ **6.** 264 square units

7. In any triangle, the sum of the lengths of any two sides must be greater than the length of the remaining side.

8. $|\mathbf{v}| = 10$; $\theta \approx 126.9°$ **9.** $\langle 1, -3 \rangle$ **10.** $\langle -6, 18 \rangle$ **11.** -20 **12.** 2.7 mi   **13.** 43 ft   **14.** $\langle -346, 451 \rangle$

15. 1.91 mi

CHAPTER 8   COMPLEX NUMBERS, POLAR EQUATIONS, AND PARAMETRIC EQUATIONS

Connections *(page 335)*

1. A beginning algebra student might have difficulty with the concept because no *real* number (positive, negative, or zero) has a square that is negative. The student might wonder, "How can a square be a negative?"

2. $i^4 = 1$ since $(i^2)^2 = (-1)^2 = 1$; $i^3 = -i$ since $i^3 = i^2 \cdot i = -1 \cdot i = -i$

8.1 Exercises *(page 341)*

1. B   **3.** D   **5. (a)** Imaginary Numbers   **(b)** Rational Numbers   **7.** $6i, -6i$ **9.** $4i\sqrt{3}, -4i\sqrt{3}$

11. $-\dfrac{3}{4} + \dfrac{\sqrt{7}}{4}i, -\dfrac{3}{4} - \dfrac{\sqrt{7}}{4}i$ **13.** $-2 + i\sqrt{7}, -2 - i\sqrt{7}$ **15.** $\dfrac{1}{3} + \dfrac{\sqrt{6}}{3}i, \dfrac{1}{3} - \dfrac{\sqrt{6}}{3}i$ **17.** $1 + i, 1 - i$ **19.** -3

21. $-\sqrt{30}$ **23.** $\dfrac{\sqrt{6}}{2}$ **25.** $\dfrac{\sqrt{3}}{3}i$ **27.** 4   **29.** $-3 - 9i$ **31.** $x = 2, y = -3$ **33.** $x = \dfrac{1}{2}, y = 15$

35. $x = 14, y = 8$ **37.** $5 - 3i$ **39.** $-5 + 2i$ **41.** $-4 + i$ **43.** $8 - i$ **45.** $-14 + 2i$ **47.** 5   **49.** $-5i$

51. $2 - 2i$ **53.** $\dfrac{13}{20} - \dfrac{1}{20}i$ **57.** 1   **59.** -1 **61.** i **63.** -1 **65.** 0   **67.** $E = 30 + 60i$

69. $Z = \dfrac{233}{37} + \dfrac{119}{37}i$ **71.** $110 + 32i$ **73.** $\left(\dfrac{\sqrt{2}}{2} + \dfrac{\sqrt{2}}{2}i \right)^2 = i$ **75.** The step $\sqrt{-1} \cdot \sqrt{-1} = \sqrt{(-1)(-1)}$ is invalid,

since the rule $\sqrt{a \cdot b} = \sqrt{a} \cdot \sqrt{b}$ does not apply when both a and b are negative.   **77.** $b = 0$

Connections *(page 344)*

1. $\langle 10, 1 \rangle$ or $10 + i$ **2.** Answers will vary.

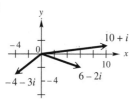

8.2 Exercises *(page 348)*

1. magnitude (length)

3.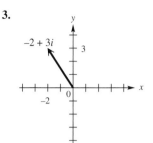

5.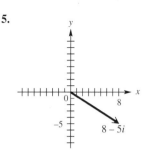

7.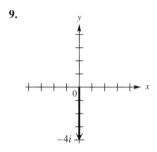

9.

11.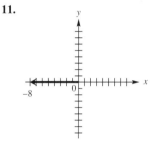

13. $1 - 4i$ **15.** $3 - i$ **17.** $3 - 3i$ **19.** $-3 + 3i$ **21.** $2 + 4i$ **23.** $7 + 9i$ **25.** $\sqrt{2} + i\sqrt{2}$ **27.** $10i$

29. $-2 - 2i\sqrt{3}$ **31.** $\dfrac{\sqrt{3}}{2} + \dfrac{1}{2}i$ **33.** $\dfrac{5}{2} - \dfrac{5\sqrt{3}}{2}i$ **35.** $-\sqrt{2}$ **37.** $3\sqrt{2}(\cos 315° + i \sin 315°)$

39. $6(\cos 240° + i \sin 240°)$ **41.** $2(\cos 330° + i \sin 330°)$ **43.** $5\sqrt{2}(\cos 225° + i \sin 225°)$

45. $2\sqrt{2}(\cos 45° + i \sin 45°)$ **47.** $4(\cos 180° + i \sin 180°)$ **49.** $2(\cos 270° + i \sin 270°)$

51. $\sqrt{13}(\cos 56.31° + i \sin 56.31°)$ **53.** $-1.0260604 - 2.8190779i$ **55.** $12(\cos 90° + i \sin 90°)$

57. $\sqrt{34}(\cos 59.04° + i \sin 59.04°)$ **59.** the circle of radius 1 centered at the origin **61.** the vertical line $x = 1$

63. yes **67.** B **69.** A

8.3 Exercises *(page 353)*

1. multiply; add **3.** $-3\sqrt{3} + 3i$ **5.** $-4i$ **7.** $12\sqrt{3} + 12i$ **9.** $-\dfrac{15\sqrt{2}}{2} + \dfrac{15\sqrt{2}}{2}i$ **11.** $-3i$ **13.** $\sqrt{3} - i$

15. $-1 - i\sqrt{3}$ **17.** $-\dfrac{1}{6} - \dfrac{\sqrt{3}}{6}i$ **19.** $2\sqrt{3} - 2i$ **21.** $-\dfrac{1}{2} - \dfrac{1}{2}i$ **23.** $\sqrt{3} + i$ **25.** $.65366807 + 7.4714602i$

27. $30.858023 + 18.541371i$ **29.** $.20905693 + 1.9890438i$ **31.** $-3.7587705 - 1.3680806i$ **33.** 2

34. $w = \sqrt{2}\ \text{cis}\ 135°;\ z = \sqrt{2}\ \text{cis}\ 225°$ **35.** $2\ \text{cis}\ 0°$ **36.** 2; It is the same. **37.** $-i$ **38.** $\text{cis}(-90°)$

39. $-i$; It is the same. **43.** $1.18 - .14i$ **45.** approximately $27.43 + 11.5i$

8.4 Exercises *(page 359)*

1. $27i$ **3.** 1 **5.** $\dfrac{27}{2} - \dfrac{27\sqrt{3}}{2}i$ **7.** $-16\sqrt{3} + 16i$ **9.** $-128 + 128i\sqrt{3}$ **11.** $128 + 128i$

13. $\cos 0° + i \sin 0°$,
$\cos 120° + i \sin 120°$,
$\cos 240° + i \sin 240°$

15. 2 cis 20°,
2 cis 140°,
2 cis 260°

17. $2(\cos 90° + i \sin 90°)$,
$2(\cos 210° + i \sin 210°)$,
$2(\cos 330° + i \sin 330°)$

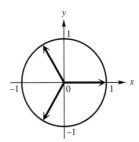

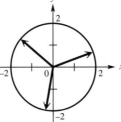

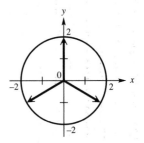

19. $4(\cos 60° + i \sin 60°)$,
$4(\cos 180° + i \sin 180°)$,
$4(\cos 300° + i \sin 300°)$

21. $\sqrt[3]{2}(\cos 20° + i \sin 20°)$,
$\sqrt[3]{2}(\cos 140° + i \sin 140°)$,
$\sqrt[3]{2}(\cos 260° + i \sin 260°)$

23. $\sqrt[3]{4}(\cos 50° + i \sin 50°)$,
$\sqrt[3]{4}(\cos 170° + i \sin 170°)$,
$\sqrt[3]{4}(\cos 290° + i \sin 290°)$

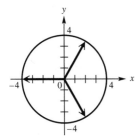

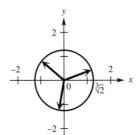

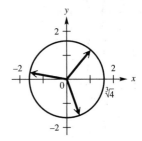

25.

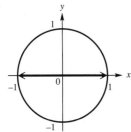

$\cos 0° + i \sin 0°$,
$\cos 180° + i \sin 180°$

27.

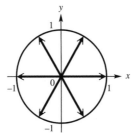

$\cos 0° + i \sin 0°$, $\cos 60° + i \sin 60°$,
$\cos 120° + i \sin 120°$,
$\cos 180° + i \sin 180°$,
$\cos 240° + i \sin 240°$,
$\cos 300° + i \sin 300°$

29.

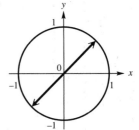

$\cos 45° + i \sin 45°$,
$\cos 225° + i \sin 225°$

31. $\cos 0° + i \sin 0°$,
$\cos 120° + i \sin 120°$,
$\cos 240° + i \sin 240°$

33. $\cos 90° + i \sin 90°$,
$\cos 210° + i \sin 210°$,
$\cos 330° + i \sin 330°$

35. $2(\cos 0° + i \sin 0°)$,
$2(\cos 120° + i \sin 120°)$,
$2(\cos 240° + i \sin 240°)$

37. $\cos 45° + i \sin 45°$,
$\cos 135° + i \sin 135°$,
$\cos 225° + i \sin 225°$,
$\cos 315° + i \sin 315°$

39. $\cos 22.5° + i \sin 22.5°$,
$\cos 112.5° + i \sin 112.5°$,
$\cos 202.5° + i \sin 202.5°$,
$\cos 292.5° + i \sin 292.5°$

41. $2(\cos 20° + i \sin 20°)$,
$2(\cos 140° + i \sin 140°)$,
$2(\cos 260° + i \sin 260°)$

43. $1, -\dfrac{1}{2} + \dfrac{\sqrt{3}}{2}i, -\dfrac{1}{2} - \dfrac{\sqrt{3}}{2}i$ **45.** $\cos 2\theta + i \sin 2\theta$ **46.** $(\cos^2 \theta - \sin^2 \theta) + i(2 \cos \theta \sin \theta) = \cos 2\theta + i \sin 2\theta$

47. $\cos 2\theta = \cos^2 \theta - \sin^2 \theta$ **48.** $\sin 2\theta = 2 \cos \theta \sin \theta$ **49. (a)** yes   **(b)** no   **(c)** yes

51. $1, .30901699 + .95105652i, -.809017 + .58778525i, -.809017 - .5877853i, .30901699 - .9510565i$

53. $-4, 2 - 2i\sqrt{3}$ **55.** $.87708 + .94922i, -.63173 + 1.1275i, -1.2675 - .25240i, -.15164 - 1.28347i,$
$1.1738 - .54083i$ **57.** false

Connections *(page 363)*
1. $x - y = 3$ **2.** $x^2 + y^2 = 4$

8.5 Exercises *(page 370)*
1. (a) II   **(b)** I   **(c)** IV   **(d)** III

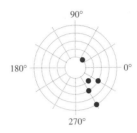

Graphs for Exercises 3, 5, 7, 9, 11

Answers may vary in Exercises 3–11.

3. $(1, 405°), (-1, 225°)$ **5.** $(-2, 495°), (2, 315°)$ **7.** $(5, 300°), (-5, 120°)$ **9.** $(-3, 150°), (3, -30°)$
11. $(3, 660°), (-3, 120°)$

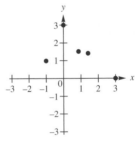

Graphs for Exercises 13, 15, 17, 19, 21

Answers may vary in Exercises 13–21.

13. $(\sqrt{2}, 135°), (-\sqrt{2}, 315°)$ **15.** $(3, 90°), (-3, 270°)$ **17.** $(2, 45°), (-2, 225°)$ **19.** $(\sqrt{3}, 60°), (-\sqrt{3}, 240°)$
21. $(3, 0°), (-3, 180°)$

23. cardioid   **25.** limaçon   **27.** four-leaved rose

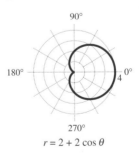

$r = 2 + 2 \cos \theta$

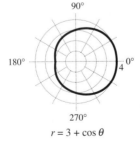

$r = 3 + \cos \theta$

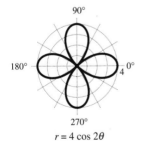

$r = 4 \cos 2\theta$

29. lemniscate

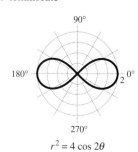

$r^2 = 4 \cos 2\theta$

31. cardioid

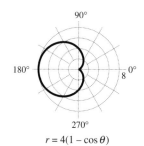

$r = 4(1 - \cos \theta)$

33.

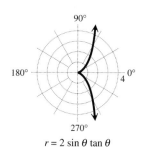

$r = 2 \sin \theta \tan \theta$

35. (a) $(r, -\theta)$ (b) $(r, \pi - \theta)$ or $(-r, -\theta)$ (c) $(r, \pi + \theta)$ or $(-r, \theta)$ **36.** $-\theta$ **37.** $\pi - \theta$ **38.** $-r; -\theta$ **39.** $-r$

40. $\pi + \theta$ **41.** the polar axis **42.** the line $\theta = \dfrac{\pi}{2}$ **43.** 4; 45°, 135°, 225°, 315°

47. $\left(\dfrac{4 + \sqrt{2}}{2}, \dfrac{\pi}{4}\right), \left(\dfrac{4 - \sqrt{2}}{2}, \dfrac{5\pi}{4}\right)$

51. $x^2 + (y - 1)^2 = 1$

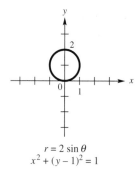

$r = 2 \sin \theta$
$x^2 + (y - 1)^2 = 1$

53. $y^2 = 4(x + 1)$

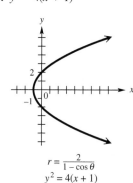

$r = \dfrac{2}{1 - \cos \theta}$
$y^2 = 4(x + 1)$

55. $(x + 1)^2 + (y + 1)^2 = 2$

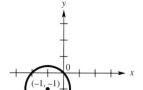

$r + 2 \cos \theta = -2 \sin \theta$
$(x + 1)^2 + (y + 1)^2 = 2$

57. $x = 2$

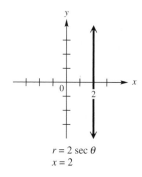

$r = 2 \sec \theta$
$x = 2$

59. $x + y = 2$

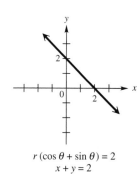

$r(\cos \theta + \sin \theta) = 2$
$x + y = 2$

61. $r = \dfrac{4}{\cos \theta + \sin \theta}$

63. $r = 4$ **65.** $r = 2 \csc \theta$ or $r = \dfrac{2}{\sin \theta}$ **67.**

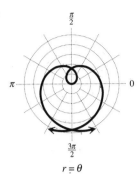

$r = \theta$

69. $r = \dfrac{2}{2 \cos \theta + \sin \theta}$

8.6 Exercises *(page 380)*

1. C   **3.** A

5. $x = 2t$
$y = t + 1$
for t in $[-2, 3]$

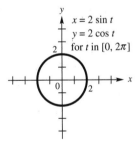

$(6, 4)$
$(-4, -1)$

$y = \dfrac{1}{2}x + 1$, for x in $[-4, 6]$

7.

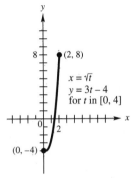

$(2, 8)$
$x = \sqrt{t}$
$y = 3t - 4$
for t in $[0, 4]$
$(0, -4)$

$y = 3x^2 - 4$, for x in $[0, 2]$

9.

$x = t^3 + 1$
$y = t^3 - 1$
for t in $(-\infty, \infty)$

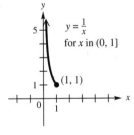

$y = x - 2$, for x in $(-\infty, \infty)$

11.

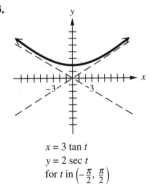

$x = 2 \sin t$
$y = 2 \cos t$
for t in $[0, 2\pi]$

$x^2 + y^2 = 4$, for x in $[-2, 2]$

13.

$x = 3 \tan t$
$y = 2 \sec t$
for t in $\left(-\dfrac{\pi}{2}, \dfrac{\pi}{2}\right)$

$y = 2\sqrt{1 + \dfrac{x^2}{9}}$, for x in $(-\infty, \infty)$

15.

$y = \dfrac{1}{x}$
for x in $(0, 1]$
$(1, 1)$

$y = \dfrac{1}{x}$, for x in $(0, 1]$

17.

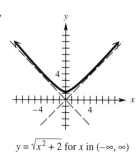

$y = \sqrt{x^2 + 2}$ for x in $(-\infty, \infty)$

19.

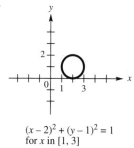

$(x - 2)^2 + (y - 1)^2 = 1$
for x in $[1, 3]$

21.

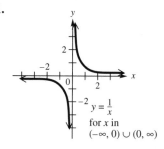

$y = \dfrac{1}{x}$
for x in
$(-\infty, 0) \cup (0, \infty)$

23.

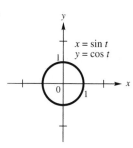

$x = \sin t$
$y = \cos t$

25.

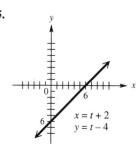

$x = t + 2$
$y = t - 4$

27.

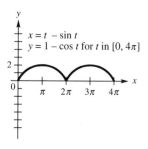

$x = t - \sin t$
$y = 1 - \cos t$ for t in $[0, 4\pi]$

29. (a) $x = 24t, \ y = -16t^2 + 24\sqrt{3}t$ **(b)** $y = -\dfrac{1}{36}x^2 + \sqrt{3}x$ **(c)** 2.6 sec; 62 ft

31. (a) $x = (88 \cos 20°)t, \ y = 2 - 16t^2 + (88 \sin 20°)t$ **(b)** $y = 2 - \dfrac{x^2}{484 \cos^2 20°} + (\tan 20°)x$ **(c)** 2 sec; 160 ft

33. (a)

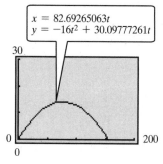

$x = 82.69265063t$
$y = -16t^2 + 30.09777261t$

(b) 20.0° **(c)** $x = (88 \cos 20.0°)t, \ y = -16t^2 + (88 \sin 20°)t$

35. Many answers are possible, two of which are $x = t, \ y = m(t - x_1) + y_1$ and $x = t^2, \ y = m(t^2 - x_1) + y_1$.

37. Many answers are possible; for example, $x = a \sec \theta, \ y = b \tan \theta$ and $x = t, \ y^2 = \dfrac{b^2}{a^2}(t^2 - a^2)$.

41. the pair $x = \cos t, \ y = -\sin t$

Chapter 8 Review Exercises (page 384)

1. $3i$ **3.** $-9i, 9i$ **5.** $-2 - 3i$ **7.** $5 + 4i$ **9.** $29 + 37i$ **11.** $-32 + 24i$ **13.** $-2 - 2i$ **15.** $\dfrac{8}{5} + \dfrac{6}{5}i$

17. $-\dfrac{3}{26} + \dfrac{11}{26}i$ **19.** i **21.** -1 **23.** $-30i$ **25.** $-\dfrac{1}{8} + \dfrac{\sqrt{3}}{8}i$ **27.** $8i$ **29.** $-\dfrac{1}{2} - \dfrac{\sqrt{3}}{2}i$

33.

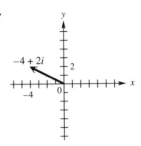

35. $5 + 4i$

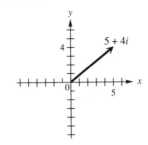

37. $3i$ **39.** $8(\cos 120° + i \sin 120°)$ **41.** $-2 - 2i\sqrt{3}$ **43.** a circle of radius 2 with the origin as center

45. $\sqrt[10]{8}(\cos 27° + i \sin 27°)$, $\sqrt[10]{8}(\cos 99° + i \sin 99°)$, $\sqrt[10]{8}(\cos 171° + i \sin 171°)$, $\sqrt[10]{8}(\cos 243° + i \sin 243°)$,

$\sqrt[10]{8}(\cos 315° + i \sin 315°)$ **47.** one   **49.** $5(\cos 60° + i \sin 60°)$, $5(\cos 180° + i \sin 180°)$, $5(\cos 300° + i \sin 300°)$

51. $\cos 135° + i \sin 135°$, $\cos 315° + i \sin 315°$ **53.** $\left(\dfrac{5\sqrt{2}}{2}, -\dfrac{5\sqrt{2}}{2}\right)$ **55.** a circle

57. cardioid   **59.** B   **61.** C   **63.** $y^2 = -6\left(x - \dfrac{3}{2}\right)$ or $y^2 + 6x - 9 = 0$ **65.** $x^2 + y^2 = 4$

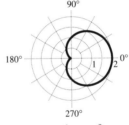

$r = -1 + \cos \theta$

67. $r = \tan \theta \sec \theta$ or $r = \dfrac{\tan \theta}{\cos \theta}$ **69.** $r = 2 \sec \theta$ or $r = \dfrac{2}{\cos \theta}$ **71.**

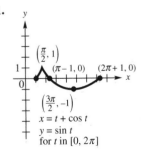

$x = t + \cos t$
$y = \sin t$
for t in $[0, 2\pi]$

73. $y = \sqrt{x^2 + 1}$, for x in $[0, \infty)$ **75.** $y = 3\sqrt{1 + \dfrac{x^2}{25}}$, for x in $(-\infty, \infty)$ **77.** $\left(\dfrac{v_0^2 \sin \theta \cos \theta}{32}, \dfrac{v_0^2 \sin^2 \theta}{64}\right)$ or

$\left(\dfrac{v_0^2 \sin 2\theta}{64}, \dfrac{v_0^2 \sin^2 \theta}{64}\right)$

Chapter 8 Test *(page 386)*

1. (a) $7 - 3i$
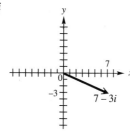
(b) $-3 - 5i$ **(c)** $14 - 18i$ **(d)** $\dfrac{3}{13} - \dfrac{11}{13}i$ **2. (a)** $-i$ **(b)** $2i$

3. $\dfrac{1}{4} + \dfrac{\sqrt{31}}{4}i, \dfrac{1}{4} - \dfrac{\sqrt{31}}{4}i$ **4. (a)** $3(\cos 90° + i \sin 90°)$ **(b)** $\sqrt{5}$ cis $63.43°$

(c) $2(\cos 240° + i \sin 240°)$ **5. (a)** $\dfrac{3\sqrt{3}}{2} + \dfrac{3}{2}i$ **(b)** $3.06 + 2.57i$ **(c)** $3i$

6. (a) $16(\cos 50° + i \sin 50°)$ **(b)** $2\sqrt{3} + 2i$ **(c)** $4\sqrt{3} + 4i$

7. 2 cis $67.5°$, 2 cis $157.5°$, 2 cis $247.5°$, 2 cis $337.5°$

Answers may vary in Exercise 8.

8. (a) $(5, 90°), (5, -270°)$ **(b)** $(2\sqrt{2}, 225°), (2\sqrt{2}, -135°)$ **9. (a)** $\left(\dfrac{3\sqrt{2}}{2}, -\dfrac{3\sqrt{2}}{2}\right)$ **(b)** $(0, -4)$

10. cardioid **11.** three-leaved rose **12.** $x - 2y = -4$

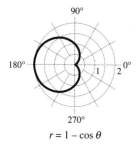

$r = 1 - \cos\theta$

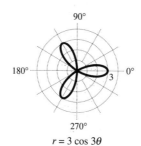

$r = 3\cos 3\theta$

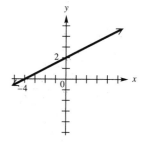

13.

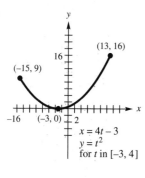

$x = 4t - 3$
$y = t^2$
for t in $[-3, 4]$

14.

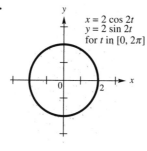

$x = 2\cos 2t$
$y = 2\sin 2t$
for t in $[0, 2\pi]$

15. $z^2 - 1 = -1 - 2i; r = \sqrt{5}$ and $\sqrt{5} > 2$

CHAPTER 9   EXPONENTIAL AND LOGARITHMIC FUNCTIONS

Connections *(page 397)*

1. 2.717 **2.** $.9512$ **3.** $\dfrac{x^6}{6 \cdot 5 \cdot 4 \cdot 3 \cdot 2 \cdot 1}$

9.1 Exercises *(page 399)*

1. 9 **3.** $\dfrac{1}{9}$ **5.** $\dfrac{1}{16}$ **7.** 16 **9.** 5.196152423 **11.** $.0390103297$ **13.** yes; an inverse function

14.

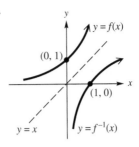

15. $x = a^y$ **16.** $x = 10^y$
17. $x = e^y$ **18.** (q, p)

In Exercises 19–29, we give only a traditional graph.

19.

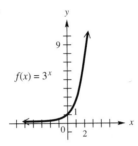

$f(x) = 3^x$

21.

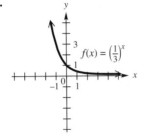

$f(x) = \left(\frac{1}{3}\right)^x$

23.

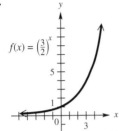

$f(x) = \left(\frac{3}{2}\right)^x$

25.

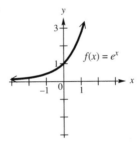

$f(x) = e^x$

27.

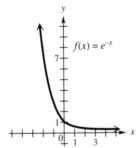

$f(x) = e^{-x}$

29.

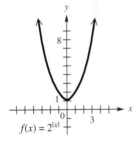

$f(x) = 2^{|x|}$

31.

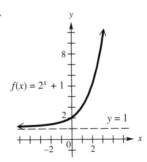

$f(x) = 2^x + 1$

$y = 1$

33.

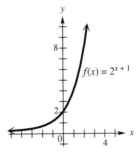

$f(x) = 2^{x+1}$

35.

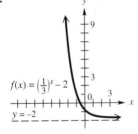

$f(x) = \left(\frac{1}{3}\right)^x - 2$

$y = -2$

37.

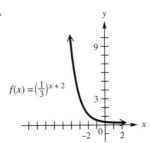

$f(x) = \left(\frac{1}{3}\right)^{x+2}$

39. 2.3
41. .75
43. .31

45.

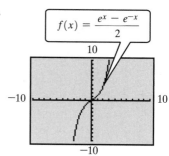
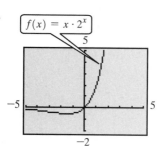

47.

49. $\dfrac{1}{2}$ **51.** -2 **53.** 0

55. 3   **57.** $-8, 8$ **59.** $\dfrac{1}{5}$ **61.** $-\dfrac{2}{3}$ **63.** $13,891.16   **65.** $21,223.33   **67.** 1.0%

69. (a)

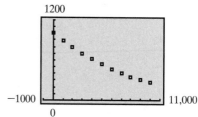

(b) exponential   **(c)**

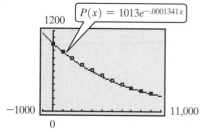

(d) $P(1500) \approx 828$ mb; $P(11,000) \approx 232$ mb

71. (a) The function gives approximately 5238 million, which differs by 82 million from the actual value.   **(b)** 5662 million
(c) 6618 million   **73. (a)** about 207   **(b)** about 235   **(c)** about 249

Connections *(page 409)*

1.
$$\log_{10} 458.3 \approx 2.661149857$$
$$+ \log_{10} 294.6 \approx 2.469232743$$
$$\approx 5.130382600$$
$$10^{5.130382600} \approx 135,015.18$$
A calculator gives $(458.3)(294.6) = 135,015.18$.

2. Answers will vary.

9.2 Exercises *(page 410)*

1. E   **3.** G   **5.** C   **7.** F   **9.** $\log_3 81 = 4$ **11.** $\log_{2/3}\left(\dfrac{27}{8}\right) = -3$ **13.** $6^2 = 36$ **15.** $(\sqrt{3})^8 = 81$

19. -4 **21.** -3 **23.** 9   **25.** $\dfrac{1}{5}$ **27.** 64   **29.** $\dfrac{2}{3}$ **33.**

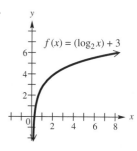

35.

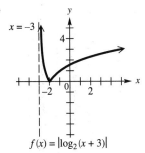

$f(x) = |\log_2(x + 3)|$

37.

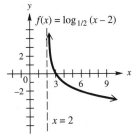

$f(x) = \log_{1/2}(x - 2)$

$x = 2$

39.

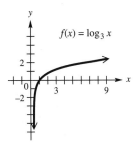

$f(x) = \log_3 x$

41.

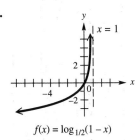

$f(x) = \log_{1/2}(1 - x)$

43.

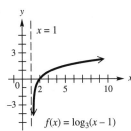

$f(x) = \log_3(x - 1)$

45. E   **47.** B   **49.** F

51.

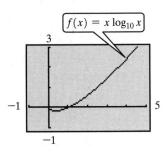

$f(x) = x \log_{10} x$

55.

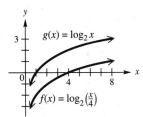

$g(x) = \log_2 x$

$f(x) = \log_2\left(\frac{x}{4}\right)$

53. $\log_a x - \log_a y$

54. Since $\log_2\left(\dfrac{x}{4}\right) = \log_2 x - \log_2 4$ by
the quotient rule, the graph of
$y = \log_2\left(\dfrac{x}{4}\right)$ can be obtained by
translating the graph of $y = \log_2 x$
downward by $\log_2 4 = 2$ units.

56. 0; 2; 2; 0; By the quotient rule, $\log_2\left(\dfrac{x}{4}\right) = \log_2 x - \log_2 4$. Both sides should
equal 0. Since $2 - 2 = 0$, they do.

57. $\log_2 6 + \log_2 x - \log_2 y$ **59.** $1 + \left(\dfrac{1}{2}\right)\log_5 7 - \log_5 3$

61. cannot be simplified   **63.** $\left(\dfrac{1}{2}\right)(\log_m 5 + 3 \log_m r - 5 \log_m z)$

65. $\log_a\left(\dfrac{xy}{m}\right)$ **67.** $\log_m\left(\dfrac{a^2}{b^6}\right)$ **69.** $\log_a[(z - 1)^2(3z + 2)]$ **71.** .7781   **73.** .3522

75. (a)

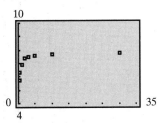

(b) The interest rates increase with time but not at
a constant rate. Therefore, a linear function
would not model the data well. The interest
rates gradually level off. This resembles a
translated logarithmic function. An
(increasing) exponential function does not
level off, but rather continues to increase at an
even faster rate.

77. (a) -4 **(b)** 6

9.3 Exercises *(page 419)*

1. increasing   **3.** $f^{-1}(x) = \log_5 x$ **5.** natural; common   **7.** There is no power of 2 that yields a result of 0.

9. 1.5563   **11.** −1.3768   **13.** 4.3010   **15.** 3.5835   **17.** −3.1701   **19.** 4.6931   **21.** 3.2   **23.** 1.8

25. 2.0×10^{-3} **27.** 1.6×10^{-5} **29.** poor fen   **31.** rich fen   **33.** 2.3219   **35.** −.2537   **37.** 1.9376

39. −1.4125   **41.** D   **43.** $4v + \frac{1}{2}u$ **45.** $\frac{3}{2}u - \frac{5}{2}v$ **47. (a)** 3   **(b)** 5^2 or 25   **(c)** $\frac{1}{e}$ **49. (a)** 5   **(b)** ln 3

(c) 2 ln 3 or ln 9   **51.** domain: $(-\infty, 0) \cup (0, \infty)$; range: $(-\infty, \infty)$; symmetric with respect to the y-axis

53. When $x \geq 4$, $4 - x \leq 0$, and we cannot obtain a real value for the logarithm of a nonpositive number.

55. (a) 20   **(b)** 30   **(c)** 50   **(d)** 60   **(e)** about 3 decibels   **57. (a)** 3   **(b)** 6   **(c)** 8   **59.** about $126,000,000I_0$

61. about 68 million visitors; We must assume that the rate of increase continues to be logarithmic.

63. (a) 2   **(b)** 2   **(c)** 2   **(d)** 1   **65.** 1   **67.** between 7°F and 11°F

69. (a)

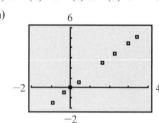

Let $x = \ln D$ and $y = \ln P$ for each planet. From the graph, the data appear to be linear.

(b)

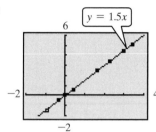

The points (0, 0) and (3.40, 5.10) determine the line $y = 1.5x$ or $\ln P = 1.5 \ln D$. (Answers will vary.)

(c) $P \approx 248.3$ yr

9.4 Exercises *(page 428)*

1. $\log_7 19$; $\frac{\log 19}{\log 7}$; $\frac{\ln 19}{\ln 7}$ **3.** $\log_{1/2} 12$; $\frac{\log 12}{\log(\frac{1}{2})}$; $\frac{\ln 12}{\ln(\frac{1}{2})}$ **5.** 1.6309   **7.** −.0803   **9.** 2.2694   **11.** 2.3863

13. −.1227   **15.** no solution   **17.** 2   **19.** 140.0112   **21.** 25   **23.** 4   **25.** 4.5   **27.** −17.5314   **29.** 8

31. 4   **33.** 1, 100   **37.** $f^{-1}(x) = \ln(x + 4) - 1$; domain: $(-4, \infty)$; range: $(-\infty, \infty)$ **39.** 1.52   **41.** 0

43. 2.45, 5.66   **45.** during 2001   **47.** about 24%

49. (a) $P(T) = 1 - e^{-.0034 - .0053T}$

(b)

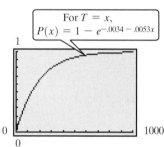

For $T = x$,
$P(x) = 1 - e^{-.0034 - .0053x}$

51. 1.8 yr

53. 6.48%

55. (a) $17,742.41   **(b)** $6191.93

(c) $11,550.48   **(d)** They are the same.

(c) $P(60) \approx .275$ or 27.5%. The reduction in carbon emissions from a tax of $60 per ton of carbon is 27.5%.

(d) $T = 130.14

Chapter 9 Review Exercises *(page 431)*

1. B   **3.** C   **5.** $\log_2 32 = 5$ **7.** $\log_{3/4}\left(\frac{4}{3}\right) = -1$

9.

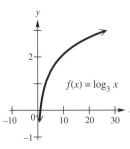

$f(x) = \log_3 x$

11. 2   **12.** 3

13. It lies between 2 and 3.

14. By the change-of-base theorem,

$$\log_3 16 = \frac{\log 16}{\log 3} = \frac{\ln 16}{\ln 3} \approx 2.523719014.$$

15. $-1; 0$

16. It lies between -1 and 0. $\frac{1}{5} = .2 < .68 < 1$, so $-1 = \log_5 .2 < \log_5 .68 < \log_5 1 = 0$;

$\log_5 .68 = \dfrac{\log .68}{\log 5} = \dfrac{\ln .68}{\ln 5} \approx -.2396255723$ **17.** $9^{3/2} = 27$ **19.** $e^{3.8067} \approx 45$ **21.** 2

23. $2 \log_5 x + 4 \log_5 y + \frac{1}{5}(3 \log_5 m + \log_5 p)$ **25.** 1.6590   **27.** 6.1527   **29.** 6.0486   **31.** $\dfrac{5}{3}$ **33.** 2.1152

35. 1.3026   **37.** 2   **39.** $\dfrac{1}{2}$ **41.** -3 **43.** (a) about $4,000,000 I_0$ (b) about $3,200,000 I_0$ (c) about 1.25 times as great

45. 89 decibels is about twice as loud as 86 decibels. This is a 100% increase.   **47.** 4.0 yr   **49.** \$25,149.59   **51.** 1999

53. (a)

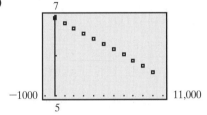

There appears to be a linear relationship.

55. (a)

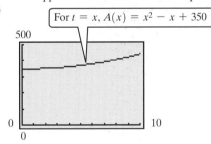

For $t = x$, $A(x) = x^2 - x + 350$

(b)

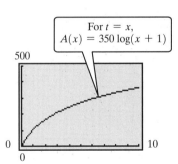

For $t = x$, $A(x) = 350 \log(x + 1)$

(c)

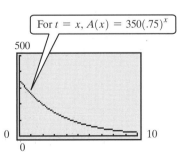

For $t = x$, $A(x) = 350(.75)^x$

(d)

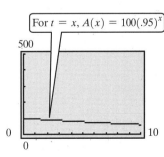

For $t = x$, $A(x) = 100(.95)^x$

Function (c) best describes $A(t)$.

57. (a) $\log_4(2x^2 - x) = \dfrac{\ln(2x^2 - x)}{\ln 4}$ **(b)**

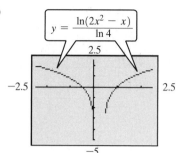

(c) $-\dfrac{1}{2}, 1$ **(d)** $x = 0, x = \dfrac{1}{2}$

Chapter 9 Test *(page 434)*

1. (a) f and g are inverse functions. **(b)** The graph of g is the reflection of the graph of f across the line $y = x$. **(c)** $(9, 2)$

2. (a) B **(b)** A **(c)** C **(d)** D   **3.** $\dfrac{1}{2}$ **4. (a)** $\log_4 8 = \dfrac{3}{2}$ **(b)** $8^{2/3} = 4$

5. They are inverses.

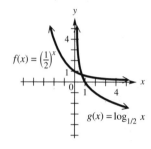

6. $2 \log_7 x + \dfrac{1}{4} \log_7 y - 3 \log_7 z$ **7.** 2.3755

8. -3.0640 **9.** 1.1674   **10.** -2.6038 **11.** 5

12. $\dfrac{5}{2}$ **13.** 2   **14.** 4.7833   **15.** 20.1246   **17.** 10 sec

18. (a) about 18.89 yr **(b)** about 18.84 yr   **19.** about 329.3 g

20. near the end of 2015

Index

A

Absolute value of a complex number, 345
Acute angles, 17
  trigonometric functions of, 62
Acute triangle, 26
Addition
  of complex numbers, 338
  of ordinates, 177
Adjacent side to an angle, 62
Aerial photography, 281
Airspeed, 321
al-Biruni, 304
Alternate exterior angles, 24
Alternate interior angles, 24
Alternating current, 342
Ambiguous case of the law of sines, 294
Amplitude, 143
  of cosine function, 143
  of sine function, 143
Angle(s), 15
  alternate exterior, 24
  alternate interior, 24
  between vectors, 316
  complementary, 17
  corresponding, 25
  coterminal, 20
  of depression, 86
  of elevation, 86
  of inclination, 216
  initial side of, 15
  measure of, 16, 104
  negative, 16
  positive, 16
  quadrantal, 20
  reference, 71
  side adjacent to, 62

side opposite, 62
significant digits for, 84
standard position of, 20
supplementary, 17
terminal side of, 15
types of, 17
vertex of, 16
vertical, 24
Angle sum of a triangle, 26
Angular velocity, 127, 207
  applications of, 128
Applied trigonometry problems, steps to
  solve, 86
Approximately equal to, 67
  symbol for, 67
Arc length, 109
Arccos, 243
Arccot, 245
Arccsc, 245
Archimedes, spiral of, 357, 367
Arcsec, 244
Arcsin, 241
Arctan, 244
Area of a sector, 112
Area of a triangle, 286, 304
  Heron's formula for, 304
Argument, 154
  of a complex number, 345
Arrhenius, Svante, 416
Aryabhata the Elder, 124
Asymptote, vertical, 166
Axis
  horizontal, 2
  imaginary, 343
  polar, 361
  real, 343
  vertical, 2

B

Babbage, Charles, 409
Bearing, 91
Bernoulli, Jakob, 361
Bessel, Friedrich, 1
Bezier curves, 376
Braking distance, 71
British nautical mile, 278
British thermal unit, 427

C

Calculator graphing of polar equations, 363
Cardano, Girolamo, 335
Cardioid, 364
Cartesian equations, 363
Cayley, Sir Arthur, 333
Change-of-base theorem for logarithms,
  417
Chuquet, Nicolas, 335
Circle
  arc length of, 109
  circumference of, 104
  sector of, 111
  unit, 38, 119
Circular trigonometric functions, 118
  applications of, 122
  domains of, 120
  evaluating, 120
  graphs of, 137
Circumference of a circle, 104
Cis θ, 345
Clinometer, 52
Closed interval, 6
Cofunction identities, 63, 205
Cofunctions of trigonometric functions, 63
Combined functions, graph of, 177

Common logarithms, 413
  applications of, 413
Complementary angles, 17
Complex number(s), 334
  absolute value of, 345
  argument of, 345
  conjugate of, 339
  De Moivre's theorem for, 355
  graph of, 343
  imaginary part of, 334
  modulus of, 345
  nth root of, 356
  nth root theorem for, 357
  operations on, 337
  polar form of, 344
  powers of, 355
  product theorem for, 351
  quotient theorem for, 352
  real part of, 334
  rectangular form of, 334, 343
  roots of, 356
  standard form of, 334, 343
  trigonometric form of, 344
Complex plane, 343
  imaginary axis of, 343
  real axis of, 343
Components of a vector, 310
  horizontal, 310, 313
  vertical, 310, 313
Compound interest, 395
Conditional trigonometric equations, 253
  factoring method for solving, 255
  graphing calculator methods for solving, 255
  with half-angles, 262
  intersection-of-graphs method for solving, 254
  linear methods for solving, 254
  with multiple angles, 263
  quadratic formula method for solving, 258
  steps to solve algebraically, 258
  steps to solve graphically, 259
  x-intercept method for solving, 255
Congruence axioms, 282
Congruent triangles, 27, 282
Conjugate of a complex number, 339
Converse of the Pythagorean theorem, 12
Coordinate plane, 2
  locating points in, 3
Coordinate system
  polar, 138, 361
  rectangular, 2
Coordinates, photographic, 288
Copernicus, 1
Corresponding angles, 25
Cosecant function, 35, 62, 119, 166, 186
  characteristics of, 167
  domain of, 120, 166
  graph of, 166
  inverse of, 245
  period of, 167
  range of, 46, 166

steps to graph, 168
Cosine function, 35, 62, 119, 140, 186
  amplitude of, 143
  characteristics of, 141
  difference identity for, 202
  domain of, 120, 141
  double-angle identity for, 217
  graph of, 141
  half-angle identity for, 226
  horizontal translation of, 156
  inverse of, 242
  period of, 145
  range of, 46, 141
  steps to graph, 146, 160
  sum identity for, 203
  translating graphs of, 156
  vertical translation of, 158
Cosine wave, 141
Cosines, law of, 299
Cotangent function, 35, 62, 119, 170, 186
  characteristics of, 172
  domain of, 120, 172
  graph of, 171
  horizontal translation of, 176
  period of, 172
  range of, 46, 172
  steps to graph, 172
  vertical translation of, 176
Coterminal angles, 20
Cross-multiplication property of proportions, 28
Curve fitting, 161
  exponential, 399
  logarithmic, 419
  sinusoidal, 161
Cycloid, 376

D

Damped pendulum, 279
Decay, exponential, 398
Decibel, 415
Decimal degrees, 19
Degree measure, 16
  converting to radian measure, 106
Degree/radian relationship, 104
  table of, 106
De Moivre, Abraham, 355
De Moivre's theorem, 355
Denominator, rationalizing, 44
Dependent variable, 7
Depression, angle of, 86
Derivatives
  of natural logarithmic functions, 415
  of parametric equations, 85
  of trigonometric functions, 194
Difference identity
  application of, 206
  for cosine, 202
  for sine, 210
  for tangent, 211
Digits, significant, 83
Diophantus, 335

Direction angle for a vector, 312
Discriminant, 337
Distance formula, 4
Division
  of complex numbers, 340, 351
  of imaginary numbers, 337
Domain(s)
  of circular trigonometric functions, 120
  of inverse trigonometric functions, 246
  of a relation, 8
Dot product of vectors, 315
  geometric interpretation of, 316
  properties of, 316
Double-angle identities, 217
  application of, 222
  simplifying expressions using, 219
  verifying, 220

E

e, 396
Eccentricity, 363, 373
Elevation, angle of, 86
Equal vectors, 310
Equation(s)
  Cartesian, 363
  with complex solutions, 336, 358
  conditional trigonometric, 253
  decay, 398
  exponential, 393, 423
  growth, 398
  with inverse trigonometric functions, 268
  logarithmic, 404, 425
  parametric, 85, 373
  parametric form of, 149
  polar, 363
  rectangular, 363
  trigonometric, 253
Equilateral triangle, 26
Equilibrant vector, 319
Eratosthenes, 117
Euclid's *Elements,* 303
Euler, Leonhard, 361, 396
Even function, 141
Exact number, 84
Exponential decay, 398
Exponential equations, 393, 423
  solution of, 393, 423
  summary of solving methods, 427
Exponential expressions, evaluating, 389
Exponential functions, 388
  graphs of, 388, 390, 392
Exponential growth, 398
Exponential to logarithmic form, 404
Exponents
  fractional, 388
  negative, 388
  properties of, 389
  rational, 388

F

Factoring method for solving trigonometric equations, 255

Faraday, Michael, 185
Faraday's law, 185
Finke, Thomas, 124
Four-leaved rose, 366
Fractals, 333, 347
Fractional exponents, 388
Frequency, fundamental, 265
Function(s), 7
  circular, 118
  even, 141
  exponential, 388
  horizontal translations of, 154
  inverse, 238
  inverse cosine, 242
  inverse sine, 240
  inverse tangent, 244
  inverse trigonometric, 238
  logarithmic, 404
  odd, 140
  one-to-one, 238
  periodic, 138
  trigonometric, 35, 186
  vertical line test for, 9
  vertical translations of, 158
Function notation, 7
Function values of special angles, 65, 66
Fundamental frequency, 265
Fundamental identities, 187
Future value, 395
$f(x)$ notation, 7

Grade resistance, 79
Graph(s)
  of circular functions, 137
  of combined functions, 177
  of complex numbers, 343
  of cosecant function, 166
  of cosine function, 141
  of cotangent function, 171
  of exponential functions, 388, 390, 392
  of inverse cosecant function, 245
  of inverse cosine function, 243
  of inverse cotangent function, 245
  of inverse functions, 239
  of inverse secant function, 245
  of inverse sine function, 241
  of inverse tangent function, 244
  of logarithmic functions, 405
  parametric, 374
  of polar coordinates, 362
  of polar equations, 363
  of secant function, 166
  of sine function, 139
  slope of tangent line to, 397
  summary of polar, 368
  of tangent function, 171
Graphical addition of complex numbers, 344
Graphing calculator methods for solving trigonometric equations, 255
Groundspeed, 321

Growth, exponential, 398
Gunter, Edmund, 124

Half-angle identities, 226
  application of, 232
  simplifying expressions using, 229
  verifying, 230
Half-open interval, 6
Harmonic motion, 153
Hellmich, Eugene W., 335
Hermann, Jacob, 361
Heron of Alexandria, 304, 335
Heron's area formula, 304
  application of, 304
Hipparchus, 1
Horizontal axis, 2
Horizontal component of a vector, 310, 313
Horizontal line test for one-to-one functions, 238
Horizontal translations, 154
Hypotenuse, 3

i, 334
  powers of, 340
  simplifying powers of, 341
Identities, 43, 185 *See Trigonometric Identities*
i, j form of a vector, 315
Imaginary axis, 343
Imaginary numbers, 334
  operations on, 337
Imaginary part of a complex number, 334
Impedance, 343, 355
Inclination, angle of, 216
Independent variable, 7
Initial point of a vector, 309
Initial side of an angle, 15
Inner product of vectors, 315
Interest
  compound, 395
  simple, 394
Intersection-of-graphs method for solving equations, 254
Interval notation, 6
Intervals, types of, 6
Inverse cosecant function, 245
  graph of, 245
Inverse cosine function, 242
  evaluating, 243
  graph of, 243
Inverse cotangent function, 245
  graph of, 245
Inverse functions, 238
  general statements about, 239
  graph of, 239
  notation for, 238
Inverse secant function, 244
  graph of, 245
Inverse sine function, 240
  evaluating, 241

  graph of, 241
Inverse tangent function, 244
  graph of, 244
Inverse trigonometric equations, 268
Inverse trigonometric functions, 238
  domains of, 246
  equations with, 268
  evaluating, 247
  graphs of, 241, 243, 244, 245
  notation for, 238
  ranges of, 246
  summary of, 246
Inverses theorem, 409
Isosceles triangle, 26

Julia set, 347

Latitude, 110
Law of cosines, 299
  applications of, 301
  derivation of, 299
Law of sines, 283
  ambiguous case of, 294
  applications of, 284
  derivation of, 283
Law of tangents, 309
Legs of a right triangle, 3
Lemniscate, 365
Length of an arc, 109
Limaçons, 368
Line(s), 15
  angle of inclination of, 216
  parallel, 24
  segment, 15
  slope of, 216
Line segment, 15
Linear methods for solving trigonometric equations, 254
Linear velocity, 127
  applications of, 128
Locating points in a plane, 3
Logarithmic equations, 404, 425
  solution of, 404, 425
  summary of solving methods, 427
Logarithmic functions, 404
  characteristics of graph of, 406
  graph of, 405
Logarithmic to exponential form, 404
Logarithms, 402
  change-of-base theorem for, 417
  common, 413
  definition of, 403
  natural, 415
  properties of, 407, 423
Longitude, 112

Mach number, 232, 252
Magnitude of a vector, 309

Mahavira, 335
Mandelbrot, Benoit B., 333
Mandelbrot set, 360
Mathews, Max, 237
Measure of an angle, 16, 104
Midpoint formula, 5
Minute, 18
Modulus of a complex number, 345
Mole, 413
Multiplication
    of complex numbers, 338, 350
    of imaginary numbers, 337

N

Napier, John, 409
Natural logarithms, 415
    derivatives of, 415
Nautical mile, 278
Negative angle, 16
    identities, 187
Negative exponents, 388
Negative number, square root of, 336
Newton, Sir Isaac, 361
Notation
    function, 7
    interval, 6
    inverse trigonometric function, 238
    set-builder, 6
nth root of a complex number, 356
nth root theorem for complex numbers, 357
Number(s)
    complex, 334
    exact, 84
    imaginary, 334
    mach, 232, 252

O

Oblique triangle, 283
    data required for solving, 283
    solving procedures for, 303
Obtuse angle, 17
Obtuse triangle, 26
Odd function, 140
One-to-one function, 238
    horizontal line test for, 238
Open interval, 6
Opposite of a vector, 310
Opposite side to an angle, 62
Orbital period of a planet, 14
Ordered pair, 2
Ordinates, addition of, 177
Origin, 2
Orthogonal vectors, 318

P

Pacioli, Luca, 335
Pair, ordered, 2
Parallax method of measure, 1
Parallel lines, 24
    transversal of, 24

Parallelogram rule for vectors, 310
Parameter, 373
Parametric equations, 85, 373
    derivatives of, 85
    of a plane curve, 373
Parametric form of equations, 149
Parametric graphs, 374
    applications of, 377
    converting to rectangular equivalent, 374
Pascal, Blaise, 409
Path of projectile, 377
Pendulum, damped, 279
Period
    of cosecant function, 167
    of cosine function, 145
    of cotangent function, 172
    of periodic functions, 138
    of secant function, 167
    of sine function, 145
    of tangent function, 172
Periodic functions, 138
    period of, 138
pH, 413
Phase shift, 154
Photographic coordinates, 288
Pi (π), 105
Plane
    complex, 343
    coordinate, 2
    quadrants of, 3
Plane curve, 373
    parametric equations of, 373
Planet
    orbital period of, 14
    sidereal period of, 14
Polar axis, 361
Polar coordinate system, 138, 361
    polar axis of, 361
    pole of, 361
Polar coordinates of a point, 361
    converting to rectangular coordinates, 369
    graph of, 362
Polar equations, 363
    calculator graphing of, 363
    graph of, 363
Polar form of a complex number, 344
Polar graphs, summary of, 368
Pole of a polar coordinate system, 361
Polya, George, 87
Polya's four-step process for problem
    solving, 87
Positive angle, 16
Powers
    of complex numbers, 355
    of i, 340
Present value, 395
Product theorem for complex numbers, 351
Product-to-sum identities, 221
Properties
    of exponents, 389
    of logarithms, 407, 423
Proportions, cross-multiplication property of, 28

Protractor, 18
Pseudo-asymptotes, 167
Ptolemy, 304
Pythagorean identities, 47, 48, 186
Pythagorean theorem, 3
    converse of, 12
Pythagorean triple, 12
Pythagoras, 237

Q

Quadrantal angles, 20
    trigonometric values of, 40
Quadrants of a plane, 3
Quadratic equation with complex solutions, 336
Quadratic formula method for solving
    trigonometric equations, 258
Quotient identities, 48, 186
Quotient theorem for complex numbers, 352

R

Radian, 104
Radian/degree relationship, 104
    table of, 106
Radian measure, 104
    applications of, 109
    converting to degree measure, 106
Radiative forcing, 401, 416
Range(s)
    of circular trigonometric functions, 46, 166
    of inverse trigonometric functions, 246
    of a relation, 8
    of trigonometric functions, 46
Rates, related, 85
Rational exponents, 388
Rationalizing the denominator, 44
Ray, 15
r cis θ, 345
Real axis, 343
Real part of a complex number, 334
Reciprocal, 43
Reciprocal identities, 43, 186
Rectangular coordinates, 2
    converting to polar coordinates, 369
Rectangular equations, 363
    converting to polar equations, 369
Rectangular form of a complex number, 334, 343
    converting to trigonometric form, 346
Reference angles, 71
    table of, 72
Regression, sine, 162
Related rates, 85
Relation, 6
    domain of, 8
    range of, 8
Resultant vectors, 310
Richter scale, 421
Right angle, 17
Right triangles, 26
    applications of, 86, 91

legs of, 3
solving, 84
Right-triangle-based trigonometric functions, 62
Roots of a complex number, 356

S

Scalar product of a vector, 310, 315
Scalars, 309
Scalene triangle, 26
Scatter diagram, 161
Secant function, 35, 62, 119, 166, 186
  characteristics of, 167
  domain of, 120, 166
  graph of, 166
  period of, 167
  range of, 46, 166
  steps to graph, 168
Second, 18
Sector of a circle, 111
  area of, 112
Segment of a line, 15
Semiperimeter, 304
Set-builder notation, 6
Side adjacent to an angle, 62
Side opposite an angle, 62
Sidereal period of a planet, 14
Significant digits, 83
  for angles, 84
  calculating with, 84
Signs of trigonometric functions, 45
Similar triangles, 26
  conditions of, 27
Simple harmonic motion, 153
Simple interest, 394
Simplifying trigonometric expressions, 189
Sine function, 35, 62, 119, 140, 186
  amplitude of, 143
  characteristics of, 140
  difference identity for, 210
  domain of, 120, 140
  double-angle identity for, 217
  graph of, 139
  half-angle identity for, 226
  horizontal translation of, 155
  inverse of, 240
  model, 148, 161
  period of, 145
  range of, 46, 140
  steps to graph, 146, 160
  sum identity for, 210
  translating graphs of, 155
Sine regression, 162
Sine wave, 139
Sines, law of, 283
  ambiguous case of, 294
Sinusoid, 139
Slant height, 13
Slope of a line, 216
Snell's law, 81, 278
Solar constant, 109
Solving triangles, 282

Sound waves, 210
Special angles, trigonometric function values of, 65, 66
Spiral of Archimedes, 367
Square root of a negative number, 336
Standard form of a complex number, 334, 343
Standard position of an angle, 20
Straight angle, 17
Substitution to verify identities, 199
Subtense bar method, 95
Subtraction
  of complex numbers, 338
  of vectors, 310
Sum identity
  for cosine, 203
  for sine, 210
  for tangent, 211
Sum of vectors, 310
Supplementary angles, 17

T

Tangent function, 35, 62, 119, 170, 186
  characteristics of, 172
  difference identity for, 211
  domain of, 120, 171
  double-angle identity for, 218
  graph of, 171
  half-angle identity for, 226
  inverse of, 244
  period of, 172
  range of, 46, 171
  steps to graph, 172
  sum identity for, 211
  vertical translation of, 175
Tangents, law of, 309
Terminal point of a vector, 309
Terminal side of an angle, 15
Theorem on inverses, 409
Thermal unit, British, 427
Three-dimensional vectors, 312
Threshold sound, 415
Translations
  horizontal, 154
  of trigonometric functions, 155, 156, 158
  vertical, 158
Transversal, 24
Triangle(s)
  angle sum of, 26
  applications of, 28
  area of, 286, 304
  congruent, 27, 282
  oblique, 283
  semiperimeter of, 304
  side length restriction of, 299
  similar, 26
  solving, 282
  solving procedures for, 303
  types of, 26
Triangulation, 307
Trigonometric equations, 253
  conditional, 253

graphing calculator methods for solving, 255
  identity, 185
  inverse, 268
Trigonometric expressions, simplifying, 189
Trigonometric form of a complex number, 344
  converting to rectangular form, 345
Trigonometric functions, 35, 186
  of acute angles, 62
  circular, 118
  cofunctions of, 63
  combinations of translations of, 159
  definitions of, 35, 62, 186
  derivatives of, 194
  domains of, 120
  finding values of, 188
  inverses of, 238
  ranges of, 46
  right-triangle-based, 62
  signs of, 45
  of special angles, 65, 66
  translation of, 155, 156, 158
  using a calculator, 77
Trigonometric identities, 43, 185
  cofunction, 63, 205
  difference, 202, 210, 211
  double-angle, 217
  fundamental, 187
  half-angle, 226
  negative angle, 187
  product-to-sum, 221
  Pythagorean, 47, 48, 186
  quotient, 48, 186
  reciprocal, 43, 186
  sum, 203, 210, 211
  verifying, 193
  verifying by calculator, 195
  verifying by substitution, 199
Trigonometric values of quadrantal angles, 40
Trigonometry problems, applied, 86
  steps to solve, 86
Triple, Pythagorean, 12
Trochoid, 376

U

Undersampling, 386
Unit circle, 38, 119
Unit vector, 315
Upper harmonics, 265

V

Variable
  dependent, 7
  independent, 7
Vector(s), 309
  algebraic interpretation of, 312
  angle between, 316
  applications of, 319
  component of, 310
  direction angle for, 312

Vector(s), (*continued*)
  dot product of, 315
  equal, 310
  equilibrant of, 319
  horizontal component of, 310, 313
  i, j form of, 315
  initial point of, 309
  inner product of, 315
  magnitude of, 309
  operations with, 314
  opposite of, 310
  orthogonal, 318
  parallelogram rule for, 310
  resultant of, 310
  scalar product of, 310, 315
  subtraction of, 310
  sum of, 310
  terminal point of, 309
  three-dimensional, 312
  unit, 315
  vertical component of, 310, 313
  x-component of, 312
  y-component of, 312
  zero, 310
Vector quantities, 309
Velocity
  angular, 127, 207
  linear, 127
Verifying trigonometric identities, 193
Vertex of an angle, 16
Vertical angles, 24
Vertical asymptote, 166
Vertical axis, 2
Vertical components of a vector, 310, 313
Vertical line test for functions, 9
Vertical translations, 158
von Helmholtz, Hermann, 237

W
Wave
  cosine, 141
  sine, 139
  sound, 210

X
x-axis, 2
x-component of a vector, 312
x-intercept method for solving trigonometric
    equations, 255

Y
y-axis, 2
y-component of a vector, 312

Z
Zero-factor property, 255
Zero vector, 310

Index of Applications

Astronomy

Angle of a star, 23
Angle of elevation of the sun, 87, 90, 122–123, 126
Angular and linear velocities of Earth, 131
Communications satellite coverage, 253
Distance between a satellite and a tracking station, 307
Distance between the sun and a star, 43
Distance from a satellite to the horizon, 59
Distance to the moon, 291
Distance traveled by a satellite, 129
Distances to stars, 1–2
Earth's diameter, 117
Eclipse on a planet, 34
Height of a satellite, 102
Hours in a Martian day, 131
Moon's diameter, 115
Orbital velocity of Jupiter, 136
Orbits of a space vehicle, 108
Orbits of planets, 14, 373
Period of revolution, 422
Planets' distances from the sun, 4, 422
Sizes and distances in the sky, 34
Spacecraft coordinate systems, 209–210
Sun's diameter, 90
Total solar eclipse, 28–29
Velocity of a star, 324
Viewing field of a telescope, 23

Automotive

Accident reconstruction, 261
Area cleaned by a windshield wiper, 116
Braking distance, 71
Car banking a curve, 330
Car's speed at collision, 71
Design of highway curves, 82
Engine specifications, 136
Grade resistance, 79–80, 82
Highway grades, 82
Pickup truck speedometer, 115
Rotating tire, 23, 59
Rotation of a gas gauge arrow, 113
Sag curve of a highway, 77
Speed limit on a curve, 82
Stopping distance on a curve, 61, 91

Biology and Life Sciences

Activity of nocturnal animals, 151
Deer population, 402
Diversity of species, 418, 422
Fish's view of the world, 81
Lynx and hare populations, 182

Business

Clothing manufacturers, 13
Employee training, 402
Paper manufacturers' measuring paper curl, 117–118
Personal computer shipments, 429
Software exports, 433

Chemistry

Distance between atoms, 290
pH of a solution, 413–414, 419
Radioactive decay, 435

Construction

Angle between a beam and cables, 307
Carpentry technique, 31
Distance between points on a crane, 307
Expansion of a bridge, 13
Height of the Eiffel Tower, 429–430
Land required for a solar-power plant, 117
Length of a tunnel, 88, 327
Testing for square corners, 13
Mounting a video camera, 90
Native American structure, 116
Playhouse layout, 306
Railroad track expansion, 13
Required amount of paint, 308, 329
Size of the threads of a bolt, 90
Surveying, 23, 102, 299, 307

Engineering

Aerial photography, 281–282, 288, 291, 330
Alternating current, 342
Amperage, 225
Angle of depression of a light, 90
Angular velocity, 129, 131, 136
Depth of field, 273
Distance across a canyon, 290, 327
Distance across a lake or river, 89, 285–286, 290, 301, 306
Distance between radio direction finders, 290, 332
Distance between two factories, 307
Distance from a fire to a house, 4–5
Distance from the ground to the top of a building, 90
Distance of a rotating beacon, 178
Electrical current, 185, 209, 268, 274, 354
Electromotive force, 261
Gears, 111, 131
Guy wire, 89
Height of a mountain, 87, 90–91, 97
Height of an object or structure, 28, 32, 33, 34, 52, 86, 90, 93–94, 97, 101, 102, 133, 327

Impedance, 343, 355
Length of a brace, 327
Length of a road, 57
Length of a rope, 306
Length of a shadow, 57, 90
Linear velocity, 129, 135
Oil in a submerged storage tank, 253
Programming language, 274
Railroad engineering, 108
Revolutions of an object, 21, 23
Rope winding around a drum, 111
Rotating pulley or wheels, 23, 56, 108, 115
Sound waves, 210
Tension of a rope, 322
Transistors on computer chips, 434
Viewing angle to an object, 182
Voltage, 152, 207, 217, 225, 236, 261
Wattage consumption, 222, 225
Weight of an object, 322

Environment
Area of a solar cell, 43
Atmospheric carbon dioxide, 152, 153, 387, 398–399
Atmospheric effect on sunlight, 135
Atmospheric pressure, 401, 433
Classifying wetlands, 414–415, 420
Cloud ceiling, 89
Coal consumption in the United States, 427–428
Global temperature increase, 416–417, 422
Landscaping formula, 252–253
Location of a forest fire, 286
Pollution trends, 182
Radiative forcing, 401–402, 416, 430
Solar energy, 103, 109
Sunset times, 183–184
Temperatures, 126, 137–138, 148, 152, 161–162, 164, 165, 181, 183
Tides for Kahului harbor, 151

Finance
Compound interest, 394–396, 401, 430, 433
Growth of an account, 433, 435
Interest rates of treasury securities, 412
Investment, 430
Retirement planning, 430

General Interest
Calories in a recipe, 88
Daylight hours in New Orleans, 268
Error in measurement, 90
Height of a balloon, 291, 331
Ladder leaning against a wall, 89
Length of a day, 126
Longitude, 112–113
Measurement of a folding chair, 290
Position of a moving arm, 152, 274
Radius of an Indian artifact, 118
Rotating hour hand on a clock, 108, 134

Viewing angle of an observer, 278
Volunteerism among college freshmen, 421

Geology
Age of rocks, 416
Angle the celestial North Pole moves, 59
Depth of a crater on the moon, 59
Earthquake intensity, 421, 432
Height of a lunar peak, 59
Volcano movement, 308

Geometry
Angle measure of a star on the American flag, 23
Arc length, 133
Area of a lot, 117
Area of a quadrilateral, 328
Area of a sector, 133
Area of a triangular object or region, 292, 308, 328
Areas of the colors of the U.S. flag, 293
Circumference of a circle, 88
Diagonals of a parallelogram, 306, 327
Dimensions of a triangle, 89
Dimensions of the Great Pyramid, 13
Distance between an arc and a chord, 91
Fractals, 333
Heron triangles, 308
Julia set, 347–348, 350, 386
Length of a diagonal, 101
Length of a side of a piece of land, 98
Lengths of the sides of a quadrilateral, 33
Lengths of the sides of a triangle, 32, 101
Mandelbrot set, 360, 385
Perfect triangles, 308
Radius of a spool of thread, 131
Slant height of a pyramid, 13
Volume of a solid, 117

Government
Carbon dioxide emissions tax, 430
Encoding and decoding information, 240
Inflation rate, 433
Vietnam Veterans' Memorial, 328

Medical
Blood pressure variation, 151
Double vision, 52
Drug level in the bloodstream, 434
Force on back muscles, 217
Reading an electrocardiogram, 324

Music
Hearing beats in music, 267–268
Hearing difference tones, 268
Musical sound waves, 152
Musical tone, 237–238, 259, 261
Pressure of a plucked string, 267
Pressure on the eardrum, 261

Pressures of upper harmonics, 265–266
Tone heard by a listener, 273

Physics
Angle between forces, 322, 329
Angle formed by radii of gears, 290
Angle of a hill, 322, 329
Ball rolling down a plane, 102
Damped pendulum, 279–280
Decibel levels, 415, 421, 433
Distance a dropped object falls, 225
Harmonic motion, 153–154
Incline angle, 320
Magnitude of forces, 312, 318, 319–320, 322, 329
Movement of a particle, 261
Object propelled, 225
Path of a projectile, 377–380, 382
Pulley raising a weight, 114
Required force, 320, 322, 329
Speed of light, 81, 278

Sports and Entertainment
Distance on a softball field, 331
Distances on a baseball diamond, 328
Flight of a ball, 381, 385
Leading NFL receiver, 87
Movement of a runner's arm, 279
Observation of a painting, 252, 274
Person in a Ferris wheel, 136
Race speed, 429
Rotation of a seesaw, 113
Shot-putter, 83, 249, 252
Skydiver fall speed, 435
Swimmer in distress, 60
Top Women's National Basketball Association scorer, 88
Weight of a sled and passenger, 329

Statistics and Demographics
Fatherless children, 429
Median home cost in the United States, 402
Population growth, 402, 435
Racial mix in the United States, 433

Travel
Airspeed of a plane, 322, 323
Altitude of a plane, 97
Bearing, 321, 323, 329
Bicycle chain drive, 115
British nautical mile, 278
Course of a plane, 322, 323
Distance between two cities, 33, 96, 110, 114, 134
Distance between two structures or objects, 55, 92–93, 96, 101, 102, 291, 306, 307, 323, 327, 328

Distance, 90, 96, 97, 101, 115, 290, 306, 323
Groundspeed of a plane, 322, 323
Length of a train, 115
Length of an oil tanker, 115

Mach number for the speed of a plane, 232, 252
Movement of a motorboat, 323
Railroad curves, 116, 232

Rotating propeller, 23, 56
Speed, 131, 329
Time to move along a railroad track, 131
Visitors to United States National Parks, 421

5.1 Fundamental Identities

$$\cot \theta = \frac{1}{\tan \theta} \qquad \sec \theta = \frac{1}{\cos \theta} \qquad \csc \theta = \frac{1}{\sin \theta}$$

$$\tan \theta = \frac{\sin \theta}{\cos \theta} \qquad \cot \theta = \frac{\cos \theta}{\sin \theta}$$

$$\sin^2 \theta + \cos^2 \theta = 1 \qquad \tan^2 \theta + 1 = \sec^2 \theta \qquad 1 + \cot^2 \theta = \csc^2 \theta$$

$$\sin(-\theta) = -\sin \theta \qquad \cos(-\theta) = \cos \theta \qquad \tan(-\theta) = -\tan \theta$$

$$\csc(-\theta) = -\csc \theta \qquad \sec(-\theta) = \sec \theta \qquad \cot(-\theta) = -\cot \theta$$

5.3, 5.4 Sum and Difference Identities

$$\cos(A - B) = \cos A \cos B + \sin A \sin B$$

$$\cos(A + B) = \cos A \cos B - \sin A \sin B$$

$$\sin(A + B) = \sin A \cos B + \cos A \sin B$$

$$\sin(A - B) = \sin A \cos B - \cos A \sin B$$

$$\tan(A + B) = \frac{\tan A + \tan B}{1 - \tan A \tan B}$$

$$\tan(A - B) = \frac{\tan A - \tan B}{1 + \tan A \tan B}$$

5.3 Cofunction Identities

$$\cos(90° - \theta) = \sin \theta$$

$$\sin(90° - \theta) = \cos \theta$$

$$\tan(90° - \theta) = \cot \theta$$

$$\cot(90° - \theta) = \tan \theta$$

$$\sec(90° - \theta) = \csc \theta$$

$$\csc(90° - \theta) = \sec \theta$$

5.5, 5.6 Multiple-Angle and Half-Angle Identities

$$\cos 2A = \cos^2 A - \sin^2 A \qquad \cos 2A = 1 - 2 \sin^2 A$$

$$\cos 2A = 2 \cos^2 A - 1 \qquad \sin 2A = 2 \sin A \cos A$$

$$\tan 2A = \frac{2 \tan A}{1 - \tan^2 A} \qquad \cos \frac{A}{2} = \pm \sqrt{\frac{1 + \cos A}{2}}$$

$$\sin \frac{A}{2} = \pm \sqrt{\frac{1 - \cos A}{2}} \qquad \tan \frac{A}{2} = \pm \sqrt{\frac{1 - \cos A}{1 + \cos A}}$$

$$\tan \frac{A}{2} = \frac{\sin A}{1 + \cos A} \qquad \tan \frac{A}{2} = \frac{1 - \cos A}{\sin A}$$

6.1 Graphs of Inverse Trigonometric Functions

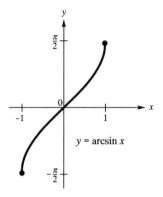

$y = \arcsin x$

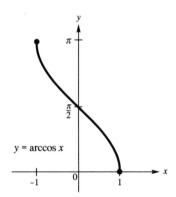

$y = \arccos x$

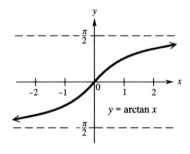

$y = \arctan x$